APPLICATION OF HIGH-PERFORMANCE COMPUTING TO EARTHQUAKE-RELATED PROBLEMS

APPLICATION OF HIGH-PERFORMANCE COMPUTING TO EARTHQUAKE-RELATED PROBLEMS

Editor

Muneo Hori

Japan Agency for Marine-Earth Science and Technology, Japan

World Scientific

NEW JERSEY • LONDON • SINGAPORE • BEIJING • SHANGHAI • HONG KONG • TAIPEI • CHENNAI • TOKYO

Published by

World Scientific Publishing Europe Ltd.
57 Shelton Street, Covent Garden, London WC2H 9HE
Head office: 5 Toh Tuck Link, Singapore 596224
USA office: 27 Warren Street, Suite 401-402, Hackensack, NJ 07601

Library of Congress Control Number: 2023025360

British Library Cataloguing-in-Publication Data
A catalogue record for this book is available from the British Library.

APPLICATION OF HIGH-PERFORMANCE COMPUTING TO EARTHQUAKE-RELATED PROBLEMS

ISBN 978-1-80061-462-8 (hardcover)
ISBN 978-1-80061-463-5 (ebook for institutions)
ISBN 978-1-80061-464-2 (ebook for individuals)

For any available supplementary material, please visit
https://www.worldscientific.com/worldscibooks/10.1142/Q0432#t=suppl

Desk Editors: Logeshwaran Arumugam/Shi Ying Koe

Typeset by Stallion Press
Email: enquiries@stallionpress.com

https://doi.org/10.1142/9781800614635_fmatter

Preface

This major reference summarizes the theories, analysis methods, and computational results of various earthquake simulations that take advantage of supercomputers. Specifically, it covers simulations in the following three fields. The first field is earth sciences related to earthquakes. The second field is earthquake engineering for the evaluation of structure and ground seismic responses. The third field is social sciences related to post-earthquake disasters such as traffic flow and economic recovery. High-resolution simulations of the three fields provide a bird's-eye view of earthquake hazards, earthquake disasters, and post-earthquake recoveries.

This book consists of 10 chapters. The first two chapters are for the simulation in the earth sciences field, namely, the simulations of earthquake cycles and plate boundary behavior. The numerical methods are the finite element method (FEM) and the discrete element method (DEM). The spatial and temporal scales of this problem are large and long, respectively, as it involves plates and the order of million years. Therefore, it is important to increase the spatial and temporal resolution for more reliable analysis, and for this the use of high-performance computation such as on supercomputers is essential.

Simulations of tsunamis caused by earthquakes are included in Chapter 3. The simulation of offshore tsunamis entering a harbor and hitting a city uses smoothed particle hydrodynamics (SPH) as the analysis method. A multi-scale analysis is developed to relate coarse modeling of the tsunami occurrence to fine modeling of tsunami inundation.

In the field of earthquake engineering, Chapter 4 of this major reference presents simulations of structures subjected to extremely strong ground motions that have not been experienced in the past. Due to the difficulty of conducting experiments, the numerical analysis of FEM must be used to evaluate the safety of structures subjected to such strong ground motions. A high-fidelity numerical analysis model of the ground and structures must be constructed, and large-scale calculations using FEM are required. In addition to earthquake ground motions, surface earthquake faults are also a threat to structures. A simulation of the generation process of surface earthquake faults is also presented. The analysis method is the FEM.

Chapters 5 and 6 present simulations of the crust and ground to numerically analyze seismic wave propagation and surface earthquake fault occurrences. Both simulations use FEM to evaluate the topographical effects of the underground structure on the seismic waves and faults. A numerical analysis model of the highest fidelity ought to be constructed, which inevitably increases the degree of freedom of the model. The use of supercomputers is thus essential. The parallel use of graphics processing units (GPU) is explained in Chapter 5.

The remaining four chapters are simulations in the field of social sciences, namely, the simulations of traffic in a damaged urban area and the simulations of regional and national economic activities after an earthquake. The traffic simulation analyzes the traffic flow in the entire region. The simulation of a regional economy analyzes the optimal economic activities after a disaster. The simulation of a national economy is a very large-scale numerical analysis using agent models for each economic entity. Parallel computation, which is rarely used in the social sciences, enables large-scale and high-speed numerical analysis.

Unlike the fields of earth sciences and earthquake engineering, the social sciences use various digital information about cities, regions, and nations as input data for simulations. This major reference introduces the digital information used and its processing and treatment for numerical simulation in the last chapter.

Forecasting of hazards, disasters, and responses based on scientific simulations enhanced with high-performance computing will increasingly be used to prepare for and respond to earthquakes. This is because of the increasing performance of computers and the expansion of digital information in cities. Research and development of more advanced numerical analysis methods in earth sciences, earthquake engineering, and social sciences related to post-earthquake disasters that effectively use computers and digital information are needed. We hope that this major reference will be of some help.

https://doi.org/10.1142/9781800614635_fmatter

About the Editor

Muneo Hori received his Ph.D. from the University of California, San Diego, in 1986. He became a professor at the Earthquake Research Institute, The University of Tokyo, in 2001. He was appointed as the Director-General of the Research Institute for Value-Added-Information Generation, Japan Agency for Marine-Earth Science and Technology, in 2019. He also works as a research professor at the School of Engineering, The University of Tokyo, and as a program officer in the field of civil engineering at the Japan Science and Technology Agency. Muneo Hori served as a managing editor of the *Journal of Earthquake and Tsunami* from 2006 to 2022. His primary research topics are computational earthquake engineering for developing large-scale simulation systems and applied mechanics and mathematics for deformation and fracture of solids. He received the Prime Minister's Award for Contribution to Natural Disaster Mitigation, 2022. He was a finalist for the ACM Gordon Bell Prizes four times.

https://doi.org/10.1142/9781800614635_fmatter

Contents

https://doi.org/10.1142/9781800614635_0001

Chapter 1

Earthquake Generation Cycle and Crust Deformation in Subduction Zones

Takane Hori* and Ryoichiro Agata

Japan Agency for Marine-Earth Science and Technology, Kanazawa-ku, Yokohama, Japan

**horit@jamstec.go.jp*

In this chapter, we introduce a model and a numerical simulation method based on repeated megathrust earthquakes at a subducting plate boundary, as well as a calculation method for crustal deformation associated with slip at a plate boundary with realistic topography and subsurface structure. As an application example, we will evaluate the effects of changes in the magnitude and recurrence interval of megathrust earthquakes caused by a nearby earthquake along the Nankai Trough, southwest Japan. Furthermore, the calculation of the slip response Green's functions at the plate boundaries along the Nankai Trough and the Japan Trench will also be presented.

1. Introduction

There are three major types of earthquakes that can cause damage in Japan: megathrust earthquakes that occur at the boundary between a subducting plate and an inland plate carrying the Japanese islands, large shallow earthquakes that occur inland (shallower than 20 km) where traces of active faults remain in the crust, and large earthquakes that occur inside a subducting plate. An example of megathrust earthquakes is the 2011 earthquake off the Pacific coast of Tohoku (magnitude (M) 9.0). It is the largest of the three types and is commonly accompanied by a tsunami, causing widespread damage by both shaking and tsunami. To prepare for such damage mitigation, it is necessary to study possible future earthquake scenarios as a hazard assessment. To do so quantitatively, it is necessary to utilize numerical simulations based on earthquake occurrence models and observation data. In this chapter, we introduce a numerical simulation method of earthquake generation and a crustal deformation analysis method necessary for data analysis for monitoring the preparation process of earthquakes and its application examples, focusing on megathrust earthquakes, which are more widespread and large-scale and for which modeling of earthquake generation is relatively advanced.

Plate subduction is actually a complex dynamic process involving phenomena such as mantle convection and crustal rupture, and the plate geometry that causes plate boundary earthquakes changes on geological time scales of a million years or more. However, such long-term processes are not the subject here. So, subduction is modeled as a displacement discontinuity at the boundary under a given geometry. The rate of displacement discontinuity is estimated using global navigation satellite system (GNSS) and other satellite data. If subduction proceeds at a uniform and constant rate at the plate boundary, plate boundary earthquakes will not occur. However, if a part of the plate boundary is locked or slips slower than its surroundings, the stress acting on the boundary increases (Fig. 1). Here, slip deficit rate (SDR) is defined as $V_{\text{pl}} - V$, V_{pl} is the relative plate motion and V is the slip rate at a point on the plate boundary. If we ignore the stress change due to the uniform steady sliding with V_{pl}, the slip deficit is the source of shear stress increase at the plate boundary. When the stress exceeds the strength of the boundary there, sliding occurs to release the slip deficit. If the slip is sufficiently fast, elastic waves propagate around the surface. This high-speed slip at the plate boundary is an earthquake, and the propagating elastic waves are seismic waves.

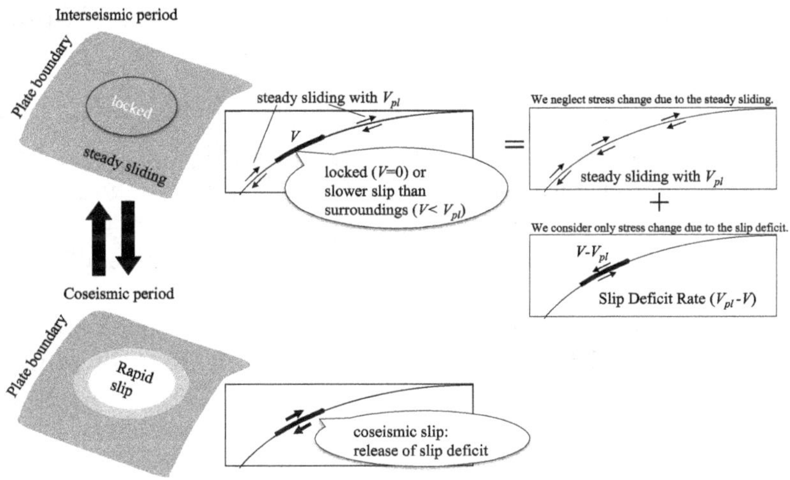

Fig. 1. Conceptual model of earthquake generation at the plate boundary in subduction zone.

Thus, the generation of megathrust earthquakes is treated here as a repetition of plate boundary locking (accumulation of slip deficit) and its release at the plate boundary. In the following, we first introduce a simulation method for repeated occurrence of earthquakes as a means of understanding the pattern of occurrence (changes in magnitude and recurrence intervals) of megathrust earthquakes and examining possible future scenarios (Section 2). As an example, we will discuss the repetition of the Nankai Trough megathrust earthquakes and the evaluation of the effects of stress changes due to nearby earthquakes that may cause disturbances to the occurrence pattern of megathrust earthquakes. On the other hand, in order to examine the feasibility of such scenarios, it is necessary to estimate the current status and past history of plate boundary locking and sliding conditions. In order to estimate more accurately the slip and slip deficit, analysis methods that take into account the complex topography and subsurface structure of the subduction zone are expected to play an important role. In Section 3, we introduce a method to calculate the slip response function, which is necessary for estimating fault slip and slip deficit at a plate boundary from surface crustal deformation data, taking into account more realistic topography and subsurface structure, and examples of its application.

2. Earthquake Generation Cycle Simulations for Nankai Subduction Zone in Southwest Japan

Among megathrust earthquakes, the Nankai Trough has a relatively well-known past history and has repeatedly occurred with a recurrence interval of about 100–200 years, while the extent of the epicenter has changed (Fig. 2). The formulation of how to model the repeated occurrence of such giant earthquakes is explained in Section 2.1. We also discuss how to handle changes in the timing and magnitude of occurrence due to earthquakes occurring in adjacent areas. The method for their numerical simulation is also explained in Section 2.2. Then, the parameter setting of the model targeting the Nankai Trough mega earthquake, its simulation results, and a

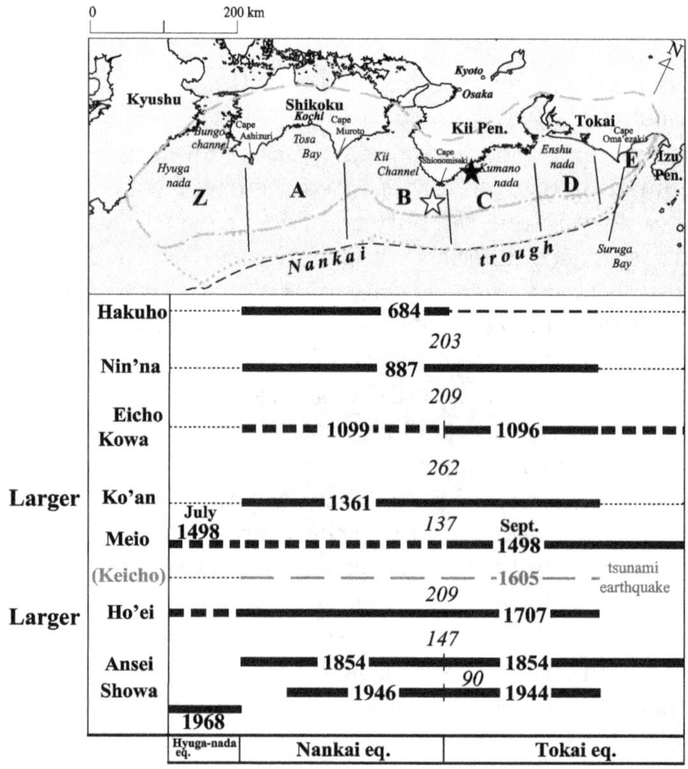

Fig. 2. Historical sequence of Nankai Trough earthquakes (after Ref. 1).

discussion of the evaluation of the impact of earthquakes in adjacent areas are presented in Sections 2.3–2.5.

2.1. *Formulation of earthquake generation cycle*

The repetition of earthquakes at plate boundaries can be described by a system of equations that specifies the increase or decrease of shear stress due to slip deficit and its release, the relationship between stress, strength, and slip (constitutive relationship), and the change in strength (strength evolution law) under a given plate boundary geometry and relative velocity between plates.

2.1.1. *Stress variation with spatio-temporal variation of slip*

The following approximate treatment of the changes in stress at the plate boundary caused by spatio-temporal changes in slip on the boundary is given. First, we treat only the shear stress component acting on the fault plane in the direction parallel to subduction. Slip is also restricted to only directions parallel to this. The medium surrounding the plate boundary is assumed to be elastic and dynamic stress changes are not included in the slip response function, and there is no time variation of the slip response function. Furthermore, stress changes due to steady-state slip at the plate boundary are neglected. Under the above approximation, the shear stress change can be written as follows:

$$\tau(x,t) = \int_{\Omega} G(x : x_0)(u(x_0,t) - V_{\mathrm{pl}}(x_0)t)dx_0 - \eta V(x,t). \quad (1)$$

Equation (1) is an approximated expression of fault shear stress τ in the subduction direction using the static slip response to forward slip G, slip deficit on the plate interface $V_{\mathrm{pl}}t-u$, and fault slip rate V. The first term on the right-hand side of this equation represents a quasi-static fault shear stress as the spatial integration of the product of G and slip deficit over the target plate interface Ω, where u and V_{pl} are the slip on the plate interface and relative plate velocity, respectively. V and u have the relation of $du/dt = V$. The second term represents the effect of radiation damping introduced by Ref. 2, the coefficient η controls the degree of the effect of the seismic radiation, and $\eta = \mu/2\beta$ is assumed following Ref. 2, where μ and β are the rigidity and shear wave velocity of the material surrounding the source region, respectively. In the following, we set $\mu = 30$ GPa and $\beta = 3.27$ km/s.

2.1.2. *Constitutive relationship of fault and evolution law of strength*

Here we use the rate- and state-dependent friction law, which is a friction law introduced on the basis of rock experiments. This friction law consists of Eq. (2),[3] which specifies the constitutive relationship of the fault (the relationship between slip rate, stress, and strength), and Eq. (3), which specifies the strength evolution.

$$V = V_* \exp\left[\frac{\tau - (\tau_s^* + \Delta\tau_s)}{A}\right]. \tag{2}$$

$$\frac{d\Delta\tau_s}{dt} = \frac{B}{L/V_*}\exp\left(-\frac{\Delta\tau_s}{B}\right)\exp\left(-\frac{V}{V_c}\right) - \frac{V}{L}\left\{\Delta\tau_s + B\ln\left(\frac{V}{V_*}\right)\right\}. \tag{3}$$

In Eq. (2), V_*, τ_s^*, and $\Delta\tau_s$ are an arbitrary reference slip velocity on the frictional surface, the strength of a fault sliding with a velocity of V_*, and the differential fault strength measured from τ_s^*, respectively. Because Eq. (2) implies the slip rate V is accelerated or decelerated from V_* depending on the difference between the fault shear stress τ and the variable $\tau_s(= \tau_s^* + \Delta\tau_s)$, the variable τ_s is regarded as a threshold to acceleration of the fault with respect to the fault shear stress τ (Fig. 3(a)). Hence, we call the variable τ_s as the "strength" of the fault. A frictional parameter $A(= a\sigma)$ controls the variation of V with respect to the certain difference of stress and strength, where parameters a (and b, as will be described in what follows) are frictional coefficients, and σ is the effective normal stress. The differential strength $\Delta\tau_s$ here is defined as $\Delta\tau_s = B \ln (V_*\theta/L)$, where θ is the state variable representing a contact state of the interface. It should be noted that we can obtain the ordinary rate- and state-dependent friction equation by substituting the definition of $\Delta\tau_s$ in Eq. (2).

Equation (3) is an equivalent differential equation to the composite equation describing the evolution of the state variable θ,[4]

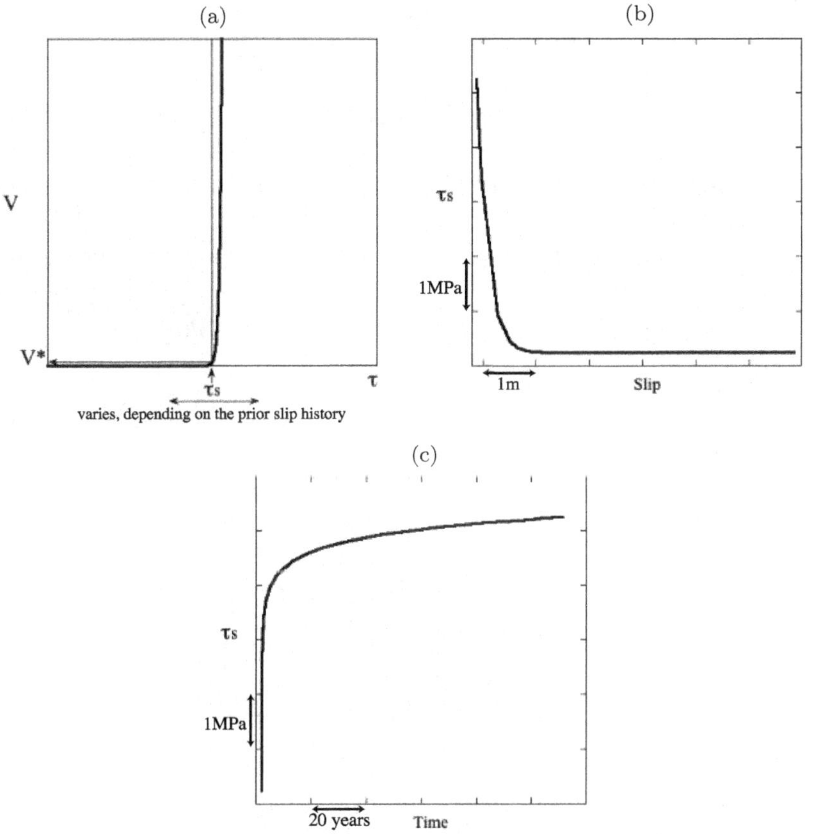

Fig. 3. (a) Schematic diagram for slip velocity V and shear stress τ (modified from Ref. 3, where τ_s is the interfacial strength), (b) slip dependence of strength τ_s for $V > 1$ cm/s, and (c) temporal variation of strength τ_s for $V > 1$ cm/s.

although the integrand is $\Delta\tau_s$ instead of θ. The first term on the right-hand side of Eq. (3) expresses the time-dependent increase of $\Delta\tau_s$ (healing effect) if the slip rate V is small enough ($V \sim V_c$) (Fig. 3(c)). The second term represents the slip-dependent decrease of $\Delta\tau_s$ (slip-weakening effect) as shown in Fig. 3(b). So, $B(= b\sigma)$ and L are frictional parameters controlling strength recovery and slip weakening. V_c switches the behavior of Eq. (3), depending on the slip rate V, and is called the cutoff velocity. Following Ref. 4, we assume $V_c = 10^{-8}$ m/s in Eq. (3) to fit the theoretical frictional property to the experimental data of friction on granite surfaces obtained by

Ref. 5. Though $A(= a\sigma)$ and $B(= b\sigma)$ depend on the normal stress σ which could vary with seismic cycles, we neglect the temporal variation of σ for simplification.

Note that we abbreviate the functional expressions as variables in Eqs. (2) and (3): the variables V_*, τ_{*s}, A, B, and L are functions with respect to only the spatial coordinate (i.e., constant in time), while $V_c(= 10^{-8}$ m/s) is constant with respect to time and space. V, τ, and $\Delta\tau_s$ are unknown variables with respect to time and space.

2.1.3. *Evaluation of the effect of stress changes due to surrounding earthquakes*

When dealing with changes in the timing and magnitude of the target mega earthquake due to earthquakes in adjacent areas, we assume for simplicity that only stress changes affect the source region of the target mega earthquake, ignoring changes in strength before and after the earthquake in the adjacent area. In this case, the slip velocity change in the source region of the earthquake is given by the following equation based on (2):

$$V(x, T_E^+) = V(x, T_E^-) \exp\left[\frac{\Delta\tau_E(x)}{A}\right], \tag{4}$$

where $V(x, T_E^+)$ and $V(x, T_E^-)$ are slip velocity at x just after and before the occurrence of a surrounding earthquake, respectively. $\Delta\tau_E(x)$ is the elastic stress change distribution due to the earthquake in the adjacent area. The slip velocity is accelerated for $\Delta\tau_E > 0$ or decelerated for $\Delta\tau_E < 0$. By calculating the continuation of the earthquake cycle using the modified velocity and stress distributions (with the strength distribution intact) as initial conditions, the effect of the surrounding earthquake on the occurrence of the target megathrust earthquake can be evaluated for any time period.

2.2. *Numerical simulation method of earthquake generation cycle*

By spatially discretizing the above governing equations, the plate interface Ω of interest is partitioned into a set of N small finite sub-faults. Here, the subscript i denotes the index of the sub-fault

$(1 < i < N)$; the differential equation for $\Delta\tau_{s,i}$ is directly obtained from Eq. (3), and it is clear that the evaluation of $d\Delta\tau_{s,i}/dt$ for each sub-fault i can be performed using only variables related to that sub-fault. On the other hand, the specific expression for the change in the slip rate V_i can be derived from Eqs. (1) and (2) as follows:

$$\frac{dV_i}{dt} = \frac{V_i}{A_i\left(1 + \eta\frac{V_i}{A_i}\right)}\left\{\sum_{j=1}^{N} G_{ij}(V_i - V_{pl,j}) - \frac{d\Delta\tau_{s,i}}{dt}\right\}, \quad (5)$$

where the subscripts i and j indicate the index of the sub-faults $(1 < i, j > N)$.

It should be noted that direct evaluation of dV_i/dt for all i requires $O(N^2)$ operations because G_{ij}, the slip response of sub-fault i due to unit slip of sub-fault j, is an $N \times N$ dense matrix. Therefore, in quasi-dynamic earthquake cycle simulations with a large number of sub-faults N, the direct evaluation of Eq. (5) is the most time-consuming process, requiring a large memory for G_{ij}.

To efficiently compute seismic cycles, many attempts to reduce the memory size and computation time of Eq. (5) have been considered.[6–8] Among them, we applied the hierarchical matrix (H-matrix) method to calculate the multiplication of the $N \times N$ slip response function matrix and slip deficit rate vector. We adopted the procedure of Ref. 8 because, unlike other attempts, quasi-dynamic simulations using the H-matrix can be easily applied to seismic cycles on a 3D surface, since the H-matrix is simply a mathematical approximation as a low-rank compressed representation of the original dense matrix G_{ij}. Thus, regardless of the geometry of the fault of interest or the direction of fault slip, the H-matrix approximation is applicable to quasi-dynamic cycle simulations and suitable for our target problem: following Ref. 8, for the $N(10^5–10^6)$ treated here, the original computation time $O(N^2)$ is reduced to $O(N\log N)$ and the memory size is efficiently reduced to about $O(N)$.

For more efficient computation of seismic cycles, simulations using the H-matrix are run on massively parallel computers. In the parallelized simulation code, all sub-faults are divided into groups according to the number of available CPUs. The velocity V_i and differential strength $\Delta\tau_{s,i}$ of each group are then assigned on the CPU responsible for each group. In addition, the CPUs store only a subset of the H-matrix used to evaluate the velocity rate dVi/dt in each group. Although only the velocities of each group in charge are stored on each CPU, the current velocities of all sub-faults are needed to evaluate Eq. (5). Therefore, before evaluating Eq. (5) at each CPU, the speeds of all sub-faults are collected through Message Passing Interface (MPI) communication, and the evaluation calculation of Eq. (5) is performed at each CPU. The calculated sliding velocity rate and strength rate are integrated over time by the adaptive stepping fifth order Runge–Kutta method.[9]

2.3. *Problem setting*

As shown in Fig. 4(a), we model the boundary between the upper continental plate and the subducting Philippine Sea Plate along the Nankai Trough and Suruga Trough from southern Kyushu to Suruga Bay. Here, we adopt a discretized version of the 3D Philippine Sea Plate boundary geometry of Ref. 10. Next, we define a 3D coordinate system as shown in Fig. 3(b) to divide the plate interface into sub-faults. Using this coordinate system, we divide the plate interface into sub-faults so that the projection of the sub-faults onto the $Y-Z$ plane forms a homogeneous rectangular lattice cell (Fig. 4(b)). For these sub-faults, we prepare the slip response G_{ij} in Eq. (5) as the stress change that enhances the slip of sub-fault i due to unit slip of sub-fault j. Since sub-faults generally form non-planar shapes, the sub-faults are approximated by a set of three triangles, as shown in Fig. 4(b). G_{ij} is then evaluated as the sum of the stress changes due to unit slip on the set of three triangles representing sub-fault j at the center of sub-fault i (marked with X in Fig. 4). The direction of unit slip on each triangle of a sub-fault is determined from the direction of subduction and the shape of the triangle (strike and dip) as follows. First, the horizontal unit vector with the subduction direction is divided into two components: the strike of the triangle and the direction orthogonal to it. Next, the latter component of the unit vector is rotated in the direction of inclination of this triangle, and this rotated vector shall be the inclined component of the unit slip. Finally, the tangential slip that can occur on the triangle is defined as a dislocation along the unit vector consisting of the strike and dip components described previously.

By discretizing the model plate interface into sub-faults with sizes less than 1 km^2 by setting $\Delta Y = 1$ km and $\Delta Z = 0.1$ km, the total

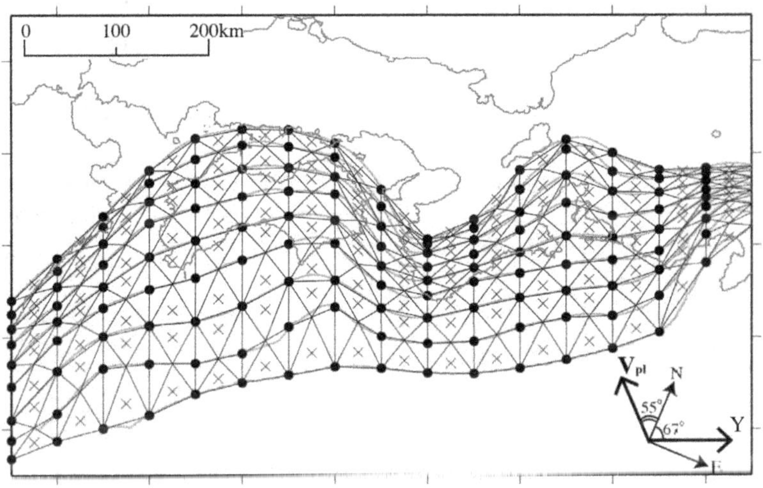

(a)

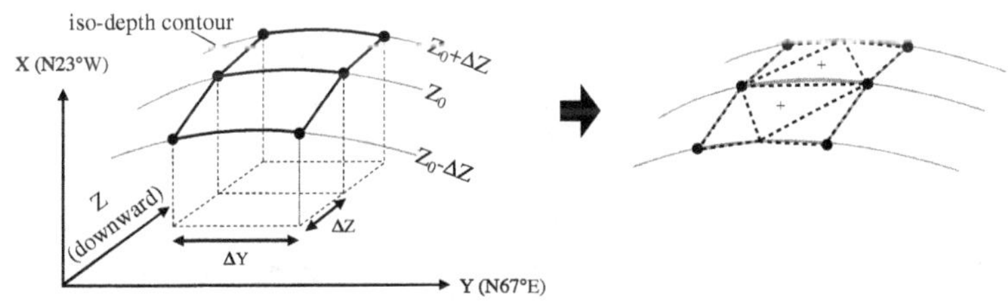

(b)

Fig. 4. (a) Schematic diagram of spatial discretization for the calculation of the slip response function on a 3D fault. The isobath (interval 5 km) on the top of the subducting Philippine Sea plate is shown as a gray curve.[10] Along the isobath, solid lines are distributed at regular intervals (50 km) along the Y-axis (N67°E). Conversely, across the isobath, solid lines are distributed with a constant interval of 5 km depth in the X-axis direction (N23°W). Here, the solid line spacing in this figure is an example of how to discretize the plate interface, and a finer resolution distribution is used in the actual calculation, and (b) schematic illustration of the definition of sub-faults. The left-hand side figure shows division of the plate interface into sub-faults with non-planar shape using the projection to Y–Z plane of the coordinate system in (a). The right-hand side figure shows approximation of a non-planar sub-fault with a set of three triangles.

number of sub-faults treated here is 308,736. For these discretized sub-faults, the set of slip responses G_{ij} in Eq. (5) is evaluated in a homogeneous elastic half-space and compressed as an H-matrix as described previously. The vertical position of the free surface in the elastic half-space is then set to the 5-km depth of the interface geometry.[10] This means that the upper limit of the model interface does not reach the seafloor (i.e., the free surface). It has been shown in Ref. 11 that once the seafloor is reached, the shallow slip becomes too large to stop the rupture.

The distribution of parameters is shown in Fig. 5; the distribution of V_{pl} follows Ref. 12. In addition to the depth dependence, the heterogeneous distribution of frictional properties is given for heterogeneous subsurface structures known from structural surveys,[13] and for the epicentral regions of Bungo slow slip event

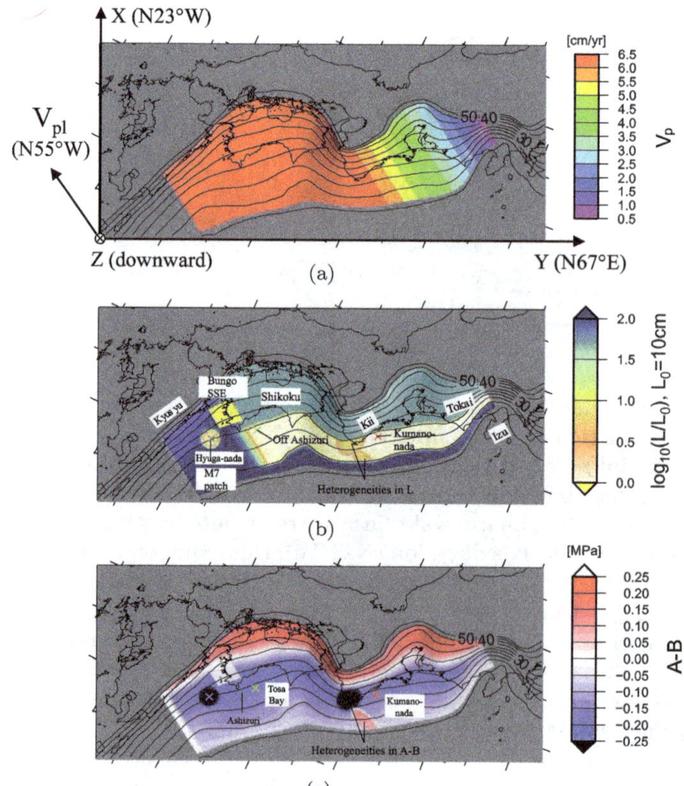

Fig. 5. Distribution of model parameters assumed in this study. (a) Relative plate velocities between the Philippine Sea plate and continental plates, (b) characteristic slip distance L, and (c) A–B. The three X-marks indicate the location of the cumulative slip time series evaluated in Figs. 6 and 8.

(SSE, a square shaped patch) and M7 class Hyuga-nada earthquake (a circle patch).

2.4. *Results*

When the two patches, Hyuga-nada earthquake and Bungo SSE, are included in the Nankai earthquake cycle model, the Hyuga-nada earthquake (shown by the solid line in Fig. 6(a)) occurs with a shorter recurrence time interval than the Nankai Trough earthquake (shown

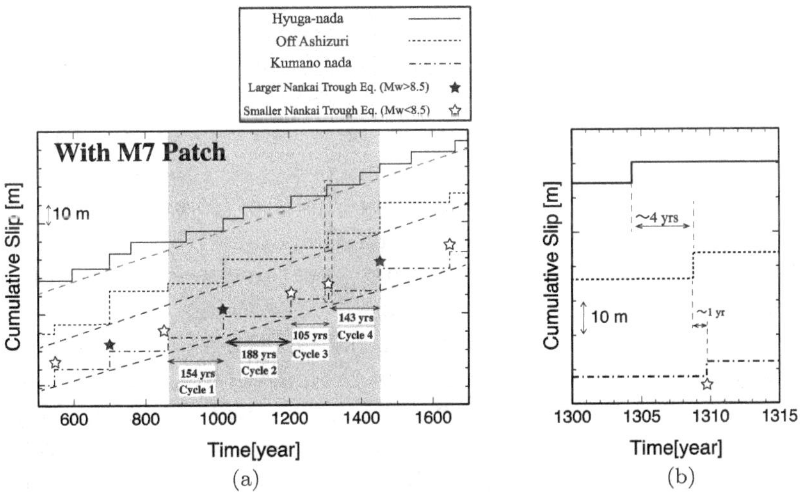

Fig. 6. (a) Cumulative slip in the simulation of the Nankai Trough earthquake cycle including the M7 patch in the Hyuga-nada and the Bungo SSE. Each cumulative slip profile is evaluated at the crosses shown in Fig. 4(b). Solid and open stars represent Nankai Trough's larger earthquakes (Mw $>$ 8.5) and smaller earthquakes (Mw $<$ 8.5), respectively. Except for the step around time = 1310, the larger and smaller Nankai Trough earthquakes occur alternately, and the larger Nankai Trough earthquake is accompanied by a slip of about 10 m in the Kumano-nada Sea (large step in the single-dotted line) and a slip of more than 10 m off Ashizuri-oki (large step in the dotted line). Except for the smaller event at time = 1310, smaller Nankai Trough earthquakes are accompanied by smaller slips than larger earthquakes both in Kumano-nada and Ashizuri-oki, but the relationship of slip magnitude between Kumano-nada and Ashizuri-oki is opposite to that of larger earthquakes. As for the Hyuga-nada earthquake (Mw $\sim$ 7.5), several slip events have occurred in Hyuga-nada, independent of the Nankai Trough earthquake, and (b) enlarged view of the dashed line box in (a). The Hyuga-nada earthquake triggers the Nankai earthquake 4 years later, followed by the Tokai earthquake with a time lag of 1 year.

by the dotted line and single-dotted line in Fig. 6(a)). With the exception of the Nankai Trough earthquake around time = 1310, large and small Nankai Trough earthquakes occur alternately. Around time = 1310, a small Nankai earthquake occurs at only 105 years recurrence interval, as shown in the dashed rectangle in Figs. 6(a) and 6(b). The slip distribution shows that a few years after the Hyuga-nada earthquake (Fig. 7(f)), a Nankai earthquake occurred at the western end of Shikoku and then propagated eastward. The rupture stops due to frictional heterogeneity under the Kii Peninsula (Fig. 7(g)). It then propagates to the eastern Kii Peninsula as a Tokai earthquake about one year later (Fig. 7(h)). Since the Nankai earthquake occurred several years after the Hyuga-nada earthquake (Mw 7.5), it is considered to be a trigger event due to coseismic slip or afterslip after the Hyuga-nada earthquake. However, the frequency of Nankai earthquake triggering events is low (several times in the 2000-year simulation), and it is difficult to examine the conditions for a Nankai earthquake to be triggered by a Hyuga-nada earthquake from observational data. Therefore, we conducted a numerical experiment to evaluate the impact of the Hyuga-nada earthquake on any timing of a representative Nankai Trough cycle.

Using Eq. (4), we evaluate the impact of the Mw7.5 Hyuga-nada earthquake on the Nankai Trough earthquake cycle. We consider the behavior of the Nankai Trough earthquake cycle when the Hyuga-nada M7 patch is removed from the above earthquake cycle model as the representative one. As a result, the representative cycle behavior is that two types of Nankai Trough earthquakes with large magnitude differences occur alternately, as shown by the black star (Mw $>$ 8.5) and white star (Mw $<$ 8.5) in Fig. 8. As shown in Figs. 8 and 9, the time interval immediately preceding a large Nankai Trough earthquake is shorter than the time interval immediately preceding a small earthquake, and the variation in the recurrence interval is about 150–180 years, which is smaller than that of historical earthquakes (90–262 years). In the Hyuga-nada, the afterslip of about 5 m occurs only after larger Nankai Trough earthquakes (shown by the solid line in Fig. 8). In other interseismic periods of Nankai Trough earthquakes, a region of stable slip dominates (solid line in Fig. 8).

Both types of Nankai Trough earthquakes occur at the plate boundary beneath the Kumano-nada Sea. The initiated rupture then propagates eastward and ruptures the eastern segment. At the same

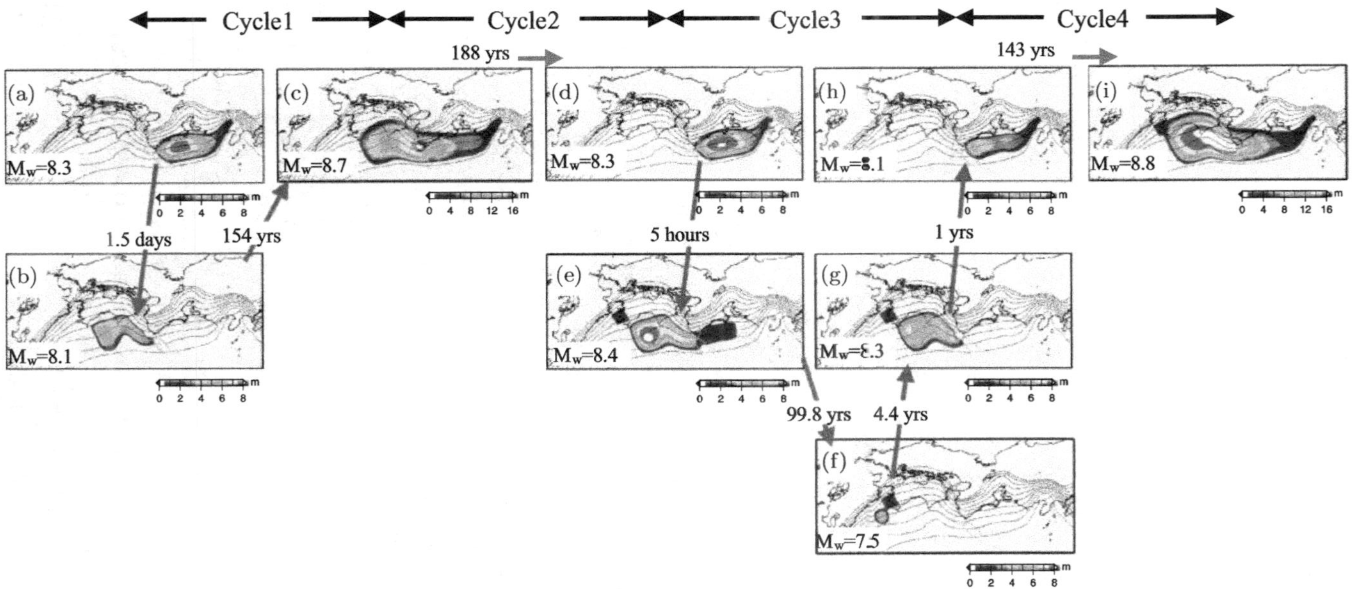

Fig. 7. Seismic slip distribution of the Nankai Trough and Hyuga-nada earthquakes in the gray area of Fig. 6(a). As shown in (c) and (f), the Nankai Trough's larger earthquakes rupture almost simultaneously from western Shikoku to the Tokai region. On the other hand, in smaller earthquakes, the eastern and western parts of the Nankai Trough earthquakes occur separately with a time lag. At the end of the third cycle, the Nankai earthquake occurs about 4 years after the preceding Hyuga-nada earthquake. The Tokai earthquake will then occur alone 1 year later.

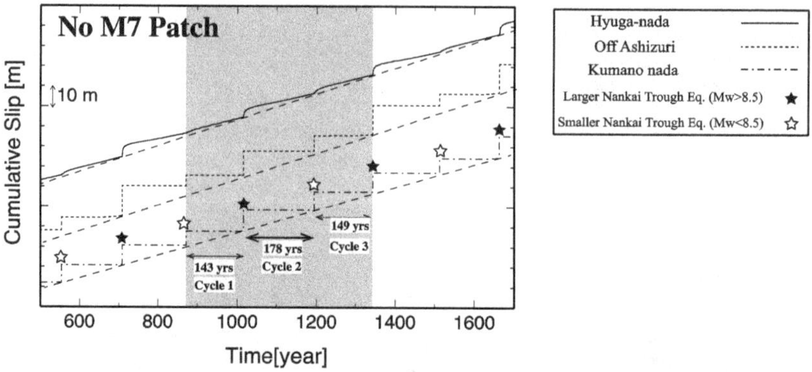

Fig. 8. Cumulative slip in the Nankai Trough earthquake simulation in the absence of the M7 patch in the Hyuga-nada and Bungo SSE. Each cumulative slip profile is evaluated at the location of the crosses as shown in Fig. 5. Solid and open stars indicate the occurrence of a larger Nankai Trough earthquake (Mw $<$ 8.5) and a smaller Nankai Trough earthquake (Mw $>$ 8.5), respectively. It can be seen that Nankai Trough earthquakes alternate between larger and smaller, with larger Nankai Trough earthquakes accompanied by a slip of about 10 m in the Kumano-nada (large step in the single-dotted line) and a slip of more than 10 m off Ashizuri-oki (large step in the dotted line). Smaller earthquakes in the Nankai Trough have smaller slip than larger earthquakes in both Kumano-nada and Ashizuri-oki, but the relationship of slip magnitude between Kumano-nada and Ashizuri-oki is opposite to that of larger earthquakes. In the Hyuga-nada, afterslip of about 5 m corresponds to the occurrence of a larger earthquake.

time, the westward rupture also begins to cross the Kii Peninsula barrier (heterogeneity shown in Figs. 5(b) and 5(c)) and crosses the barrier at different time intervals after the eastward rupture, as shown in Fig. 9. However, the third cycle of breakdown (Fig. 9(f)) begins at the western end of Shikoku and propagates eastward. This type of rupture occurs only once every few thousand years and is seldom seen. Thus, seismic cycles can be divided into three types according to their magnitude and pattern of destruction.

Thus, the behavior of the Nankai Trough giant earthquakes (shortening of the recurrence interval) at the end of the third cycle in Figs. 6 and 7 is typical of the trigger effect of the Mw7.5 Hyuga-nada earthquake. In order to quantitatively evaluate how the trigger effect changes the recurrence interval of Nankai Trough earthquakes, we conducted numerical experiments by applying the velocity disturbance caused by the Mw7.5 earthquake shown in Eq. (4) to the three

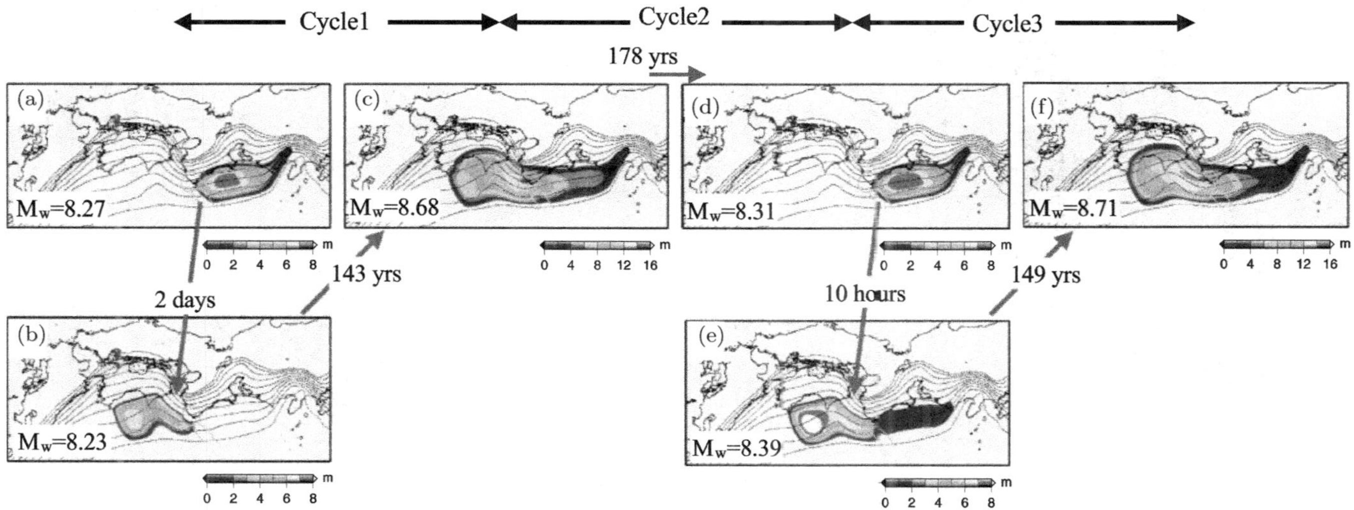

Fig. 9. Seismic slip distribution of the Nankai Trough earthquakes in the gray area of Fig. 8. The Nankai Trough earthquakes occur alternately, and the interval between the earthquake cycles preceding a larger earthquake is shorter than that preceding a smaller earthquake. Small earthquakes at the beginning of the first cycle and at the end of the second cycle rupture the east and west sides of the Nankai Trough separately, with a time lag of about 2 days and 10 hours, respectively. On the other hand, for larger earthquakes (c) and (f), the western part of Shikoku and the Tokai region are ruptured almost simultaneously.

types of earthquake cycles in Figs. 8 and 9. The results show that the trigger effect changes the recurrence interval of the Nankai Trough earthquake cycles. The slip distribution of the Hyuga-nada earthquake shown in Fig. 7(f) was employed to evaluate Eq. (4).

Numerical experimental results showing the relationship between trigger timing (horizontal axis) and the ratio of the modified recurrence interval to the original recurrence interval (vertical axis) are presented in Fig. 10.

For Cycle 1, Fig. 10(a) shows that the resulting earthquake cycle interval changes significantly at 80% of the original cycle time scale.

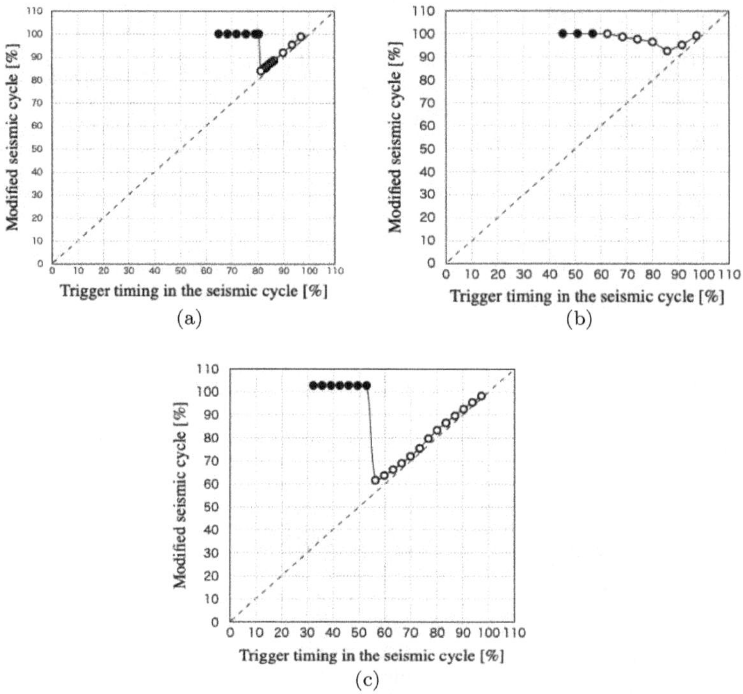

Fig. 10. Changes in the seismic cycle of the Nankai Trough according to the timing of the Hyuga-nada earthquake (Mw = 7.5). In each figure, the horizontal axis indicates the timing of the Hyuga-nada earthquake in the original seismic cycle, and the vertical axis indicates the modified seismic cycle interval relative to the original seismic cycle; values less than 100 indicate a shorter seismic cycle. (a) Case of Cycle 1 in Fig. 8, (b) case of Cycle 2 in Fig. 8, and (c) case of Cycle 3 in Fig. 8. The open circles indicate that the epicenter of the Nankai Trough earthquake is off Ashizuri-oki. Solid circles indicate earthquakes with epicenters in the Kumano-nada.

For triggering before 80% of Cycle 1, neither the location nor the timing of the onset of the next Nankai Trough earthquake rupture is affected. For timing after 80%, the Nankai earthquake starts 3–4 years after the trigger, off Ashizuri-oki, and the earthquake rupture then propagates eastward. The maximum shortening of the seismic cycle is about 23 years, whereas the original Cycle 1 interval was 143 years. This is when it triggered at 81% of the cycle.

In the case of Cycle 2, the interval between triggered earthquake cycles changes gradually (Fig. 10(b)), with a maximum shortening of only 8%, with a trigger timing of 86%. As the original interval of Cycle 2 is 178 years, the maximum shortening of the earthquake cycle is about 14 years, and the interval between the trigger timing and the next Nankai Trough earthquake is about 11 years. 60% or earlier triggers initiate rupture in Kumano-nada, while later triggers initiate rupture in western Shikoku.

In Cycle 3 of Fig. 10(c), the trigger effect changes significantly as in Cycle 1, but the trigger timing that minimizes the interval between modified cycles is much earlier than in Cycle 1, about 57% of the original cycle in Cycle 3. For triggers earlier than 57%, rupture begins in the Kumano-nada, but without the trigger effect, the rupture begins at Ashizuri-oki in Cycle 3. On the other hand, for triggers after 57%, a Nankai earthquake occurs off Ashizuri-oki 2–8 years after the trigger. The maximum shortening of the recurrence interval is about 57 years, compared to the original recurrence interval of 149 years, which corresponds to a reduction of about 40% of the original recurrence interval. Such a significant time reduction is seen only in Cycle 3. Triggered earthquakes off Ashizuri-oki tend to stop at the barrier below the Kii Peninsula, and then the eastern part of the Kii Peninsula is ruptured with a time lag, as shown in the simulation results in Figs. 7(g) and 7(h).

Many triggered earthquakes stop at the segment boundary at the tip of the Kii Peninsula, and the time lags of triggered Nankai earthquakes in the three types of cycles are shown in Fig. 11, which is closely correlated with the shortening of the cycles shown in Fig. 10. The shorter the recurrence time is by the trigger, the larger the lag time between the eastern and western segment ruptures. On the other hand, as shown in Fig. 12, the shorter the recurrence time, the smaller the rupture of the western segment. This is because the shorter the recurrence time, the smaller the accumulated slip deficit.

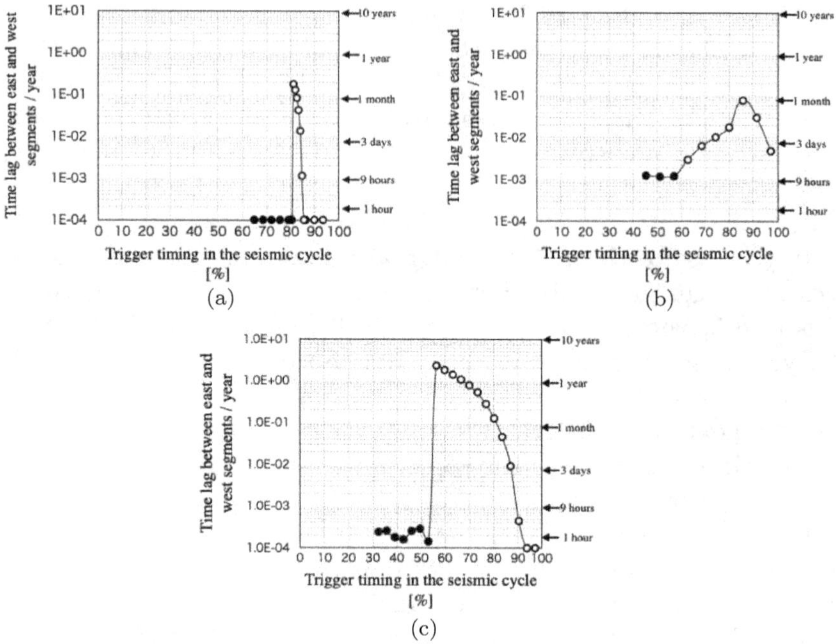

Fig. 11. Variability of possible delay of rupture on the west and east sides of the Nankai Trough earthquake due to the timing of the Hyuga-nada earthquake (Mw = 7.5). In each figure, the horizontal axis indicates the timing of the Hyuga-nada earthquake, and the vertical axis indicates the delay of rupture in the western and eastern sections. (a) Case of Cycle 1 in Fig. 8, (b) case of Cycle 2 in Fig. 8, and (c) case of Cycle 3 in Fig. 8. The open circles indicate that the epicenter of the Nankai Trough earthquake is off Ashizuri-oki. Solid circles indicate earthquakes with epicenters in the Kumano-nada.

Smaller rupture of the west segment results in smaller stress concentrations at the barrier under the Kii Peninsula (see Figs. 5(b) and 5(c)). As a result, the time to induce failure in the eastern segment should be longer.

2.5. *Discussion*

As shown in the results, during the seismic cycle following the smaller Nankai Trough earthquake, the triggered Nankai earthquake occurs in the latter half of the original recurrence interval. The eastward propagation of rupture of the early triggered Nankai earthquake tends to stop at the barrier below the Kii Peninsula, and the rupture

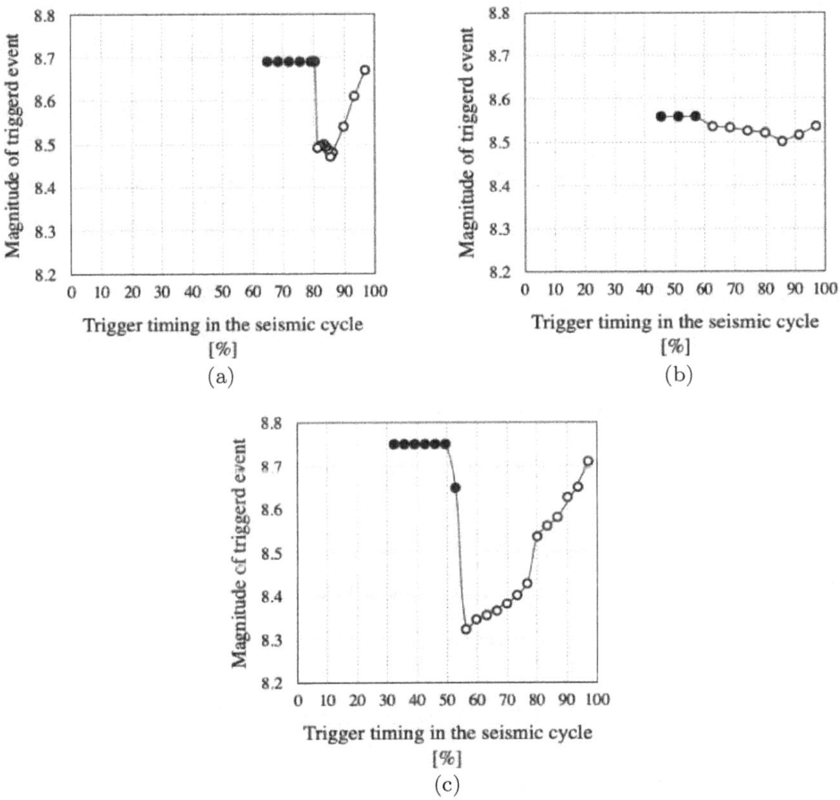

Fig. 12. Variation of the magnitude of Nankai Trough earthquakes according to the timing of the Hyuga-nada earthquake (Mw = 7.5). In each figure, the horizontal axis indicates the timing of the Hyuga-nada earthquake, and the vertical axis indicates the total magnitude of earthquakes that occurred in the western and eastern sections. (a) Case of Cycle 1 in Fig. 8, (b) case of Cycle 2 in Fig. 8, and (c) case of Cycle 3 in Fig. 8. The open circles indicate that the epicenter of the Nankai Trough earthquake is off Ashizuri-oki. Solid circles indicate earthquakes with epicenters in the Kumano-nada.

of the eastern segment occurs separately within a few years after the triggered Nankai earthquake. As a result, the magnitudes of the early triggered Nankai Trough earthquakes will be small. In addition, the occurrence of smaller Nankai Trough earthquakes in the early stage can postpone the occurrence of larger Nankai Trough earthquakes to the next earthquake cycle (Fig. 13).

As shown in Figs. 10–12, the three types of cycles respond quite differently to the Hyuga-nada earthquake. Therefore, we consider

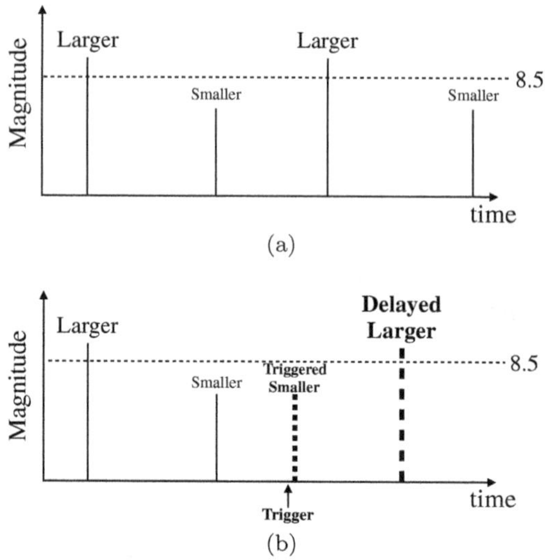

Fig. 13. Schematic representation of the seismic cycles proposed in this study. (a) Earthquake cycle with only nested asperities, and (b) possible modification of (a) by triggering.

the reason for the peculiar behavior of each cycle by focusing on the temporal variation of the slip velocity at the triggered Nankai Trough earthquake site. Figure 14 shows the time variation of the slip velocity in the three earthquake cycles, simulated in the absence of the M7 patch. The time series in Fig. 14 is evaluated at Ashizuri-oki, where many triggered earthquakes occur. As shown in this figure, the trigger start times of 80% and 55% for Cycles 1 and 3, respectively, correspond to slip velocity levels of about 2×10^{-3} V/V_{pl}. On the other hand, in Cycle 2, that level is reached just before 100%. A high slip velocity means a high stress level relative to the strength, and this velocity level is a kind of threshold for triggering in this model although slip velocities on the order of 10^{-3} V/V_{pl} are difficult to resolve by observation. It can be seen that the first cycle reaches the order of 10^{-1} V/V_{pl} in 85% and in 70% in the third cycle. The latter may be resolvable from observation.

Based on the results of numerical experiments, the smaller Nankai Trough earthquakes caused by the Hyuga-nada earthquake will significantly advance in time only when the interplate coupling in western Shikoku is low and the slip rate reaches a certain level, as

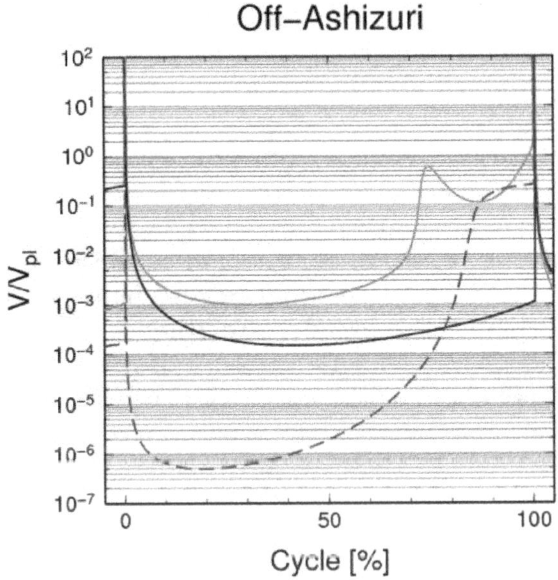

Fig. 14. Temporal variation of slip velocity at the simulated Nankai Trough earthquake site off Ashizuri-oki during the three types of earthquake cycles shown in Fig. 8.

in Cycle 3 (Fig. 14). Therefore, the significance of the Hyuga-nada earthquake for the next Nankai Trough earthquake will strongly depend on the current status of that earthquake cycle.

2.6. *Summary*

To explore the scenario of the next Nankai Trough earthquake, we focused on the Nankai earthquake triggered by the preceding Hyuga-nada earthquake. Detailed numerical experiments showed that after smaller Nankai Trough earthquakes such as the 1946 and 1944 earthquakes, the outer extension of the rupture zone is stressed and slides at a higher slip rate than after larger Nankai Trough earthquakes. Therefore, the Hyuga-nada earthquake (Mw 7.5), which is an extension of the western margin of the Nankai Trough earthquake, tends to trigger a Nankai earthquake, especially in the latter half of the seismic cycle after a Nankai Trough earthquake. Whether the triggering is actually likely to occur depends on the current slip deficit rate near Hyuga-nada. In Section 3, we will introduce GF calculations with more realistic structures, which are necessary to estimate the slip deficit rate more accurately.

3. Calculation of Green's Functions for the Subduction Zones in Japan

Accurate estimates of seismic rupture and its subsurface preparation processes (e.g., aseismic slip and interplate coupling) are important to better understand the process of earthquake generation and to assess earthquake and tsunami hazards. However, in most cases, surface measurements are the only observations available for such estimates. In general, by assuming a linear elastic medium, the surface displacement due to any fault slip on a known fault plane embedded in the subsurface is calculated by superimposing the surface response Green's function (GF) on the elastic deformation due to unit slip on the fault plane. This allows linear slip inversion of the geophysically observed surface displacement and displacement velocity, which can be used for seismic slip,[14–16] aseismic slip,[17–20] slip deficit rates.[21–24] The tsunami waveform inversion for seismic slip also uses GFs to relate seafloor deformation to fault slip distribution.[25]

GFs computed in elastic homogeneous half space are widely used for such purposes because their analytical formulas[26–29] do not require high computational costs and can be calculated with familiar computational programs. Recent geodetic observation techniques, such as the nationwide continuous observation network of the Global Navigation Satellite System (GNSS) over land,[30,31] high-resolution land deformation data from interferometric synthetic aperture radar (InSAR),[32–34] GNSS-acoustic (GNSS-A) technology,[24,35–37] and seafloor pressure (OBP) instruments[38,39] provide more collective information on seafloor deformation. Combining these observations can improve the analysis of crustal deformation over large areas with large spatial variations in topography and elastic properties. Notably, consistent analysis of geodetic data over such large areas can be achieved using finite element (FE)[40–42] or spectral element (SE) methods,[43,44] where multiple studies have reported that incorporating a realistic Earth model based on the elastic homogeneous half-space often requires a more realistic Earth model. For example, the effects of topography and elastic properties are discussed.[40,45–49]

These results suggest that GFs calculated by incorporating realistic topographic and subsurface models are essential for estimating fault slip and interplate coupling based on recent geodetic data, which are observed not only on land but also on the seafloor.

Ignoring the elliptical shape of the Earth (usually approximated by a projected plane, or at best a sphere) is another source of error when comparing computed and observed data. This method is problematic in two respects. One is the influence of elliptical shapes in the calculation of response due to fault slip, which has been studied;[50] for GNSS observation data, the geodetic reference system 1980 (GRS80) adopted by the International Union of Geodesy and Geophysics[51] and the Japan Geodetic System 2000[52] are applied according to the International Earth Reference Coordinate System (http://itrf. ensg.ign.fr/), while numerical models are usually defined in the local projected Cartesian or symmetric spherical coordinate system. Less attention has been paid to this latter aspect, but it is a more fundamental issue involving the former. If the area to be modeled is small, any coordinate transformation method will not cause significant distortion. However, if the model covers a relatively large area where the ground is quite curved, as mentioned above, the difference in coordinate systems can cause non-negligible errors in the calculation results. Such errors can be avoided by making the coordinate system used for model calculation the same as that of GNSS observation data. Therefore, in order to calculate a fairly wide range of GFs, it is desirable to consider the Earth as an ellipsoid according to the definition of the GRS80 coordinate system.

The regions covered here are the Nankai Trough and the subduction zone of the Japan Trench, which are known for megathrust earthquakes.[53,54] For the Japanese Islands, the Japan integrated velocity structure model version 1 (JIVSM),[55,56] a unified velocity structure model proposed by the Japanese government for seismic hazard assessment, was deemed suitable. We used a high-fidelity FE model that faithfully reproduces the resolution and layer boundaries of the JIVSM subsurface structure model. Coordinate transformations necessary for data processing were performed faithfully to the definition of the coordinate system used in the geodetic reference system. After coordinate transformation, a high-resolution FE mesh was constructed from the structured data using an automatic mesh generator developed especially for stratified layer structures.[42,57] FE analysis is then performed using a high-performance FE solver[58] that can calculate GFs incorporating high-resolution digital elevation model (DEM) data from JIVSM.

3.1. *Coordinate transformation between the geographic and a Cartesian coordinate system*

In order to calculate the elastic GF, it is necessary to assume elastic properties of the subsurface structure. JIVSM is composed of DEM data consisting of multiple layers defined in a geographic coordinate system (longitude, latitude, and elevation) as well as a seismic wave velocity structural model defined over a wide area including the source region of a magnitude 9-class earthquake, and seismic wave velocities in each layer are included. In contrast, elastic deformation calculations are usually based on a Cartesian coordinate system. Therefore, to construct a numerical simulation model (FE model), it is necessary to transform the DEM data from a geographic coordinate system to a Cartesian coordinate system. However, constructing a numerical simulation model for GF calculations in the Cartesian coordinate system usually involves work that requires user-dependent parameters to define the transformation, such as projection onto a hemisphere (e.g., the Mercator diagram method) or spherical approximation. This section takes this into account and describes the procedure for transforming the coordinates of DEM data in a manner faithful to the original definition of the coordinate system.

We performed coordinate transformations based on GRS80, a compliant ellipsoid of GRS80 that is practically equivalent to the World Geodetic System 1984 (WGS84) (https://www.nga.mil/ProductsServices/GeodesyandGeophysics/Pages/WorldGeodeticSystem.aspx),[59] which is practically equivalent to the WGS84, and GRS80 was used as a compliant ellipsoid (Geodetic System). A target Cartesian coordinate system was defined with the representative point (center point) of the target area Pc as the origin. The y-axis is defined in the north direction and the z-axis in the upward direction (the x-axis is then automatically determined by the right-hand system). Hereafter, unless otherwise noted, the unit of length is the meter (m). First, the coordinates of point P in the input DEM data were transformed to the Earth-centered and Earth-fixed (ECEF) coordinate system as follows (Fig. 15(a)).

$$X = (N + h)\cos\varphi\cos\lambda, \tag{6}$$

$$Y = (N + h)\cos\varphi\sin\lambda, \tag{7}$$

$$Z = \{N(1 - e^2) + h\}\sin\varphi, \tag{8}$$

where φ, λ, and h are latitude, longitude, and ellipsoidal height (the sum of elevation and geoid height), and X, Y, and Z are the coordinates in the ECEF system. $N(= a/\sqrt{1 - e^2\sin^2\varphi})$, a, and e are the prime vertical radius of curvature, equatorial radius, and first eccentricity of the Earth ellipsoid. For the GRS80 reference ellipsoid, $a = 6378137m, e = 0.081819191042815791$. We use the geoid model from Earth Gravitational Model 2008 (EGM2008),[60] a high-resolution global geoid model that includes both land and sea areas.

Next, for the convenience of defining the region of interest for FE mesh generation, the coordinates of the ECEF system were transformed and rotated to the local Cartesian coordinate system of the region of interest (Fig. 14(b)) as follows.

$$\begin{pmatrix} x \\ y \\ z \end{pmatrix} = \begin{pmatrix} -\sin\lambda_c & \cos\lambda_c & 0 \\ -\sin\varphi_c\cos\lambda_c & -\sin\varphi_c\sin\lambda_c & \cos\varphi_c \\ \cos\varphi_c\cos\lambda_c & \cos\varphi_c\sin\lambda_c & \sin\varphi_c \end{pmatrix} \begin{pmatrix} X - X_c \\ Y - Y_c \\ Z - Z_c \end{pmatrix}, \tag{9}$$

where φ_c and λ_c are the latitude, longitude, and ellipsoidal height of Pc, X_c, Y_c, and Z_c are the coordinates of Pc in the ECEF system, and x, y, and z are the coordinates in the local Cartesian coordinate system. The mapping matrix in Eq. (9) is the product of a rotation matrix with $\frac{\pi}{2} + \lambda_c$ around the Z-axis, one with $\frac{\pi}{2} - \varphi_c$ around the x-axis. The coordinate transformation was performed for each point in the original DEM data set defined in the geographic coordinate system; the construction of the FE model requires DEM data equally spaced in the $x-y$ plane in the Cartesian coordinate system. The data points in the original DEM data are usually defined on a structural lattice of equal intervals in the geographic coordinate system, but when the coordinates of the data points are transformed to the local Cartesian coordinate system based on Eq. (9), the structural lattice is not preserved. Therefore, data points with equally spaced structural lattices on the $x-y$ plane of the local Cartesian

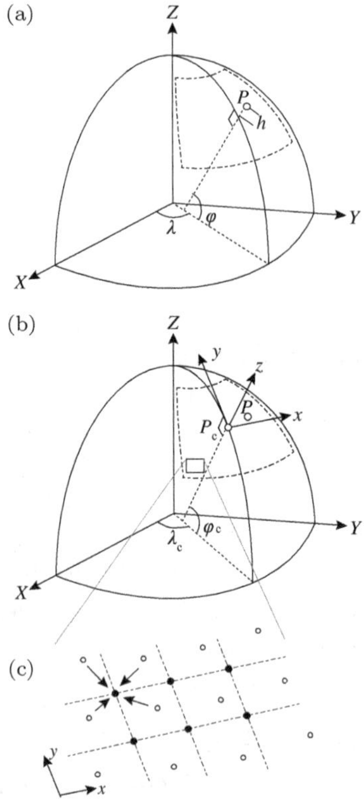

Fig. 15. Schematic representation of the coordinate transformation. For simplicity, 1/8 of the GRS 80 reference ellipsoid is visualized. (a) The coordinates of P are transformed from the geographic coordinate system to the ECEF system. The area of the input data is defined as a rectangular region in the geographic coordinate system, so it appears distorted in the ECEF system as if it were surrounded by a dotted line, (b) the coordinates of P are transformed from the ECEF system to the local Cartesian coordinate system with Pc as the origin, and (c) generation of new data points, indicated by equally spaced black circles in the $x-y$ plane in the local Cartesian coordinate system. The white circles are data points in the original DEM data that are not placed on a structured grid equally spaced in the local Cartesian coordinate system. The z-components of the points in the black circles are interpolated from the points in the white circles using a near neighbor algorithm as indicated by the arrows.

coordinate system were newly generated, and their z coordinates were interpolated from the neighboring points of the original data (Fig. 15(c)). Neighborhood point interpolation is used, in which each point is assigned an average value weighted by the reciprocal of the distance between the four neighboring points.

From the response displacement field calculated by the FE method (see the following section), the displacement at any point on the ground surface can be extracted. The xyz component of displacement is defined in a local Cartesian coordinate system characterized by Pc, but GNSS and GNSS-A geodetic observation data are usually provided as ENU (east-west, north-south, up-down) coordinates. Therefore, observed and calculated displacements cannot be directly compared. The displacements calculated in the local Cartesian coordinate system must be transformed into ENU components at P. Consistent with the method described above, a coordinate transformation of the computed displacements based on the GRS80 ellipsoid is performed. The coordinate transformation is defined as follows:

$$\begin{pmatrix} u_E \\ u_N \\ u_U \end{pmatrix} = \begin{pmatrix} -\sin\lambda & \cos\lambda & 0 \\ -\sin\varphi\cos\lambda & -\sin\varphi\sin\lambda & \cos\varphi \\ \cos\varphi\cos\lambda & \cos\varphi\sin\lambda & \sin\varphi \end{pmatrix} \times \begin{pmatrix} -\sin\lambda_c & \cos\lambda_c & 0 \\ -\sin\varphi_c\cos\lambda_c & -\sin\varphi_c\sin\lambda_c & \cos\varphi_c \\ \cos\varphi_c\cos\lambda_c & \cos\varphi_c\sin\lambda_c & \sin\varphi_c \end{pmatrix}^{\mathrm{T}} \begin{pmatrix} u_x \\ u_y \\ u_z \end{pmatrix}, \tag{5}$$

where u_E, u_N, u_U, u_x, u_y, and u_z are displacements in the ENU and xyz components at P. In this method, the ECEF system components are first converted and then converted again to ENU components in P.

3.2. *Calculation method of Green's function*

In calculating GF, we aimed to use a numerical simulation model that is faithful to the reference subsurface database, i.e., an FE mesh with layer boundaries of the same resolution as JIVSM (about 1 km). If such a mesh were constructed for a large area with a side length of several thousand kilometers and more than 10 layers of different seismic intensity, the typical degrees of freedom (DOF) would be on the order of billions. Robustly constructing a high-quality mesh with complex layer boundary geometry in such a large model is not an easy task. In general, the creation of the FE mesh itself can take much more time than the FE analysis of a large-scale problem.[58] We constructed a high-resolution FE mesh from coordinate-transformed DEM data using a robust and fast automatic tetrahedral mesh generation tool developed specifically for layered structures.[42] In this method, the mesh is generated based on uniform background cells covering the entire area of interest, and the resolution of the layer interface *ds* is defined. The geometry of the ground surface and layer interface was slightly simplified to preserve the quality of the elements. Unnecessary elements were integrated to generate larger elements elsewhere. This method enabled the automatic and robust construction of high-resolution tetrahedral meshes directly from DEM data of crustal structures, without the need to create computer-aided design (CAD) models.

Normalized third-order bicubic B-splines[14] were used as basis functions for the unit slip (see Ref. 61 for definition). For sub-faults located on the plate boundary, B-spline style slip was input along the structured grid. Slip in the direction of subduction and orthogonal to it, established based on previous studies,[12,62] was taken into account. Slip beyond the trench or trough axis is assumed to be zero.

The response displacements due to unit slip on the sub-fault are calculated using the FE method. By applying the FE formulation to the governing equations of elastic deformation and placing boundary conditions, the problem becomes the following linear equation: $Ku = f$, where K, u, and f are the global stiffness matrix, the displacement vector, and the outer force vector. The fault slip is transformed into a force vector using the split-node method.[63] Dirichlet conditions are applied to the sides and bottom of the modeling domain. Specifically, the displacement in the direction of the surface normal vector is fixed to zero and no displacement is imposed in other directions.[41] As mentioned earlier, FE models that faithfully reproduce seismic wave velocity structure data typically have billions of DOF. Therefore, a large set of simultaneous equations must be solved to obtain u. For this purpose, we have introduced a state-of-the-art FE solver for computing crustal deformation.[58] The basic algorithm is a conjugate gradient (CG) method with the addition of a memory-efficient matrix–vector product calculation. The convergence of the CG solver was improved by using an adaptive preconditioner that combines geometric and arithmetic multigrid methods. The resulting solver is scalable to 105 computational cores and in some cases obtained a 20-fold speedup from widely used solvers.[58] For more information on FEM large-scale fast computation methods, see Chapter 5.

3.3. *Calculation of Green's functions for the subduction zones in Japan*

Following the method described in the previous section, a numerical simulation model of the Nankai Trough and Japan Trench subduction zone was constructed based on JIVSM. JIVSM is a seismic wave velocity structure model covering most of the Japanese Islands and was developed to simulate long-period seismic motions and to evaluate their seismic hazards. The structure, which includes shallow soil to the oceanic and continental mantle, is modeled as a 23-layer structure, and its geometry and elastic properties are provided in the data set. The shallowest 14 layers in the model are thin, and seismic velocity changes between these layers are assumed to be unaffected by crustal deformation over relatively long wavelengths, so they are integrated into a layer that includes the accretionary wedge and shallower layers and the remaining layers for the calculation. JIVSM is basically based on the Bessel ellipsoid since it is defined by the former Japan Geodetic System (Tokyo Datum), the coordinates were first converted to GRS80 and then applied to the aforementioned procedure.

The deep part of the plate boundary where interplate fault slip is expected to occur is considered by JIVSM to be located at the boundary between the upper part of the subducting oceanic crust and the lower part of the continental crust or mantle. However, in the shallow part of the megathrust, fault slip does not necessarily occur at the boundary between continental and oceanic crust, nor does it necessarily occur on a single plane. Since this study does not aim to calculate such complex GFs, we simply assume that the relative motion between the two plates is accommodated by a single fault. By smoothly connecting the axes of troughs and trenches with the boundaries between continental and oceanic crust, a new surface for applying shallow fault slip is generated. The connection point on the crustal boundary is located at a depth of 15 km, thus generating a smooth surface. We generated surfaces using the adjustable tension continuous surface grid method.[64] The newly generated surfaces are shown in Fig. 16.

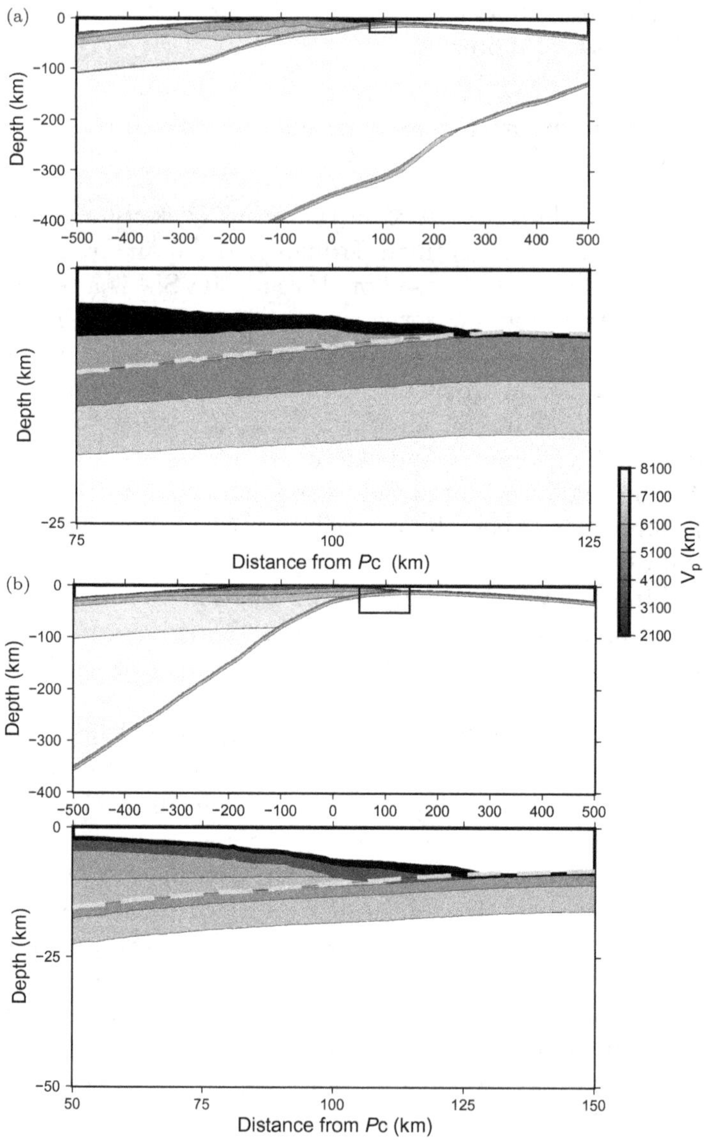

Fig. 16. Seismic wave velocity (P-wave) profiles on the cutting plane. Along the line in Fig. 18, profiles on the cutting plane of (a) Nankai Trough, and (b) Japan Trench subduction zones are shown. The upper figure is an overview view. The lower figure is a close-up of the area around the axis of the trough and trench, indicated by the solid rectangle in the upper figure. The dashed line is a newly generated surface created by interpolating the top surface of the subducting plate and the axis of the trough and trench.

The parameters for mesh construction, such as domain size and mesh resolution, were selected based on the results of an evaluation of numerical solutions for elastic deformation caused by a Mw = 9 earthquake in the Japan Trench region.[65] As a result, the horizontal element size of the layer interface is $ds = 1$ km (ds corresponds to the edge length here) and the domain size is 2500 km × 2500 km × 1100 km. The model domain is kept large enough that approximation errors due to boundary conditions at the model boundaries are negligible in the displacement calculations for observation points in the region of interest, i.e., -1000 km $\leq x \leq 1000$ km and -1000 km $\leq y \leq 1000$ km, with Pc as the origin. The vertical thickness of each layer is assumed to be 0 in the FE model if it is less than 0.125 ds in the original data. This is because the presence of such a thin layer would cause difficulties in the generation of the FE mesh. Based on the results of several preliminary calculations with different element sizes ds, the elements of the sediment layers in the upper plate were further refined down to $ds = 0.5$ km.

We calculated GF in the Nankai Trough subduction zone in southwestern Japan (Fig. 17(a)); Pc is located 80 km offshore from Cape

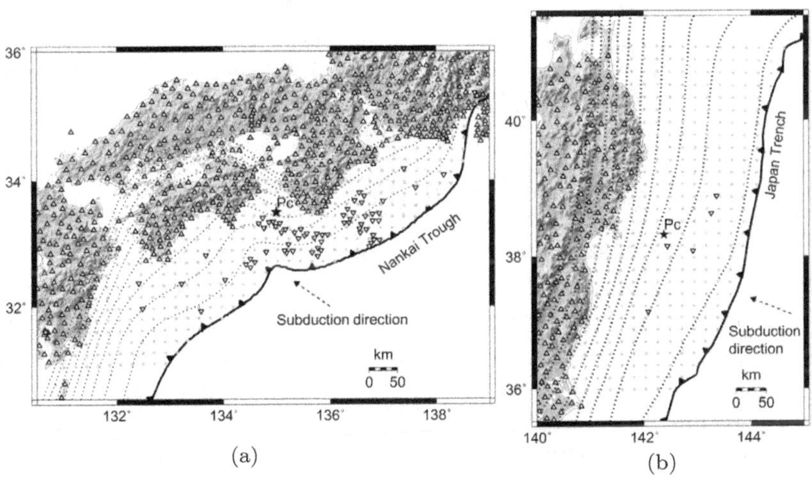

Fig. 17. Overview of the area subject to GF calculations. (a) is the Nankai Trough, and (b) is the subduction zone of the Japan Trench. Gray dots indicate the center points of the B-spline function-type slip distribution applied to the subfaults. Triangles and inverted triangles represent GEONET and representative seafloor stations for geodetic observations,[24,36,39] respectively.

Table 1. Location of *Pc* in the subduction zones of the Nankai Trough and Japan Trench, area for calculating GF, and direction of subduction.

	Pc (Lon., Lat., ellipsoidal height)	Region for GFs SW-NE edges	Subduction direction
Nankai Trough	(135.0,33.5,0)	(131,31)–(139,35.5)	N55W
Japan Trench	(142.369, 38.322, 0)	(140.5,36)–(145, 41)	N68.5W

Note: The subduction directions are based on Ref. 12 for the Nankai Trough and Ref. 61 for the Japan Trench, respectively.

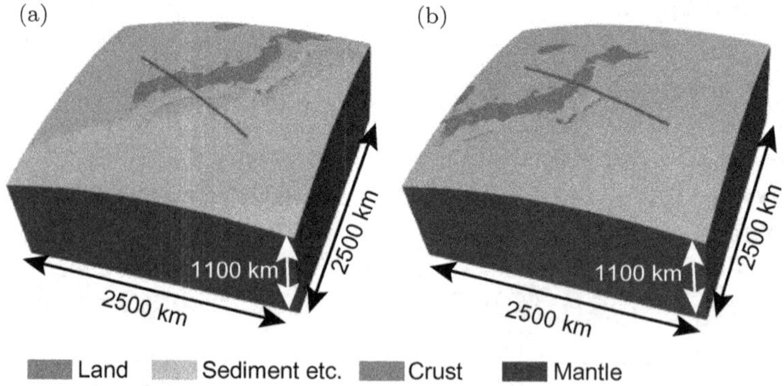

Fig. 18. Overview of FE model. (a) FE model of the Nankai Trough, and (b) FE model of the subduction zone of the Japan Trench. Seismic wave velocity profiles on the solid lines are shown in Fig. 16.

Muroto (Table 1). Figure 18(a) shows the FE mesh of the target area. The meshed model surface well represents the elliptical shape of the Earth centered east of Cape Muroto. Figure 16(a) shows a cross-section of the velocity structure incorporated into the FE model. The plate boundary of the constructed model has 460 structured grid points that introduce sub-faults with third-order B-spline functions at 20 km node intervals, and the upper limit of the slip depth is 50 km. The slip direction of each sub-fault was set to two directions, and a total of 920 GFs were calculated. Figure 19(a) shows an example of GF (surface displacement) calculation when B-spline type slip is applied to a sub-fault on a plate boundary in the Nankai Trough subduction zone. Note that in this example the unit slip at locations beyond the trough axis is forced to be set to zero.

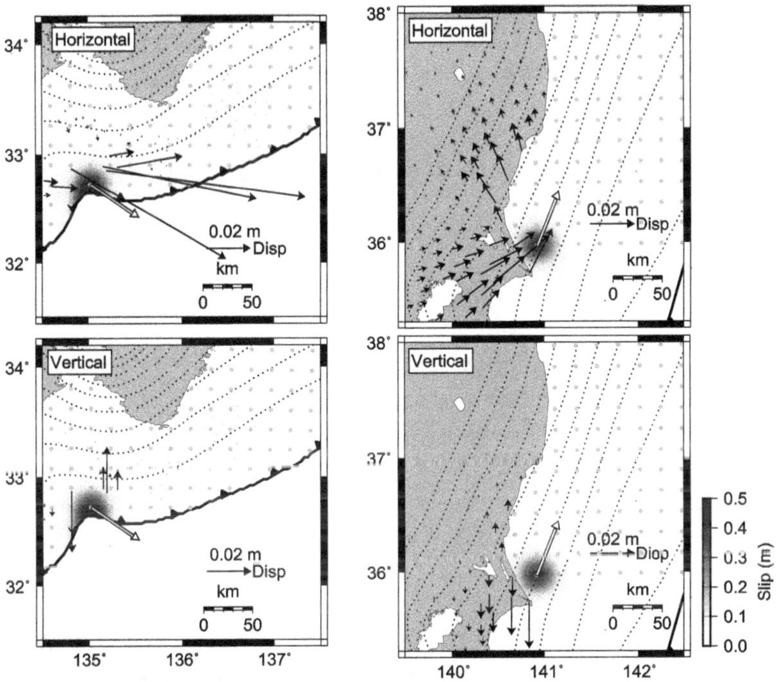

(a) (b)

Fig. 19. Examples of GFs calculated based on FE modeling. (a) Unit slip in the direction opposite to plate convergence and its response displacement in the Nankai Trough, and (b) unit slip in the direction orthogonal to plate convergence in the Japan Trench subduction zone. The gray-scale map shows the slip distribution. The black arrows are the horizontal (top) and vertical (bottom) surface displacements due to fault slip. White arrows indicate the direction of fault slip.

Similarly, GF was calculated for the subduction zone of the Japan Trench in Northeast Japan (Fig. 17(b)), with *Pc* placed near the epicenter of the 2011 Tohoku-Oki earthquake (Table 1). Figure 18(b) shows that the meshed model surface well represents the elliptical shape of the earth centered on the epicenter. Figure 16(b) shows a cross-section of the velocity structure incorporated into the FE model. In the constructed model, 405 structured grid points were placed at 20 km intervals with a slip depth limit of 100 km. In total, 810 GFs were calculated. Figure 19(b) shows an example of the calculated displacement (surface displacement) when B-spline type slip is applied to a sub fault on a plate boundary in the Japan Trench subduction zone.

3.4. *Application to slip deficit rate estimation in the Nankai Trough*

Using the calculated GFs, we estimated the interseismic SDR of the Nankai Trough subduction zone and compared it with the results obtained based on homogeneous elastic half-space.[24] Both used the interseismic displacement rate data: the GNSS-A seabed geodetic data and GNSS land-based continuous data to estimate the interplate SDR of the Nankai Trough subduction zone. They used the method of Ref. 14 (imposing a smoothness constraint on the slip distribution as prior information and determining the hyperparameters of the constraint using Akaike's Bayes Information Criterion (ABIC)) on these data for inversion analysis. We used the same data set and methodology to estimate the interplate SDR of the Nankai Trough subduction zone, except that we incorporated 3D heterogeneous elastic properties and the elliptical shape of the Earth into our analysis.

Figure 20(a) shows the estimated SDR using the GFs calculated here. A large SDR is estimated for Cape Ashizuri, similar to the preferred model of Ref. 24 (Fig. 21(a)) and other SDR models.[66,67] However, the region of large SDR in our model is considerably smaller than in the model of Ref. 24. Our result is more consistent with the fact that the maximum SDR in this subduction zone is not expected to exceed the reported plate convergence rate, i.e., about 7 cm/year.[12] In particular, our model can effectively explain the displacements observed at the ASZ2 station without requiring a large slip deficit at the plate boundary immediately below the station. Comparing Figs. 20(b) and 21(b), we also observe a better fit in our SDR model at stations TOK1 and TOK2. This difference in goodness-of-fit to the observations can be attributed to the fact that we considered the consistency of fault geometry, trough axis, and surface geometry in our numerical model. Such consistency is expected to significantly improve the prediction accuracy of the model if the database of subsurface structures used in the analysis is reliable. Furthermore, this consistency in modeling structures around the trough axis allows for

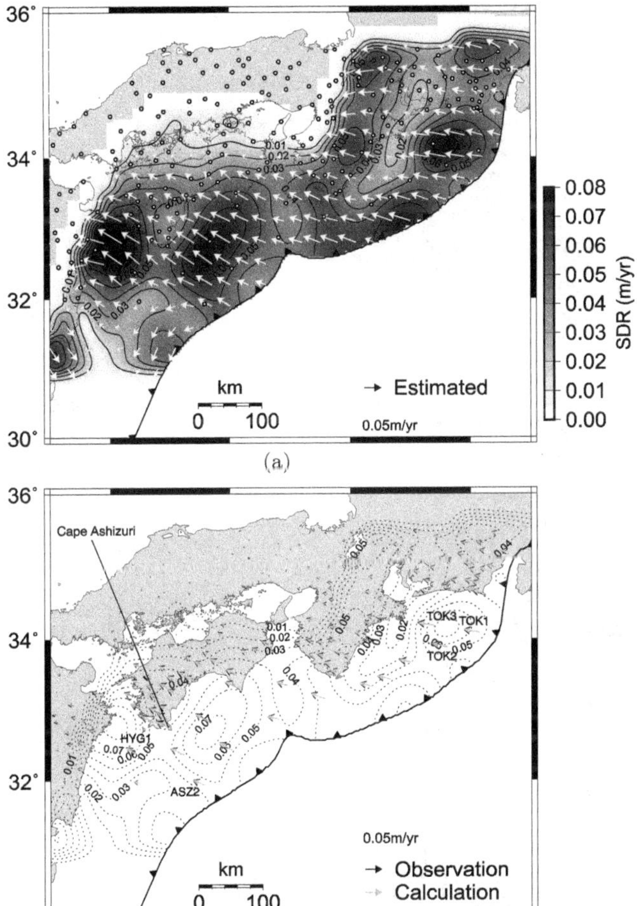

Fig. 20. SDR distribution estimated using the GFs calculated here. (a) Gray-scale contours and vectors of the estimated SDR. Open circles indicate observation points, and (b) comparison of observed and calculated velocity at observation points. Dotted contours indicate the norm of the estimated SDR.

the placement of fault patches close to the trough axis. This is in contrast to the traditional approach based on a combination of planar faults and flat ground surface, where the location of such patches cannot be determined when considering curved trough axes and ground surface topography. It is not surprising that fault-slip models that

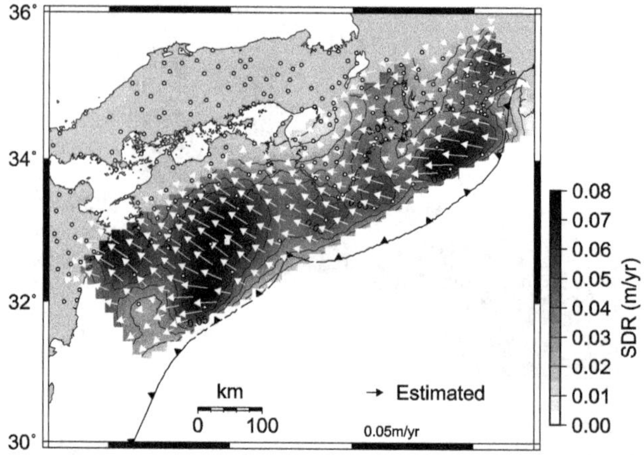

(a)

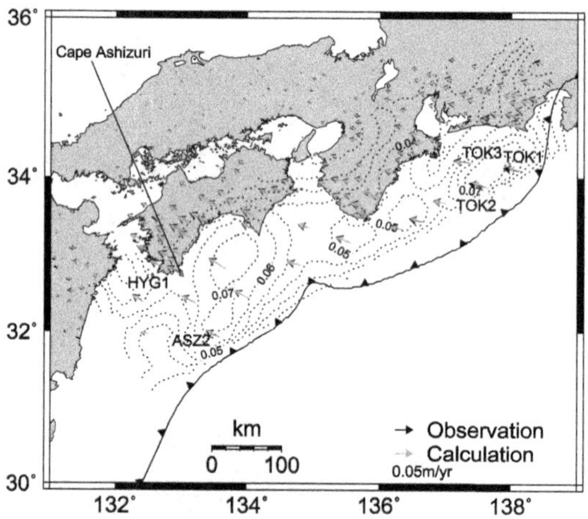

(b)

Fig. 21. SDR distribution estimated by Ref. 24 (a) Gray-scale contours and vectors of the estimated SDR. Open circles indicate observation points, and (b) comparison of observed and calculated speeds at observation points. Dotted contours indicate the norm of the estimated SDR.

account for such fault patches can better explain observations near the trough axis.

3.5. *Summary*

When calculating GF, we proposed the concept of matching the coordinate system and geometry of the calculation model with the observed data. In other words, we introduced the world geodetic system (ITRF and compliant ellipsoid GRS80) used in GNSS for observation of geodetic data. We also introduced the 3D topography and heterogeneous subsurface structure, which have been pointed out as necessary in recent years.The resolution of the layer boundaries of the subsurface structure is the same as the DEM data of the JIVSM Version 1,[55,56] and we used an FE model that is faithful to the Japanese unified velocity structure model. The coordinate transformation of the velocity structure DEM data from the geographic coordinate system to the local Cartesian coordinate system was performed faithfully to the definition of each coordinate system, without projection onto a hemisphere or approximation with a spherical shape. The computational costs of FE mesh generation and FE analysis associated with high-resolution FE models were overcome by introducing robust and fast automatic tetrahedral mesh generation tools developed especially for stratified layer structures[42] and state-of-the-art FE solvers,[58] respectively, into the FE analysis. An empirical application of the computed GF to the estimation of interseismic SDR in the Nankai Trough subduction zone was also presented. Comparison with estimation results based on a simplified Earth model showed that the computed GFs provide more consistent estimation results for both geodetic data and understanding of plate tectonics.

3.6. *Future directions*

In the future, higher resolution modeling will be required as the seismic velocity structure model is updated to introduce more detailed structures, for example, splay faults. DEM data for subsurface structures in other subduction zones around the world, such as CRUST1.0,[68] can also be applied to the FE mesh generation procedure described here, provided the dataset is defined in a geographic coordinate system. In addition, viscoelastic GFs are in demand for the analysis of geodetic data over longer periods of time. The computational cost required to create a viscoelastic GF is expected to be significantly higher than that for an elastic GF. Therefore, it is essential to introduce more efficient algorithms for FE analysis that are compatible with new computational environments.[69]

In addition, performing cycle simulations in Section 2 on a realistic structural model in Section 3 is one of the next important issues. To do so, calculations with sufficient spatial resolution to converge stress changes on the fault plane must be performed for a large number of GFs, which is a challenging task for computational science.

Furthermore, estimation of the slip loss rate, which is important for scenario feasibility assessment, requires Uncertainty Quantification, which takes into account errors in observation data and model errors. This also requires more extensive calculations, including the introduction of ensembles to represent the distribution of errors.

Acknowledgment

This study was supported by Post K computer project (Priority issue 3: "Development of Integrated Simulation Systems for Hazard and Disaster Induced by Earthquake and Tsunami"). Some of the results were obtained using the K computer at the RIKEN Advanced Institute for Computational Science (Proposal numbers hp150215, hp180207, and hp190177). This study was also supported by MEXT as "Program for Promoting Researches on the Supercomputer Fugaku" (Large-scale numerical simulation of earthquake generation, wave propagation, and soil amplification: Proposal number hp200126) and used computational resources of Oakforest-PACS provided by The University of Tokyo.

References

1. K. Ishibashi, Status of historical seismology in Japan, *Ann. Geophys.* **47**, 339–368 (2004).
2. J. R. Rice, Spatio-temporal complexity of slip on a fault, *J. Geophys. Res.* **98**, 9885–9907 (1993).
3. M. Nakatani, Conceptual and physical clarification of rate and state friction: Frictional sliding as a thermally activated rheology, *J. Geophys. Res.* **106**, 13347–13380 (2001), doi:10.1029/2000JB900453.
4. N. Kato and T. E. Tullis, A composite rate- and state-dependent law for rock friction, *Geophys. Res. Lett.* **28**, 1103–1106 (2001).
5. M. L. Blanpied, T. E. Tullis, and J. D. Weeks, Effects of slip, slip rate, and shear heating on the friction of granite, *J. Geophys. Res.* **103**, 489–511 (1998).

6. N. Kato, Numerical simulation of recurrence of asperity rupture in the Sanriku region, northeastern Japan, *J. Geophys. Res.* **133**, B06302 (2008).
7. T. E. Tullis, J. Salmon, N. Kato, and M. Warren, The application of fast multipole methods to increase the efficiency of a single fault numerical earthquake model, *AGU Fall Meeting Suppl.* **80**, F924 (1999).
8. M. Ohtani, K. Hirahara, Y. Takahashi, T. Hori, M. Hyodo, H. Nakashima, and T. Iwashita, Fast computation of quasi-dynamic earthquake cycles simulation with hierarchical matrices, *Proc. Com. Sci.* **4**, 1456–1465 (2011).
9. W. H. Press, S. A. Teukolsky, W. T. Vetterling, and B. P. Flannery, Numerical *Recipes in Fortran 77: The Art of Scientific Computing, 2nd edition.* New York: Cambridge University Press (1996).
10. T. Baba, Y. Tanioka, P. R. Cummins, and K. Uhira, The slip distribution of the 1946 Nankai earthquakes estimated from tsunami inversion using a new plate model, *Phys. Earth Planet. Int.* **132**, 59–73 (2002).
11. M. Hyodo and T. Hori, Re-examination of possible great interplate earthquake scenarios in the Nankai Trough, southwest Japan, based on recent findings and numerical simulations, *Tectonophys.* **600**, 175–186 (2013).
12. K. Heki and S. Miyazaki, Plate convergence and long-term crustal deformation in central Japan, *Geophys. Res. Lett.* **28**(12), 2313–2316 (2001).
13. S. Kodaira, T. Hori, A. Ito, S. Miura, G. Fujie, J.O. Park, T. Baba, H. Sakaguchi, and Y. Kaneda, A cause of rupture segmentation and synchronization in the Nankai Trough revealed by seismic imaging and numerical simulation, *J. Geophys. Res.* **111**, B09301 (2006).
14. T. Yabuki and M. Matsu'ura, Geodetic data inversion using a Bayesian information criterion for spatial distribution of fault slip, *Geophys. J. Int.* **109**(2), 363–375 (1992).
15. M. Chlieh M, J. P. Avouac, V. Hjorleifsdottir, T. R. A. Song, C. Ji, K. Sieh, A. Sladen, H. Hebert, L. Prawirodirdjo, Y. Bock *et al.*, Coseismic slip and after-slip of the great Mw 9.15 Sumatra-Andaman earthquake of 2004, *Bull. Seismol. Soc. Am.* **97**(1A), 152–173 (2007).
16. T. Iinuma, R. Hino, M. Kido, D. Inazu, Y. Osada, Y. Ito, M. Ohzono H. Tsushima, S. Suzuki, H. Fujimoto, and S. Miura, Coseismic slip distribution of the 2011 off the Pacific Coast of Tohoku Earthquake (M9.0) refined by means of seafloor geodetic data, *J. Geophys. Res.* **117**(B7), 1–18 (2012).

17. J. J. McGuire and P. Segall, Imaging of aseismic fault slip transients recorded by dense geodetic networks, *Geophys. J. Int.* **155**(3), 778–788 (2003).
18. S. Miyazaki, P. Segall, J. Fukuda, and T. Kato, Space time distribution of afterslip following the 2003 Tokachi-oki earthquake: Implications for variations in fault zone frictional properties, *Geophys. Res. Lett.* (2004), doi:10.1029/2003GL019410.
19. R. Nakata, H. Hino, T. Kuwatani, S. Yoshioka, M. Okada, and T. Hori, Discontinuous boundaries of slow slip events beneath the Bungo Channel, southwest Japan, *Sci. Rep.* **7**(1), 6129 (2017).
20. Q. Qiu, J. D. Moore, S. Barbot, L. Feng, and E. M. Hill, Transient rheology of the Sumatran mantle wedge revealed by a decade of great earthquakes, *Nat. Commun.* **9**(1), 995 (2018).
21. H. Jin, T. Kato and M. Hori, Estimation of slip distribution using an inverse method based on spectral decomposition of Green's function utilizing Global Positioning System (GPS) data, *J. Geophys. Res.* (2007), doi:10.1029/2004JB003378.
22. C. Hashimoto, A. Noda, T. Sagiya, and M. Matsu'ura, Interplate seismogenic zones along the Kuril-Japan trench inferred from GPS data inversion, *Nat. Geosci.* **2**(2), 141–144 (2009).
23. J. P. Loveless and B. J. Meade, Geodetic imaging of plate motions, slip rates, and partitioning of deformation in Japan, *J. Geophys. Res.: Solid Earth* **115**(B2) (2010).
24. Y. Yokota, T. Ishikawa, S. Watanabe, T. Tashiro, and A. Asada, Seafloor geodetic constraints on interplate coupling of the Nankai Trough megathrust zone, *Nature* **534**(7607), 374 (2016).
25. K. Satake, Inversion of tsunami waveforms for the estimation of a fault heterogeneity: Method and numerical experiments, *J. Phys. Earth* **35**(3), 241–254 (1987).
26. M. Comninou and J. Dundurs, The angular dislocation in a half space, *J. Elast.* B(3–4), 203–216 (1975).
27. Y. Okada, Surface deformation due to shear and tensile faults in a half-space, *Bull. Seismol. Soc. Am.* **75**(4), 1135–1154 (1985).
28. Y. Okada, Internal deformation due to shear and tensile faults in a half-space, *Bull. Seismol. Soc. Am.* B(2), 1018–1040 (1992).
29. B. J. Meade, Algorithms for the calculation of exact displacements, strains, and stresses for triangular dislocation elements in a uniform elastic half space, *Comput. Geosci.* **33**(8), 1064–1075 (2007), doi:10.1016/j.cageo.2006.12.003.
30. S. Miyazaki and Y. Hatanaka, The outlines of the GEONET, *Meteorol. Res. Note.* **192**, 105–131 (1998).

31. J. Beavan, L. M. Wallace, N. Palmer, P. Denys, S. Ellis, N. Fournier, S. Hreinsdottir, C. Pearson, and M. Denham, New Zealand GPS velocity field: 1995–2013. *NZ J. Geol. Geophys.* **59**(1), 5–14, (2016).
32. D. Massonnet, M. Rossi, C. Carmona, F. Adragna, G. Peltzer, K. Feigl, and T. Rabaute, The displacement field of the landers earthquake mapped by radar interferometry, *Nature* **364**(6433), 138 (1993).
33. T. Wright, B. Parsons, J. Jackson, M. Haynes, E. Fielding, P. England, and P. Clarke, Source parameters of the 1 October 1995 dinar (Turkey) earthquake from SAR interferometry and seismic bodywave modelling, *Earth Planet. Sci. Lett.* **172**(1–2), 23–37 (1991).
34. T. Kobayashi, Earthquake rupture properties of the 2016 Kumamoto earthquake foreshocks (M j 6.5 and M j 6.4) revealed by conventional and multiple-aperture InSAR, *Earth Planets Space* **69**(1), 7 (2017).
35. C. D. Chadwell and F. N. Spiess, Plate motion at the ridge-transform boundary of the south Cleft segment of the Juan de Fuca Ridge from GPS-Acoustic data, *J. Geophys. Res.* **113**(B4), 04415 (2008).
36. M. Sato, T. Ishikawa, N. Ujihara, S. Yoshida, M. Fujita, M. Mochizuki, and A. Asada, Displacement above the hypocenter of the 2011 Tohoku-Oki earthquake, *Science* **332**(6036), 1395 (2011).
37. F. Tomita, M. Kido, Y. Osada, R. Hino, Y. Ohta, and T. Iinuma, First measurement of the displacement rate of the Pacific Plate near the Japan Trench after the 2011 Tohoku-Oki earthquake using GPS/acoustic technique, *Geophys. Res. Lett.* **42**(20), 8391–8397 (2015).
38. Y. Ito, T. Tsuji, Y. Osada, M. Kido, D. Inazu, Y. Hayashi, H. Tsushima, R. Hino, and H. Fujimoto, Frontal wedge deformation near the source region of the 2011 Tohoku-oki earthquake, *Geophys. Res. Lett.* (2011), doi:10.1029/2011GL048355.
39. Y. Kaneda, K. Kawaguchi, E. Araki, H. Matsumoto, T. Nakamura, S. Kamiya, K. Ariyoshi, T. Hori, T. Baba, and N. Takahashi, Development and application of an advanced ocean floor network system for megathrust earthquakes and tsunamis. in: P. Favali, L. Beranzoli, and A. De Santis (eds.), *Seafloor Observatories,* pp. 643–662. Berlin: Springer, (2015).
40. T. Masterlark, Finite element model predictions of static deformation from dislocation sources in a subduction zone: sensitivities to homogeneous, isotropic, Poisson-solid, and half-space assumptions, *J. Geophys. Res.* **108**(B11), 2540 (2003).
41. B. T. Aagaard, M. G. Knepley, and C. A. Williams, A domain decomposition approach to implementing fault slip in finite-element models of quasi-static and dynamic crustal deformation, *J. Geophys. Res. Solid Earth* **118**(6), 3059–3079 (2013).

42. T. Ichimura, R. Agata, T. Hori, K. Hirahara, C. Hashimoto, M. Hori, and Y. Fukahata, An elastic/viscoelastic finite element analysis method for crustal deformation using a 3-D island-scale high-fidelity model, *Geophys. J. Int.* **206**(1), 114–129 (2016).
43. D. Komatitsch and J. Tromp, Spectral-element simulations of global seismic wave propagation-I. Validation, *Geophys. J. Int.* **149**(2), 390–412 (2002).
44. H. N. Gharti, L. Langer, and J. Tromp, Spectral-infinite-element simulations of coseismic and post-earthquake deformation, *Geophys. J. Int.* **216**(2), 1364–1393 (2018).
45. T. Ichimura, R. Agata, T. Hori, K. Hirahara, and M. Hori, Fast numerical simulation of crustal deformation using a three-dimensional high-fidelity model, *Geophys. J. Int.* **195**(3), 1730–1744 (2013).
46. C. Kyriakopoulos, T. Masterlark, S. Stramondo, M. Chini, and C. Bignami, Coseismic slip distribution for the Mw 9 2011 Tohoku-Oki earthquake derived from 3-D FE modeling, *J. Geophys. Res. Solid Earth* **118**(February), 3837–3847 (2013).
47. C. A. Williams and L. M. Wallace, Effects of material property variations on slip estimates for subduction interface slow-slip events, *Geophys. Res. Lett.* **42**(4), 1113–1121 (2015).
48. C. A. Williams and L. M. Wallace, The impact of realistic elastic properties on inversions of shallow subduction interface slow slip events using seafloor geodetic data, *Geophys. Res. Lett.* **45**(15), 7462–7470 (2018).
49. L. Langer, H. N. Gharti, and J. Tromp, Impact of topography and three-dimensional heterogeneity on coseismic deformation, *Geophys. J. Int.* **217**(2), 866–878 (2019).
50. H. Cheng, B. Zhang, L. Huang, H. Zhang, and Y. Shi, Calculating coseismic deformation and stress changes in a heterogeneous ellipsoid earth model, *Geophys. J. Int.* **216**(2), 851–858 (2019).
51. H. Moritz, Geodetic reference system 1980, *J. Geodesy.* **74**(1), 128–133 (2000).
52. S. Matsumura, M. Murakami, and T. Imakiire, Concept of the new Japanese geodetic system, *Bull. Geograph. Surv. Inst.* **51**, 1–9 (2004).
53. M. Ando, Source mechanisms and tectonic significance of historical earthquakes along the Nankai Trough Japan, *Tectonophys.* **27**(2), 119–140 (1975).
54. S. Ozawa, T. Nishimura, H. Suito, T. Kobayashi, M. Tobita, and Imakiire T (2011) Coseis-mic and postseismic slip of the 2011 magnitude-9 Tohoku-Oki earthquake, *Nature* **475**(7356), 373–376, (2011), doi:10.1038/nature10227.
55. K. Koketsu, H. Miyake, and Y. Tanaka, A proposal for a standard procedure of modeling 3-D velocity structures and its application to

the Tokyo metropolitan area, Japan, *Tectonophys.* **472**(1–4), 290–300 (2009).
56. K. Koketsu, H. Miyake, and H. Suzuki, Japan integrated velocity structure model version 1, in: *Proceedings of the 15th World Conference on Earthquake Engineering* (1773), (2012).
57. T. Ichimura, M. Hori, and J. Bielak, A hybrid multiresolution meshing technique for finite element three-dimensional earthquake ground motion modelling in basins including topography, *Geophys. J. Int.* **177**(3), 1221–1232 (2009).
58. K. Fujita, T. Ichimura, K. Koyama, H. Inoue, M. Hori, and L. Maddegedara, Fast and scalable low-order implicit unstructured finite-element solver for earth's crust deformation problem, in: *Proceedings of the Platform for Advanced Scientific Computing Conference*, p. 11. ACM (2017).
59. J. A. Slater and S. Malys, WGS 84–past, present and future, in: *Advances in Positioning and Reference Frames*, pp. 1–7. Springer (1998).
60. N. K. Pavlis, S. A. Holmes, S. C. Kenyon, and J. K. Factor, The development and evaluation of the Earth Gravitational Model 2008 (EGM2008), *J. Geophys. Res.* (2012), doi:10.1029/2011JB008916.
61. C. De Boor, On calculating with B-splines, *J. Approx. Theor.* **6**(1), 50–62 (1972).
62. G. F. Sella, T. H. Dixon, and A. Mao, Revel: A model for recent plate velocities from space geodesy, *J. Geophys. Res.* **107**(B4), 11 (2002).
63. H. Melosh and A. Raefsky, A simple and efficient method for introducing faults into finite element computations, *Bull. Seismol. Soc. Am.* **71**(5), 1391–1400 (1981).
64. W. Smith and P. Wessel, Gridding with continuous curvature splines in tension, *Geophysics* **55**(3), 293–305 (1990).
65. R. Agata, T. Ichimura, K. Hirahara, M. Hyodo, T. Hori, C. Hashimoto, and M. Hori, Numerical verification criteria for coseismic and postseismic crustal deformation analysis with large-scale high-fidelity model, *Proc. Comput. Sci.* **51**, 1534–1544 (2015), doi:10.1016/j.procs.2015.05.344.
66. S-i. Watanabe, Y. Bock, D. Melgar, and K. Tadokoro, Tsunami scenarios based on interseismic models along the Nankai trough, Japan, from seafloor and onshore geodesy, *J. Geophys. Res.* **123**(3), 2448–2461 (2018).
67. A. Noda, T. Saito, and E. Fukuyama, Slip-deficit rate distribution along the Nankai trough, southwest Japan, with elastic lithosphere and viscoelastic asthenosphere, *J. Geophys. Res.* **123**(9), 8125–8142 (2018).

68. G. Laske, G. Masters, Z. Ma, and M. Pasyanos, Update on CRUST1.0 – A 1-degree global model of Earth's crust, EGU General Assembly Vienna, Austria 15, 2658 (2013).
69. T. Ichimura, K. Fujita, T. Yamaguchi, A. Naruse, J.C. Wells, C.J. Zimmer, T. P. Straatsma, T. Hori, S. Puel, T. W. Becker, *et al.*, 416-pflops fast scalable implicit solver on low-ordered unstructured finite elements accelerated by 1.10-exaflops kernel with reformulated ai-like algorithm: For equation-based earthquake modeling, In: *SC19: International Conference for High Performance Computing, Networking, Storage and Analysis* (2019).

https://doi.org/10.1142/9781800614635_0002

Chapter 2

Large-Scale Simulation of the Discrete Element Method to Analyze Multi-Scale Granular Dynamics in the Practical Geosciencc Applications

Mikito Furuichi* and Daisuke Nishiura
Center for Mathematical Science and Advanced Technology,
Japan Agency for Marin-Earth Science and Technology,
Showa-machi, Kanazawa-ku Yokohama, Japan
**m-furuic@jamstec.go.jp*

The implementation and demonstration of a parallel discrete element method (DEM) simulation for practical scientific and engineering applications are presented. To efficiently calculate the pairwise interactions on the shared and distributed memory systems, preconditioning is applied using sorting techniques with the grid cell label and prefix sum. The sorted memory allocation allows for efficient computation of DEM kernels and preconditioning tasks. The benchmark test using different algorithms addresses the optimal shared memory parallelization method of DEM for different multi- and many-core devices, including graphics processing units (GPUs). For distributed memory parallelization, space-based partitioning is employed. To achieve large-scale practical DEM applications, the iterative load-balancing method over memory implementations (MPI) processes is applied to control the flexible orthogonal domain decomposition geometry. The efficiency of the proposed technique is verified and validated through numerical tests. Finally, as an example of a large-scale DEM application, a real-scale numerical sandbox experiment is presented as an analog system of accretionary prism formation. Stress chain analysis revealed the mechanism bridging the micro- and macroscale structures during thrust formation.

1. Introduction

The discrete element method (DEM) is one of the standard simulation methods originally developed for granular physics. Recently, DEM has been extensively utilized in various fields of study, such as geotechnics, geoscience, industry, and disaster prevention problems.

A high computational cost is a major difficulty in DEM. A clear solution is parallel computing. Massively parallel supercomputer systems are commonly used in high-performance computing (HPC). However, the difficulty of parallelization prevented us from performing a large-scale DEM simulation using HPC. Thus, typical DEM simulation studies employ less than a few million particles. Several challenges have recently been reported to parallelize particle simulations.[1–7]

Previous studies involved the development of a workload balancer for smoothed particle hydrodynamics (SPH) and molecular dynamics (MD), particle in a cell, and N-body simulations. The moving Lagrangian particle method easily suffers from a load imbalance problem in a fixed domain, and dynamic load-balancing is a key technique for overcoming such imbalance problems.

The major difference between the DEM and other particle simulations is the calculation of the tangential force. Therefore, the cost of the DEM is bounded by the number of contacts rather than the number of particles. Thus, the optimization of computing contact particle pairs is important, although it has not been well considered in previous studies, particularly for distributed memory parallelization. The naive implementation of tangential forces may result in a significant amount of memory usage and slow computation.

Herein, we present a hybrid parallelization method for shared and distributed memory systems for DEM. For parallel particle simulation with short-range interactions, the workload can be divided into computational units in two ways: data-based partitioning that divides the workloads with the particle labels reflecting the data location (device and memory address), and space-based partitioning that divides the workloads with spatial domain decomposition (DD) reflecting the physical location of particles. We implemented data-based partitioning with shared-memory parallelization and space-based partitioning with distributed memory parallelization.

Modern parallel computing systems use multi-core central processing units (CPUs), graphics processing units (GPUs), or many integrated core (MIC) processors. Although various parallel algorithms for the particle simulation method for shared memory have been proposed, the performance of these algorithms on different processing units is unknown. Thus, it is important to compare the different numerical algorithms for each processing unit. We present a performance comparison test to select the optimal parallelization algorithms of the DEM for a target processing unit.

A limitation of shared memory parallelization is the size of the system, which is generally up to approximately ten million DEM particles with less than four CPUs (or GPUs) or 64 GB memory of HPC systems.

Thus, distributed memory parallelization with MPI is required to manage over a billion DEM particles. In contrast to shared memory parallelization, it is natural to employ space-based partitioning with DD that only requires communication with neighboring processes, because alternative all-to-all communication required for data-based partitioning is slow and memory-intensive with MPI. Our implementation of DD can compute DEM kernels involving tangential forces without explicit MPI synchronization of the particle contacts of ghost particles. We also applied a dynamic load balancer based on an iterative nonlinear solver to control the geometries of flexible rectangular DD. The proposed method is designed to achieve converged parallel performance iteratively for problems with non-uniform particle distribution and non-uniform calculation costs.

Our implementation was verified and validated using a reproducibility test and a series of parallel scaling tests. The beneficial strong and weak scalabilities of up to 2.4 billion particles were obtained using the K-computer. These performance tests demonstrate parallel scaling and the credibility of the code.

We also present an algorithm for characterizing stress chains using hybrid parallelization. Stress chains are important for analyzing the results of DEM simulations. However, the general algorithm is difficult to parallelize, particularly when selecting networks longer than several particles. The proposed algorithm can count the number of particles that are tightly connected iteratively with operations using the nearest-neighbor particles.

We present a numerical simulation study of a sandbox experiment as an application of a large-scale DEM for solid Earth science. This experiment represents a numerical model of accretionary prism formation. This study used the world record of the number of DEM particles in published scientific papers. The results demonstrated that large-scale DEM simulation is a powerful tool for scientific discoveries.

The remainder of this chapter is organized as follows. In the next section, we describe the DEM equations. In Section 3, the implementation of shared memory is explained with a performance comparison test between different algorithms, implementations, and device units. In Section 4, parallelization for distributed memory systems is presented, focusing particularly on the treatment of load-balancing. A new MPI DD is also presented. The proposed parallelization techniques are demonstrated through performance and reproducibility tests. In Section 5, the parallel implementation of the numerical stress chain analysis is explained to analyze large-scale DEM simulation data. In Section 6, a numerical sandbox experiment to imitate the accretionary prism formation is presented. Section 7 concludes the chapter.

2. Basics of DEM

In an ordinary DEM simulation, the Newton–Euler equations of motion are solved for the translational and rotational motions of each particle, as follows:

$$M\frac{\partial \boldsymbol{u}}{\partial t} = \sum(\boldsymbol{F}_c^n + \boldsymbol{F}_c^t) + M\boldsymbol{g}, \tag{1}$$

$$I\frac{\partial \boldsymbol{\omega}}{\partial t} = \sum\left(r_p\boldsymbol{n}\times\boldsymbol{F}_c^t - \mu_r r_p\{1-(\boldsymbol{n}\cdot\boldsymbol{\omega})^2/\boldsymbol{\omega}^2\}|F_c^t|\frac{\boldsymbol{\omega}}{|\boldsymbol{\omega}|}\right), \tag{2}$$

where $\boldsymbol{u}$, $\boldsymbol{\omega}$, and r_p are the velocity, angular velocity, and radius of the particle, respectively, M and I are the mass and moment of the inertial tensor, respectively. $\boldsymbol{n}$ denotes the unit normal vector to the contact point between the particles from the particle center, and μ_r is the rolling friction coefficient.[8] $\boldsymbol{F}_c^n$ and $\boldsymbol{F}_c^t$ are the normal and tangential forces induced by the contact between two particles or between a particle and a wall, respectively. The position of particle $\boldsymbol{x}$ was updated with $\boldsymbol{u}$ using the Euler time-stepping method. Further details on the model descriptions can be found in Refs. 9–11.

These contact forces are represented using the Voigt model, which comprises a nonlinear spring and dashpot, and stems from the Hertz–Mindlin model.[12–14]

$$\boldsymbol{F}_c^n = K_{ij}^n \boldsymbol{\delta}_{ij}^n + \eta_{ij}^n \boldsymbol{v}_{ij}^n, \tag{3}$$

$$\boldsymbol{F}_c^t = \min(K_{ij}^t \boldsymbol{\delta}_{ij}^t + \eta_{ij}^t \boldsymbol{v}_{ij}^t,\ \mu|\boldsymbol{F}_\mathrm{c}^\mathrm{n}|\boldsymbol{t}), \tag{4}$$

where $\boldsymbol{\delta}_{ij}$ is the relative displacement vector and $\boldsymbol{v}_{ij}$ is the relative velocity vector between i- and jth particles. In the normal direction, the spring constant of each contact point between particles i and j is determined according to Hertzian contact theory:

$$K_{ij}^n = \frac{2E}{3(1-m^2)}\sqrt{\frac{(r_p)_i(r_p)_j}{(r_p)_i + (r_p)_j}|\boldsymbol{\delta}_{ij}^n|}, \tag{5}$$

and depends on the normal compressed length $|\delta_{ij}^n|$ of the spring. m is Poisson's ratio and E is Young's modulus of the particles used in the DEM. The spring constant in the tangential direction is expressed as

$$K_{ij}^t = \frac{4G}{(2-m)}\sqrt{\frac{(r_p)_i(r_p)_j}{(r_p)_i + (r_p)_j}|\boldsymbol{\delta}_{ij}^n|}, \tag{6}$$

where G is the shear modulus, which is related to Young's modulus and Poisson's ratio as follows:

$$G = \frac{E}{2(1+m)} \tag{7}$$

and consequently, it is expressed as follows:

$$K_{ij}^t = \frac{3(1-m)}{(2-m)} K_{ij}^n. \tag{8}$$

The dumping coefficient η is represented as follows:

$$\eta_{ij} = \frac{1}{2.2}\ln\left(\frac{2}{e} - 1\right)\sqrt{\frac{M_i M_j}{M_i + M_j} K_{ij}}, \tag{9}$$

where e is the restitution coefficient, and Eq. (9) is associated with the nonlinear spring model.[14] If the tangential force by the Voigt model satisfies the following relation:

$$|\boldsymbol{F}_c^t| > \mu |\boldsymbol{F}_c^n|, \tag{10}$$

particles slide between each other, and the tangential force is expressed as follows:

$$\boldsymbol{F}_c^t = \mu |\boldsymbol{F}_c^n| \boldsymbol{t}, \tag{11}$$

where the kinetic friction force is based on the Coulomb friction law, instead of the spring and dashpot. Furthermore, μ is the friction coefficient and $\boldsymbol{t}$ is the unit tangential vector. The restitution (viscous damping) force owing to the dashpot and kinetic friction force act as dissipative forces in terms of the inter-particle contact force.

3. Shared Memory Parallelization

Shared memory parallelization is based on data-based partitioning, which generally shows better workload balance with finer-grained workload division than spatial DD. The shared memory parallelization (e.g., OpenMP, CUDA) can easily handle complicated operations for data-based partitioning without significant cording tasks with the help of hardware design for global data access. We investigated an efficient implementation method of shared-memory parallelization for the particle method with short-range interactions, including DEM and SPH.[7]

There were three major modules of particle simulations with short-range interactions. These include the creation of the neighbor particle contact pair list, calculation of interactions, and updating of particle information. The contact pair list is created as a preconditioning for computing kernels, consisting of the calculation of interactions and update of particle information. In the following sections, we introduce shared-memory parallelization schemes for each module.

3.1. *Efficient computation of preconditioning*

For an efficient search of potentially interacting particle pairs for preconditioning, we generated a list of neighbor particle pairs.

In this study, the pair list was updated when the cumulative displacement of a particle exceeded $\alpha(r_p)_i$, where α was a constant set to 0.1. This technique is similar to that proposed in Ref. 4, based on the Verlet list method.

First, every particle is identified by a particle label (i or j) and belongs to a cell labeled by *cell_label*. The grid cells are defined on a uniform Cartesian grid with a size of $2r_{\max}(1+\alpha)$, where $r_{\max}$ is the maximum particle radius. To assign each particle to a corresponding cell in parallel, we first sorted the particle labels using *cell_label* as a sortkey. The minimum and maximum particle labels in each cell were then recorded in arrays *ID*[1, *cell_label*] and *ID*[2, *cell_label*], respectively. This type of sorting is known as an efficient algorithm for creating a neighboring particle list in parallel computing.[4, 15] This algorithm is shown in Sub_C, where N_{p} is the total number of particles and *IB* is the array of cell labels corresponding to the positions of all particles.

Sub_C Storing particle labels to corresponding cell data

Call sort {particle label is sorted by cell label}
for $i = 1$ to N_{p} **do** {enable thread parallelism}
  $ibl = IB[i\text{-}1]$
  $ibi = IB[i]$
  $ibr = IB[i+1]$
  if $ibl < ibi$ **then**
    $ID[1, ibi] = i$
  end if
  if $ibr > ibi$ **then**
    $ID[2, ibi] = i$
  end if
end do

In the next step, we created a neighbor particle pair list. From the array *ID*, we can determine whether particles belong to the same or adjacent cells. Particle pairs with a distance less than $(1+\alpha)[(r_p)_i + (r_p)_j]$ are listed as neighbor particle pairs. The pair list is created as one-dimensional arrays $List_{\mathrm{i}}[l] = i$ and $List_{\mathrm{j}}[l] = j$, which indicates that the lth pair consists of the ith and jth particles. The procedure for searching for pairs is performed using the outer and inner loops. The outer loop considers the ith particle index, and the inner loop searches for the index of the jth particle, which is a neighbor of the

ith particle. We labeled the particle pairs obtained in this manner using a sequential index number that is used for efficient memory access. In the simplest method, such labeling tasks are conducted sequentially, although pair-searching tasks can be parallelized in the outer loop. Because the number of the ith particle's neighbors is unknown before the inner loop calculation, it is not straightforward to parallelize the labeling of the pair list array with a sequential index in the outer loop.

In our method, we counted the number of neighboring particles for $j > i$ and $j < i$, defined as arrays *ipair* and *jpair*, respectively. In addition, the prefix sum *ipl* of the pair list for the ith defined as follows:

$$ipl[i] = \sum_{n=1}^{i} ipair[n] \quad \text{with } ipl[0] = 0, \tag{12}$$

was calculated to parallelize the pair list labeling task. This preconditioning allows the neighboring particles of i and $j > i$ to be listed in parallel to the arrays of the particle pair list, $List_{\mathrm{i}}[l]$ and $List_{\mathrm{j}}[l]$, because the pair list label l for particle i ranges from $ipl[i-1]+1$ to $ipl[i]$.

We considered seven implementation methods for particle pair creation, named "Sub_ L#," where # is the label of the subroutine. Table 1 provides an overview of these subroutines. The details of the subroutines used to calculate the pair list are summarized as follows:

- Sub_L1, Sub_L1s: In these subroutines, two pair list arrays (i.e., $List_{\mathrm{i}}$ and $List_{\mathrm{j}}$) are used to calculate the interactions between particles by parallelizing each index of the particle pair list, as shown in Section 3.2. The difference between Sub_L1 and Sub_L1s lies in the prefix sum operation; Sub_L1 computes this in parallel, whereas Sub_L1s computes this sequentially.
- Sub_L2: This subroutine uses only the $List_{\mathrm{j}}$ pair list. The difference between Sub_L1 and Sub_L2 in calculating the interparticle interactions is explained in Section 3.2.
- Sub_L3: This stores all neighbor j-th particles for each i-th particle in the pair list (i.e., $List_{\mathrm{j}}$ of Sub_L3 contains particles with $j > i$ and $j < i$, whereas Sub_L2 only creates a list for $j > i$). This pair list is used to calculate the interactions between each pair twice, first for the ith particle from j, and second for the jth particle from i.

Table 1. Summary of subroutines for particle pair list creation.

Subroutine name	Major aspects
Sub_L1s	• Pair list is constructed by two arrays of $List_i$ and $List_j$ that contain particle i and particle $j > i$ • Pair list is labeled using the sequential prefix sum • Reference table array of *Ref* is created to sum up interactions
Sub_L1	• Pair list is constructed by two arrays of $List_i$ and $List_j$ that contain particle i and particle $j > i$ • Pair list is labeled using parallel prefix sum • Reference table array of *Ref* is created to sum up interactions
Sub_L2	• Pair list is constructed by one array of $List_j$ which contains particle $j > i$ • Pair list is labeled using parallel prefix sum • Reference table array of *Ref* is created to sum up interactions
Sub_L2a	• Pair list is constructed by one array of $List_j$ that contains particle $j > i$ • Pair list is labeled using parallel prefix sum • Reference table array of *Ref* is created to sum up interactions • Pair list data are aligned to allow coalescent memory access on GPU
Sub_L3	• Pair list is constructed by one array of $List_j$ that contains all neighbor particles • Pair list is labeled using parallel prefix sum
Sub_L3a	• Pair list is constructed by one array of $List_j$ that contains all neighbor particles • Pair list is labeled using parallel prefix sum • Pair list data are aligned to allow coalescent memory access on GPU
Sub_L4	• Pair list is constructed by one array of $List_j$ that contains particle $j > i$ • Pair list is labeled using parallel prefix sum • The number of searched cells is half that of the other subroutines

```
    ipair[i] = Iij
#if defined Sub_L1, Sub_L2, or Sub_L2a
    jpair[i] = Iji
#endif
    end do
#if defined Sub_L1s
     Call sequential_prefix_sum {sequential prefix sum for ipair and
      jpair}
#elif
Call parallel_prefix_sum {parallelized prefix sum for ipair and jpair}
#endif
for i = 1 to Np do {enable thread parallelism}
  for k = 1 to ipair[i] do
    l = ipl[i-1] + k
#if defined Sub_L1
     List_i[l] = i
#endif
#if defined Sub_L1, Sub_L2, Sub_L3, or Sub_L4
     List_j[l] = tempij[k, i]
#elif defined Sub_L2a or Sub_L3a
     List_j[i+N16_p*(k-1)] = tempij[k, i]
     {where N16_p = int((N_p+15)/16)*16}
#endif
     end do
end do
#if defined Sub_L1, Sub_L2, or Sub_L2a
     Call reference_table {create reference table Ref using jpair and
      tempji}
#endif
```

3.2. *Kernel calculation with action–reaction law with many-core computing*

We first calculated the interactions for each lth pair $F[l]$ in parallel using the pair list in Section 3.1. The interaction forces $F[l]$ exerted on the i-th particle are then gathered as the force term $Force[i]$. In our study, we employed seven subroutines named "Sub_I#." These differ in their choice of pair list from Section 3.1 and the summation algorithm for the interactions as summarized in Table 2. The details of these subroutines are as follows:

- Sub_I1: This subroutine uses the pair list from Sub_L1 to compute $F[l]$ simultaneously. By utilizing the reference table $Ref[l]$, the

Table 2. Summary of subroutines for interaction calculation.

Subroutine name	Major aspects
Sub_I1	• Interaction calculation is parallelized in the pair list index by utilizing $List_{\mathrm{i}}$ and $List_{\mathrm{j}}$ • Summation of interactions are parallelized by utilizing *Ref*
Sub_I1a	• Interaction calculation is parallelized in the pair list index by utilizing $List_{\mathrm{i}}$ and $List_{\mathrm{j}}$ • Summation of interactions are parallelized by utilizing *Ref* • Temporal arrays of interactions (*Ftemp*) are read with coalescent memory access on GPU
Sub_I2	• Interaction calculation is parallelized in the particle index by utilizing $List_{\mathrm{j}}$ • Summation of interactions are parallelized by utilizing *Ref*
Sub_I2a	• Interaction calculation is parallelized in the particle index by utilizing $List_{\mathrm{j}}$ • Summation of interactions are parallelized by utilizing *Ref* • $List_{\mathrm{j}}$ and *Ref* are read with coalescent memory access on GPU
Sub_I3	• Interaction calculation is parallelized in the particle index by utilizing $List_{\mathrm{j}}$ • Summation of interactions is parallelized by double calculation without action–reaction law
Sub_I3a	• Interaction calculation is parallelized in the particle index by utilizing $List_{\mathrm{j}}$ • Summation of interactions is parallelized by double calculation without action-reaction law • $List_{\mathrm{j}}$ is read with coalescent memory access on GPU
Sub_I4	• Interaction calculation is parallelized in the particle index by utilizing $List_{\mathrm{j}}$ • Summation of interactions is parallelized using atomic-add instruction

interaction forces exerted on particle i are then combined as

$$\text{Force}[i] = \sum_{l=ipl[i-1]+1}^{ipl[i]} F[l] - \sum_{l=jpl[i-1]+1}^{jpl[i]} F[Ref[l]], \qquad (12)$$

where $jpl[i]$ is the prefix sum of $jpair[i]$, and $jpair[i]$ is the number of neighboring jth particles ($j < i$) for the ith particle. The first

Sub_I#: Calculation of interactions, except for Sub_I1 and Sub_I1a

```
#define Sub_I# {subroutine name, Sub_I2, Sub_I2a, etc.}
  Force = 0
———————Calculation of interactions between particle pair———————
for i = 1 to N_p do {enable thread parallelism for particle index}
AFtemp = 0
  for l = ipl[i-1]+1 to ipl[i] do
#if defined Sub_I2, Sub_I3, or Sub_I4
      j = List_j[l]
#elif defined Sub_I2a or Sub_I3a
      l16 = i + N16_p *(l − ipl[i-1])
      j = List_j[l16]
#end if
      Ftemp = 0
       if distance between particle i and particle j is less than (r_p)_i + (r_p)_j
       then
      Ftemp = {forces}
       endif
#if defined Sub_I2
      F[l] = Ftemp
      AFtemp=AFtemp+Ftemp
#elif defined Sub_I2a
      F[l16] = Ftemp
      AFtemp=AFtemp+Ftemp
#elif defined Sub_I3 or Sub_I3a
      Force[i] = Force[i] + Ftemp
#elif defined Sub_I4
!$OMP ATOMIC
      Force[j] = Force[j] - Ftemp
      AFtemp=AFtemp+Ftemp
#end if
      end do
#if defined Sub_I2 or Sub_I2a
      Force[i] = AFtemp
#elif defined Sub_I4
!$OMP ATOMIC
      Force[i] = Force[i] + AFtemp
#end if
end do
———————Summation of interactions———————
#if defined Sub_I2 or Sub_I2a
for i = 1 to N_p do {enable thread parallelism}
  for k = 1 to jpair[i] do
#if defined Sub_I2
```

```
      l = jpl[i-1] + k
#elif defined Sub_I2a
      l = i + N16_p*k
#end if
      Force[i] = Force[i] - F[Ref[l]]
    end do
  end do
#end if
```

3.3. *Performance test of shared memory parallelization methods*

Here, we show the performance analysis of the SPH simulation as a representative particle method with short-range interactions, including DEM. We tested our code implementation by using several com puter architectures. To examine the shared memory performance, we used a single-node of many/multi-core architectures, including the GPU, as shown in Table 3.

Table 3. Salient architecture of the shared-memory parallel computers employed.

System	GPU	FX10	2CPU	MIC
OS	Windows	Linux	Linux	Linux
Compiler	Microsoft C++	Fujitsu Fortran	Intel Fortran 13	Intel Fortran 13
Parallelization	CUDA 5.0	OpenMP	OpenMP	OpenMP
		Processor		
Product line	GK110	SPARC64	Xeon	XeonPhi
Product model	Titan	IXfx	E5-2680	SE10P
Number of cores	2688 (14 SMX)	16	8	61
Peak performance [GFlops]	1300	236.5	172.8	1073
L1 data cache	48 KB/SMX	32 KB/Core	32 KB/Core	32 KB/Core
Read-only cache	48 KB/SMX	—	—	—
L2 cache	1536 KB	12 MB	2048 KB	31 MB
L3 cache	—	—	20 MB	—
Power	250	110	130	300
		Memory		
Granularity [GB]	6	32	32	8
Band width [GB/s]	288.4	85	51.2	352

the elapsed time was measured. The computation time per step is averaged over 100 iterations.

To discuss the performance of each program code on each parallel processor, Fig. 2 shows the average computation time per time step for the creation of the particle pair list (T_L) by Sub_L#, the calculation of interactions between particles (T_I) by Sub_I#, and the update of particle information (T_U). The performance test used all the cores on each processor. In multi-core CPU systems, the number of parallel threads is equal to that of the cores. To ameliorate the performance degradation caused by data access latency, tests on the MIC and GPU systems were performed using four threads per core and 256 threads per streaming multiprocessor (SM), respectively (Table 3).

Figures 2(a) and 2(b) show the performance on two Xeon processors (2CPU) and the SPARC64 processor (FX10), respectively.

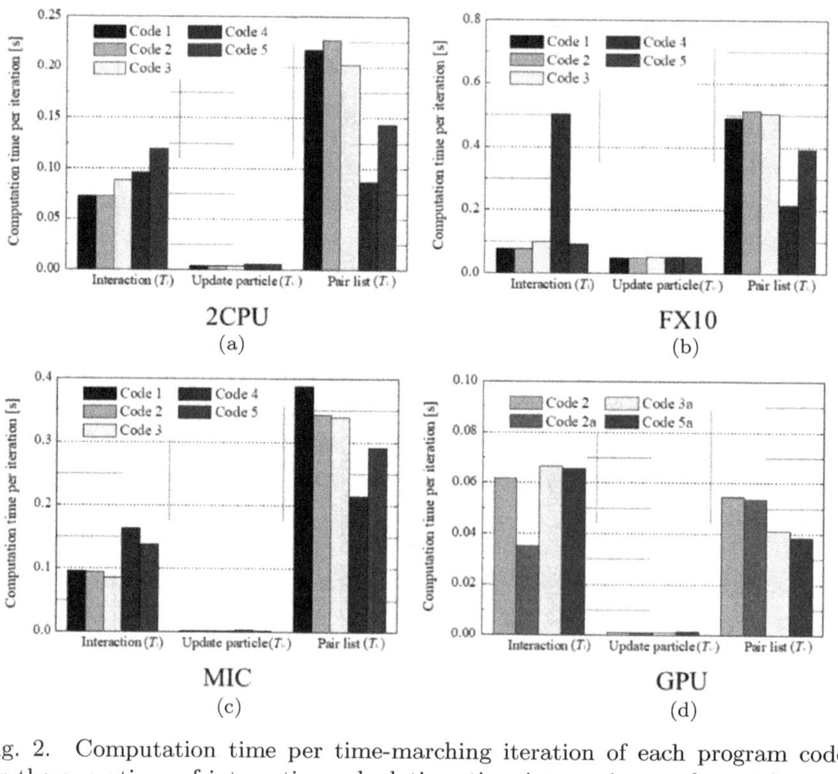

Fig. 2. Computation time per time-marching iteration of each program code for the operations of interaction calculation, time integration, and particle-pair-list creation on each shared-memory parallel computer. (a) 2 CPU, (b) FX10, (c) MIC, and (d) GPU.

We first compared the times for Codes 1 and 2, which use subroutines Sub_L1s and Sub_L1, respectively. Sub_L1 is parallelized for the prefix sum, whereas Sub_L1s is not. We found that T_L for Code 1 was smaller than that for Code 2. This implies that parallelizing the prefix sum does not improve the computational performance of multi-core CPUs because the 16 cores of the multi-core CPU are not sufficient to overcome the parallelization overhead. The parallel prefix sum performs $O(N_{\mathrm{p}}\log_2 N_{\mathrm{p}})$ additive operations, whereas the sequential prefix sum performs only $O(N_{\mathrm{p}})$.[16] Thus, the overhead of parallelization is $\log_2 N_{\mathrm{p}}$. In this case, $\log_2 N_{\mathrm{p}}$ is 20.7 for $N_{\mathrm{p}} = 1{,}687{,}016$. This indicates that at least 21 cores are required to accelerate the prefix sum operation using parallelization. In terms of calculating interactions, both Codes 1 and 2 are parallelized in a single do-loop and show the same T_{I}.

For Code 3, T_{L} is smaller than that for Code 2 because Code 3 creates only $List_{\mathrm{j}}$ instead of both $List_{\mathrm{i}}$ and $List_{\mathrm{j}}$. However, the T_{I} value of Code 3 was greater than that of Code 2. The interactions in Code 3 are calculated using a double do-loop, with the particle index as the outer loop and the pair list index as the inner loop. Although we can parallelize the outer loop, the parallel acceleration is degraded by the imbalance in the inner loop length (i.e., the number of interacting particles).

Code 4 uses atomic instructions to calculate the summation of interactions. This produces a greater T_{I} than that of Codes 1–3. Atomic instructions require synchronization of the thread barrier, which degrades parallel performance. In particular, the performance of the interaction calculation is extremely slow for FX10 because SIMD is unavailable with the thread barrier of the atomic instructions. SIMD plays a critical role in the efficient performance of FX10. However, atomic instructions have some benefits as only half the number of cells must be searched for neighboring particles, and the creation of a reference table is not required. Therefore, Code 4 results in the best T_{L} value among all codes.

In Code 5, the arithmetic cost of calculating interactions is twice that of the other codes, because the interactions are calculated twice. Therefore, the T_{I} value of Code 5 is the worst among all the codes on the 2CPU. However, Code 5 shows a good T_{I} for FX10 because SIMD can be used for the calculation of interactions. The SIMD acceleration on FX10 renders the computation time comparable to of the code that can be parallelized. In other words, the computational cost of the remaining fraction $(1 - S)$ cannot be reduced through parallelization. The parallel fraction for each value of n for $m = 240$ is shown in Fig. 3(b). The parallel fraction decreased as n increased. From this result, it was confirmed that the increase in overhead owing to atomic instructions significantly degrades the performance of Code 4. The performance of Code 5 compared to that of Code 3 was the same as that of the multi-core CPUs.

Figure 2(d) illustrates the performance of each code on the GPU. First, we can observe that the computational performance of the interaction calculation in Code 2(a) significantly improved compared to that in Code 2 because of the coalescent address treatment. Code 2(a) showed the best performance in terms of calculating interactions.

The $T_{\rm I}$ result for Code 3(a) was worse than that for Code 2(a). This performance degradation is caused by an imbalance of the inner loop length, as in multi-core CPUs. In addition, the decrease in the parallelized loop length of Code 3(a), which is shorter than that of Code 2(a), could degrade the parallel efficiency in many-core computing. However, the interaction calculation in Code 5(a) is slightly faster than that in Code 3(a) for the same reason as that for FX10. Thus, we found that the double calculation of interactions is effective in SIMD architectures, such as GPU and FX10. We note that the order relation of $T_{\rm L}$ between Codes 3(a) and 5(a) is equal to that between Codes 3 and 5 in other processors.

In terms of updating particle information, the computational cost was the worst for FX10. The cache hierarchy of FX10 is two-level, rather than the three-level system of 2CPU, and the total cache per core is the lowest among the multi-CPUs and MIC. However, the memory bandwidth per flop (B/F) of FX10 was greater than that of 2CPU and MIC. The large B/F value of FX10 can efficiently supply data to the processor from the local memory when the data overflow from the small cache. However, the access speed of local memory is less than that of cache memory, even if the B/F of local memory is large. To update the particle information, data, such as position, velocity, and force, are required. Thus, we believe that the particle data used in the arithmetic operations overflow from the small cache on FX10. In comparison with the 2CPU and MIC systems, the amount of overflow data from the cache is greater, because the cache size is smaller. Hence, the computational performance of

FX10 was lower than that of the other systems, regardless of the large B/F value. By contrast, the computational performance of the GPU is considered the best of all the processors because the hiding of memory access latency is perfectly implemented by reading and writing from/to coalescent memory addresses aligned on the 128-byte boundary.

Therefore, no single code provided the best performance for both pair list creation (T_{L}) and interaction calculation (T_{I}). The cost balance of our simulation depends on the target problem characterized by the update frequency of the particle pair list. To determine the best simulation code, we examined the effect of the pair-list update frequency on the overall computation time. We then plotted the relationship between the update frequency f and the computation time per iteration T_{all} for each code, as shown in Fig. 4. T_{all} is defined as follows:

$$T_{\mathrm{all}} = T_{\mathrm{I}} + T_{\mathrm{U}} + T_{\mathrm{L}} f, \tag{14}$$

where f is defined by the number of updates per iteration, as well as T_{I}, T_U, and T_L are obtained from the results shown in Fig. 2. For 2CPU, the computation times of Codes 1 and 4 are the smallest in the ranges $0 < f < 0.189$ and $0.189 < f < 1$, respectively. For FX10, the performance of Code 1 was the best when $0 < f < 0.186$. However, in contrast to the 2CPU result, Code 5 exhibits the best performance when $0.186 < f < 1$. For MIC, the computation time of Code 4 is the smallest in the range of $0.626 < f < 1$, and for Code 3, it is the smallest outside this range. For the GPU, Code 2a exhibited the best performance for all values of f.

To compare the performance between processors, the performance of the best code on each processor is summarized in Fig. 5. This figure illustrates the shorter computation time given by the two best codes at $f = 1$ and $f = 0.01$. When the update frequency f is high, the computation time is found to have the order "Code 2(a) on GPU" < "Code 4 on 2CPU" < "Code 4 on MIC" < "Code 5 on FX10". However, when f is small, the order is "Code 2a on GPU" < "Code 1 on 2CPU" < "Code 3 on MIC" < "Code 1 on FX10". We note that MIC and 2CPU exhibit similar computation times.

From these results, if we target the computational speed on a single-node system, we recommend using Code 2(a) on the GPU.

a solid line in the figure. Generally, a high B/F results in a highly effective performance because the fast data supply reduces the idle time of the arithmetic units. The effective performance correlated well with B/F, except for MIC. The effective performance of MIC is the worst, even though its B/F is as high as that of multi-core processors. We believe that there are two reasons for this performance loss on MIC: the way of the data supply and the low effective flop performance. In our implementation, random access to read particle data is inevitable, resulting in frequent local memory access owing to cache misses. We believe that such local memory access may significantly degrade computational performance because the memory bandwidth of the MIC in practical use (i.e., the effective bandwidth) is not as good as that shown by the theoretical performance. We implemented the STREAM benchmark to estimate effective memory bandwidth. The triad test showed that the effective bandwidth of the MIC was 40% of its theoretical value. This is considerably smaller than the effective bandwidth of other processors, which is more than 70% of the theoretical bandwidth. Figure 6(a) depicts a plot of the effective B/F based on the obtained effective memory bandwidth as a dashed line. The figure shows a high correlation between the effective B/F and the effective performance. The low effective flop performance comes from our loop calculation with indirect access which disables the SIMD instructions.

Figure 6(b) shows the computational speed divided by the processor's peak watt for the code used in Fig. 5, which we define as the "power efficiency". The theoretical peak performance per watt (GF/W) of each processor is also presented as a reference. Similar to the effective performance in Fig. 7(a), the power efficiency is proportional to GF/W for the GPU, FX10, and 2CPU. The power efficiency of MIC was the lowest, although the GF/W of MIC was the second highest following that of GPU. This means that the MIC cannot efficiently utilize its power supply because the effective performance of the MIC is lower than that of the other processors.

Therefore, if we compare parallel computer systems with the same theoretical peak performance, the fastest computational speed for our simulation code is obtained on a multi-core CPU system. If we compare parallel computer systems with the same peak power consumption, the highest computational speed is obtained for the GPU system. Thus, the multi-node system of multi-core CPUs and GPUs is a suitable HPC system to perform particle simulation methods with short-range interactions, such as DEM and SPH.

In the next section, we discuss the performance of DEM with multiple nodes of multi-core devices. In general, the update frequency of the particle pair list is less than 10^{-1} and double calculation of the DEM may require a large memory size. Code1 using the action-reaction law is selected as a high-performance DEM implementation that saves memory size for our target computer systems in Section 4.

4. Distributed Memory Parallelization

Distributed memory parallelization with MPI is required to handle over ten million DEM particles with more computational resources than a single shared memory system. Space-based partitioning with DD, which requires the communication of ghost particles with neighbor processes (Fig. 7(a)), is commonly used for distributed memory parallelization. One of the problems of MPI parallelization is the load balance among the different MPI processes. Moving Lagrangian particle methods inherently suffer from workload imbalance problems in fixed subdomains because particles move around and change the distribution of the workload during the simulation run. Load imbalances degrade the parallel performance of the DEM simulations. To overcome this problem, we developed an iterative dynamic load balancer.[17] The imbalances in the execution time between parallel logical processes could be iteratively minimized by controlling DD during the simulation run. Because the computational cost of a DEM particle depends on the number of contacts and interaction types rather than the number of particles commonly used for SPH, MD, and N-body, the evaluation of workload by execution time is suitable for DEM applications.

Here, we present the basic framework of the iterative load-balancing method utilizing the local change in the DD. The multigrid-level technique was also used to improve the performance of the iterative load-balancing method. The multigrid technique is

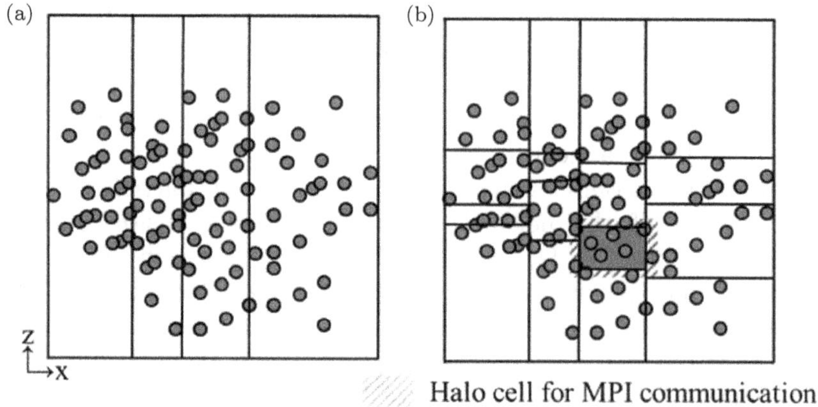

Fig. 8. Schematic image of (a) the first domain division in the x-direction, and (b) the final DD after the second domain division in the z-direction. Configuration of particles in the MPI subdomains. The circles represent the particles.

method.[1–3] First, the columns are created by dividing the entire domain in the x-direction (Fig. 8(a)). Each column is then divided in the z-direction to create rectangular subdomains. Figure 8(b) presents an example of our DD for 4×4 subdomains. Because the areas of the subdomains can be arbitrarily controlled, a perfect load balance can be attained in principle, although it is limited in practice by the grid size. Each MPI process holds a subdomain and adjacent halo cells, as shown in Fig. 8(b). The particles at the halo cells are updated by the particles received from the neighboring subdomain via MPI communication, whereas the particles at the outermost cells of the subdomain are sent to other MPI processes for contacting subdomains. Our algorithm guarantees that a particle located within a subdomain interacts only with particles that are in the subdomain and halo cells in the prior preconditioning step. Thus, the same cell-to-particle reference table used in shared memory parallelization can be used for distributed memory parallelization until the next preconditioning time step.

Here, we present an algorithm for changing the subdomains for load-balancing. The design of the algorithm is based on an iterative nonlinear solver, which gradually improves the solution to satisfy the target equation with the Jacobian.[18,19] In our implementation, a host process gathers the elapsed time T_i from ith MPI process to create $\boldsymbol{T}$ and calculates the optimal subdomain changes at the preconditioning time step to update the cell-to-particle table. The Jacobian is estimated from the changes in the elapsed time of each MPI process by the recent domain changes, which reflect the workload distribution inside the subdomain. The algorithm of our iterative load balancer is summarized in the BALANCER algorithm.

Algorithm BALANCER ($\boldsymbol{T}$)

Step 0. If $n = 0$, set $\boldsymbol{x}^0$, $\hat{J}$ and $\delta\boldsymbol{x} = \boldsymbol{0}$
Step 1. Update $\hat{J}_{ii}$ for $\{i : \delta x_i \neq 0\}$
Step 2. $\delta\boldsymbol{x} = \boldsymbol{0}$
Step 3. For $L = 0$ to L_{top}
Step 4. For $i \in (G_L \setminus \sum_0^{L-1} G_i)$
Step 5. Compute residual $(f_i^n)_L$
Step 6. Solve $\delta x_i = -\beta(\hat{J}_{ii})^{-1}(f_i^n)_L$
Step 7. Update $\boldsymbol{x}^{n+1} = \boldsymbol{x}^n + \delta\boldsymbol{x}$
Step 8. If $\delta\boldsymbol{x} = 0$, control γ with line search
Step 9. Return

For simplicity, we consider the 1D problem shown in Fig. 9, in which the space domain Ω_N is divided for N MPI processes by $\{x_i : i = 0, 1, 2, 3, \ldots (N-1)\}$ with $x_0 = 0$. T_i^n is the workload for the cell defined between x_{i-1}^n and x_i^n at the nth preconditioning step. The deviation of workload $\boldsymbol{f}$ discretized at the cell boundaries is expressed as follows:

$$f_i^n = T_{i+1}^n - T_i^n. \tag{15}$$

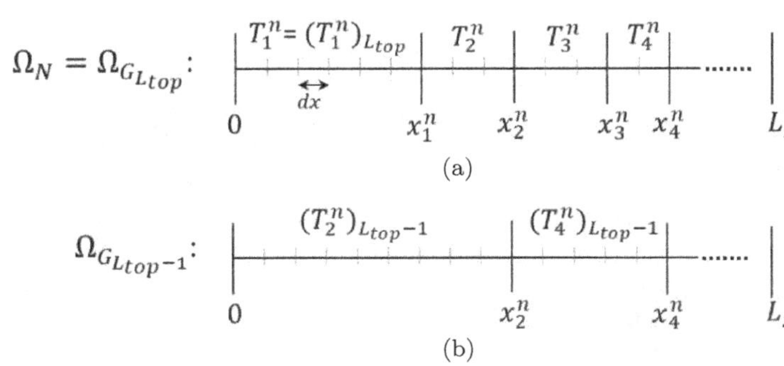

Fig. 9. Example of multi-level DD for the 1D problem (a) original level DD, and (b) coarser level DD.

The workload T_i^n is a nonlinear function of x_i^n, mainly because of the non-uniform particle distributions in the subdomain and the communication cost for each ghost particle. Thus, we can regard f_i^n as the residual vector component of the nonlinear problem of $\|\boldsymbol{f}\| = 0$. The goal $\|\boldsymbol{f}\| = 0$ indicates a perfect load balance. The load balancer can be defined as an iterative operation of the Newton-like nonlinear solver to minimize $\|\boldsymbol{f}\|$. The domain boundary is updated using

$$\delta x_i = -\beta \left(\hat{J}_{ii}\right)^{-1} f_i^n, \tag{16}$$

where $\delta x_i = x_{i+1}^n - x_i^n$, $\beta = 0.95$ is a constant parameter empirically obtained, and $\hat{J}_{ii}$ is the diagonal matrix of the inexact Jacobian defined by

$$\hat{J}_{ii} = \frac{f_i^n - f_i^{n-1}}{x_i^n - x_i^{n-1}}. \tag{17}$$

In a practical implementation of the BALANCER algorithm, $\hat{J}$ averaged over the last five evaluations was used for smooth changes in workload balance. In addition, the domain change at each load-balancing time step was represented by the grid size dx (i.e., $\delta x/dx = \{-1, 0, 1\}$).

4.3. *Multigrid level relaxation*

To improve the relaxation for large wavelength modes, inexact Jacobian calculation incorporates a multigrid technique.[20] The iteration of local relaxation is difficult to reduce the residual $\boldsymbol{f}$ with a large wavelength mode. Multigrid is a well-established technique to address this limitation. For simplicity, we consider the 1D DD $\{x_i : i \in G_L\}$ in Fig. 9, where $G_L = \{M, 2M, 3M, \dots (N-M) : M = 2^{L_{\text{top}}-L}\}$ with $N_0 \cdot 2^{L_{\text{top}}} = N$ and $N_0 \geq 2$ is a constant. The L_{top}-th and 0th grids represented the finest (original) and coarsest grid levels, respectively. The workload on the L ($\neq L_{\text{top}}$)th coarse grid is defined by the average as follows:

$$(T_i)_L = \frac{(T_{i-1})_{L+1} - (T_i)_{L+1}}{2}. \tag{18}$$

Smoothing on the Lth grid is applied by the inexact Jacobian expressed by

$$\hat{J}_{ii} = \frac{(f_i)_L^n - (f_i)_L^{n-1}}{x_i^n - x_i^{n-1}}, \tag{19}$$

where $(f_i)_L = (T_i)_L - (T_{i-M})_L$: $M = 2^{l_{top}-1}$. Coarser grid relaxation is recursively applied from the top to the bottom level.

Our iteration for calculating the domain change $\delta \boldsymbol{x}$ is designed to simultaneously smooth low-frequency and high-frequency errors using a set of coarse grids. We first calculated δx_i at the domain boundary of the coarsest level G_0 using the Newton-like procedures of Eqs. (16) and (19). The smoothing process then moves to the 1st grid G_1 and calculates δx_i in $\{x_i : i \in (G_1 \backslash G_0)\}$. With this method, the priority of relaxation by a smoother is provided to a large wavelength. These processes were repeated until the top level was reached.

4.4. *Flexible rectangular DD*

For DD geometry, we applied new types of 2D flexible rectangular DD. Flexible rectangular DD is commonly used in improved ORB and the slice-grid method.[1–3] First, the entire domain is divided in the x-direction to create columns, and then each column is divided in the z-direction to create rectangular subdomains. In the sandbox experiment, which was characterized by a rectangular-shaped geometry, this method showed good parallel scaling.[17] However, one of the difficulties of the original DD method was the aspect ratio.

For example, Figs. 10(a) and 10(b) show the DD obtained with the load balancer for a circular particle distribution using 64 spatial MPI subdomains. The result with the original method of Fig. 10(a), in which each of the eight rows was divided into eight columns, shows a large aspect ratio in the outer regions. A subdomain with a large aspect ratio results in a large number of ghost particles to communicate and degrade the parallel performance. The original flexible rectangular DD suffers from the aspect ratio problem, with a geometry deviating from a rectangular shape in practical engineering applications. To solve this problem, we considered a new flexible

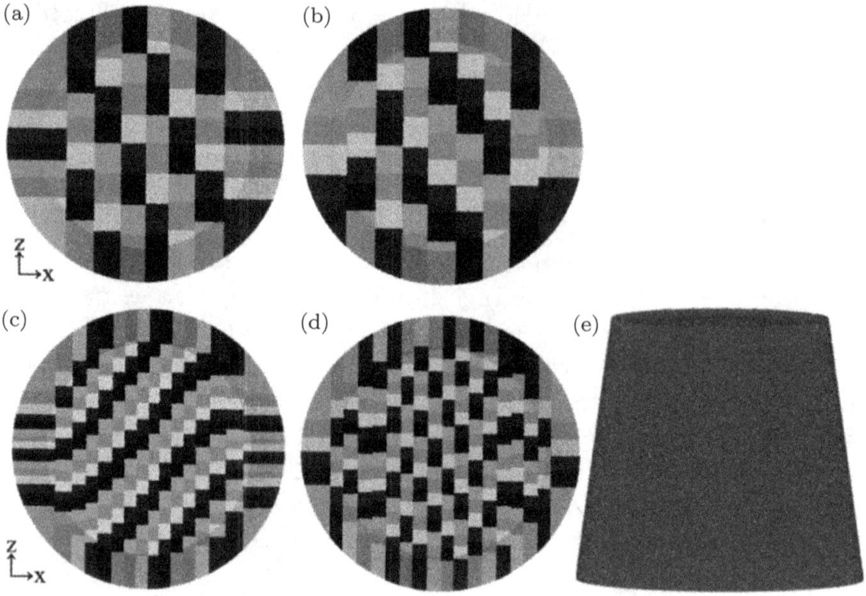

Fig. 10. Top views of particles used in the slump flow experiment in Section 4.6 from (a, c) original DD, and (b, d) FR-DD methods. The different shades indicate the total 64 MPI subdomains in (a) and (b) and 256 MPI subdomains in (c) and (d). (e) Side view of the initial setup for 3D slump flow test of the Abrams cone with 13M particles in Section 4.6.

rectangular DD (FR-DD) that allows each row to have a different number of subdomains, as shown in Fig. 10(b). Therefore, the maximum aspect ratio of the rectangular domain is improved from 3.9 in Fig. 10(a) to 2.1 in Fig. 10(b), using the same 64 subdomains. The performance improvement observed with the FR-DD technique is discussed in Section 4.6.

4.5. *Verification of DEM implementation for the distributed memory system*

To verify our parallel implementation of the DEM code, we performed reproducibility tests as shown in Fig. 11. Because verification of the OpenMP implementation is intrinsic, the test focuses on distributed memory parallelization with dynamic load balance. The dynamic option of shared-memory parallelization (OpenMP) was switched off to prevent changes in the order of summation calculation.

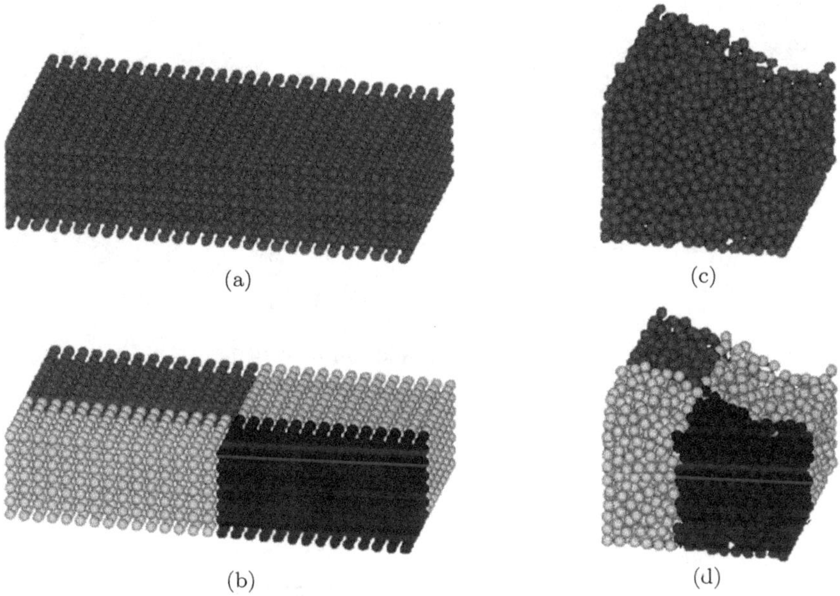

Fig. 11. Results after pushing the left wall from uniformly distributed particles with (a, c) single and (b, d) four MPI processes. Different shades identify different MPI processes. In the DD of (b), columns are divided by the same number of rows (i.e., (2, 2)).

First, the particles were filled in the box by gravity and pushed to the left by moving the left wall. The results obtained using a single MPI process are shown in Figs. 11(a) and 11(c). The DD for MPI parallelization is a flexible orthogonal grid domain in which each column is divided by the same number of rows using the four MPI processes, as shown in Figs. 11(b) and 11(d). We also performed a verification test with the FR-DD by dividing each column by a different number of rows, as shown in Fig. 12. In this test, we computed the dropping particles initially placed in the lattice positions in a cylinder with different parallelization: (a) a single MPI process and (b) 10 MPI processes. Figures 12(a) and 12(b) show identical solutions that verify our parallel implementation. Our code implementation successfully showed a perfectly identical solution from the same initial setup but was performed with a different number of MPI processes and different choices of DD.

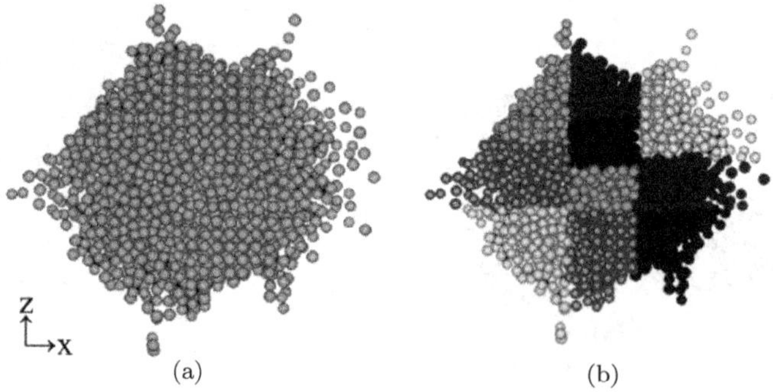

Fig. 12. Top views of falling particle simulation with (a) single, and (b) 10 MPI processes. Different shades identify different MPI processes. In the FR-DD of (b), columns are divided by the different number of rows (i.e., (3, 4, 3)).

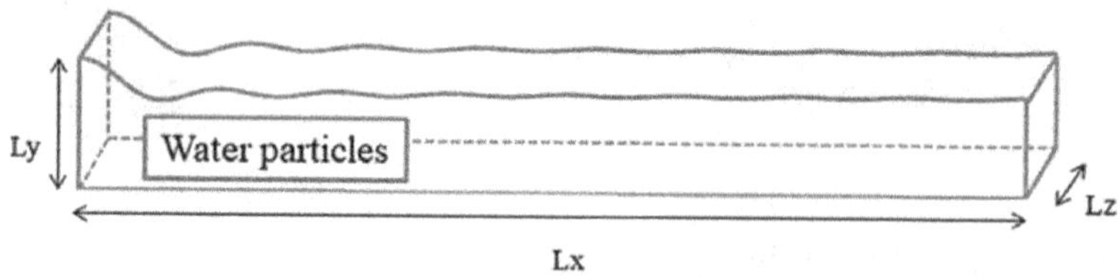

Fig. 13. Initial setting of wavy water problem.

4.6. *Validation test of iterative load-balancing scheme*

The "wavy water" problem was solved using the SPH method to examine the basic performance of the proposed load balancer. In this test, an Intel Xeon cluster system (XE) consisting of four nodes (eight CPUs) connected with InfiniBand was used. As shown in Fig. 13, the box-sized water particles ($L_x = 128$, $L_y = 1.8$, $L_z = 1$) are initially placed with a wavy surface shape, expressed as follows:

$$y = 1 + \frac{0.8}{16} \sum_{n=1}^{16} \cos\left(n\pi \frac{x}{L_x}\right). \tag{20}$$

The smoothing operation at the multigrid level is a key technique in our load-balancing algorithms. To confirm the effectiveness of this technique, we performed a wavy water test with approximately 27 million particles. This simulation used 64 MPI processes for 1D DD in the x-direction. Initially, the simulation starts with

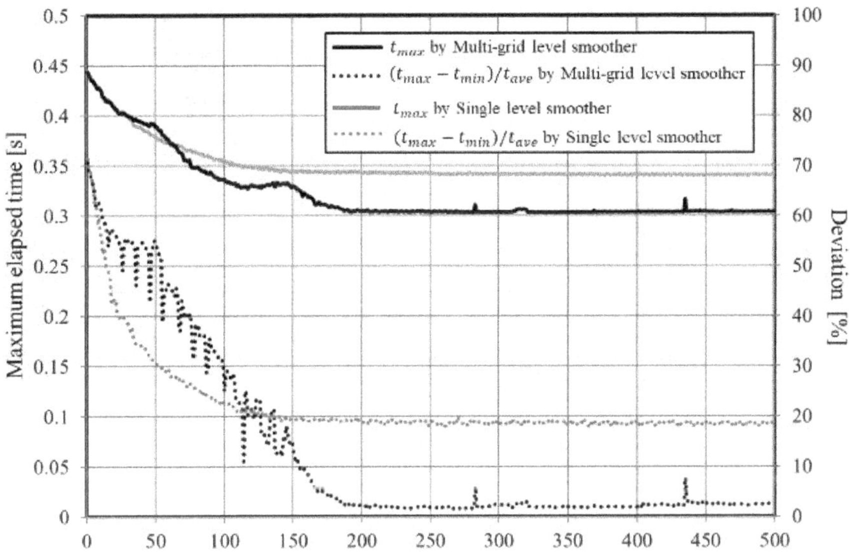

Fig. 14. Results with and without multigrid level smoothing method for the wavy water test. The maximum time and imbalance are plotted against the number of the domain change.

a significant load imbalance because the space domain is uniformly divided into subdomains, although the particle distribution depends on x from Eq. (20). Figure 14 shows the parallel performance using two types of smoother: simple single-grid (finest-grid) level and multi-grid. Both methods improve the parallel performance iteratively with local domain changes. The quality of each load balancer can be analyzed with the performance converged after approximately 200 domain changes. The performance curve of the single-grid level was trapped by a local minimum, although the global deviation of the performance remained approximately 20%. By contrast, the multigrid-level method successfully reduced the maximum deviations by approximately 2%. To investigate the detailed relaxation behavior, we plotted the elapsed time as a function of the MPI process number, as shown in Fig. 15. The single-grid level smoother can dampen the load imbalance from the initial state. However, the local deviation in cost was extremely small to change the subdomains to improve the performance. The multigrid level method showed nearly the same workload across all MPI processes. The single-grid level smoother failed to reduce the low-frequency modes of

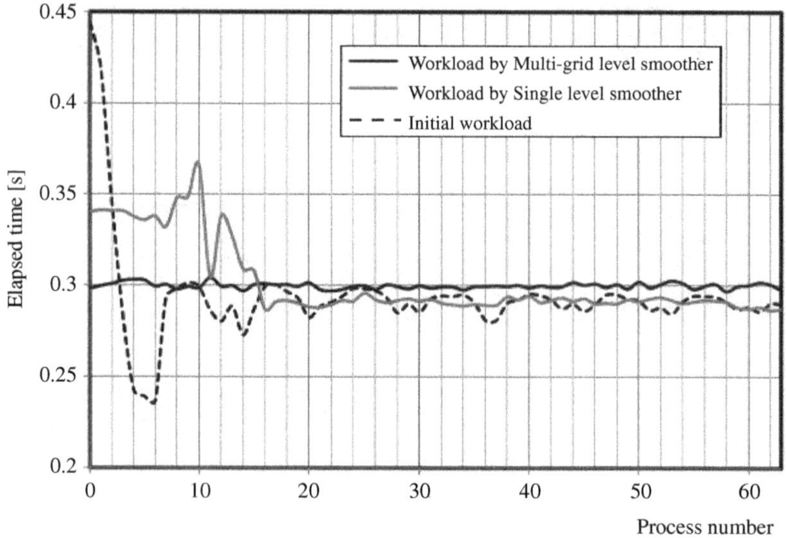

Fig. 15. Workload for each MPI process at 500 domain changes of Fig. 14. The imbalance of the elapsed time at the initial domain is also shown as a reference.

workload imbalance, although it dumped the high-frequency modes. The observed trends are consistent with the inherent nature of local relaxation schemes, such as the Jacobi method.[20] The multigrid level technique was introduced to overcome this problem, with global communication of information at each grid level. The load imbalance applied by the multigrid level smoother exhibits a flat line, as shown in Fig. 15. This confirms the robustness of our iterative multigrid level smoother against varying frequency modes of the load imbalance.

Next, we focus on improving the performance with the FR-DD method, which allows for a different number of columns for each row to reduce the aspect ratio, as shown in Fig. 16. We report the parallel performance of the DEM particles filled in the cone used for the slump flow test in a geotechnical study. A third-generation Earth Simulator (ES) based on NEC SX-ACE with a fat-tree network was used. MPI-OpenMP hybrid parallelization was applied for each node, and four OpenMP threads were launched for one CPU on the ES.

To associate with the reduction of aspect ratio from 3.9 to 2.1 (Fig. 10(a) → Fig. 10(b)) and from 9.3 to 3.7 (Fig. 10(c) → Fig. 10(d)), 4.1% and 15.4% performance improvements were observed with the 256 cores (64 MPI processes) and 1024 cores

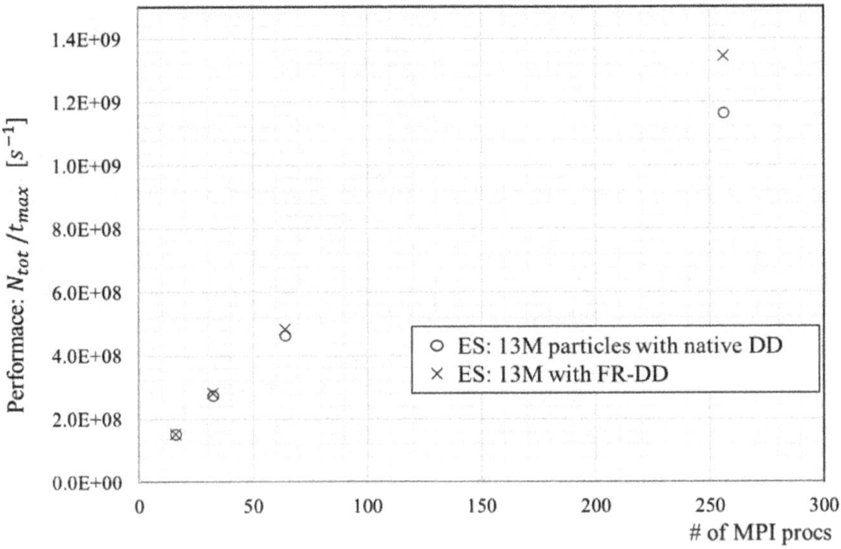

Fig. 16. Scaling performance of the slump flow with the native DD and FR-DD using up to 64 MPI processes of ES.

(256 MPI processes), respectively, in Fig. 16. Because the aspect ratio of the original domain division and the relative cost of ghost particles against total computation increases with the number of total MPI processes, the improvement with 256 MPI processes was larger than that with 64 MPI processes.

4.7. *Parallel scalability test using the HPC machines*

The parallel scalability of our algorithms was examined for the direct 3D numerical simulation of sandbox experiments. This numerical experiment with 0.2 billion and 2.4 billion (B) particles involves the horizontal shortening of a layer of sand. The simulation with 2.4 B uses approximately 100 μm sized particles in a $1 \times 1 \times 0.02$ m box. More details are provided in Section 6. For HPC parallel scaling testing, an ES and a K-computer (K) operated by SPARC64 processors with a Tofu network topology were used. In both systems, MPI–OpenMP hybrid parallelization was applied for each node: four and eight OpenMP threads were launched for one CPU on the ES and K-computer, respectively. An alternative approach is to use only MPI, which is considered a flat MPI implementation.

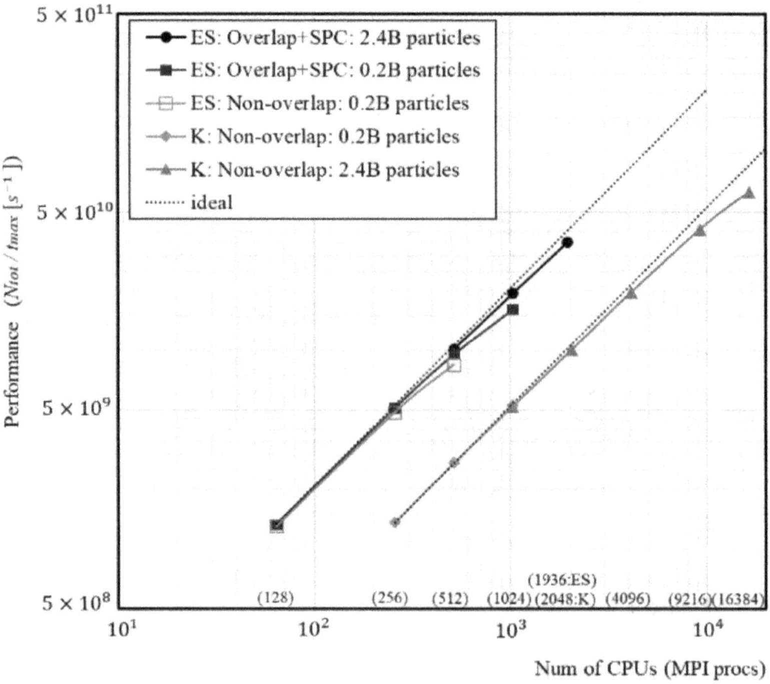

Fig. 17. Strong and weak scalability of DEM on ES and K-computer.

Recently, many-core architectures, such as Fugaku, GPGPU, and PEZY-SC, have significantly accelerated computing performance.[21] Because MPI implementation uses memory for the ghost particles, the overhead to launch MPI processes as many as the number of cores (e.g., >64) in a single many-core device should be significant. Therefore, hybrid parallelization of the DEM is essential for current or next-generation HPC systems. Basic optimizations, such as the insertion of directives, loop unrolling, and adjustment of chunk size, were applied for efficient SIMD operations on the K-computer and efficient vectorization on the ES. All floating-point calculations were performed with double-precision.

The results of the scaling tests are shown in Fig. 17. The figure shows two types of scaling performances: strong and weak. When the problem size (N_{tot}) is fixed, strong scaling performance can be measured by increasing the number of cores ($\# \, of \, cores$). When $N_{\text{tot}}/(\# \, of \, cores)$ is fixed, weak scaling can be measured with increasing N_{tot}. Thus, Fig. 17 shows that the strong and weak scaling of our DEM code was confirmed with up to 16,384 CPUs (131,072 cores) of the K-computer and 1,936 CPUs of the ES.

5. Stress Chain Analysis

DEM simulations can handle simulation sizes of over a billion particles owing to the development of the parallelization techniques described in Sections 3 and 4. Large-scale DEM simulations generate large amounts of data to reproduce complex granular phenomena. Stress chain analysis is a powerful tool for recording scientific discoveries from these datasets. Stress chains are one of the most prominent features of granular dynamics because they control most of the load applied to the system and probe the non-continuum features of granular materials. Because stress chains are constructed using only particles that form a strong network of contact forces, the number of particles that must be analyzed can be significantly reduced. Therefore, identifying stress chains is critical for extracting the catachrestic features of complex 3D granular dynamics from large DEM simulation datasets using parallel computing.

The stress chains were quantitatively identified using the method proposed in Ref. 22. Thus, we first calculated the most compressive principal stress vector (σ_3) from the particle stress tensor as follows:

$$\sigma_{IJ} = \frac{1}{V}\sum_{c=1}^{n_c} f_I^c r_J^c, \tag{21}$$

where $1 < c < n_c$ denotes all particle pairs in contact, V is the volume of a particle, f_I is the Ith component of the force between two contacting particles, and r_J is the Jth component of the radius vector from the center of the particle to the point of contact with the paired particle. Tension is a positive value in Eq. (21); thus, the most compressive vector is the minor principal stress $\boldsymbol{\sigma}_3$. To form stress chains, ith particles should have $|\boldsymbol{\sigma}_3^i|$ greater than the average and should contact jth particles. In addition, the stress vectors of the ith and jth particles (i.e., $\boldsymbol{\sigma}_3^i$ and $\boldsymbol{\sigma}_3^j$) should be aligned in the relative position vector δ_{ij}^n within a 45° angle. Finally, the strongly connected particles should be longer than a given number of particles, which is four in this case.

Identifying the group of candidate particles $i \in \Omega_C$ to form stress chains and compute Eq. (21) in parallel using the particle number as a loop counter is straight-forward. By contrast, the parallel implementation of counting the length of the particle connection of $i \in \Omega_C$ requires some techniques. We developed an iterative algorithm with a counter n which is summarized by Algorithm STRESS_CHAIN, to identify the stress chains that are longer than a given number of particles in parallel.

The length number L_i is introduced for each candidate particle $i \in \Omega_C$. These length numbers are updated using the sum of the maximum L_{c1} and second maximum length number L_{c2} of the adjacent particles, where $c \in \boldsymbol{N}_i$. $\boldsymbol{N}_i$ denotes a group of particles connected to the ith particle. Because the length number at the edge of the chain $L_{i_{ed}}$ shows the integration of the connecting particles, the length of the chain can be obtained with a certain magic number $\boldsymbol{M}$, as shown in Fig. 18. The iterative operations are based on nearest-neighbor communications and can be parallelized using the same shared (OpenMP) and distributed memory implementations (MPI) as the DEM code.

Algorithm STRESS_CHAIN

Step 0.	Set the length number $L_i = 1$ for particles $i \in \Omega_C$ and the iterative counter $n = 0$.
Step 1.	Update $L_i = L_{c1} + L_{c2}$ where $L_{c1} = max\{L_c : c \in N_i\}$ *and* $L_{c2} = max\{L_c : c \in N_i \setminus c1\}$ ($L_{c2} = 0$ *when* $N_i \setminus c1 = \emptyset$).
Step 2.	If $L_{i\_ed} < M(n)$, then the length of the chain is the iteration counter n.
Step 3.	Update the counter with $n = n + 1$ and return to Step 1.

6. Large-Scale DEM Application in Geoscience: Numerical Simulation of a Real-Scale Numerical Sandbox Experiment

The proposed DEM technique enables us to perform large-scale DEM simulations that open the door to new scientific and engineering

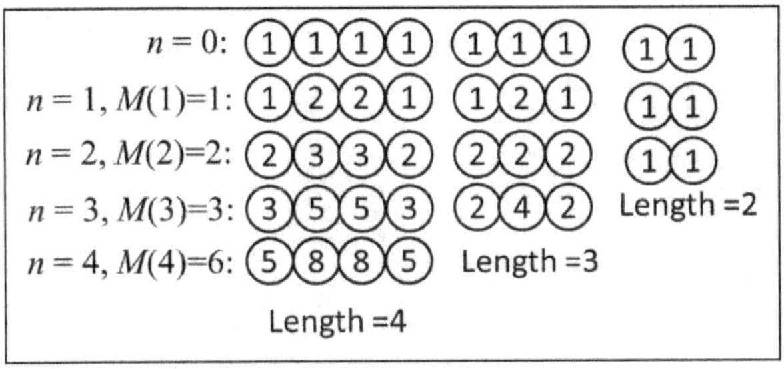

Fig. 18. Example of the nth iterative operation to count the length and criteria number $M(n)$ until $n = 4$.

applications. Here, we introduce one such application in solid Earth science.[23]

Stress states in accretionary prisms are important for understanding the building and release of seismic energy. Numerous researchers have conducted sandbox experiments using scaled physical analog models to understand the formation of accretionary prisms (Fig. 19(a)). However, laboratory experiments cannot fully explore the 3D state inside granular media. To overcome this challenge, we performed direct 3D numerical simulations with 2.4 B particles (approximately 100 μm particle radius and $1 \times 1 \times 0.02$ m domain size). The bottom, fore, and backstop walls had low friction coefficients, whereas the side walls in the z-direction had free-slip conditions. When pushed against the backstop, the sand layer deformed into pop-up structures with new frontal thrusts forming sequentially, a behavior of typical taper wedges formed in the sandbox experiments. The situation after the formation of the first pop-up structure is illustrated in Fig. 19(b). At the end of the experiment, the undulation of the frontal thrust in Fig. 19(c) was similar to the geometries observed in accretionary prisms.

Figure 20 shows the amplitude of the contact force in the thin middle layer during a thrusting event. Before and after the thrusting in Figs. 20(a) and 20(c), respectively, many small waves can be observed in the incoming layer. These waves arise from local changes in the particle contacts induced by the compressive stress in the pushed layer. By contrast, during thrusting in Fig. 20(b), such noisy waves are significantly reduced because the rearrangement of particle contacts concentrated in the fault region releases stress energy. These results demonstrate the ability of our DEM simulation to capture

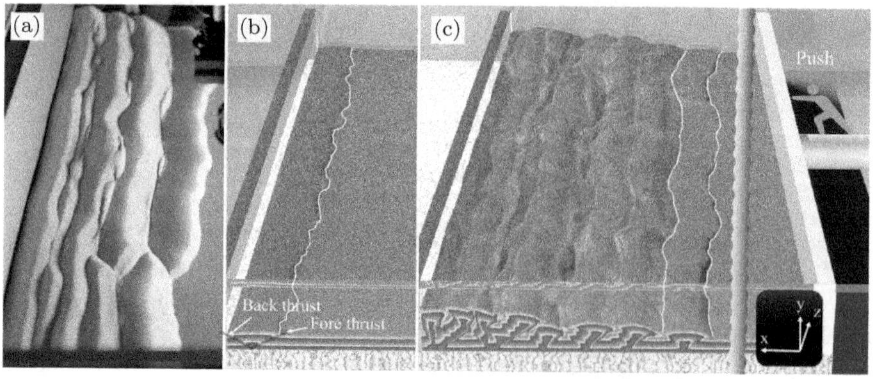

Fig. 19. Horizontal shorting of the sand box to mimic accretionary prism in a (a) laboratory experiment and snapshots of the numerical simulation results at (b) 1.4%, and (c) 40% shortening. The sand particles are colored according to their initial depth (y-coordinate).

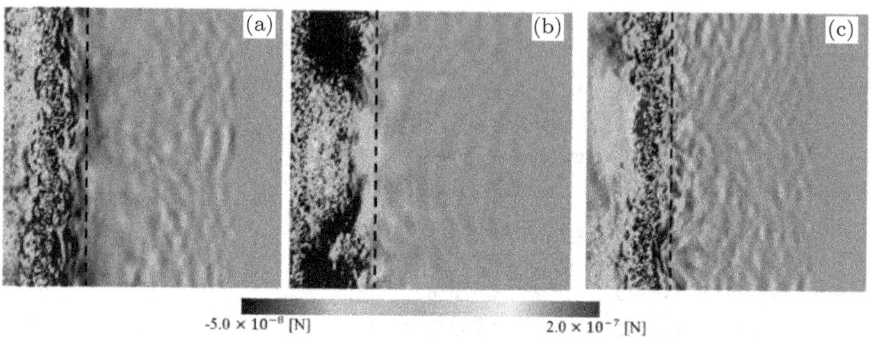

Fig. 20. Snapshots of the compression force in the thin layer at the height of the middle of the initial thickness. The plotted data are the following: (a) before the new thrust is formed, (b) during an active thrust event, and (c) after the thrust formation. The dotted lines in (a), (b), and (c), represent the position of the frontal thrust.

the complexity of the deformation mechanism of the granular layer sequentially, ranging from the local to regional scales.[24]

The number of particles is important for understanding the generation process of a fault because the macroscale thrust formation

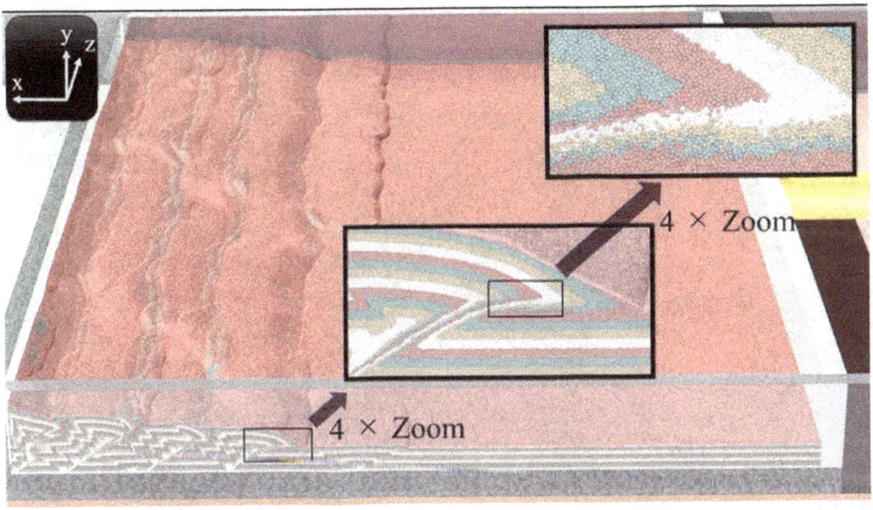

Fig. 21. Result of the real-scale numerical sandbox experiment with 2.4 B DEM particles.

starts from the microscale event of each grain motion. Figure 21 shows a snapshot of the simulation results with zoomed figures.

6.1. *Horizontal undulation of thrust and arcuate stress state inside the granular layer*

The numerical sandbox simulation reproduces the horizontal shortening of a layer of sand by which the characteristic structures of accretionary prisms, such as thrusts and décollements, are formed. In the simulation, macroscale undulations of frontal faults were found from the first pop-up event. Fault undulations are commonly found in the trenches of accretionary prisms. Both the amplitude and characteristic length of the thrust undulation were several orders of magnitude greater than the grain size. To understand the way the initial microscale perturbations of the layer thickness produce macroscale undulation structures, we performed a stress chain analysis, as explained in Section 5.

Soon after the beginning of the simulation, the stress chains intensified in the shortening direction (x-direction). A multitude of stress bands is oriented $\sim 30°$ from the x-direction in the xz-plane (Fig. 22(a)). This angle is attributed to the macroscopic friction

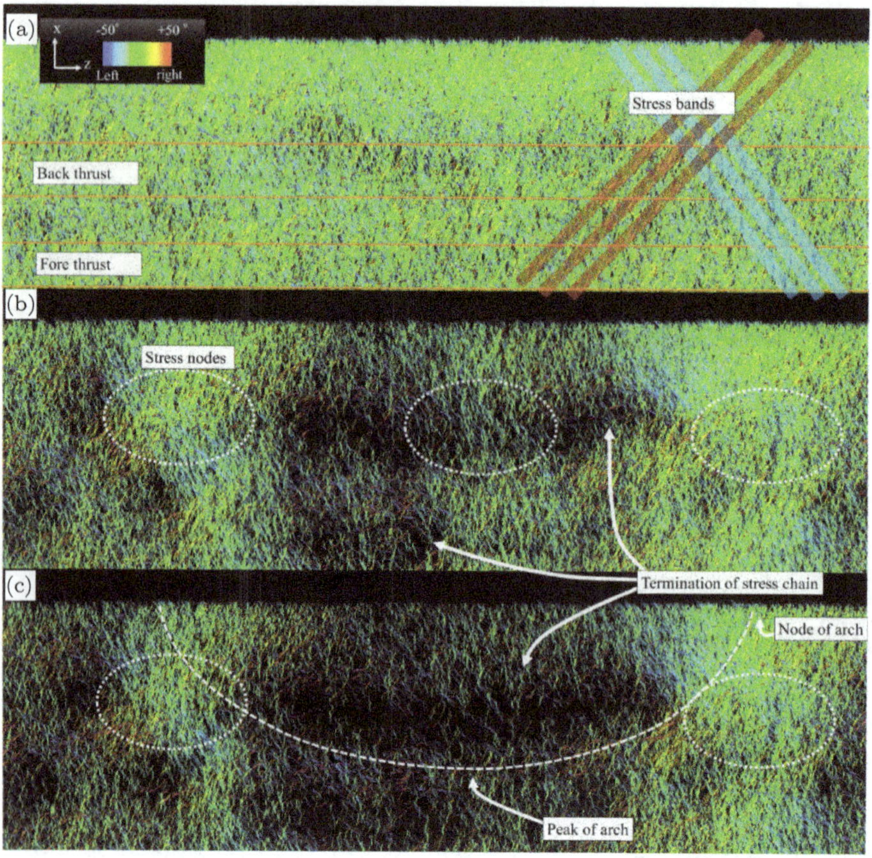

Fig. 22. Visualization results of stress chains in the surface thin layer (0.95,0.005, 0.4)(1, 0.01, 0.6). The degree of the shortening is (a) 0.55% before thrusting, (b) 0.65% at the beginning of the rearrangement of the stress chain, and (c) 0.68% at the end of the rearrangement of the stress chain. The most compressible principal stress vectors in the stress chain were rendered as colored lines. The color shows the angle of the vector from the x-axis in the horizontal plane. The background is colored back.

angle of the sand. The first pop-up structure is formed with fore and back thrust, as shown in Fig. 22(b). The stress chains have largely disappeared around the faulted region in comparison to those in Fig. 22(a) because the shear motion of the particles breaks the existing short stress chains. Simultaneously, the reconnection of the particle contacts induces the rearrangement of the stress chains in and around the pop-up. At the beginning of the rearrangement stage,

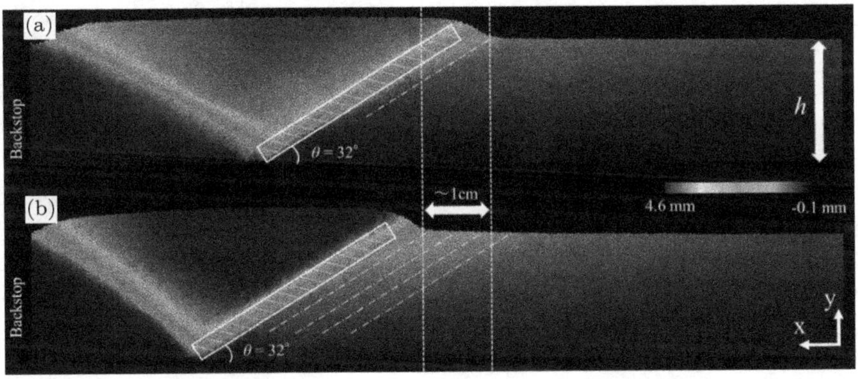

Fig. 23. Cross-sectional view of the particles in the thrust at the (a) peak, and (b) node of the stress arch. The colors denote the commutative displacement of the particle height from the initial position. The shape of the shear band is delineated by the white box to measure the angle of thrust θ. The weak shear bands are shown by the dot-dashed lines.

thick bands are observed. Then, the positively and negatively tilted thick bands cross the back thrust and generate the nodes. In the pop-up structure, pairs of positively and negatively tilted bands were connected. The stress chain structure in the rearrangement stage was shaped as an arch. Horizontal undulations emerge simultaneously with the stress arch. The undulation of the frontal thrust is consistent with the shape of the arch. As shown in the cross-sectional views in Fig. 23, the positions of the thrusts differ in the arch region. The light-shaded region in the frontal thrust in Fig. 23 shows weak shear bands that were generated before the thrusting event. The weak shear bands at the node position (Fig. 23(b)) are wider than those in the peak region (Fig. 23(a)) because the stress localization to develop the fault is affected by the arch structure. The compressive strength was at its maximum at the peak of the arch. After the rearrangement stage, the arch structure gradually changed but maintained its signature, as shown in Fig. 22(c). The undulation of thrust survives and characterizes the pop-up structure in the lateral direction.

6.2. *Multi-scale mechanism for generating arcuate geological-scale structure*

In earlier studies, the horizontal undulations of thrusts were mainly attributed to geological-scale perturbations, such as sea mountains, layer properties, erosion, and multiple plate convergence. By contrast, our numerical sandbox experiment and internal stress chain analysis show that the undulation of faults emerges from microscale perturbations owing to the nonlinear rheological behavior that forms the stress arch. The discontinuous change in stress orientation during the rearrangement of the stress chain causes a multi-scale arch formation mechanism. With an analogy to the deformations between the upper crust and granular material, this observation leads to a new feasible mechanism for generating arcuate geological-scale (km-scale or higher) structures from microscale perturbations.

7. Conclusion

Implementation techniques and algorithms of DEM are presented for hybrid parallelization in shared and distributed memory environments. Our proposed preconditioning method enables the efficient parallel computation of tangential contact forces using a sorting technique. We used the action-reaction law to save memory usage for each particle pair without degrading the total performance in comparison to the double calculation method. In addition, we proposed a new DD method to improve the performance of particle distributions characterized by nonrectangular geometries found in practical geotechnical and engineering problems.

The proposed techniques were verified and validated using a series of performance tests. Our hybrid parallel DEM implementation enabled the utilization of HPC with a good scaling performance. The computational performance is accelerated by shared memory systems, and over billions of particles can be handled by the huge computational resources in the distributed memory system.

Large-scale DEM simulation technology was a powerful tool for identifying hidden multi-scale mechanisms in granular systems. For instance, we introduced a direct numerical simulation of a sandbox experiment using over a billion particles. Despite the nearly uniform initial conditions, the macroscale undulations of faults were similar to those observed in the trenches of an accretionary prism. Our stress chain analysis showed that these undulations were caused by the formation of stress arches generated from microscale perturbations inherent in a granular layer.

From a technical viewpoint, the remaining issue of parallelization is the further increase in the flexibility of DD geometries. For example, it would be useful to dynamically adjust the number of columns in each row according to the changing particle distributions in the load-balancing scheme.

Parallelization of the coupled large-scale multi-physics simulation with the DEM is also challenging. In this direction, we developed an application named DEM-based parallel multi-physics simulator (DEPTH) by extending the interaction force model and performing DEM–SPH coupled simulation.

The data analysis of DEM simulations is also a challenging task. Large-scale simulations can reproduce complex phenomena in huge amounts of unstructured and discretized point data. It is difficult to analyze such data using ordinal procedures, particularly for the physical interpretation of the complex network of particle contacts. Post-processing technology should be improved using advanced stress chain analysis and topological data analysis methods.

With the proposed parallelization technologies and modern HPC resources, we can handle over 10 billion DEM particles. This number of particles covered a special length scale of 10^3–10^5 in one dimension. For some scientific and engineering problems, this number is sufficient to perform real-scale applications of granular materials. Granular physics has a long-term history based on the theoretical treatment and small-scale numerical simulations. However, it is still difficult to fully understand the highly nonlinear dynamics with a large number of frictional elements. Thus, direct numerical simulations are a promising approach for assessing and handling the granularity of multi-scale dynamics hidden in various practical applications.

References

1. M. Furuichi and D. Nishiura, Iterative load-balancing method with multigrid level relaxation for particle simulation with short-range interactions, *Comput. Phys. Commun.* **219**, 135–148 (2017), doi:10.1016/j.cpc.2017.05.015.
2. S. Tsuzuki and T. Aoki, Large-scale particle simulations for debris flows using dynamic load balance with space filling curves on a gpu-rich supercomputer, in: *European Congress on Computational Methods in Applied Sciences and Engineering (ECCOMAS2016)*, E5457, (2016).

3. M. Iwasawa, A. Tanikawa, N. Hosono, K. Nitadori, T. Muranushi, and J. Makino, Implementation and performance of FDPS: A framework for developing parallel particle simulation codes, *Publ. Astron. Soc. Jpn.*, **68**(4), 54 (2016), doi:10.1093/pasj/psw053.
4. J. M. Dominguez, A. J. Crespo, and M. Gomez-Gesteira, Optimization strategies for CPU and GPU implementations of a smoothed particle hydrodynamics method, *Comput. Phys. Comm.* **184**(3) (2013) 617–627.
5. P. Incardona, A. Leo, Y. Zaluzhnyi, R. Ramaswamy, and I. F. Sbalzarini, OpenFPM: A scalable open framework for particle and particle-mesh codes on parallel computers, *Comput. Phys. Commun.*, **241**, 155–177 (2019), doi:10.1016/j.cpc.2019.03.007.
6. B. Jelinek, G. Mason, J. Peters, D. Johnson, M. Brumfield, A. Carrillo, C. Goodman, and F. Vahedifard, Large scale MPI-Parallelization of LBM and DEM systems: Accelerating research by using HPC. in *The International Conference for High Performance Computing, Networking, Storage and Analysis.* Dallas, TX (2018), https://sc18.supercomputing.org/proceedings/tech_poster/poster_files/post260s2-file3.pdf.
7. D. Nishiura, M. Furuichi, and H. Sakaguchi, Computational performance of a smoothed particle hydrodynamics simulation for shared-memory parallel computing, *Comput. Phys. Comm.* **194** 18–32 (2015), doi:10.1016/j.cpc.2015.04.006.
8. H. Sakaguchi and T. Igarashi, Plugging of the flow of granular materials during the discharge from a silo, *Int. J. Mod. Phys. B.* **7**, 1949–1963 (1993).
9. J. Chen, M. Furuichi, and D. Nishiura, Discrete element simulation and validation of a mixing process of granular materials, *Materials* **13**, 1208 (2020).
10. J. Chen, M. Furuichi, and D. Nishiura, DEM study of the influences of the geometric and operational factors on the mechanical responses of an underwater mixing process, *Powder Technol.* **392**, 251–263 (2021).
11. J. Chen, A. Kitamura, E. Barbieri, D. Nishiura, and M. Furuichi, Analyzing effects of microscopic material parameters on macroscopic mechanical responses in underwater mixing using discrete element method, *Powder Technol.* **401**, 117304 (2022).
12. H. Hertz, Über die berührung fester elastischer körper, *Jurnal für die Reine und Angewandte Mathematik*, **92**, 156–171 (1881).
13. R. D. Mindlin, Compliance of elastic bodies in contact, *J. Appl. Mech.* **16**, 259–268 (1949).
14. Y. Tsuji, T. Tanaka, and T. Ishida, Lagrangian numerical simulation of plug flow of cohesionless particles in a horizontal pipe, *Powder Technol.* **71**(3), 239–250 (1992).

15. D. Nishiura and H. Sakaguchi, Parallel-vector algorithms for particle simulations on shared-memory multiprocessors, *J. Comput. Phys.* **230**(5) 1923–1938 (2011), doi:10.1016/j.jcp.2010.11.040.
16. W. D. Hillis and G. L. Steele Jr, Data parallel algorithms, *Commun. ACM.* 29(12), 1170–1183 (1986).
17. M. Furuichi, D. Nishiura, M. Asai, and T. Hori, The first real-scale DEM simulation of a sand-box experiment using 2.4 billion particles, in: *The International Conference for High Performance Computing, Networking, Storage and Analysis.* Denver, CO (2017), http://sc17.supercomputing.org/SC17%20Archive/tech_poster/tech_poster_pages/post113.html.
18. D. A. Knoll and D. E. Keyes, Jacobian-free Newton–Krylov methods: A survey of approaches and applications, *J. Comput. Phys.* **193**(2), 357–397 (2004).
19. M. Furuichi, Dave A. May, Implicit solution of the material transport in Stokes flow simulation: Toward thermal convection simulation surrounded by free surface, *Comput. Phys. Comm.* **192**, 1–11 (2015).
20. U. Trottenberg, C. W. Oosterlee, A. Schuller (eds.), *Multigrid.* Academic Press, San Diego (2000).
21. N. Hosono and M. Furuichi, Implementation of SPH and DEM for a PEZY-SC heterogeneous many-core system, *ICCES 2019: Mech. Mach. Sci.*, **75** (2020), doi:10.1007/978-3-030-27053-7_60.
22. J. F. Peters, M. Muthuswamy, J. Wibowo, and A. Tordesillas, Characterization of force chains in granular material, *Phys. Rev. E.* **72**, 041307 (2005).
23. M. Furuichi, D. Nishiura, O. Kuwano, A. Bauville, T. Hori, and H. Sakaguchi, Arcuate stress state in accretionary prisms from real-scale numerical sandbox experiments, *Sci. Rep.* **8**, 8685 (2018).
24. D. Nishiura, M. Furuichi, and H. Sakaguchi, Real-scale DEM simulations on the fault evolution process observed in sandbox experiments, *Adv. Powder Technol.* **32**(11), 4432–4441, (2021).

https://doi.org/10.1142/9781800614635_0003

Chapter 3

Multi-Scale Tsunami Simulation with High-Fidelity Modeling and Visualization

Mitsuteru Asai[*,‡] and Nur'Ain binti Idris[†]

[*]Department of Civil Engineering, Kyushu University, Motooka, Nishi-ku, Fukuoka, Japan

[†]Department of Civil Engineering, Universiti Tun Hussein Onn (UTHM), Malaysia

[‡]asai@doc.kyushu-u.ac.jp

This chapter introduces a multi-scale tsunami simulation technique and its high-fidelity modeling and visualization. This simulation technology[1] is based on the particle simulation method[2] originally developed by the first author. Multi-scale tsunami simulators perform to solve tsunami run-up phenomena in a city scale. However, to accurately predict the inundation area, a 3D high-fidelity model that resolves buildings and houses in the town is required. Therefore, we proposed a particle generator for multi-scale tsunami simulation.[3] The key technology for a multi-scale tsunami simulator is a virtual wave generator using a fixed ghost boundary method. This boundary process has been updated to a Simplified version of the Generalized Fixed Ghost Particle method (SGFGP),[4] which is easier to implement and gives more general boundary conditions. In this chapter, we summarize our past research and show new numerical examples of multi-scale tsunami simulations that have been updated with the SGFGP boundary treatment.

1. Introduction

The 2011 off the Pacific coast of the Tohoku earthquake was the largest earthquake ever recorded in Japan, and the massive tsunami that accompanied the earthquake caused extensive damage to the coastal areas of the Tohoku region. Based on the experience of this unprecedented tsunami damage, there is an urgent need to study disaster prevention and mitigation technologies for coastal areas. In order to create a safe coastal zone, it is necessary to construct disaster countermeasures that include not only hard infrastructure such as levees and bridges but also soft measures such as disaster prevention education and the development of hazard maps. As one of such measures, the tsunami run-up analysis is expected to play an important role.

Currently, tsunami analysis widely practiced at the practical level is a numerical analysis based on shallow-water long wave theory, in which the vertical acceleration of water particles is assumed to be very small and the flow velocity is averaged in the vertical direction. Since this analysis method stimulates the phenomenon in two dimensions, it is effective for understanding a wide range of tsunami propagation phenomena from the epicenter to the coast, but it is difficult to determine the free surface shape of the local flow, where three-dimensionality dominates. In particular, when tsunamis run up to land, detailed prediction of the flow near buildings is difficult because the three-dimensionality of the wave cannot be ignored. Numerical analysis methods using meshes, such as the finite difference method and the finite element method, complicate the 3D flow analysis of waves with large waveform changes under the following conditions. The mesh may collapse during large deformations, the shape of the free surface may change significantly from time to time, and discontinuities may occur.

In this study, we developed a multi-scale simulation based on the smoothed particle hydrodynamics (SPH) Method.[5] SPH has been widely used in fluid and solid mechanics in recent years. There are two types of SPH for free surface flow problems: weakly compressible SPH (WCSPH) proposed by Monaghan,[6] and incompressible SPH (ISPH) proposed by Cummins.[7] Our group developed a multi-scale tsunami run-up simulation using ISPH in combination with the 2D

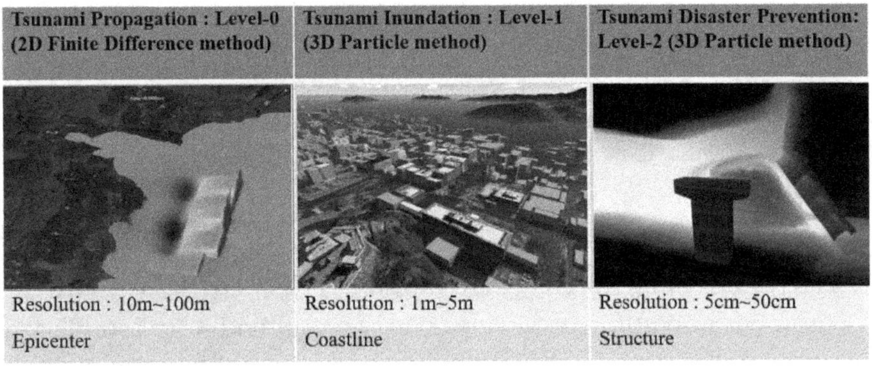

Fig. 1. Coupled simulations between a 2D shallow-water-equation-based simulation and a 3D particle simulation.

difference method based on the conventional shallow-water long wave theory and the 3D ISPH.

Our concept of multi-scale tsunami analysis is summarized in Fig. 1. The first Level-0 analysis is the same as a practical tsunami simulation using 2D shallow-water equations solved mainly by the finite difference method. The main objective of the Level-0 simulation is to predict the magnitude of a tsunami immediately. As mentioned earlier, countermeasures against tsunami disasters need to be developed for tangible infrastructures such as breakwaters and bridges as well as intangible consider aspects such as disaster prevention education and hazard maps. Tsunami run-up analysis is expected to play an important role in tsunami countermeasures. The multi-scale tsunami analysis aims to perform 3D tsunami inundation simulations with high accuracy by taking advantage of the fast computation speed of 2D finite difference simulations. With the above background, we aim to establish a tool to efficiently analyze tsunami run-ups in urban areas using the SPH method, which is a type of particle method. For this purpose, a geographic information system (GIS), we developed a series of procedures to develop a detailed topographic analysis model that reflects the topography, elevation, and exterior shape of buildings by referring to 3D location information (SHP) and digital elevation map (DEM) data obtained from

In the 3D analysis of tsunami run-up, regardless of the analysis method, such as the mesh method or the particle method, it is necessary to determine in advance the relationship between the spatial resolution (or the mean particle distance in the case of the particle method) and the accuracy of the analysis. In particular, for 3D analysis, the resolution needs to be as low as possible because of the high cost of analysis. In this study, analytical models with different particle distances developed using a single CAD data were prepared. Numerical experiments were conducted to determine the minimum spatial resolution required to obtain a valid solution by comparing the analysis results under the same conditions.

Finally, a photo-realistic visualization method was established to provide an intuitive understanding of tsunami phenomena. This makes it possible to apply the results of tsunami run-up analysis in urban areas to soft measures such as tsunami disaster prevention education.

2. Overview of a Stabilized ISPH for 3D Tsunami Inundation Analysis

2.1. *Incompressible smoothed particle hydrodynamics*

The following incompressible Navier–Stokes governing equations are utilized as the governing equations of fluid domain.

$$\nabla \cdot \boldsymbol{u} = \boldsymbol{0}, \tag{1}$$

$$\frac{D\boldsymbol{u}}{Dt} = -\nabla p + \nu \nabla^2 \boldsymbol{u} + \tfrac{1}{\rho} \nabla \cdot \boldsymbol{\tau} + \boldsymbol{g}, \tag{2}$$

where ρ and ν are the density and kinematic viscosity of the fluid, $\boldsymbol{u}$, p, and $\boldsymbol{g}$ are the velocity vector, pressure, and gravity vector, respectively. The turbulence stress $\boldsymbol{\tau}$ is necessary to represent the effects of turbulence with a coarse spatial discretization. In the general incompressible flow simulation, the density is assumed by a constant value with its initial value, and then the pressure is determined by the pressure Poisson equation (PPE), which is derived from the continuity Eq. (1). In the sense of physical observation, physical density should keep its initial value for incompressible flow. However, during numerical simulation, the *particle* density may change slightly from the initial value because the particle density is strongly dependent on particle locations in the SPH method. If the particle distribution can keep almost uniform, the difference between *physical* and *particle* density may be vanishingly small. In other words, accurate SPH results in incompressible flow need to keep the uniform particle distribution. For this purpose, the different source terms in the PPE can be derived using the *particle* density. The SPH interpolations are introduced into the original mass conservation law before the perfect compressibility condition is applied.

$$\langle \nabla \cdot \boldsymbol{u}_i^{n+1} \rangle = -\frac{1}{\rho^0} \frac{\langle \rho_i^{n+1} \rangle - \langle \rho_i^* \rangle}{\Delta t}. \tag{3}$$

Then, the PPE is reformulated as follows:

$$\langle \nabla^2 p_i^{n+1} \rangle = \frac{\rho^2}{\Delta t} \langle \nabla \cdot \boldsymbol{u}_i^* \rangle + \alpha \frac{\rho^0 - \langle \rho_i^n \rangle}{\Delta t^2}, \tag{4}$$

where α is the relaxation coefficient, $\boldsymbol{u}_i^*$ is temporal velocity given by the following predictor calculation without pressure gradient term in the original Navier–Stokes equations,

$$\boldsymbol{u}_i^* = \boldsymbol{u}_i^n + \Delta t \left(\nu \langle \nabla^2 \boldsymbol{u}_i^n \rangle + \frac{1}{\rho^0} \langle \nabla \cdot \boldsymbol{\tau}_i^n \rangle + \boldsymbol{g} \right). \tag{5}$$

After calculating the above temporal velocity, the current pressure is implicitly determined by solving the modified PPE. And then, the current velocity can be updated from its temporal value by taking account of the pressure gradient term as

$$\boldsymbol{u}_i^{n+1} = \boldsymbol{u}_i^* - \Delta t \langle \nabla p_i^{n+1} \rangle. \tag{6}$$

2.2. *Spatial discretization by SPH*

In this chapter, the triangle bracket $\langle \cdot \rangle$ means an SPH approximation for the function ϕ, the gradient $\nabla \phi$, and the Laplacian $\nabla^2 \phi$ as follows:

$$\phi_i^n(\boldsymbol{x}_i, t) \approx \langle \phi_i^n \rangle = \sum_j \frac{m_j}{\rho_j^0} \phi W(|\boldsymbol{r}_{ij}|, h) \tag{7}$$

$$\nabla \phi(\boldsymbol{x}_i, t) \approx \langle \nabla \phi_i^n \rangle = \frac{1}{\rho_i^0} \sum_j m_j (\phi_j - \phi_i) \nabla W(|\boldsymbol{r}_{ij}|, h) \tag{8}$$

$$= \rho_i^0 \sum_j m_j \left(\frac{\phi_i}{\rho_i^0} + \frac{\phi_j}{\rho_j^0} \right) \nabla W(|\boldsymbol{r}_{ij}|, h) \quad (9)$$

$$\nabla^2 \phi(\boldsymbol{x}_i^n, t) \approx \langle \nabla^2 \phi_i^n \rangle = \rho_i^0 \sum_j m_j \left(\frac{\rho_i^0 + \rho_j^0}{\rho_i^0 \rho_j^0} \frac{\boldsymbol{r}_{ij} \nabla W(|\boldsymbol{r}_{ij}|, h)}{\boldsymbol{r}_{ij}^2 + \eta^2} \right) (\phi_i - \phi_j), \quad (10)$$

where $W(|\boldsymbol{r}_{ij}|, h)$ is called by the kernel function in the SPH, and it should be a function of influence radius h distance between the target particle i and its neighbor particle j. The distance $|\boldsymbol{r}_{ij}|$ is given by the relative coordinate vector $\boldsymbol{r}_{ij}(= \boldsymbol{x}_i - \boldsymbol{x}_j)$.

Note that the gradient models have a manifold property that can convert each other in the analytical sense, but a different formula in the numerical simulation has been applied. Only the pressure gradient term $\langle \nabla p_i^{n+1} \rangle$ in (6) is approximated by the upper formulation (8), and the other gradient terms like $\langle \nabla \cdot \boldsymbol{u}_i^* \rangle$ in Eq. (4) and *abc* in (5) are given by the lower formulation (9). The other notice in our writing is the differences between *physical* density ρ_i^0 and *particle* density $\langle \rho_i^n \rangle$. The physical density never changes during our numerical simulation because of the incompressible property, but the particle density in the modified PPE (4) is determined by the above SPH weight averaging (7) in order to know the uniformity of the particle location.[a]

2.3. *Boundary treatment using a modified mirror boundary method*

The boundary treatment of the particle method, and other meshless methods, is a negligible issue. This chapter uses a modified mirror boundary treatment with a virtual marker to satisfy a reasonable slip or no-slip boundary condition in the velocity and the Neuman condition in the pressure.

Many conventional methods of boundary treatment have been introduced within the SPH framework, such as the fixed wall particle

[a]The particle density is utilized to evaluate the pressure explicitly in the original SPH, but the stabilized ISPH solves the PPE with the particle density in order keep a suitable (or better) homogeneous particle location.

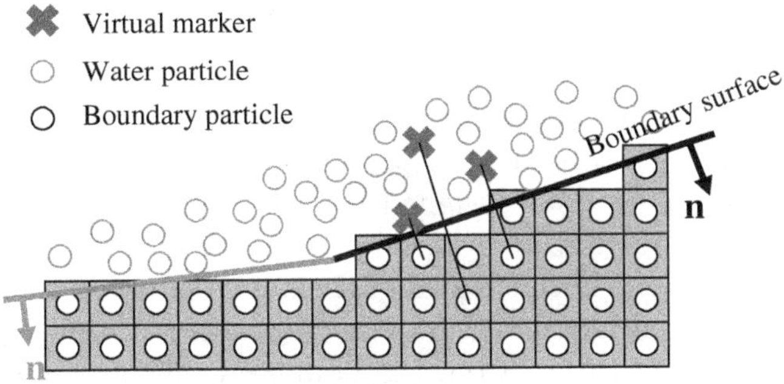

Fig. 2. Virtual marker with fixed wall particle for the incompatible boundary model.

(FWP)[8] and the ghost boundary particle (GBP).[9] Then, Marrone *et al.*[10] extended the GBP to make fixed ghost particles (FGPs). Recently, Ref. 11 introduced the virtual marker with fixed wall particle (VMFWP) that includes a special treatment of the slip and non-slip conditions with incompatible step-shape boundary models as shown in Fig. 2. Here, the compatible boundary model is defined for a case in which the boundary surface particles are located on a real solid boundary surface. The other boundary particles inside the solid region are aligned parallel to the boundary surface particles. The GBP approach was introduced to overcome the disadvantages of the FWP method by placing the wall particles into the solid wall with symmetrical points or mirroring image pairs. In addition, the method starts off by using FGP, instead of standard GBP, which provides some improvement in the uniform particle distribution because it does not depend on the position of fluid particles. This leads to a reduction in computational costs due to the repositioning of mirror particles.[10]

The use of FGP means that the symmetrical positions with the slip or no-slip condition, which respond to water particles near the wall boundary, are determined with high accuracy. However, this treatment is primarily suitable for compatible boundary models and not easy to implement for incompatible boundary models. Therefore, the VMFWP was proposed to enable consideration of both compatible and incompatible boundary models by introducing a regular grid of virtual markers.[11]

2.3.1. Virtual markers with fixed wall particles in a structured regular grid

According to the generalized VMFWP, all boundary particles are located inside the solid region within a structured regular grid. Each boundary particle is assigned its own signed distance function denoted as d_i from the nearest solid boundary surface in the normal direction $\boldsymbol{n}_i$ on the surface, and then the boundary particles are paired with virtual markers. The virtual markers are placed at mirrored symmetric locations from the boundary particles with respect to the solid boundary surface. The position of each virtual marker $\boldsymbol{X}_I$ is referenced to a boundary particle position $\boldsymbol{x}_i$, and the information contained in the signed distance function,

$$\boldsymbol{X}_I = \boldsymbol{x}_i + 2d_i\boldsymbol{n}_i. \tag{11}$$

Note that the virtual marker does not directly affect the SPH calculation but is used to satisfy adequate boundary conditions. Using this pairing system between boundary particles and their virtual markers, the slip and/or no-slip condition with respect to the velocity distribution in the vicinity of the solid boundary and the non-homogeneous pressure Neuman condition can be satisfied.

At first, the velocity $\boldsymbol{u}_I$ and pressure p_I at the virtual marker are interpolated in a manner similar to the SPH,

$$\phi(\boldsymbol{X}_I, t) \approx \langle\phi_I\rangle := \sum_j \frac{m_j}{\rho_j}\tilde{W}(|\boldsymbol{r}_{Ij}|, h)\phi_j, \tag{12}$$

$$\tilde{W} := \frac{W(|\boldsymbol{r}_{Ij}|, h)}{\sum_j \frac{m_j}{\rho_j} W(|\boldsymbol{r}_{Ij}|, h)}.$$

The only difference between the above interpolation and its original Eq. (7) is the use of a normalized weight function $\tilde{W}$. The normalized weight function is required to overcome the loss of a sufficient number of water particles to satisfy the unity condition $\sum_j W = 1.0$. This interpolation is applied to the virtual marker because the virtual marker has no explicit degrees of freedom in the SPH calculation.

Once the velocity of the virtual marker is given, adequate velocity distributions in the slip and no-slip condition can be determined by

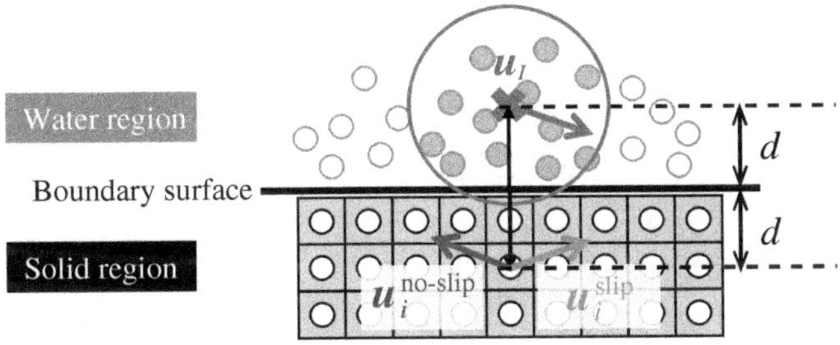

Fig. 3. Slip and no-slip boundary velocity field using the virtual marker.

the following equation:

$$\boldsymbol{u}_i^{\text{slip}} := \boldsymbol{M}\,\langle \boldsymbol{u}_I \rangle,\ M_{ij} = \delta_{ij} - 2n_i n_j \quad \text{(for slip condition)}, \tag{13}$$

$$\boldsymbol{u}_i^{\text{no-slip}} := \boldsymbol{R}\,\langle \boldsymbol{u}_I \rangle,\ R_{ij} = -\delta_{ij} \quad \text{(for no-slip condition)}. \tag{14}$$

where the second-order tensors $\boldsymbol{M}$ and $\boldsymbol{R}$ are the transportation tensors for the mirror symmetry and point symmetry, and M_{ij} and R_{ij} indicate the components in a Cartesian coordinate system. Figure 3 shows a schematic image of the velocity distribution on a boundary particle and its paired virtual marker.

2.4. *Non-homogeneous pressure Neumann condition*

Furthermore, the treatment for the non-homogeneous pressure Neumann condition can be easily formulated. The non-homogeneous pressure Neumann condition is given by

$$\frac{\partial p}{\partial \boldsymbol{n}} = \rho(\nu\Delta^2 \boldsymbol{u} + \boldsymbol{g}). \tag{15}$$

The non-homogeneous pressure Neumann condition is approximately satisfied with a virtual marker at the wall surface as

$$p_j := p_J + d_j \rho(\nu\Delta^2 \boldsymbol{u}_J + \boldsymbol{g}) \cdot \boldsymbol{n}_j,\ \boldsymbol{x}_j \in \Omega^{\text{wall}}. \tag{16}$$

Note again that the virtual marker is not the computational point but only a measuring point. Then, the pressure p_J and velocity $\boldsymbol{u}_J$

should be approximated by

$$p_j := \langle p_J \rangle + d_j \rho \langle \boldsymbol{f}_J \rangle \cdot \boldsymbol{n}_j, \; \boldsymbol{x}_j \in \Omega^{\text{wall}} \qquad (17)$$
$$\langle \boldsymbol{f}_J \rangle := \langle \nu \Delta^2 \boldsymbol{u}_J \rangle + \boldsymbol{g},$$

where $\langle p_I \rangle$ is the pressure on the virtual marker determined by the SPH weight averaging approximation (12), and d_i indicates the distance from a solid boundary toward the target wall particle.

The distance decreases from $2d$ to d. This equation means the virtual pressure in the wall domain can be determined by adding the second term from an approximated pressure on the wall surface. This approximated value is given by the SPH interpolation with Equation (12) in the original formulation of GFGP (or FGP), but there are other possibilities for this approximation. Then, the two simplified versions are proposed as follows.

The first simplification of GFGP (SGFGP) is the use of averaging neighbor particles within the influence radius from a target wall particle. The first version is called Level-1 approximation with simple averaging carried out in this later. The next simplification uses a rough approximation by copying the pressure at each neighboring particle. This next version is called Level-0 approximation.

In the first version with Level-1 approximation, pressure at the virtual marker on the wall surface is assumed by averaging the number of the particles within the compact support as follows:

$$\begin{aligned} p_J \approx \langle p_J \rangle &= \sum_{k \in \mathbb{P}_J^{\text{fluid}}} \frac{m_k}{\rho_k} \tilde{W}(|\boldsymbol{x}_{Jk}|, h) \phi_k \quad \text{in Level-2} \\ &\approx \overline{p_J} = \sum_{k \in \mathbb{P}_J^{\text{fluid}} \bigcap \mathbb{P}_i^{\text{fluid}}}^{n_J} p_k / n_J \quad \text{in Level-1} \\ &\approx p_i \quad \text{in Level-0}. \end{aligned} \qquad (18)$$

One of the above approximations is substituted into Eq. (16) in order to establish a relationship among pressure value on the wall particle $\boldsymbol{x}_j$ and pressure values on the neighbor water particles. Finally, the discretized pressure Poisson equations can only be limited to the unknown pressures for the water particles.

To make it much simpler, the Laplacian model for the pressure is rewritten by using a symbolic notation with A_{ij}, which means the component of the coefficient matrix related to particle i and j.

$$\begin{aligned}\langle \nabla^2 p_i \rangle &= \sum_{j \in \mathbb{P}_i} m_j \left(\frac{\rho_i + \rho_j}{\rho_i \rho_j} \frac{\boldsymbol{x}_{ij} \cdot \nabla W(|\boldsymbol{x}_{ij}|, h)}{|\boldsymbol{x}_{ij}|^2 + \eta^2} \right) (p_i - p_j) \\ &= \sum_{j \in \mathbb{P}_i} A_{ij}(p_i - p_j) \\ &= \sum_{j \in \mathbb{P}_i^{\text{fluid}}} A_{ij}(p_i - p_j) + \sum_{j \in \mathbb{P}_i^{\text{wall}}} A_{ij}(p_i - p_j). \end{aligned} \tag{19}$$

$$\mathbb{P}_i^{\text{wall}} = \left\{ j;\ h > |\boldsymbol{x}_{ij}|,\ \boldsymbol{x}_j \in \Omega^{\text{wall}},\ j \neq i \right\}. \tag{20}$$

Here, pressure on the boundary particle in the wall region in the second term of Eq. (19) is approximated by substituting Eq. (17) with each level approximation shown in Eq. (18). Then, the second term can be rewritten as

$$\begin{aligned}\sum_{j \in \mathbb{P}_i^{\text{wall}}} A_{ij}(p_i - p_j) &\approx \sum_{j \in \mathbb{P}_i^{\text{wall}}} A_{ij}(p_i - \langle p_j \rangle) + \sum_{j \in \mathbb{P}_i^{\text{wall}}} d_j \rho \langle \boldsymbol{f}_J \rangle \cdot \boldsymbol{n}_j \quad \text{in Level-2,} \\ &\approx \sum_{j \in \mathbb{P}_i^{\text{wall}}} A_{ij}(p_i - \overline{p_j}) + \sum_{j \in \mathbb{P}_i^{\text{wall}}} d_j \rho \langle \boldsymbol{f}_J \rangle \cdot \boldsymbol{n}_j \quad \text{in Level-1,} \\ &\approx \sum_{j \in \mathbb{P}_i^{\text{wall}}} \beta A_{ij}(p_i) + \sum_{j \in \mathbb{P}_i^{\text{wall}}} d_j \rho \langle \boldsymbol{f}_J \rangle \cdot \boldsymbol{n}_j \quad \text{in Level-0.} \end{aligned} \tag{21}$$

The first term of each level equation modifies the coefficient matrix components, and the second term should be converted to the source term in the discretized pressure Poisson equations. The difference among the levels is only the modification of the coefficient matrix and the same change in the source term. Figure 4 shows the schematic diagram of the relationships between the modifications made at each level. The Level-0 approximation of p_j generates a symmetric coefficient matrix, but the Level-1 and -2 approximations result in an asymmetric coefficient matrix after some modifications are made. As a result, only the Level-0 approximation can utilize the Conjugate Gradient solver, and BiCGstab or the other asymmetric solver is still needed for Level-1 and -2 cases. Note that the direct

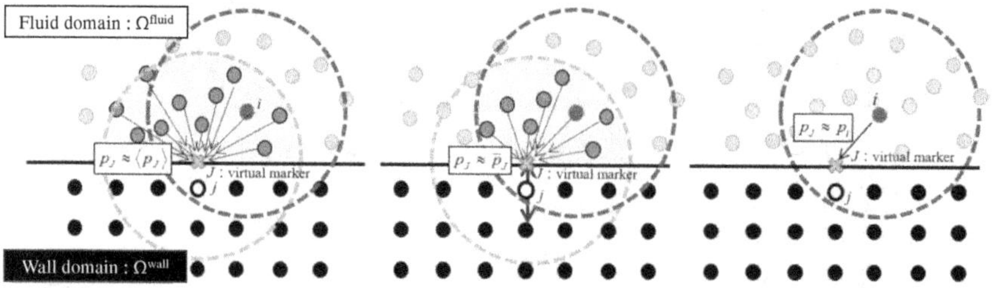

(a) Level-2 (Existing Neumann B.C.) (b) Level-1 (Simplified Neumann B.C.) (c) Level-0 (Simplified Neumann B.C.)

Fig. 4. The relationship between modifications made at each different level. (a) Level-2 (existing Neumann B.C), (b) Level-1 (Simplified Neumann B.C), and (c) Level-0 (Simplified Neumann B.C)

substitution of Eq. (16) in the Level-0 approximation may not generate any change in the coefficient matrix, but a small modification is added in the above equation with β just for the safety of solving the linear equation by the conjugate gradient method. The value of β is set around 1% in our simulations.

As shown in Eq. (18), the unknown pressure of the PPE can be limited to the water particle pressure. Solving this Poisson equation with the modified coefficient matrix and the source term automatically satisfies the non-homogeneous pressure Neumann conditions. After solving the PPE, we must first evaluate the pressure in virtual marker using Eq. (18) from the adjacent water particle pressure, and then map the wall particle pressure from the virtual marker pressure by Eq. (17). Figure 5 shows the schematic diagram of the several steps on how the boundary condition is applied with the matrix involved in solving the PPE for Level-0, -1, and -2. The non-homogenous Neumann boundary condition was applied on the solid boundary, while the Dirichlet boundary condition was applied only to the place given originally, the water surface. Note that a Dirichlet boundary will give the zero pressure of the particles of the free surface.

The advantages of Level-0 approximation are easy to implement and could also reduce the computational cost because of the symmetric coefficient matrix. A few verification tests were conducted to ensure the most efficient method and then proceed with the validation by comparing with experimental tests in the following sections.

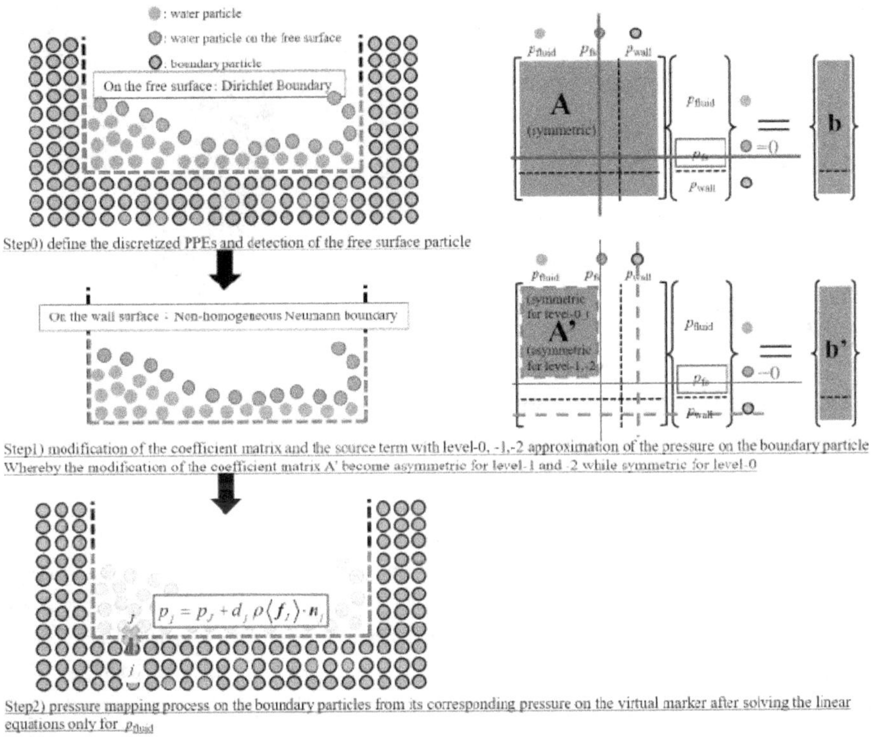

Fig. 5. Schematic diagram of the implementation of non-homogenous Neumann boundary treatments.

2.5. *Validation with 3D dam break flow with the opening gate*

The 3D dam break flow simulation is used to validate our ISPH method's versatility with SGFGP boundary treatment and compare it with the experimental results. The experimental test on the investigation of dynamic pressure loads during a dam break with an opening gate was carried out at the Technical University of Madrid (UPM) as reported by Lobovsky *et al.*[12] In order to validate the proposed SGFGP, the analysis model with the same geometry as the experimental test, including the five pressure sensors, was used as shown in the schematic diagram (Fig. 6). Figure 6 illustrates the view of the analysis model with the tank dimension and the initial water depth

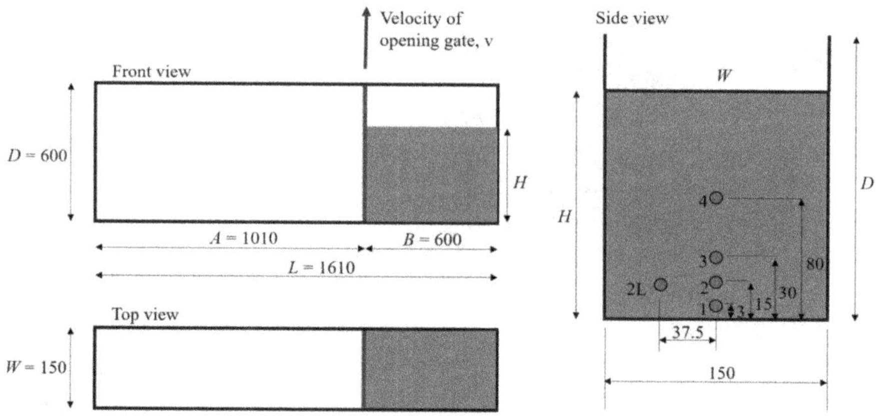

Fig. 6. The dam break simulation with an opening gate which corresponds to the experiment conducted by Lobovsky *et al.* Dimensions are in millimeters.

H in the dam reservoir from the front view, top view, and also side view. There are two different initial water depths, H, with different velocities of the opening gate, which are 300 and 600 mm of water depth and the median velocity of opening gates obtained from the experimental test are 3.46 and 4.53 m/s, respectively. In addition, the water properties give the density of 997 kg/m^3 and the kinematic viscosity of 8.9×10^{-7} m^2/s. The arrangement of the pressure sensors used in the analysis model can be seen from the side view as depicted in Fig. 6. There are four sensors located in the centerline of which the first sensor (Sensor-1) is placed 3 mm above the bed, in order to measure the impact of pressure at the lower tank corner, while the other three pressure sensors of the center line are located at 15 mm (Sensor-2), 30 mm (Sensor-3), and 80 mm (Sensor-4) between their centers and the tank bed. The additional pressure sensor (Sensor-2L), an off-centered sensor, is located at the same height as the second lowered sensor, Sensor-2, but halfway toward the back wall.

The particle simulation model of the dam break problem with the same geometry given in Fig. 6 is used. The total number of particles used for the 3D simulation is 12,541,253 and 9,160,422 for 600 mm and 300 mm, respectively. The initial particle distance d_0 is = 0.002 m, and the time increment is fixed with 0.0002 s in the analysis. The other parameters α, β are 0.005 and 0.01, respectively. An opening gate, which is included in the experimental test, also considers the design gate in the analysis model and has the same velocity as the median velocity in the experimental data. All five sensors in the analysis model were placed at the same location as the experimental test to ensure the simulation's pressure could be measured and compared with the experiment results. Since there are two cases of different heights, the simulation results of the 600 mm water height case will be discussed first, followed by the 300 mm water height case.

Figure 7 shows free surface profile evolutions by the snapshot for the 600 mm water height case. From the first observation, the simulation results show similarities in every snapshot of the free surface shape by comparing it with the experimental test. The pressure contour included in Fig. 7 shows the pressure gradient from the static condition before an opening gate and after the gate's removal and the impact after hitting the wall, which indicated the higher pressure. Furthermore, it is clear that the lower sensors are exposed to the higher pressure level in the contour is observed after the water is smashed on the wall boundary surface. Then, the pressure measured at each sensor was compared with the average value of pressure obtained from the experiment as shown in Fig. 8.

As shown in Fig. 8, the simulated pressure for each sensor is generally close values as in the experiment, although the peak values are slightly above or slightly below the peaks obtained by the experiment. Interestingly, the similar curve shape of pressure at each sensor was observed in these comparisons and it indicates a numerical convergence profile as more accurate solutions are given when the spatial resolution in the particle discretization becomes higher. However, there are small discrepancies observed between the simulation and the experiment data whereby the arrival time at each sensor from simulation results shows a bit of delay in time. Neglecting the arrival time, the value of pressure observed from the simulation using the proposed boundary treatment shows similarity and is in line with the experimental results. The opening speed in the simulation was fixed to a constant speed evaluated from the average of the experimental tests. We suggest the accurate modeling of the opening gate and slip or no-slip boundary treatment need to be modified in future work.

Next, the 300 mm water height case is simulated with the same setting. Figure 9 presents the comparison of the evolution of the free surface profile between the simulation and the experiment for the

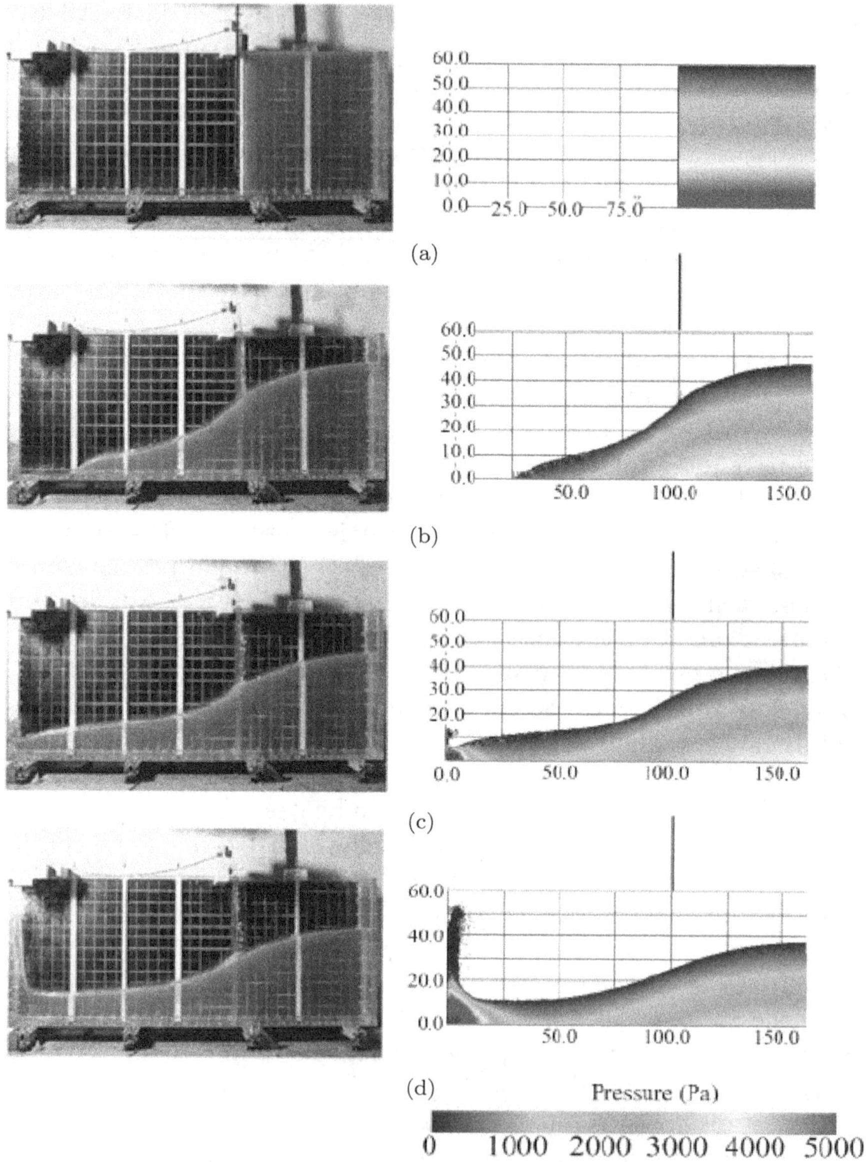

Fig. 7. The comparison of the free surface evolution between simulation and experiment for 600 mm water height case at non-dimensional times (a) 0, (b) 1.27, (c) 1.67, and (d) 1.87.

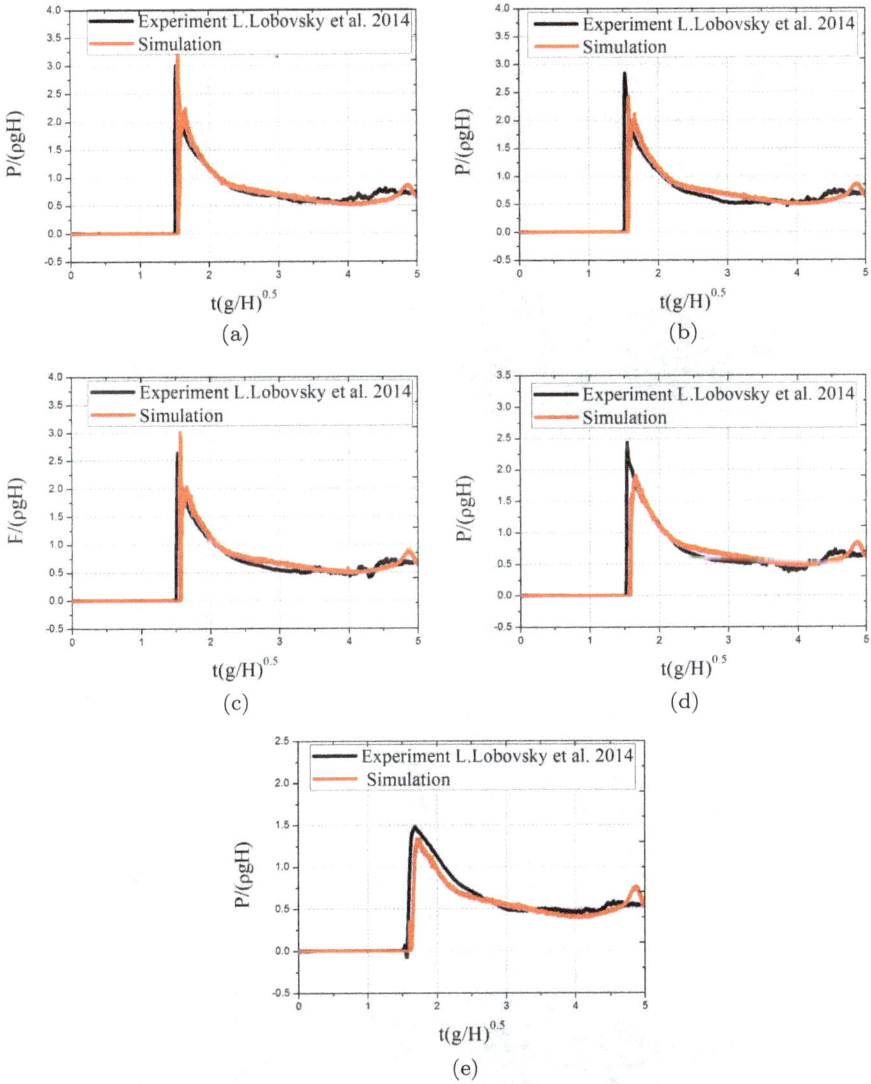

Fig. 8. The comparison of the non-dimensional pressure with the non-dimensional time between the simulation results with the average pressure from the experimental test at (a) Sensor 1, (b) Sensor 2, (c) Sensor 2L, (d) Sensor 3, and (e) Sensor 4 for the 600 mm water height case.

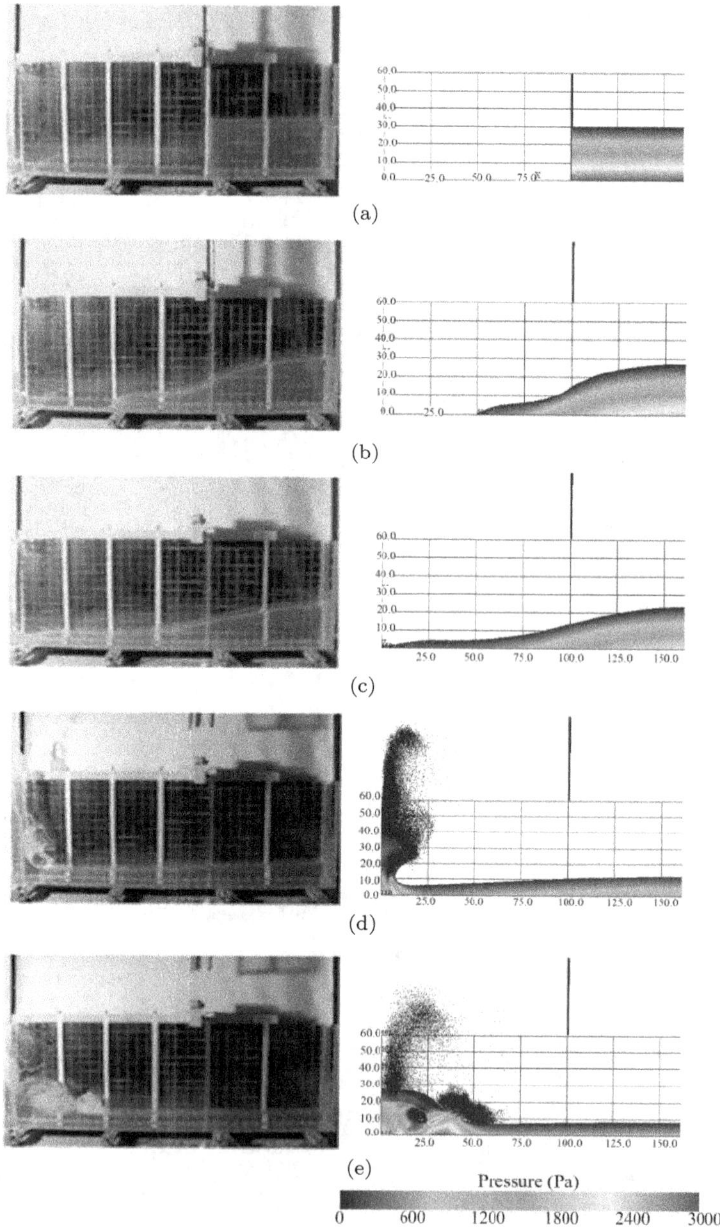

Fig. 9. The comparison of the free surface evolution between simulation and experiment for 300 mm water height case at non-dimensional times (a) 0, (b) 1.58, (c) 2.57, (d) 4.93, and (e) 6.67.

300 mm water height case at non-dimensional times (a) 0, (b) 1.58, (c) 2.57, (d) 4.93, and (e) 6.67. By comparing the snapshots taken from the simulation and experiment, a similar tendency can be observed. The snapshot taken at arrival at the edge corner and at climbing the wall indicate similar water profiles with experimental pictures as shown in Figs. 9(c) and 9(d), respectively. In addition, the static hydrostatic pressure also can be seen from the snapshot before the gate starts to open as depicted in Fig. 9(a). Similar tendency was also observed in the 600 mm water height case.

Finally, the measured pressure at each sensor obtained from the simulation is compared with the average pressure obtained from 100 conducted tests for each sensor as shown in Fig. 10. However, a bit of delay of the arrival time has also occurred at almost all sensors. Then, by neglecting the short delay in arrival time, the plotted pressure history shows similarities with the experimental data. Even though it is not completely the same as the experimental data only a small discrepancy is observed, especially at all sensors excluding Sensor 4 where the simulation results give a bit higher pressure at certain times. However, only Sensor 4 gives a bit lower value compared with the experimental results.

The tendency observed from the simulation also shows similarities with the result obtained by Cercos-Pita *et al.*[13] for Sensor 1, Sensor 2, Sensor 3, and Sensor 4. Interestingly, the upper sensor which is Sensor 4 also shows the same tendency which is lowered compared with the experimental result. In this section, the existence of the pressure difference observed at Sensor 4 is paid attention to since the same tendency is also observed by recorded pressure by Cercos-Pita *et al.*[13] Then, the comparison was made by using the same approach with the different heights as proposed in the experimental test as shown in Fig. 11. The pressure observed at the sensor located at 70 mm above the bed tank which was originally located at 80 mm shows a similarity with the experimental result. During these simulations for the dam break problem, the wavefront in the case of the 300 mm water height case is very narrow in contrast with the 600 mm case. The narrow wavefront requires a smaller particle distance.

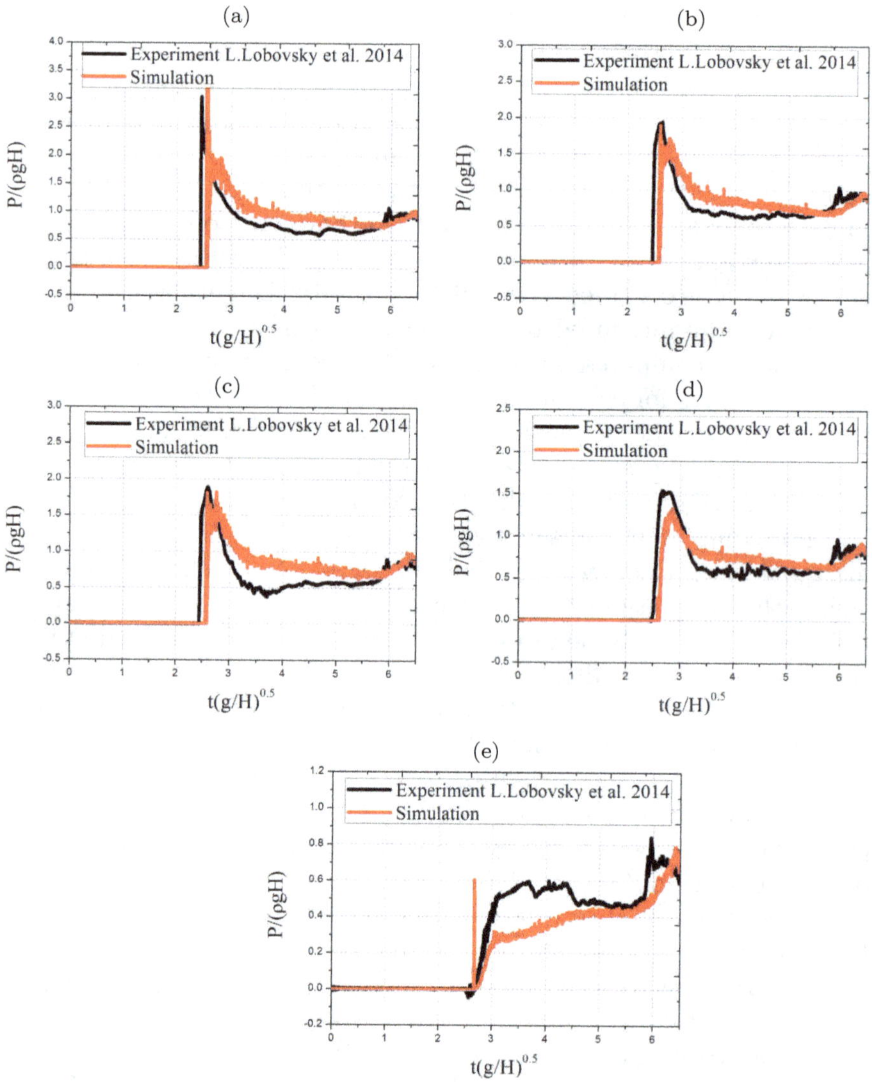

Fig. 10. The comparison of the non-dimensional pressure with the non-dimensional time between the simulation results with the average pressure obtained from 100 tests of experiment data at (a) Sensor 1, (b) Sensor 2, (c) Sensor 2L, (d) Sensor 3, and (e) Sensor 4.

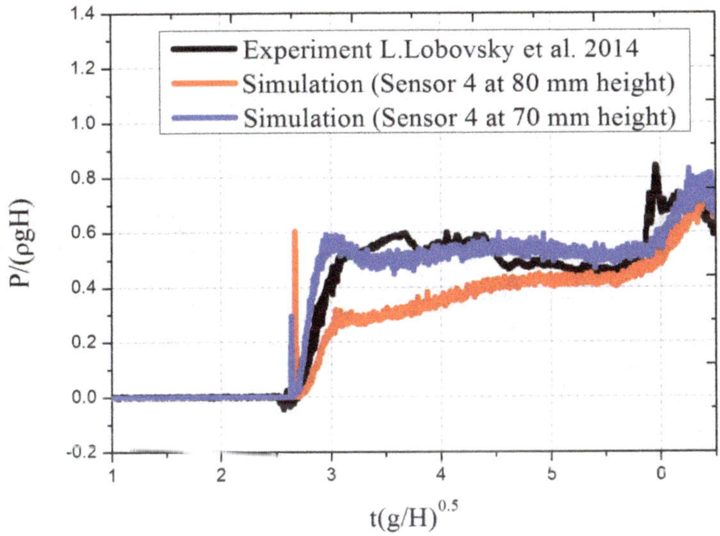

Fig. 11. The comparison of the non-dimensional pressure with the non-dimensional time between experimental result and the simulation results at the Sensor 4 with different heights.

3. Matrix-Arrayed Virtual Wave Maker for 2D–3D Coupling Simulations

Here, a matrix-arrayed virtual wave maker (VWM) is proposed to connect two different simulations. One is a pre-calculation solution (or observed data) referred to here as the Level-0 analysis. The other is a sequential particle simulation called the Level-1 analysis that produces the data following the Level-0 pre-calculation at the interface boundary.

3.1. *General functions of the virtual wave maker*

Figure 12 shows the main functions of the matrix-array shaped VWM as a schematic image for a 2D problem. In the next section, additional transitional layers will be explained.

A VWM is not a physical wave maker but rather functions as a generator of new particles and/or as a remover of unnecessary particles, dependent on pre-calculation results. At first, the cell size is assumed to be the same as the initial particle distance L_0 in the SPH.

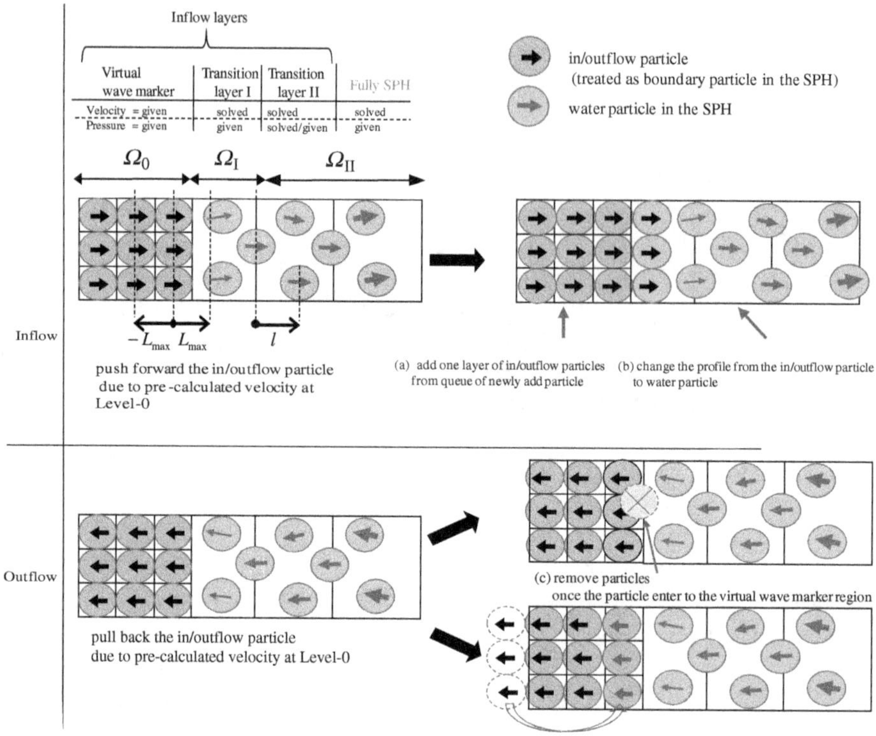

Fig. 12. General functions of the VWM and the definition of additional transitional layers

Each cell includes one in/outflow particle that becomes a candidate for the newly generated water particle in the SPH calculation. The VWM cell is moved only in the main flow direction $\boldsymbol{N}_f$ within a distance of $-L_{\max}$ to $L_{\max}$. Here, the maximum value $L_{\max}$ is the same as L_0. Then, the movement of the cell is limited to the main flow direction and the total value L^{n+1} is given,

$$L^{n+1} = L^n + dt\,\boldsymbol{U}^n_{\mathrm{lv0}} \cdot \boldsymbol{N}_f. \tag{22}$$

The incremental movement of the VWM $dt\,\boldsymbol{U}^n_{\mathrm{lv0}} \cdot \boldsymbol{N}_f$ shows a positive value in the inflow condition on the inflow boundary; the value becomes negative for outflow conditions. When the total movement reaches the maximum value $L_{\max}$, the VWM cell produces a new water particle with the current water velocity $\boldsymbol{U}^n_{\mathrm{lv0}}$, as shown in Fig. 12(b), and the cell is returned back to the last queue of the VWM cell, as shown in Fig. 12(a). In contrast, for the outflow condition, the last cell is returned back to the front cell to occupy the virtual wave maker by the in/outflow particles, as shown in Fig. 12(b), once the total movement becomes the minimum value $-L_{\max}$. In addition to the particle generation function of the virtual cell, the cell can remove unnecessary water particles in the Level-1 simulation based on the SPH, when the water particles go inside the VWM, as shown in Fig. 12(c). These functions are necessary to maintain a uniform particle distribution during a SPH calculation in the vicinity of the inflow.

Note that the in/outflow particles are treated as boundary particles in the SPH simulation at Level-1. Then, the number of the columns in the VWM is fixed at three, as shown in Fig. 12. Each in/outflow particle at the same height has the same velocity as the value at the first column from the waterside, although the velocity has a different value at greater water depths.

3.2. *VWM considering pressure from pre-analysis*

The first version of the VWM is presented here as shown in Fig. 13. The VWM introduced in this study is designed to carry over the pressure and flow velocity obtained from the preliminary analysis as in the Level-0 directly to the 3D analysis as in the Level-1 analysis. There have been many studies on multi-scale tsunami analysis and coupled simulations between different dimensions, but most of them assume hydrostatic pressure and couple the 2D and 3D analyses. The first version of the proposed VWM is the technique needed to have consistent pressure between levels. This requires the addition of several transition layers.

If the pressure is not consistent between Level-0 and Level-1, the SPH simulation of Level-1 becomes the starting point for divergence. Therefore, two types of transient regions were placed in front of the virtual wave generator (Fig. 14). As shown in this figure, the velocity from the prior analysis becomes the input data to control the movement of the VWM, and the particle velocity in the transition region I in $\Omega_{\rm I}$ and transition region II in $\Omega_{\rm II}$ are given according to the SPH method. On the other hand, the SPH method only gives the pressure for the particles located in the transition region II in $\Omega_{\rm II}$.

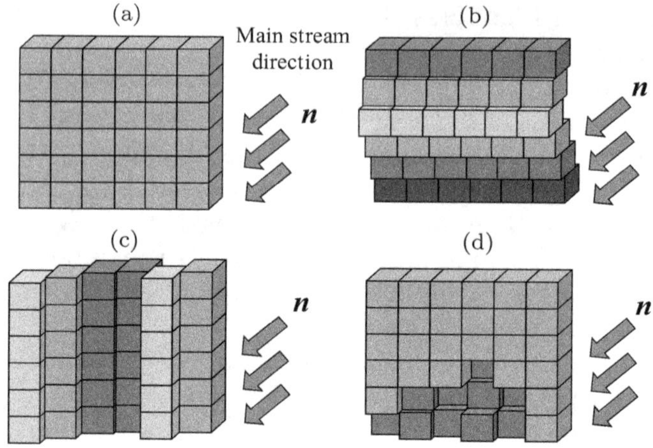

Fig. 13. Application of the virtual wave maker for several wave types. (a) Uniform velocity case, (b) 2D wave case, (c) 2D wave case (coupled with shallow water simulations), and (d) 3D wave case (coupled with 3D simulations).

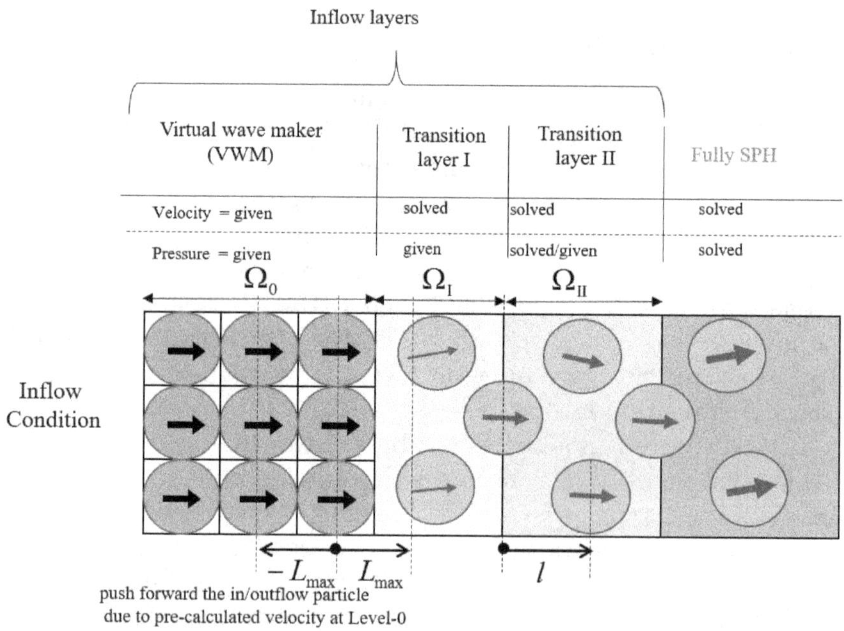

Fig. 14. The 2D schematic diagram of virtual wave maker (VWM) with additional transition region which considering the pressure from Level-0.

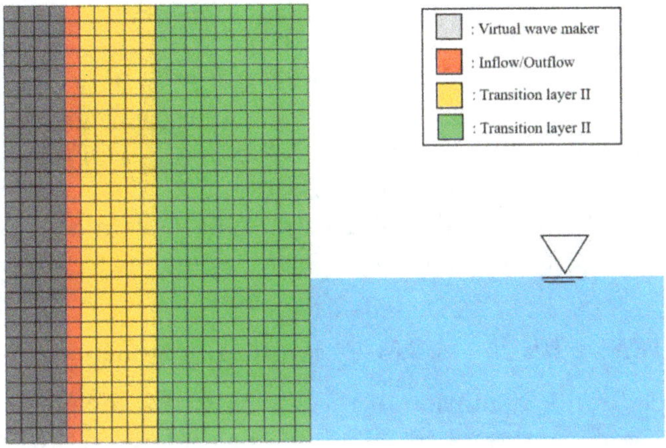

Fig. 15. The detailed overview of boundary processing cell in Level-1.

If the water particle is located inside of these transition regions, the pressure is over-written in the following manner after the pressure is given by solving the pressure Poisson equations as p^{**}.

$$p_{\mathrm{lv1}}(\boldsymbol{x}_i) := P_{\mathrm{lv0}}(\boldsymbol{x}_i) \quad \text{in}\,\Omega_0,\,\Omega_{\mathrm{I}}, \tag{23}$$

$$p_{\mathrm{lv1}}(\boldsymbol{x}_i) := \gamma(\boldsymbol{x}_i)P_{\mathrm{lv0}}(\boldsymbol{x}_i) + (1-\gamma(\boldsymbol{x}_i))p^{**}(\boldsymbol{x}_i) \quad \text{in}\,\Omega_{\mathrm{II}}, \tag{24}$$

$$\gamma(\boldsymbol{x}_i) = \frac{L_{\mathrm{II}} - l(\boldsymbol{x}_i)}{L_{\mathrm{II}}},$$

where the first equation means that both the VWM Ω_0 and the transitional layer Ω_{I} will have the same pressure as the Level-0 solution, and the second equation for the transitional layer Ω_{II} is the linear combination between the SPH calculation and Level-0 solution with respect to the distance l from the boundary between Ω_{I} and Ω_{II}. By using the above equation, we can reduce the gap between the Level-0 and Level-1 solutions without any jump in the pressure calculation. Figure 15 shows the 2D schematic diagram of the detailed overview of the boundary processing cell for the transition regions proposed. The numbers of layers for transition layer I and transition layer II are 5 layers and 15 layers, respectively.

The effects of the first version of VWM will be examined by comparing the solution with or without considering the pressure from Level-0 analysis in the transition region in Section 6. However, this VWM solution has limitations, such as its dependence on setting the number of transition layers. The most critical limitation is the need to prepare pressure time series data through pre-analysis. This required huge memories from both levels of analysis, Level-0 and Level-1. Then, in order to overcome this limitation, another version of VWM is proposed which is the VWM with SGFGP boundary treatment as already discussed in Section 2.3. The detailed implementation of the proposed VWM with SGFGP boundary treatment will be discussed in Section 3.3.

3.3. *VWM with the SGFGP boundary treatment*

The detailed implementation of the VWM with the SGFGP boundary treatment is described. The major difference between this proposed VWM with the first version of VWM as discussed in Section 3.2 is the calculation of the pressure. As mentioned previously, the pressure is dependent on the pressure calculated from the Level-0 analysis as stated in Eqs. (24) and (25). However, for the VWM with SGFGP boundary treatment, the pressure is solved using a Level-0 SGFGP approach with few modifications of the relaxation coefficient used in SPH calculation of Level-1 analysis. In Section 2.5, the proposed SGFGP boundary treatment shows the accurate pressure distribution with lower computational cost. The same safety parameter was also used in this treatment, in which the value of β is 0.01. Therefore, by using this approach, the only required data from the pre-analysis Level-0 is the velocity for the movement of VWM as stated in Eq. (22). However, two transition regions are proposed in order to control the source term in solving the PPE, especially near the boundary of VWM, as shown in Fig. 16.

The modification of the relaxation coefficient α is only necessary if the particles were located in the transition regions in the following manner:

$$\alpha_{\mathrm{lv1}}(\boldsymbol{x}_i) := 0 \quad \text{in}\, \Omega_{\mathrm{A}}, \tag{25}$$

$$\alpha_{\mathrm{lv1}}(\boldsymbol{x}_i) := \gamma(\boldsymbol{x}_i)\alpha \quad \text{in}\, \Omega_{\mathrm{B}}, \tag{26}$$

$$\gamma(\boldsymbol{x}_i) = L_{\mathrm{B}} l(\boldsymbol{x}_i),$$

where the first equation, in transition layer A Ω_{A}, means the α used is fixed to zero, and the second equation in transition layer B Ω_{B} is

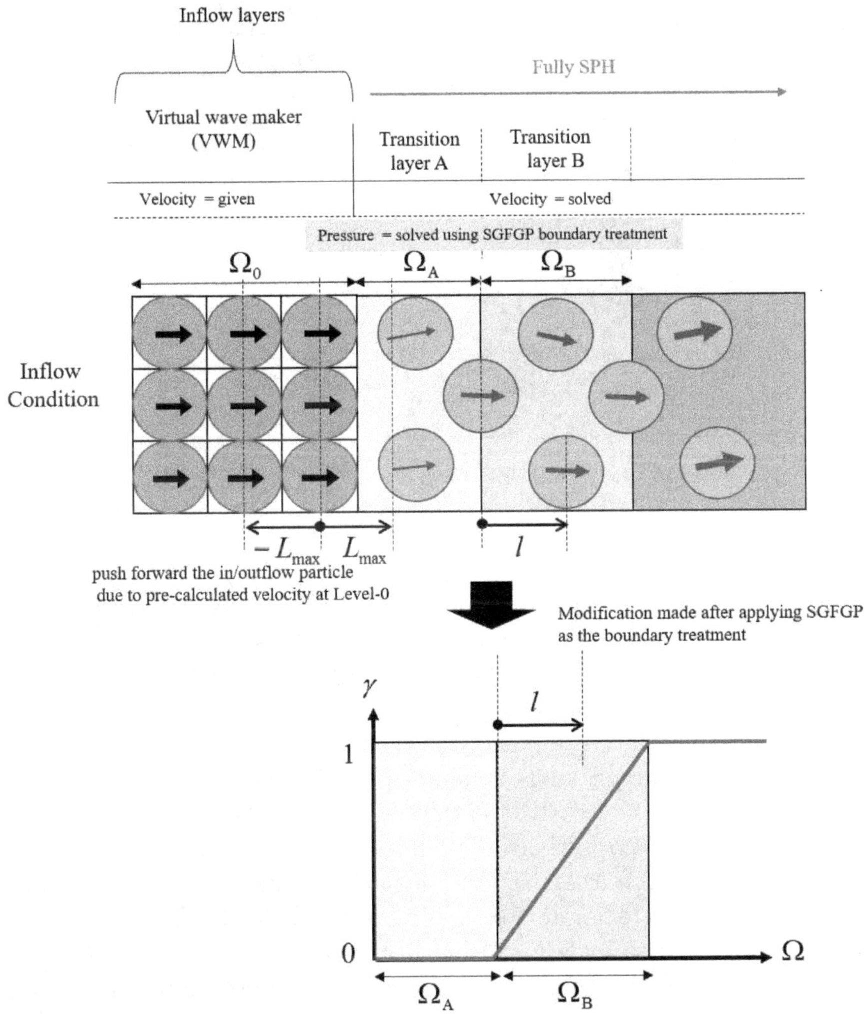

Fig. 16. The 2D schematic diagram of VWM with SGFGP boundary treatment.

the linear change of the α with respect to the distance l and L_{B} is the total length of transition layer B.

By using the above equation, we can smoothly couple the two solutions which are Level-0 and Level-1 solutions. Figure 17 shows the 2D schematic diagram of the detailed overview of the boundary processing cell for the transition regions proposed with SGFGP boundary treatment. Both versions of VWM have already been explained in the

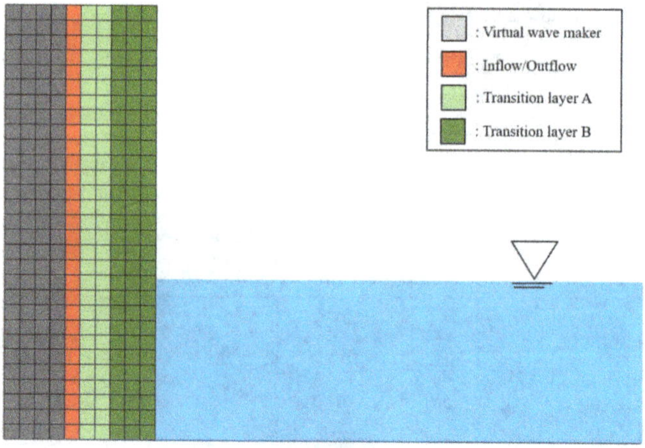

Fig. 17. The detail overview of boundary processing cell in Level-1 using VWM with SGFGP boundary approach.

handling in the transition layer. The number of layers for each transition layer shows a difference. For the proposed VWM with SGFGP boundary treatment, the number of layers is less than 3 times that of the other version. The numbers of layers for transition layer A and transition layer B are 2 layers and 3 layers, respectively.

Furthermore, by using this approach, the memory needed is less compared with the previous version of VWM and will lead to reduction in computational cost. It is well known that the proposed VWM will be used for a multi-scale tsunami analysis, which is a large-scale computational analysis, and a method with high accuracy and reasonable computational cost is always the priority. Therefore, the numerical tests conducted will be discussed in the following section in order to check the accuracy and performance before being applied in the real case of tsunami run-up simulation in the last example in Section 6.

3.4. *Validation test for VWM considering pressure from pre-analysis*

A coupled simulation between two 3D particle models was implemented as one of the validation tests for our proposed VWM. The purpose of this simulation is to show the equivalence between a single-resolution simulation and a coupled simulation with the VWM. The

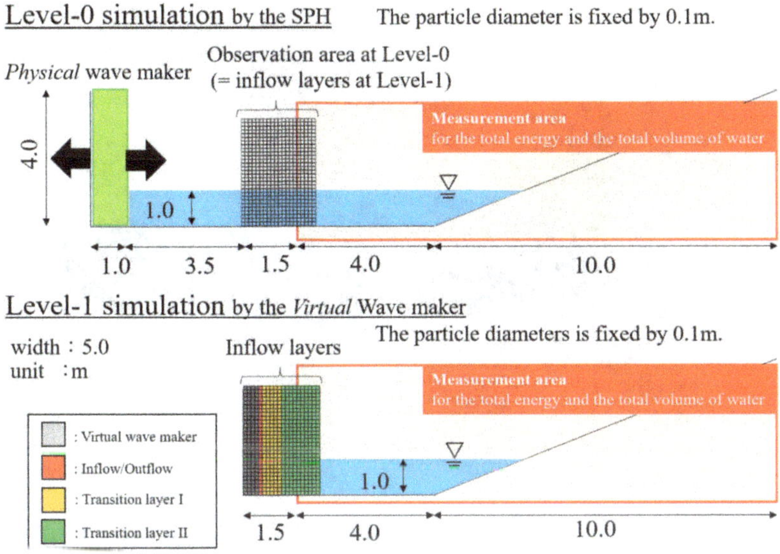

Fig. 18. Numerical validation model with a physical wave maker at Level-0.

two simulations were implemented using the same particle method, and the difference between them can only be related to the numerical error of the VWM.

The numerical test was conducted using the same resolution for both the Level-0 (pre-calculation simulation) and Level-1 (calculated) analyses. Figure 18 shows our numerical example models for the Level-0 and Level-1 simulations. The initial depth of the water was 1 m. A wave run-up problem was solved for this setting using the physical wave maker. In the Level-0 simulation, the physical wave maker generated continuous waves using five wave plates with different phases of velocity determined by $v = 1.57 \sin \pi(t + \tau)[m/s]$. The reason for having these five different phases of the five wave plates is to produce a pushing wave undertow tsunami which has 3D properties. The velocity in the x-direction of different phases of the wave-making plate is based on numbering the plate as 1, 2, 3, 4, and 5 with values of $\tau = 0$, $\tau = 2/5$, $\tau = 4/5$, $\tau = 6/5$, and $\tau = 8/5$, respectively, as shown in Fig. 19.

A solution for the whole region was found by using the SPH method as the Level-0 pre-calculation simulation. The final Level-0 simulation result was then used as the reference solution for the

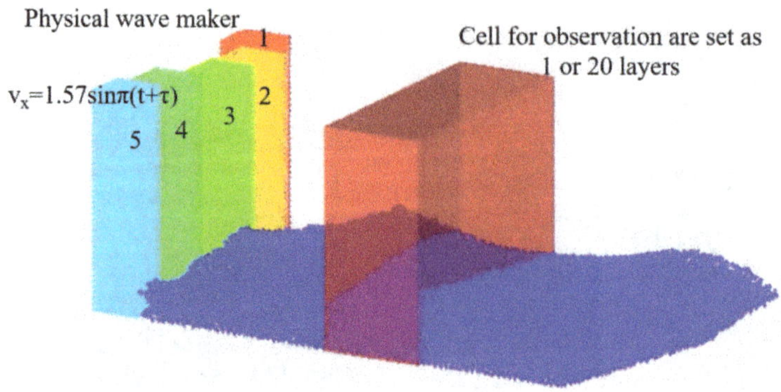

Fig. 19. The schematic diagram for generating a wave from a physical wave maker.

Level-1 simulation covering a limited region with the VWM as shown in Fig. 18. In this example, the initial particle distance was set to 0.1 m for both the Level-0 and Level-1 particle simulation models, and the time increment was fixed to 0.001 s with the slip condition applied on the solid boundary.

Here, in the case of the Level-0 particle simulation, the wave height was monitored in inflow layers consisting of the VWM and the transitional layers I and II. However, in the VWM, the water velocity was also monitored and stored to evaluate the movements of the VWM in the subsequent Level-1 simulation. In summary, the observation area for this numerical test is set to 20 layers, which will represent the inflow input data, such as the velocity and pressure that will be used for the Level-1 simulation, as shown in Figs. 18 and 19. Note that the particle locations generally differ from the center of cells in the transition layers; however, the pressure and velocity at the centers of cells can be approximated by SPH weight averaging (12).

A comparison of the wave shapes and velocity distributions obtained from the Level-0 and Level-1 models is shown in Fig. 20. In this Fig. (20), comparison is also made for Level-1 itself to check on the effect of having pressure from Level-0 or vice versa. Comparison of the waveforms shown in Fig. 20(b) and Fig. 20(c) indicates that both are similar to the reference result in Fig. 20(a), regardless of whether or not the pressure pre-evaluated in Level-0 is considered.

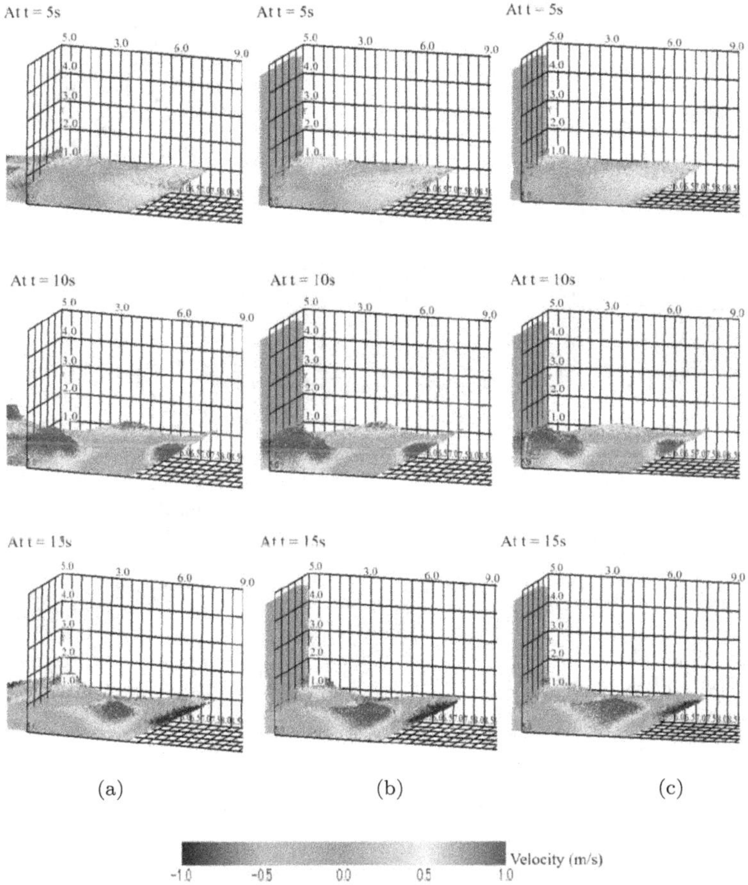

Fig. 20. The simulation results from Level-0 and Level-1 analyses in terms of velocity distribution. (a) Reference (Level-0), (b) VWM with considering pressure (Level-1), and (c) VWM without considering pressure (Level-1).

However, if the comparison is made based on the velocity distribution, there is a bit of discrepancy, especially near the boundary. It can be seen clearly in Fig. 21, where the velocity distribution is smoothly coupled between Level-0 and Level-1 without considering the pressure in the transition layers, and this is contrary to a Level-1 analysis which does not consider the pressure from the pre-analysis, Level-0.

In order to validate our simulation model quantitatively, the Level-0 and Level-1 simulation results have been compared by checking energy loss and the total volume of the fluid. The total

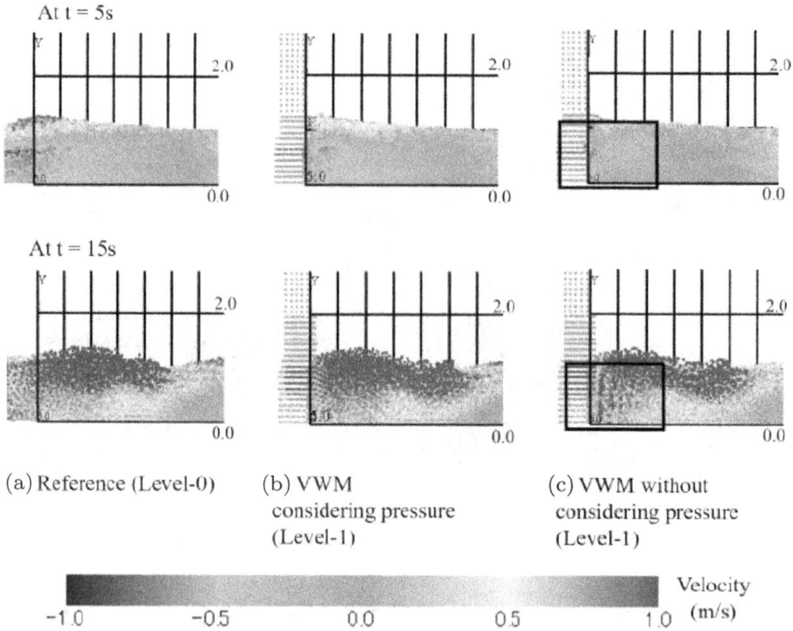

Fig. 21. The closed-up view of the velocity distribution in the vicinity of the VWM. (a) Reference (Level-0), (b) VWM considering pressure (Level-1, and (c) VWM without considering pressure (Level-1).

water head, which is the sum of the water head of all particles, and the total number of particles are measured during the simulation in the measurement area, which is defined by the box in Fig. 18. Figures 22(a) and 22(b) show the total water head and the total number of particles, respectively. In the graph, a Level-1 solution without considering the pressure distribution and additional transitional layers is added to discuss the efficiency of the pressure in transition layers. From the graphs shown in Fig. 22, it is clear that considering the pressure distribution from Level-0 analysis in transition layers may contribute to the correct coupling simulation and the precise historical records of the total water head and the total number of particles. The total water head within measured areas is calculated using the following equation.

$$\textbf{Total water head} = \sum \left(\frac{P}{\rho g} + \frac{v^2}{2g} + g \right). \tag{27}$$

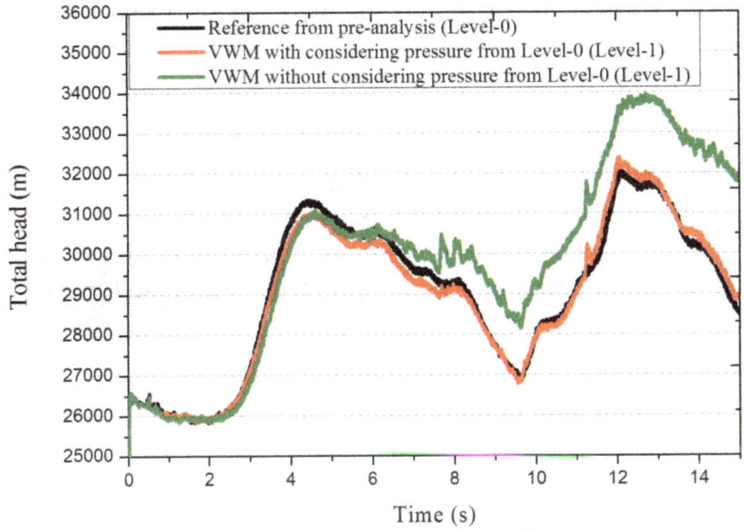

(a)

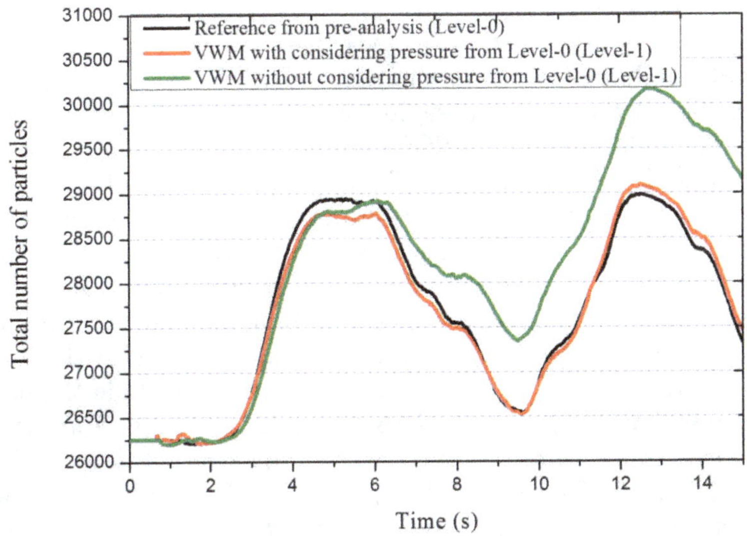

(b)

Fig. 22. The comparisons of the total energy and total volume of fluid during each particle simulation. (a) Total water head, and (b) total number of water particles.

On the other hand, the accumulated error in the time history can be seen in the case of a VWM without considering the pressure. Based on these comparisons, it indicates that the pressure distribution is one of the influencing factors necessary in order to have correct coupling analysis. Hence, as mentioned previously, the second version of the VWM with SGFGP boundary treatment was introduced in Section 3.3 in order to have correct pressure distribution without requiring huge memory of pressure data from the pre-analysis Level-0, and the validation test conducted is discussed in Section 4.

4. Pre- and Post-Processes for the Particle-Based Tsunami Inundation Simulation

4.1. *Preprocess for making a 3D high-fidelity model from Geographic Information System (GIS) data*

To understand the tsunami run-up in urban areas, detailed modeling including buildings that are the main obstacles of tsunami run-up is desirable. It goes without saying that the accuracy of 3D analysis largely depends on the reliability of the analysis model. To match the elevation and building shape in the model with those of actual geography as much as possible, we establish a procedure for developing a particle analysis model using GIS data. In this study, the particle analysis model was developed referring to DEM data for elevation data and SHP data for building information. The DEM data are the data of elevation measured by aerial survey and saved in the form of lattices with a constant interval. The DEM data exclude the heights of buildings and trees and give the elevation of the ground surface. Therefore, the DEM data were used as the elevation data, and SHP data (a format of GIS geography space data files) were used to define the geometric shapes of buildings. In the SHP data, the contour of each building is defined as 2D data, and building height information is also included. When the outlines of the ground surface and buildings are defined using general 3D CAD data, the data can be easily converted into particle model data using a general-purpose preprocessor for particle analysis (e.g., Meshman). Figure 23 shows the flow of the development of a particle model proposed in this study. In the

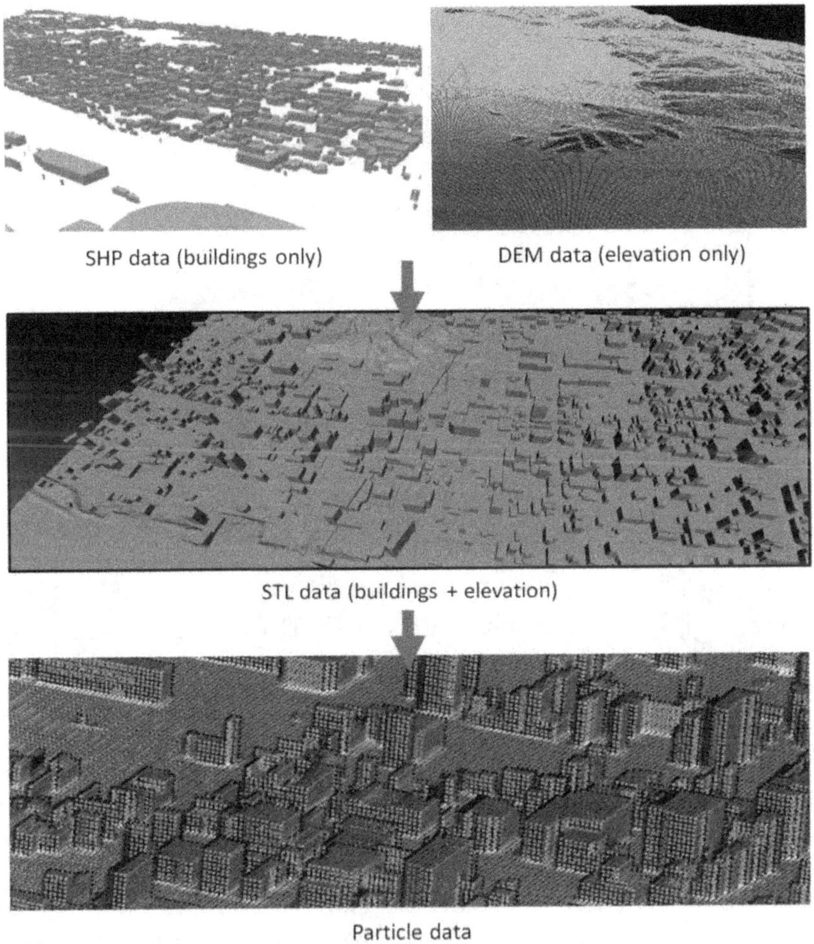

Fig. 23. Preprocess for making a high-fidelity tsunami run-up simulation model on a city scale.

following, a series of procedures to define the outline of a surface as stereolithography (STL) data (a format of 3D CAD used to define a surface as triangle patches) using the above two types of GIS data and converting the STL data into the particle data are explained in detail.

Basically, the elevation data of the location of each building are estimated from the DEM data and swept to the height of the building

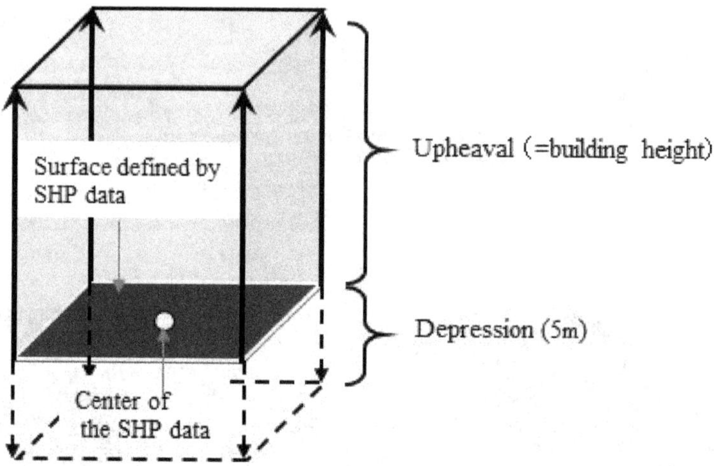

Fig. 24. Lowering of the bottom of buildings of STL data.

on the basis of the plane contour of the building using the SHP data to convert the elevation data into 3D CAD data. The ground surface generally includes slopes; thus, a gap may be generated between the bottom of a building and the ground surface. If the particle model is developed without any modification, particles may invade the gap. Therefore, the bottom of a building is lowered by a constant amount (5 m) in the preparation of CAD data of buildings, as shown in Fig. 24. The particle analysis preprocessor, Meshman, used in this study can generate particles for multiple surfaces defined in CAD data independently and develop an analysis model with a uniform particle distribution using particles with different profiles (e.g., particles for definitions of buildings and ground) by removing duplicate particles according to the order of priority if the location of a particle overlaps with that of another particle. Therefore, the STL data for the definition of the ground surface prepared from DEM data and those for the definition of buildings are saved in separate files. Next, building and ground particles are prepared as shown in Fig. 25 by prioritizing the ground particles so that the building particles are automatically removed when a building particle and a ground particle overlap in the region corresponding to the ground surface.

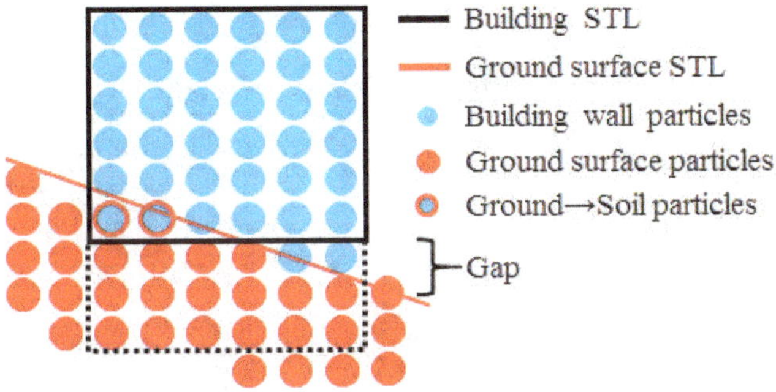

Fig. 25. Ground particles.

4.2. *Post process of particle-based tsunami simulation data as a 3D photo-realistic visualization*

4.2.1. *Smoothed surface representation from particles using the marching cubes method*

Particle location information is converted to geometric shape data used for rendering. Here, the curved surface of the implicit function is used. The influential range r is defined for each n particle on the curved surface of the implicit function, and the isosurface of density in the space is assumed to be the surface of an object.

As a method of approximating the curved isosurface of the implicit function using a polygon, a marching cubes method[14] was used. With this method, a 3D lattice is placed in space to calculate the density D, assuming that the distance from each lattice point to a vicinity particle is r. Parameters a, b, and D_0 in the following equations are changed in accordance with the particles' density and the space's scale. The Kernel function used to calculate the density is given by expressions for (a) rectangular, (b) quadric, (c) quartic, and (d) Gaussian functions.

(a) Rectangular function

$$D_{(r)} = \begin{cases} a & (|r| < b), \\ 0 & (\text{otherwise}). \end{cases} \tag{28}$$

(b) Quadric function

$$D_{(r)} = \begin{cases} D_0(1.0 - 3.0(\frac{r}{b})^2) & (0 \leq r \leq \frac{b}{3}), \\ 1.5(D_0(1.0 - (\frac{r}{b})^2) & (\frac{b}{3} < r \leq b), \\ 0 & (b < r). \end{cases} \tag{29}$$

(c) Quartic function

$$D_{(r)} = \begin{cases} D_0(1.0 - (\frac{r}{b})^2)^2 & (0 \leq r \leq b), \\ 0 & (b < r). \end{cases} \tag{30}$$

(d) Gaussian function

$$D_{(r)} = D_0 \exp(-ar^2). \tag{31}$$

Each point on the 3D lattice has two values determined on the basis of the magnitude relationship between the threshold value on the curved isosurface of the implicit function and the density calculated using the above equations. The lattice is a cube consisting of eight lattice points, and 256 different combinations exist. When duplicate combinations are removed, considering the symmetry of the the figures, 256 combinations can be classified into 15 combinations, as shown in Fig. 26. In the marching cubes method, 0–4 triangles are generated on the basis of the 15 combinations to express the curved surface of the implicit function.

4.2.2. *3D filtering to decrease polygons*

The variance of density may be large depending on the Kernel function because these values are calculated by the superposition of Kernel functions, leading to an irregular surface after polygonization. To solve this problem, a smoothing filter as shown in Fig. 27 is applied to the density values of 3D lattices. Figure 28 shows the results obtained after applying the Gaussian filter twice (Fig. 29). For smoothing, the following Gaussian filter was used.

4.2.3. *Lattice size determination*

The calculation time, memory usage, and the result of polygonization of the marching cubes method significantly differ depending on the

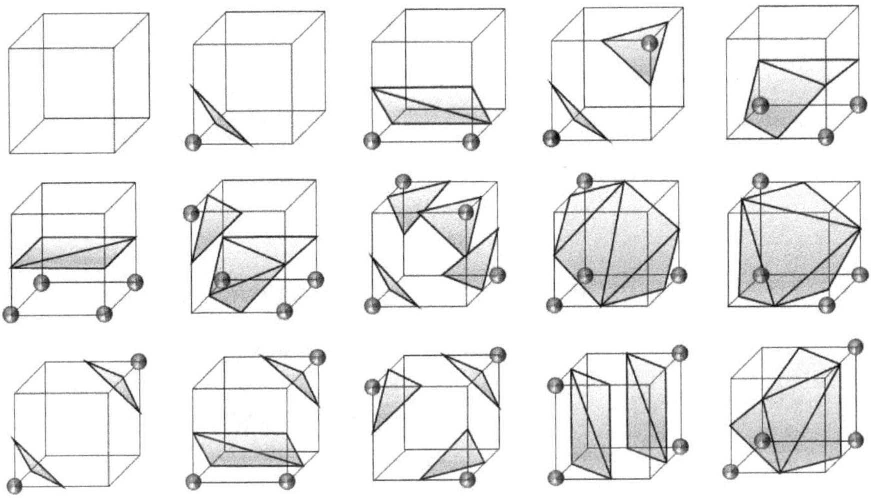

Fig. 26. Basic 15 combinations.

$\frac{1}{64}$	$\frac{2}{64}$	$\frac{1}{64}$
$\frac{2}{64}$	$\frac{4}{64}$	$\frac{2}{64}$
$\frac{1}{64}$	$\frac{2}{64}$	$\frac{1}{64}$

$\frac{2}{64}$	$\frac{4}{64}$	$\frac{2}{64}$
$\frac{4}{64}$	$\frac{8}{64}$	$\frac{4}{64}$
$\frac{4}{64}$	$\frac{8}{64}$	$\frac{4}{64}$

$\frac{1}{64}$	$\frac{2}{64}$	$\frac{1}{64}$
$\frac{2}{64}$	$\frac{4}{64}$	$\frac{2}{64}$
$\frac{1}{64}$	$\frac{2}{64}$	$\frac{1}{64}$

Fig. 27. Gaussian filter ($3 \times 3 \times 3$).

resolution used to segment the space into 3D lattices. In this study, the number of lattices along each axis was determined by designating the number of maximum cubes and using an axis-aligned bounding box (AABB) in space, including particles, so that the number of space segmentation is as high as possible in the range that the memory usage does not exceed the allowable memory (Fig. 30).

(i) Obtain the minimum and maximum position coordinates of all particles on each axis to calculate the AABB and the length of each side of the space, l_x, l_y, and l_z.

(ii) From the AABB, calculate the volume of the space, which is divided by the maximum number of cubes to calculate the volume per cube.

Fig. 28. Surface creation using rectangular function.

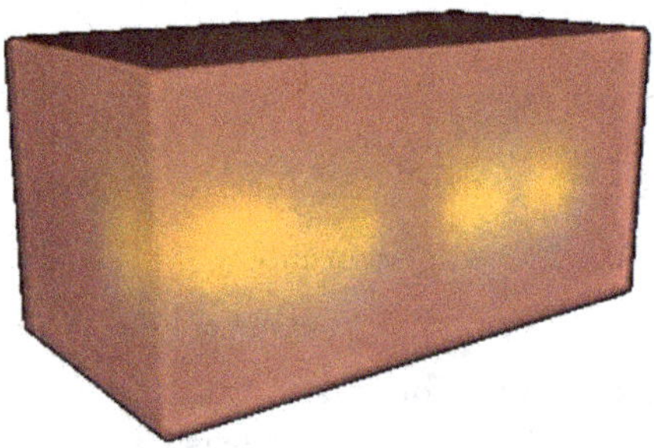

Fig. 29. Results obtained by applying Gaussian filter twice.

(iii) Calculate the cube root of the volume per cube to determine the length of a side of a lattice.
(iv) Determine the lattice size in the direction of each axis by dividing the length of each side of the space, l_x, l_y, and l_z by the length of a side of a lattice.

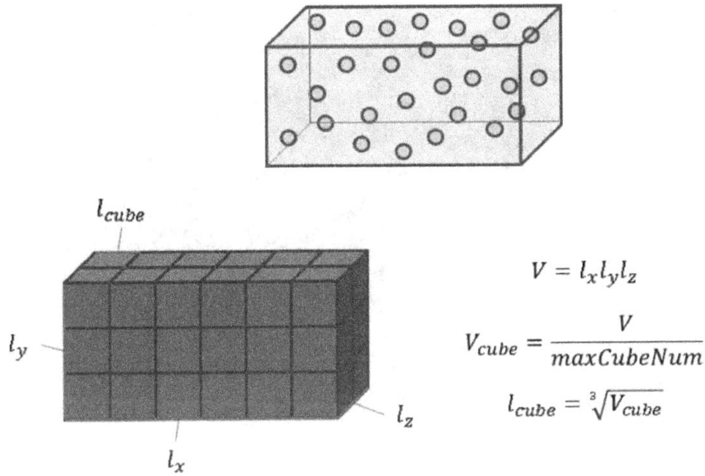

Fig. 30. Calculation of lattice size.

5. Numerical Convergence Tests for Tsunami Inundation Using the SPH

A tsunami run-up targeting a real city is simulated by the SPH method. Here, the application of the analysis model and method to disaster prevention and mitigation is examined. This is a kind of convergence test to discuss the minimum required resolution for the practical tsunami run-up analysis.

5.1. *Target city of tsunami inundation*

In tsunami countermeasures, it is important to assume the largest earthquake of any magnitude. In Japan, there has recently been concern about a massive earthquake with the Nankai Trough as its epicenter. This is a series of earthquakes with epicenters over a wide area along the Nankai Trough and is expected to cause significant damage along the Pacific coast. Therefore, in this study, tsunami run-up analysis was conducted for Kochi prefecture. Using the particle method, the number of particles required to model the land area in this region amounts to about 35.19 million particles with a diameter of 4 m and about 184.85 million particles with a diameter of 2 m.

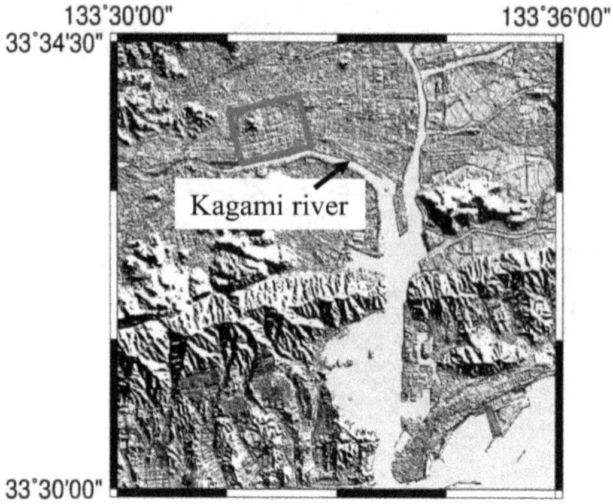

Fig. 31. Tsunami run-up simulation domain at Kochi.

Although analysis of this number of particles is possible with current supercomputers, it is practically difficult in terms of time constraints (supercomputers are shared resources, and multiple-case calculations should be avoided considering a job waiting time). Therefore, we limit the target area to the area of about 800–1,000 m^2 along the Kagami River, boxed in Fig. 31. Here, assuming that the river is inundated to the north during the tsunami run-up, the minimum spatial resolution necessary to obtain reasonable results in the numerical analysis was examined by comparing the analysis results obtained at three different resolutions.

5.2. *Simulation conditions*

Figure 32 shows the analysis model using particles with a diameter of 2 m. In this simulation model, a tsunami is simply generated with bulk water with an initial uniform flow velocity. The bulk water placed under the target area simulates the Kagami River. The Kagami River averages −4 m above sea level and assumes bulk water 8 m high at the bottom of the river. The equation for river velocity is $c = \sqrt{gh}$ based on long wave theory, and the wave velocity is assumed to be 9 m/s. Three cases with different particle sizes of 1, 2, and 4 m were analyzed. The computer used was the K computer (a supercomputer owned by RIKEN) with a time increment of 0.005

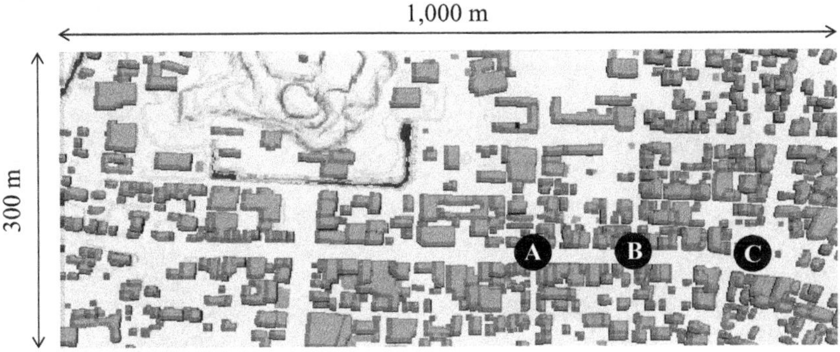

Fig. 32. Tsunami inundation simulation domain.

Table 1. Analysis conditions.

Particle diameter	Number of particles	Computer nodes	Simulation time
0.5 m	167,000,000	4,096	–
1.0 m	27,760,000	512	16 hours
2.0 m	5,200,000	256	4 hours
4.0 m	1,100,000	256	40 minutes

seconds and a total time step count of 30,000. Other numerical analysis conditions are summarized in Table 1. The nonslip boundary was applied on the ground surface.

5.3. *Evaluation of resolution*

Figure 33 shows the simulated inundation area with the water velocity distribution after 60 s. (12,000 steps) from the Tsunami input. The tsunami particles did not penetrate the building, indicating that the simulation model was successfully constructed with no gap between the building and the ground. A comparison of the analysis results for the three cases shows that the inundation area becomes larger as the spatial resolution increases. Comparing the differences in inundation areas for particle sizes of 0.5 m, 1.0 m, 2.0 m, and 4.0 m, the inundation areas appear to converge mostly with the 1.0 m resolution results when spatial resolution is increased. The enlarged image of the area circled in Fig. 33 also shows the distribution of flow velocities. Simulation results of velocity profiles with the 0.5 m and 1.0 m

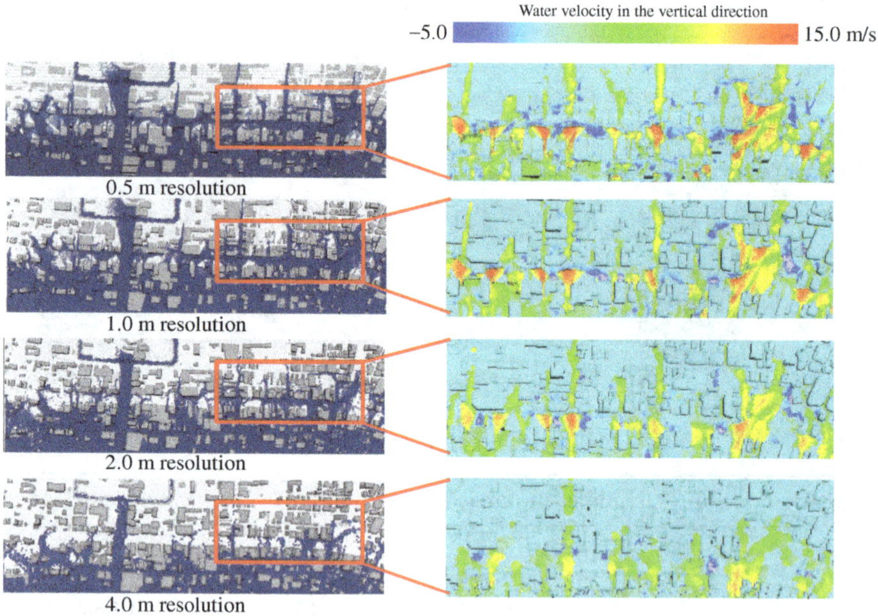

Fig. 33. Tsunami inundation area and velocity distribution after 60 s from tsunami inputs.

resolution models show similar trends. Figure 33 also shows the contour of the flow velocity in the ascending direction in the enlarged region. Even for particles generated from the same STL data, the representation of topography and building geometry differs depending on the resolution. In particular, narrow streets between buildings are not reproduced at low resolution, which would suppress tsunami run-up. In addition, the low resolution may result in poor reproduction of the slope of the ground, which may reduce the accuracy of the analysis.

To discuss more quantitatively the numerical convergence, Fig. 34 shows the simulated inundation height at points A, B, and C, indicated in Fig. 32. The tsunami inundation heights are estimated from the maximum height of water particles at the observation points. As shown in the figures, the simulated inundation heights at the three resolutions are almost the same. There is a difference between the cases with 1 m and 2 m particle resolution until the tsunami particles reach the observation point, but the case with 2 m particle resolution tends to converge with the case with 1 m particle resolution [not only

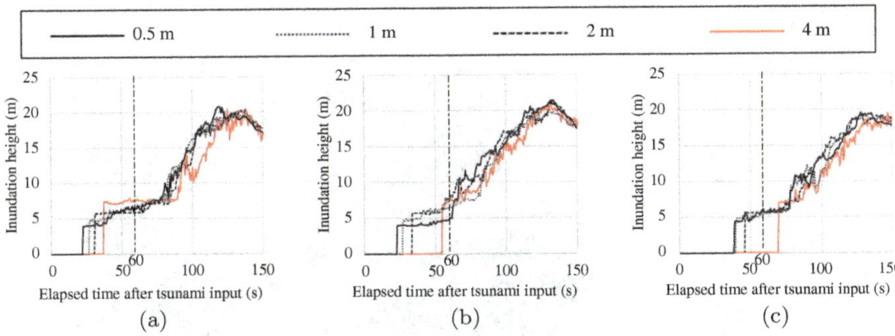

Fig. 34. Time transience of inundation water height at evaluation points A, B, and C. (a) Inundation height at point A, (b) inundation height at point B, and (c) inundation height at point C.

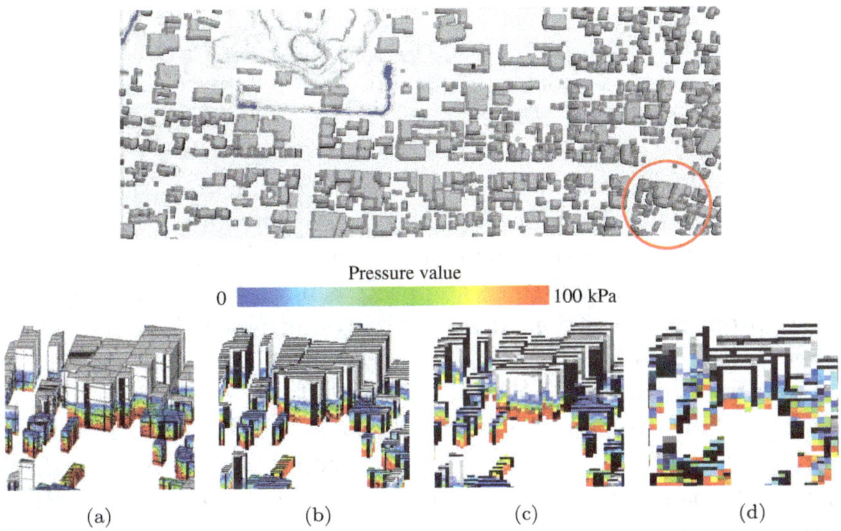

Fig. 35. Hydrodynamic pressure acting on buildings after 60 s from tsunami inputs. (a) 0.5 m resolution, (b) 1.0 m resolution, (c) 2.0 m resolution, and (d) 4.0 m resolution.

in the tsunami height but also for the pressure distribution acting on the buildings as shown in Fig. 35].

Based on the above considerations, it is desirable to adopt a simulation model with a spatial resolution of about 2 m, which can reproduce arterial roads, in order to model the tsunami run-up height in urban areas, including buildings with high accuracy.

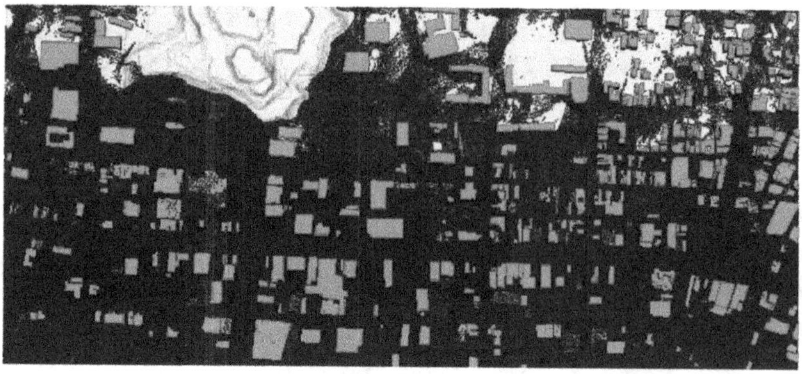

Fig. 36. Tsunami inundation area after 150 s from Tsunami inputs.

5.4. *Discussion on the simulation results*

Figure 36 shows the inundation area (at 30,000 steps) 150 s after the start of the tsunami inflow. Figure 36 shows that the tsunami did not reach the elevation of approximately 44.4 m where Kochi Castle is located and that floors above the third and fourth floors of the high-rise building were not inundated. Kochi Castle is the only designated natural high-ground evacuation site in the simulation area, and the simulation results confirm its safety. Since the inundation in the figure is assumed to occur in 150 s after the start of tsunami inflow, prompt evacuation to the area around Kochi Castle or to the upper floors of buildings is essential after an earthquake. Thus, a more elaborate disaster prevention manual can be created by conducting a tsunami run-up analysis with detailed representations of specific areas and buildings. It is also expected to be applied to disaster drills to stimulate the self-help spirit of local residents. In the next section, we propose a visualization method that focuses on the effective application of the analysis results to disaster drills. Note that the analysis results presented here are tentative in order to demonstrate the relationship between spatial resolution and accuracy and should not be used as the basis for examining the assumed inundation zones.

5.5. *Photo-realistic visualization and virtual evaluation training system*

In conventional visualization methods, particles are generally displayed as they are, as shown in Fig. 36. This visualization method can display engineering physical quantities such as flow velocity and pressure. Although these data can be applied to the formulation of building strength criteria and vulnerability assessment, a photo-realistic visualization that can convey the threat of tsunami more realistically is desired in disaster prevention training. In this study, 3D animations were developed using Lumion, an architectural visualization software widely used in the field of architecture for landscape design. Lumion has a 3D animation creation function that uses parallax effects to create 3D animations with a sense of realism. Lumion can also import general CAD data and easily add data related to landscapes, solar radiation conditions, and shadows. Therefore, 3D visualization is possible simply by preparing CAD data of a virtual city used in the analysis model and CAD data of a tsunami. In addition, the real image of the building can be added to the building data, and the texture of the water flow can be added to the tsunami data after defining the surface of the tsunami, and the CAD data, including the texture, can be imported by Lumion, thus creating a sense of reality.

The aforementioned marching cube method can create lattices and surface polygons in a stable manner in terms of memory usage, even if the area containing the particles changes. A snapshot image of a 3D animation created with this method is shown in Fig. 37. This photo-realistic visualization method allows the user to intuitively recognize the threat of a tsunami based on the relative positions of the tsunami and surrounding buildings, providing easy-to-understand information to local residents. In addition, since the 3D visualization is based on high-resolution calculations, a head-mounted display and a walking controller can be used together to conduct a virtual tsunami evacuation drill, as shown in Fig. 38.

6. Validation — Representation of the 2011 Tohoku Earthquake and Tsunami at Utatsu

In this section, we present the results of a multi-scale tsunami simulation using a finite difference method based on the 2D shallow water equation and a 3D SPH simulation. At the beginning of this chapter, we established that "multi-scale tsunami analysis is to solve tsunami phenomena at multiple levels, which can be divided into

Fig. 37. 3D animation image.

Fig. 38. Virtual evacuation training system using 3D monitor and walking controller.

Level-0, Level-1, and Level-2 analyses". However, in this example, we will focus only on Level-0 and Level-1 tsunami analysis; Level-0 represents a 2D tsunami propagation analysis and is solved by the finite difference method. Level-1, on the other hand, represents a 3D tsunami inundation simulation solved by the SPH particle method. The proposed VMW with SGF boundary treatment is used to validate the zooming analysis with different resolution particle method models and successfully shows good agreement with the reference solution. Here, a multi-scale tsunami analysis is performed by coupling two analyses with different dimensions and methods.

6.1. *2D tsunami propagation analysis based on SWE (Level-0 analysis)*

Utatsu area at Minami-sanriku-cho is selected as the target of this simulation. Utatsu area is one of the affected regions where the Utatsu Bridge totally collapsed due to the Tohoku earthquake tsunami in 2011. Figure 39 shows the analysis area of the Utatsu area, which indicates the inflow area and the 3D particle simulation area. As mentioned previously, a Level-0 analysis, which is the tsunami propagation based on the 2D shallow water equation, is required in order to continue the 3D tsunami inundation simulation in Level-1. The Level-0 simulation results based on the 2D shallow water equation were already simulated by Eric *et al.*[15] using their original seismic fault model "Tohoku University model ver.1.0." Their 2D numerical simulation results by the finite difference method are validated with various data at several sites recorded by the GPS wave gauge. The wave height and its velocity in the inflow area defined in Fig. 39 were memorized as the time history data. The measurements data are given in Fig. 40.

6.2. *Coupling between 2D shallow water (Level-0)–3D particle simulation (Level-1)*

The analysis area in Fig. 39 shows the 3D (2.15×2.6 km) location of the tsunami inundation simulation and the 2D tsunami propagation in the inflow area. When creating the topographic model, other models, such as levees and tunnels around the urban area, were taken into account based on the elevation of the 5 m mesh to reproduce the actual topographic shape. Bridges, buildings, etc., were not modeled

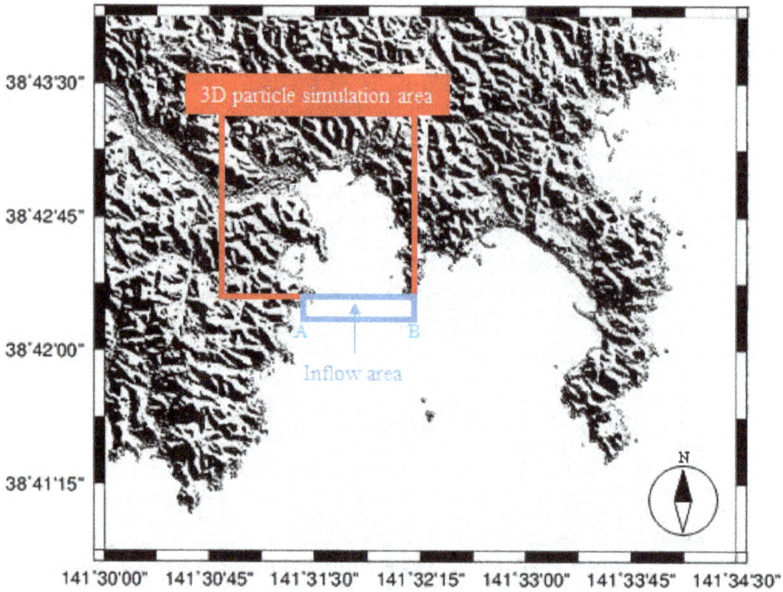

Fig. 39. The selected analysis area for tsunami inundation simulation at Utatsu.

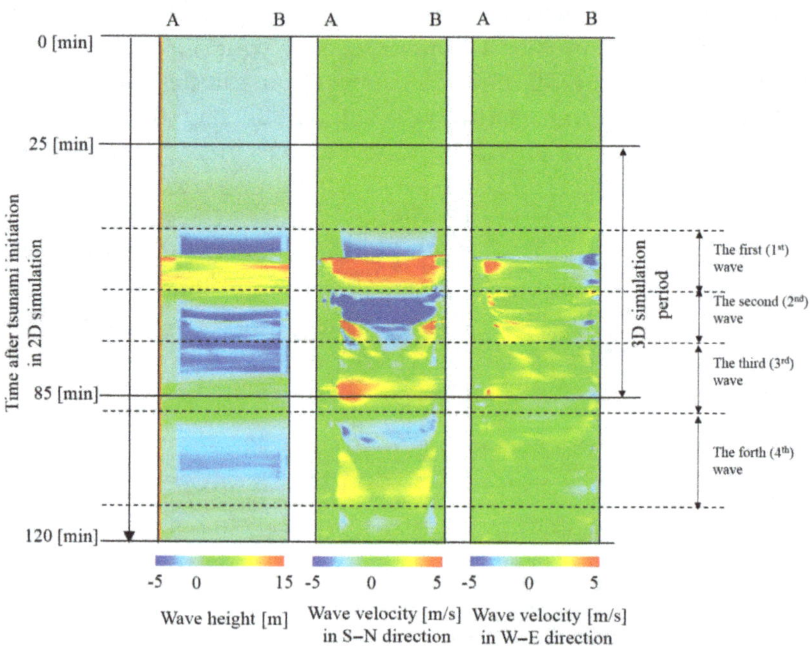

Fig. 40. Summary of the numerical solution by the 2D Finite Difference Method at inflow area.

in this analysis. Furthermore, fixed-wall particles were placed around the analysis area to create a particle model in which tsunami inflow and outflow do not occur except in the inflow region. The input information for the virtual wave generator placed in the inflow region was obtained from the 2D analysis already described in the previous section. The mean cross-sectional velocity in the depth direction was fixed because there is no velocity distribution in the inflow region.

The computation time for this tsunami simulation in 2D was more than 120 min, and it was difficult to zoom in on the 3D simulation in Level-1 due to the limitation of computation time. For the earliest numerical experiment, only the first three waves were selected in Level-1 to generate the inflow conditions, as shown in Fig. 40. The total time for the 3D calculations was limited to 60 min. The distance between particles is fixed at 2.0 m, and the time increment at 0.01 s. The total number of particles is about 17 million and are analyzed on two supercomputers, the K computer at RIKEN and the FX-10 at Kyushu University.

As shown in Fig. 40, the main flow direction of this tsunami is closer to the S–N direction. Therefore, the virtual wave generator is placed parallel to the E–W direction. Note that the pressure usually used as the inflow condition in Level-0 2D shallow-water simulations is approximated by hydrostatic pressure using the calculated wave heights. However, this approximation is not very accurate. Therefore, a virtual wave generator with appropriate boundary treatment using the SGFGP approach is proposed to couple these two levels, and the velocity distribution obtained from the 2D shallow-water equation is employed in the simulation.

The propagation of the tsunami inundation area is shown in Fig. 41 along with contour plots of the flow velocities in the S–N direction. At the bottom of each snapshot in Fig. 39, 2D finite difference results measured in the inflow area indicated in Fig. 39 are also drawn. The positive velocity value in the inflow region represents wave inundation into the shoreline region, which is the 3D particle simulation domain (Level-1). On the other hand, the negative velocity represents the undertow generated in the inflow region. This comparison confirms that our virtual wave generator can smoothly connect the results of the 3D SPH with those of the pre-computed 2D finite difference analysis. Furthermore, the fact that the velocity distribution contours in the inflow region of the pre-computed 2D shallow-water equation are similar to the velocity contours in the 3D

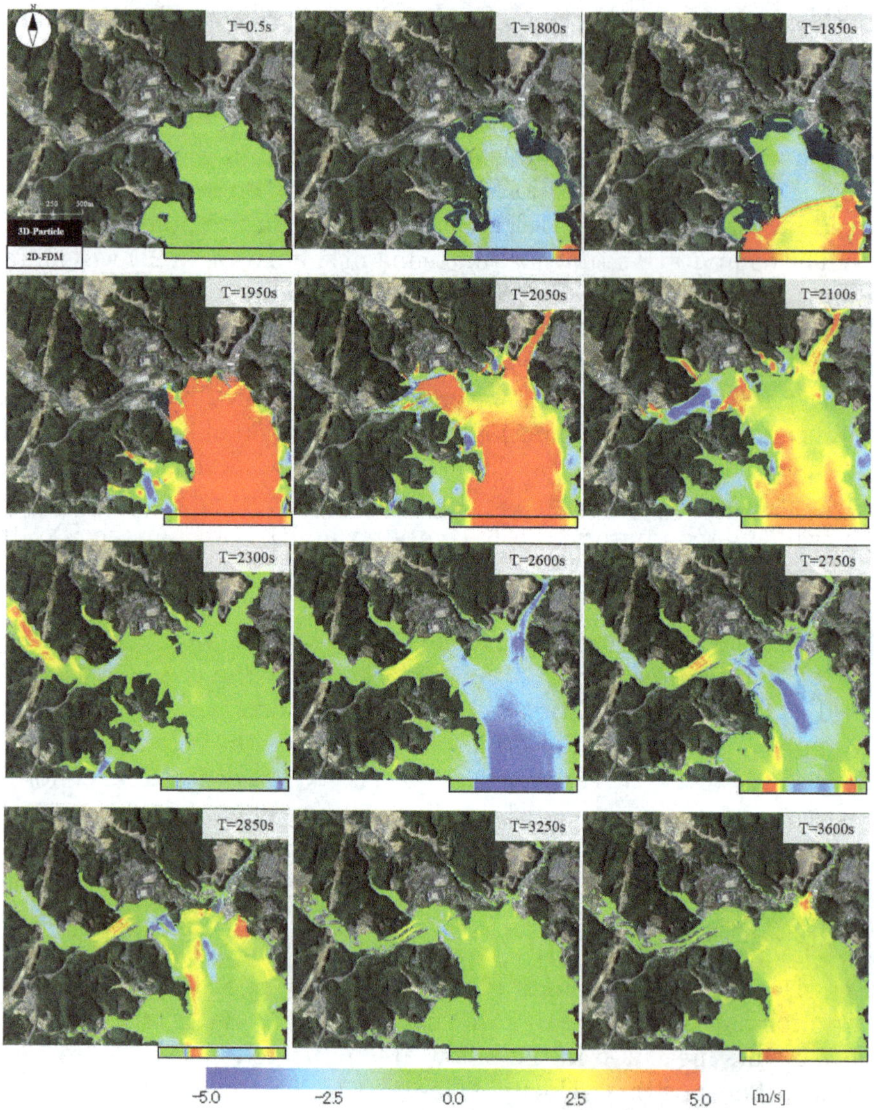

Fig. 41. Tsunami inundation result by the proposed coupling simulation at Utatsu.

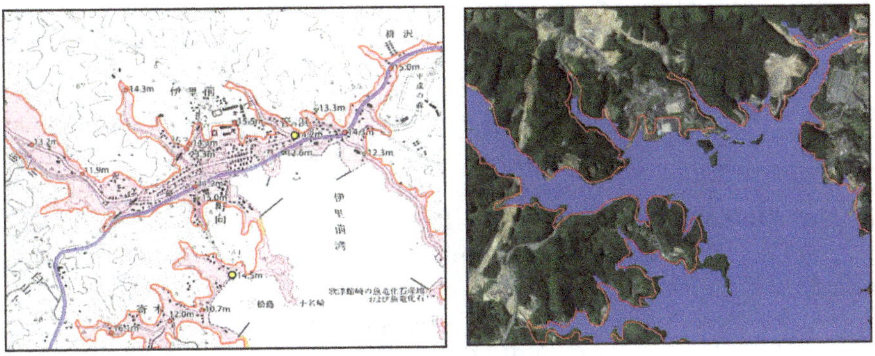

Fig. 42. Comparison in the tsunami inundation area between disaster report and 3D simulation result.

simulated region solved by SPH indicates that a smooth connection can be made.

Finally, to validate our simulation results and their accuracy, we must compare them with disaster reports in Fig. 42. Therefore, we compared the results of our tsunami inundation simulation with the disaster reports. As shown in the figure, the simulated tsunami inundation area results agree with the inundation area shown in the disaster report. While the comparison of tsunami inundation areas shows similar trends, there are differences in the details of the inundation areas. The main reason for this may be related to the low resolution. The current particle distance is 2 m, and this particle resolution model cannot resolve details of structures such as tunnels. One solution is to increase the particle resolution in the future.

The proposed virtual wave generator with SGFGP boundary processing shows outstanding trends for different levels of tsunami analysis. The virtual wave generator with SGFGP was used as a coupling method from a 2D finite difference simulation based on the shallow water equations to a 3D particle method. The main advantage of the proposed coupled simulation is its robustness to connect to various simulation tools. However, this one-way coupled simulation does not allow coupling from high-resolution solutions to low-resolution simulations. Therefore, it should be developed into a two-way coupled simulation technique for more accurate multi-scale analysis. This two-way coupled simulation would require a low-resolution to high-resolution simulation and the reverse process.

7. Conclusion

This chapter introduced a multi-scale tsunami simulation or multi-stage zoom tsunami analysis using our proposed matrix-arrayed VWM with appropriate boundary treatment (SGFGP). First, Utatsu, severely damaged by the 2011 tsunami disaster, is selected as the simulation's target area. Then, the proposed scheme, VWM with SGFGP boundary processing, was used as a one-way coupled simulation to perform a 3D tsunami inundation simulation, which was compared it with the actual disaster report. The results of the comparison between the simulations and the disaster report on the tsunami inundation area confirm that the proposed method is an accurate multi-scale analysis tool that can robustly combine various simulations. However, several recommendations and future studies are needed to ensure and improve the reliability of the proposed method.

It is well known that the procedure for multi-scale analysis coupling implemented in this chapter focuses on one-way coupling. One-way coupling is one in which data transfer is allowed only from one side, from low resolution to high resolution, or from 2D analysis to 3D analysis. There is no data transfer from high resolution to low resolution. However, due to the scale ratio limitation between low-resolution and high-resolution models, it is recommended that further research be conducted to introduce two-way coupled simulation techniques to establish a more accurate multi-step zoom tsunami analysis.

References

1. M. Asai, Y. Miyagawa, N. Idris, A. Muhari, and F. Imamura, Coupled tsunami simulations based on a 2d shallow-water equation-based finite difference method and 3d incompressible smoothed particle hydrodynamics, *J. Earthquake Tsunami.* **10**(05), 1640019 (2016).
2. M. Asai, A. M. Aly, Y. Sonoda, and Y. Sakai, A stabilized incompressible sph method by relaxing the density invariance condition, *J. Appl. Math.* (2012).
3. M. Isshiki, M. Asai, S. Eguchi, and H. O-Tani, 3d tsunami run-up simulation and visualization using particle method with GIS-based geography model, *J. Earthquake Tsunami.* **10**(05), 1640020 (2016).

4. D. Morikawa, M. Asai, N. Idris, Y. Imoto, and M. Isshiki, Improvements in highly viscous fluid simulation using a fully implicit sph method, *Comput. Part. Mech.* **6**(4), 529–544 (2019).
5. R. A. Gingold and J. J. Monaghan, Smoothed particle hydrodynamics: theory and application to non-spherical stars, *Mon. Not. R. Astron. Soc.* **181**(3), 375–389 (1977), doi:10.1093/mnras/181.3.375.
6. J. J. Monaghan, Simulating free surface flows with sph, *J. Comput. Phys.* **110**(2), 399–406 (1994).
7. S. J. Cummins and M. Rudman, An SPH projection method, *J. Comput. Phys.* **152**, 584–607 (1999).
8. J. P. Morris, P. J. Fox, and Y. Zhu, Modeling low reynolds number incompressible flows using sph, *J. Comput. Phys.* **136**(1), 214–226 (1997).
9. M. Yildiz, R. Rook, and A. Suleman, Sph with the multiple boundary tangent method, *Int. J. Numer. Meth. Eng.* **77**(10), 1416–1438 (2009).
10. S. Marrone, M. Antuono, A. Colagrossi, G. Colicchio, D. Le Touzé, and G. Graziani, δ-sph model for simulating violent impact flows, *Comput. Meth. Appl. Mech. Eng.* **200**(13), 1526–1542 (2011).
11. M. Asai, K. Fujimoto, S. Tanabe, and M. Beppu, Slip and no-slip boundary treatment for particle simulation model with incompatible step-shaped boundaries by using a virtual maker, *Trans. Japan Soc. Comput. Eng. Sci. (JSCES)*, **2013** (2013).
12. L. Lobovskỳ, E. Botia-Vera, F. Castellana, J. Mas-Soler, and A. Souto-Iglesias, Experimental investigation of dynamic pressure loads during dam break, *J. Fluids Struct.* **48**, 407–434 (2014).
13. J. Cercos-Pita, R. Dalrymple, and A. Herault, Diffusive terms for the conservation of mass equation in sph, *Appl. Math. Modell.* **40**(19–20), 8722–8736 (2016).
14. W. E. Lorensen and H. E. Cline, Marching cubes: A high resolution 3d surface construction algorithm, *ACM Siggraph Comput. Graph.* **21**(4), 163–169 (1987).
15. E. Mas, A. Suppasri, F. Imamura, and S. Koshimura, Agent-based simulation of the 2011 great east japan earthquake/tsunami evacuation: An integrated model of tsunami inundation and evacuation, *J. Nat. Disast. Sci.* **34**(1), 41–57 (2012).

https://doi.org/10.1142/9781800614635_0004

Chapter 4

Seismic Structural Response Analysis for Extremely Strong Ground Motion Using Parallel Finite Element Method

Hideaki Sonobe[*,§], Shunichi Suzuki[*], Wataru Hotta[*], Hiroki Motoyama[†,¶], and Muneo Hori[‡,||]

[*]*Taisei Co., Nuclear Facility Division, Shinjuku-ku, Tokyo, Japan*
[†]*National Defense Academy, Department of Civil and Environmental Engineering, Yokosuka, Japan*
[‡]*Japan Agency for Marine-Earth Science and Technology, Research Institute for Value-Added-Information Generation, Kanazawa-ku, Yokohama, Japan*
[§]*snbhda00@pub.taisei.co.jp*
[¶]*hiromoto@nda.ac.jp*
[||]*horimune@jamstec.go.jp*

In this chapter, seismic response analysis is performed with extremely strong earthquake motion as input. Using a parallel finite element method (FEM) tuned for seismic response analysis and enhanced with high-performance computing capability, a nuclear power plant consisting of a reactor building and ground is taken as the subject. First, a trial simulation was made using a 3D numerical analysis model of the nuclear power plant. Numerical computation of the parallel FEM was not completed when the numerical setting was improper, or when the input ground motion was magnified above two. This is mainly due to the low performance of the nonlinear analysis implemented in the parallel FEM, and the algorithm of nonlinear analysis was improved. Next, a full-scale simulation was made by using the improved algorithm. The parallel FEM was able to complete the numerical computation even for the ground motion multiplied by four times. The results showed that compressive strength remained in the concrete, and the seismic safety of the reactor building for these extremely strong ground motions was confirmed. This supports the usefulness of the parallel FEM analyzing a high-fidelity numerical model to examine the seismic safety against extremely strong seismic motion that has never been experienced.

1. Introduction

Due to the development of denser seismic observation networks, the use of stronger ground motion is required for seismic safety design and evaluation; see, for instance, the report of International Atomic Energy Agency, United Nations.[1] Numerical analysis which uses a multi-degree-of-freedom system or a 2D analysis model is often used in practice. Since certain approximations are included in the model, it generally overestimates structural seismic responses. Therefore, a 3D analysis model of high fidelity needs to be used for the stronger input ground motion. The numerical analysis of such an analysis model results in higher accuracy and resolution, which results in a more reliable estimate of seismic safety.

A parallel finite element method (FEM) is developed to realize the seismic response analysis using a 3D analysis model of high fidelity; see Gurpinar,[2] Fenves *et al.*,[3] Papadrakakis *et al.*,[4] Poursartip *et al.*,[5] and Riaz[6] for the application of large-scale numerical computatoin of FEM in earthquake engineering problems; see also Inoue *et al.*,[7] Tejchamn,[8] Kawamura, and Tanjung,[9] Bao *et al.*,[10] Usami *et al.*,[11] Isobe and Tsuda,[12] Huang *et al.*,[13] Alali,[14] Gui,[15] Hori,[16] Faria *et al.*,[17] Miyamura *et al.*,[18] Miyamura *et al.*,[19] Dobashi,[20,21] Yoshimura *et al.*,[22] Tanaka *et al.*,[23] and Motoyama *et al.*[24] for the application of high-performance computing to seismic structural response analysis. The parallel FEM is based on a general-purpose parallel FEM, called FrontISTR.[25] Implemented are special modules for the dynamic analysis, absorbing boundary conditions, and nonlinear constitutive relations of concrete and soil. Appendices A–C briefly explain the three features of the parallel FEM, namely, the material constitutive relations implemented, the capability of liquefaction analysis, and the capability of lift-up analysis.

The key feature of the parallel FEM is a *solver*, a module for solving a matrix equation of FEM, i.e.,

$$[A]\,[x] = [y], \tag{1}$$

where $[A]$ and $[y]$ are a given matrix and vector and $[x]$ is an unknown vector which corresponds to nodal displacement. The parallel FEM adapts the conjugate gradient method which solves Eq. (1) incrementally to minimize an error, i.e., the norm of $[A]\,[x] - [y]$; see, for instance, Saad,[26] Hackbush,[27] Kawai *et al.*[28] The solver is enhanced with high-performance computation capability, and high scalability is confirmed.

An extremely strong ground motion is used for the seismic safety evaluation to examine the ultimate seismic safety of a nuclear power plant. We need to know the strongest ground motion that the plant can endure, accepting the occurrence of local failure at a few structural members or connections. It is expected that the parallel FEM is used to estimate the ultimate seismic safety; higher accuracy and resolution result in the most reliable estimate of local failure when an analysis model of high fidelity is used.

In this chapter, we explain the use of the parallel FEM for the seismic structural analysis of a 3D numerical analysis model subjected to an extremely strong ground motion. A key issue is that the parallel FEM can complete the numerical computation for such a strong ground motion, which may induce local large deformation or lead to the overall failure of a structure. It is also important to examine whether the constitutive relation implemented into the parallel FEM is applicable to analyze the seismic response for the strong ground motion. It becomes necessary to improve the parallel FEM or to improve the quality of the numerical analysis model for a faster and more reliable seismic structural analysis.

The contents of this chapter are organized as follows. In Section 2, we explain a trial simulation of the seismic structural response analysis for an extremely strong ground motion. In Section 3, we improve the parallel FEM to solve issues found in the trial simulation. In Section 4, we present the results of the full-scale simulation of the seismic structural response analysis, using the improved parallel FEM. The results of the full-scale simulation are discussed from the viewpoint of numerical computation and seismic structural responses, focusing on the usefulness of the parallel FEM for the estimate of the ultimate seismic safety of a structure.

2. Trial Simulation

In this section, we explain a trial simulation of seismic structural response analysis for extremely strong ground motion. The trial simulation has the following two objectives: (1) to identify numerical computation settings for efficient numerical computation and (2) to examine the applicability of the constitutive relation implemented into the parallel FEM. The numerical computation settings are mainly concerned with solving Eq. (1), which becomes nonlinear as the matrix $[A]$ depends on the unknown vector $[x]$. That is,

$$[A]\left([x]\right)[x] = [y]. \tag{2}$$

For the second objective, we examine whether element-wise stress and strain states remain within the range of the constitutive relation implemented into the parallel FEM.

In the seismic response analysis, $[x]$, $[A]$ and $[y]$ in Eq. (2) are an increment of nodal displacement, the matrix given as the weighted sum of a global mass, damping, and stiffness matrices, and an increment of nodal force determined by the boundary conditions, respectively; it is the global stiffness matrix that depends on the nodal displacement. As will be explained later, the modified Newton method, which replaces $[A]([x])$ with a suitable matrix to make a faster convergence of the nonlinear analysis, is used as an algorithm for incrementally solving Eq. (2) by setting an initial solution as $[x] = [0]$; see Zienkiewicz and Taylor,[29] Belytschko *et al.*,[30] de Borst,[31] and Bonet and Wood[32] for the nonlinear analysis used in FEM.

In the trial simulation, we construct a 3D analysis model of a nuclear power plant, which is a set of a reactor building and the ground. Figure 1 shows the numerical analysis model constructed; (a) and (b) are the overview and the close view of the reactor building, respectively. A Cartesian coordinate of (x, y, z) is employed, with

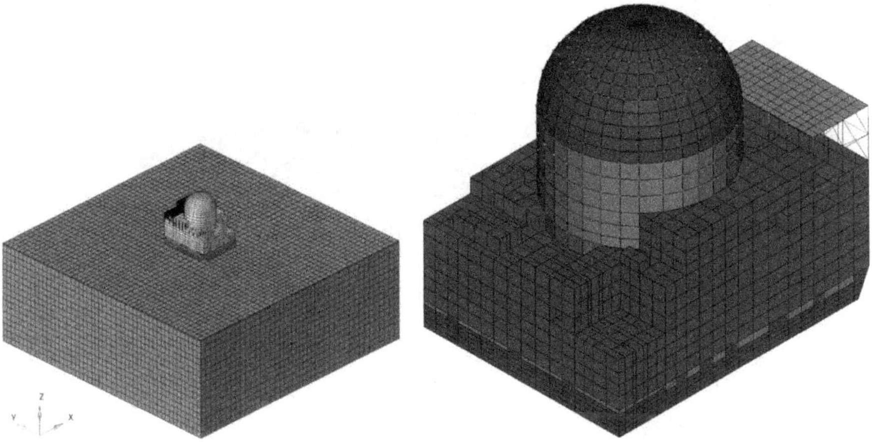

(a) (b)

Fig. 1. Numerical analysis model used in the present paper. (a) Overview, and (b) reactor building.

the x-direction parallel to the width direction of the analysis model. The dimension of the reactor building is sub-100m in the x-, y-, and z-directions, and it consists of outer shield (OS) of a cylinder with a cap, inner concrete (IC) of reinforced concrete shear walls, auxiliary equipment building (AEB), and foundation. The node and element numbers of the analysis model of the reactor building are 22,077 and 39,513, respectively.

The dimension of the ground is around several hundred meters in the x-, y-, and z-directions, and the node and element numbers are 41,115 and 42,498, respectively. It is essential to include the ground in the numerical analysis model for accurate estimate of the soil–structure interaction effects; see Wolf,[33] Hjelmstad,[34] and Pecker *et al.*[35]; see Aki[36] for the site ground motion; see also Jeremic *et al.*[37] and Wang *et al.*[38] for the soil–structure interaction of the nuclear power plant. The ground consists of a single geological layer of stiff bedrock. It should be noted that the degree-of-freedom of the whole model exceeds 100,000, which is large as a model of the seismic response analysis; the parallel FEM can analyze a model whose degree-of-freedom is larger than 10,000,000.

Joint elements proposed by Goodman *et al.*[39] are used in the interface between the foundation of the reactor building and the

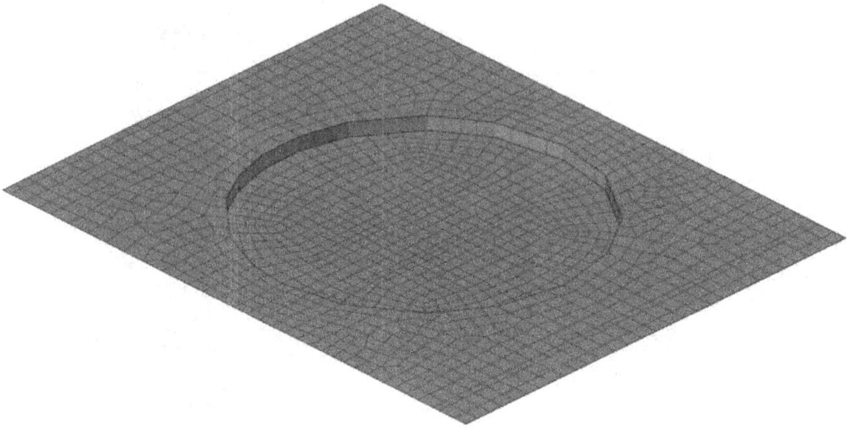

Fig. 2. Joint elements used in the numerical analysis model.

ground for the uplift analysis; subjected to strong ground motion, a part of the foundation of the reactor building is detached from the ground and then touches it, which may induce impulsive vertical ground motion in the building. Appendix C provides an example of the lift-up analysis, which models the detaching/attaching process in a more sophisticated manner than joint elements. As shown in Fig. 2, the ground beneath the reactor building is excavated, and the building bottom is set on the bedrock. The bottom of the reactor building is rectangular. The center cylinder part is excavated deeper than the other part, to confirm the attachment of the foundation to the ground. Thus, joint elements are placed on the entire rectangular bottom including the cylindrical part, i.e., the bottom and side surfaces of the cylindrical part.

In addition to the joint elements, multi-point constraint (MPC) is used in the following two interfaces: (1) the interface within the foundation; and (2) the interface between the foundation and the ground. Due to the simple configuration, only hexahedron elements are used for the ground, while IC must use tetrahedral elements for structural members of complicated geometry. The foundation compensates for this mismatch of element shapes. As the simplest solution, the foundation is decomposed into the upper part attached to IC and the lower part attached to the ground. The interface between the foundation and the ground has a mismatch in the element size since the element size of the foundation is much smaller than that of the ground. It is known that fine meshing is not needed for the ground which is far from the reactor building because it does not contribute to the soil–structure interaction effects. Thus, MPC is introduced to this interface.

Figure 3 presents the use of MPC. The foundation which is separated at the middle plane is shown in Fig. 3(a); this plane is indicated by a dashed line in the figure. Triangle facets of tetrahedral elements appear on the upper surface while only rectangular facets of hexagonal elements appear on the lower surface. MPC is introduced to

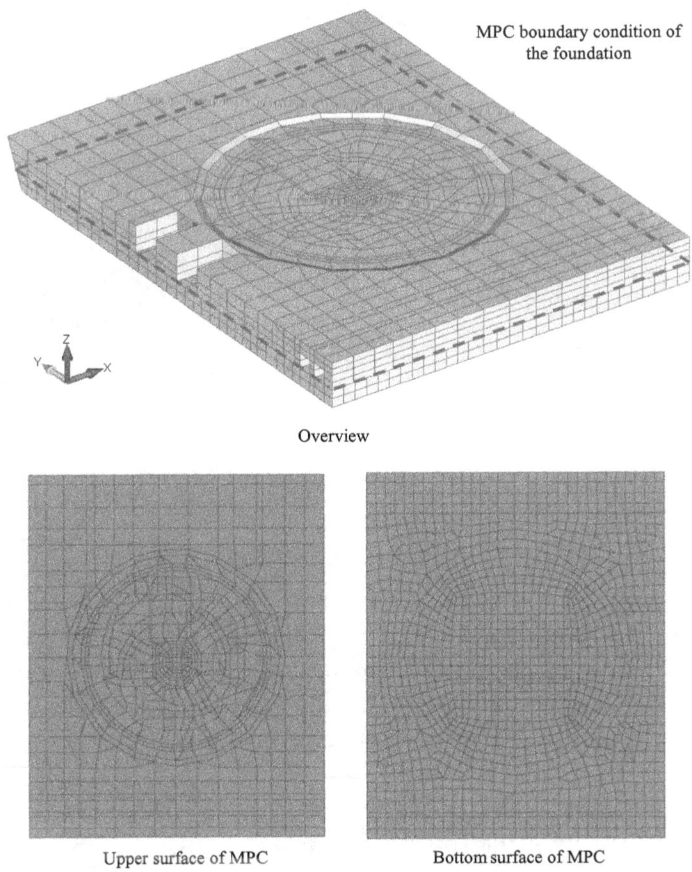

(a)

Fig. 3. The use of MPC. (a) The foundation of the reactor building. (b) The part of the ground near the foundation.

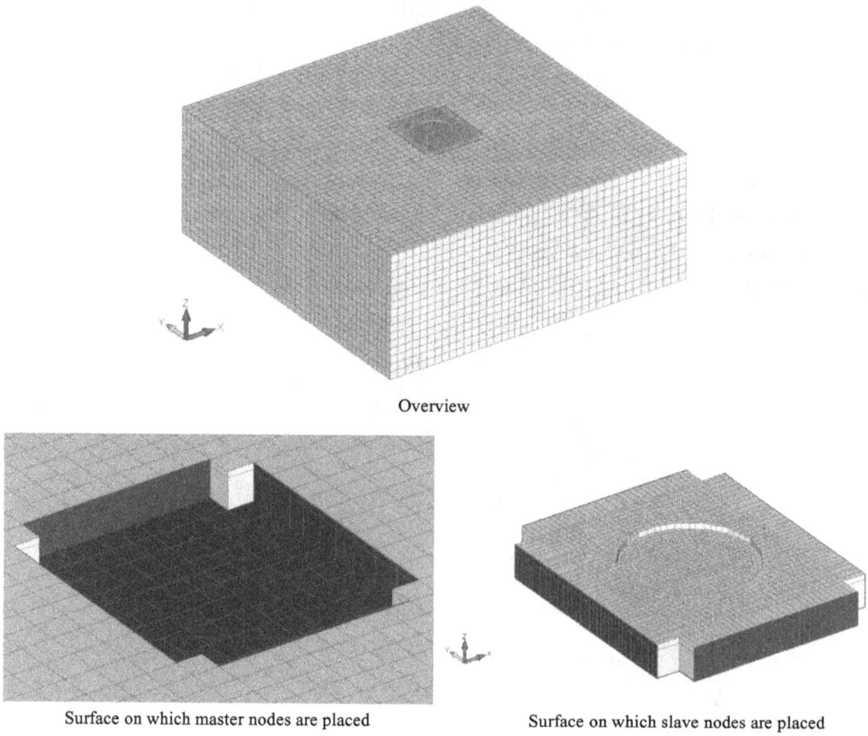

(b)

Fig. 3. (*Continued*)

connect the nodes on the upper surface to those on the lower surface. The side and bottom surfaces of the ground on which MPC is introduced are presented in Fig. 3(b). This part of the ground includes the ground attached to the foundation and a layer of one element width. The element size of this part is smaller than that of the remaining part of the ground, and MPC is introduced for the interfaces between them.

The material properties of concrete and steel of the reactor building and soil of the ground are summarized in Table 1; (a) is for the elastic parameters and density of concrete, steel, and soil, and (b) is for the nonlinear parameters of concrete and steel. The material nonlinearity is not considered for the ground since it consists of stiff bedrock.

Table 1. Material properties.

Material	Young's modulus (kN/m^2)	Poisson's ratio	Density (ton/m^3)
(a) Elastic parameters and density			
Concrete	2.88×10^7	0.2	2.3–2.4
Steel Reinforcement	2.05×10^8	—	7.8–8.0
Soil	5.44×10^7	0.34	3.0

Material	Compressive strength (kN/m^2)	Tensile strength (kN/m^2)	Shear strength (kN/m^2)	Strain at compressive strength
(b) Nonlinear parameters for concrete and steel				
Concrete	4.92×10^4	3.09×10^3	1.40×10^4	0.003417
Steel Reinforcement	—	3.45×10^5	—	—

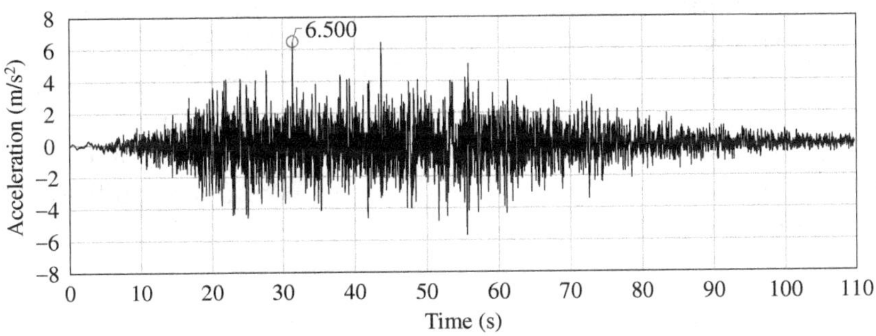

Fig. 4. Input ground motion.

The time series of input ground motion acceleration is shown in Fig. 4. The duration of this motion is 110 s, which is relatively long, and the maximum acceleration is 6.5 m/s^2; this motion is synthesized according to a method of computing a standard design seismic motion which is commonly used in Japan. It should be mentioned that the input ground motion is unidirectional; the present regulation of seismic safety uses unidirectional ground motion. The ground motion shown in Fig. 4 is magnified by the factor of 1.5, 2.0, and 4.0; the amplitude of the ground motion is simply multiplied, although it is necessary to modify the frequency characteristics of the ground motion in addition to its amplitude.

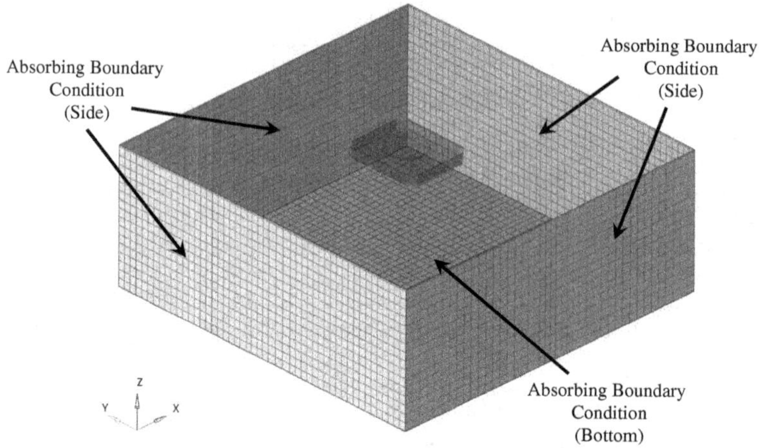

Fig. 5. Absorbing boundary condition.

In the numerical analysis model of the ground, the absorbing boundary condition is posed for the side and bottom surfaces, which are an artificial boundary of the ground, as shown in Fig. 5; see Refs. 40–43. On the boundary, a layer of a few hexagonal elements of high damping coefficients that fully absorbs waves passing through the elements at the right angle is placed. This is an ordinary numerical technique for satisfying the absorbing boundary conditions on the artificial boundary.

2.1. *Computation performance results*

In the trial simulation, we study the effects of the following four parameters on the numerical computation: (1) the amplification factor of input ground motion (1.5, 2.0, 4.0); (2) the type of modified Newton method (the use of the initial or present stiffness); (3) the time step (2.5, 1.0, 5.0×10^{-4} s; and (4) the maximum iteration number (10, 20, 50). Here, the initial or present stiffness of the modified Newton method is the stiffness computed by linear elastic response or nonlinear elastoplastic response of the present state, respectively. The iteration number is for the modified Newton method, and the iteration is terminated if the iteration number exceeds the maximum number.

For all six cases, the allowable residual error is commonly set as 1.0×10^{-9}. As explained, the residual error is for the global matrix equation which is solved by the conjugate gradient method; it is normalized with respect to the input vector. That is, the condition for solving Eq. (2) is

$$\frac{|[A]\,[x] - [y]|}{|[y]|} < 1.0 \times 10^{-9}, \tag{3}$$

where $|(.)|$ is the normal of a vector (.). We seek the most suitable numerical analysis setting of the four parameters so that the numerical computation is completed fast and with allowable accuracy.

We first summarize the results of the numerical computation of the trial simulation in Table 2. Six settings are studied, and the numerical computation is completed for the three settings with the amplification factor of 1.5, i.e., Cases 2, 3, and 4. The numerical computation is not completed for the amplification factor of 2.0 and 4.0, i.e., Cases 5 and 6. Even for the amplification factor of 1.5, the numerical computation becomes unstable when the maximum iteration number is set as 50. Although it seems odd, the numerical computation becomes stable if the maximum iteration number is limited and the solution of the modified Newton method is not fully converged. The nonlinear iterative analysis seeks an unrealistic solution if too many iterations are allowed in the modified Newton method.

Table 2. Summary of trial simulation.

	Seismic motion multiplier	Newton's method	Time step (s/step)	Maximum iteration number	Computational stability
Case 1	1.5	Initial	2.5×10^{-4}	50	Not stable
Case 2	1.5	Initial	2.5×10^{-4}	10	Stable
Case 3	1.5	Initial	1.0×10^{-4}	20	Stable
Case 4	1.5	Modified	5.0×10^{-4}	10	Stable
Case 5	2.0	Initial	2.5×10^{-4}	20	Not stable
Case 6	4.0	Initial	2.5×10^{-4}	20	Not stable

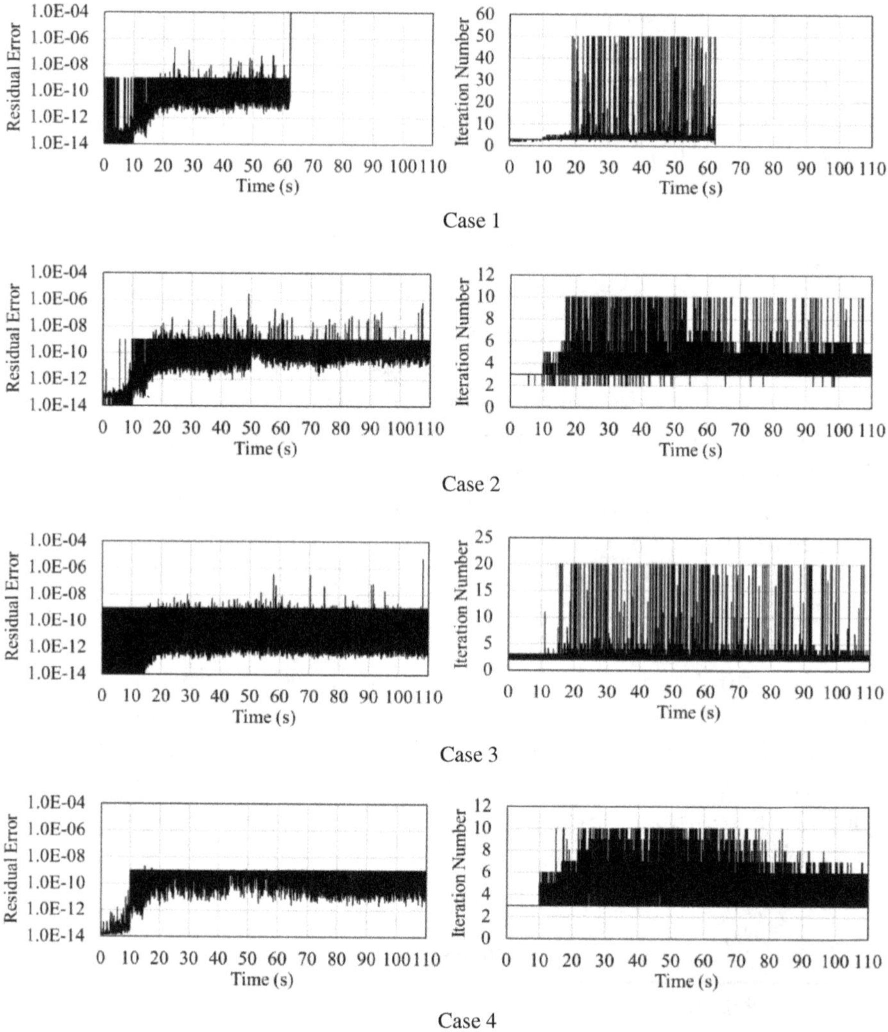

Fig. 6. Time series of residual error and iteration number.

In Fig. 6, we show the time series of the residual error and the iteration number for Cases 1–4, which commonly use the amplification factor of 1.5. As is seen, the numerical computation is not completed for Case 1. While the numerical computation is completed for Cases 2 and 3, the residual errors exceed the allowable value of 1.0×10^{-9} at many time steps; the iteration in the modified Newton method is terminated for these steps. The residual error does not exceed the allowable value only for Case 4. Regarding the iteration number, there are many steps at which the iteration number reaches the maximum number, implying that the modified Newton method does not reach a converged solution. This non-convergence of the modified Newton method happens even for Case 4.

Comparing Case 1 with Case 4, it is seen that reducing the maximum iteration number from 50 to 10 makes the numerical computation complete. It seems odd, but in the seismic structural response analysis, a residual error at one step is sent to the next step as an unbalanced force, i.e., the right-hand side of Eq. (2). The residual error can be reduced at the next step, without being accumulated if a solution that compensates for the unbalanced force is found. Comparing Case 2 with Case 3, it is seen that the iteration number is reduced by shortening the time increment from 2.5×10^{-4} to 1.0×10^{-4}. CPU time, however, may not be reduced since the total number of time steps increases 2.5 times for this setting. From the comparison of Cases 3 and 4, it is seen that the modified Newton method with the present stiffness makes a faster convergence, compared with the method with the initial stiffness; the time step of Case 4 is four times longer than that of Case 3.

The good performance of Case 4 is due to better settings of the modified Newton method with the present stiffness, a longer time increment, and a limited maximum iteration number. It should be noted that these settings do not provide an accurate solution to the material nonlinearity at all time steps; Case 4 has the largest number of time steps at which the iteration number reaches the maximum number. The residual error of solving the global matrix equation remains in the allowable range, but the level of the residual error is larger than that of Case 3.

2.2. *Seismic structural response results*

We next show the results of the structural seismic response of the trial simulation. We choose two, two, and one observation points, respectively, for OS, IC, and the foundation of the reactor building; see Fig. 7. Acceleration response spectra of the two observation points of OS and IC from Cases 1 to 4 are shown in Fig. 8; a component in the x-direction or the input ground direction is used. As is seen, the

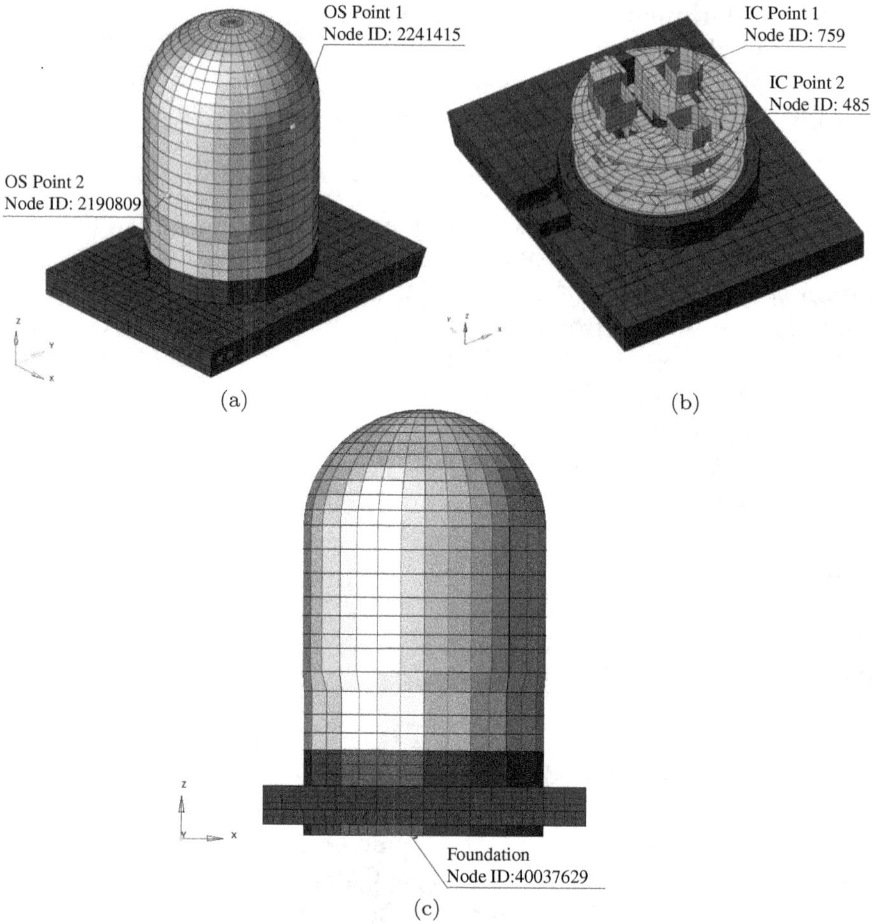

Fig. 7. Observation points of the numerical analysis model. (a) Outer shield (OS), (b) inner concrete (IC), and (c) foundation.

form of the spectra is similar for the four cases even if the numerical computation is not completed for Case 1, probably because the computation of a dominant part of the responses has been computed for this case. The similarity of the spectra for Cases 2–4 is very high, which suggests that almost the same seismic responses are computed for these three cases even though numerical computation settings are different. The agreement of the dominant frequency and its amplitude is sufficient.

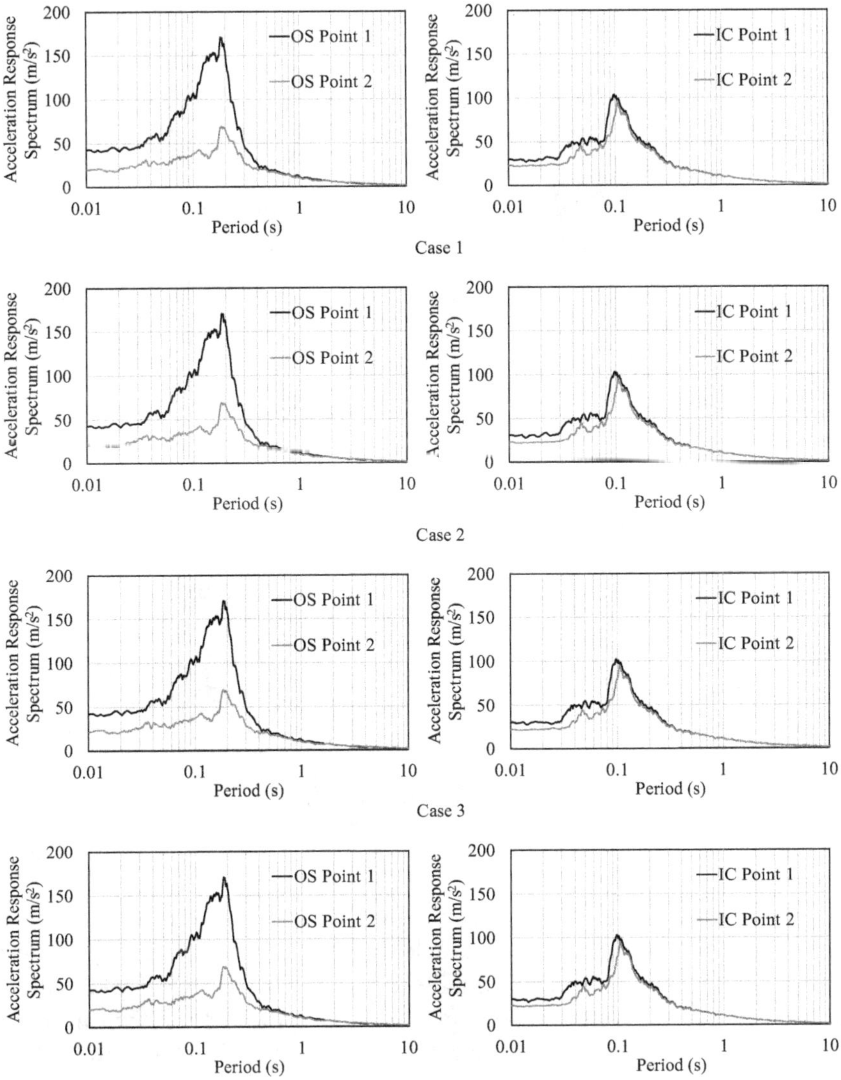

Fig. 8. Acceleration response spectra at the observation points.

Since the input ground motion is extremely strong, the concrete of the reactor building is damaged by tensile cracking. Figure 9 shows the distribution of the crack on the shear wall of the building from Cases 1 to 4; the number of cracks refers to the number of tensile

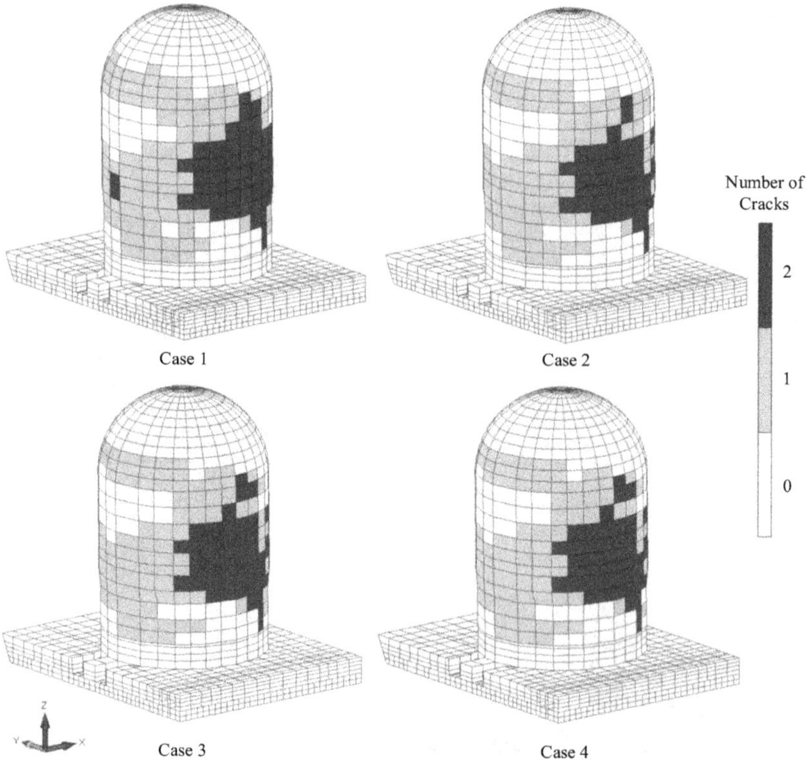

Fig. 9. Crack distribution.

cracks initiated in one element. While the numerical computation of Case 1 is not completed, the crack distribution of Case 1 is denser than that of the other three cases. While there are minor differences, the element-wise crack numbers are similarly distributed for Cases 2–4, just like Fig. 8, which again suggests that almost the same seismic responses are computed for these three cases. It should be noted that the number of cracks remains two, which are cracks in the in-plane direction of the shear wall, and that no cracking takes place in the transverse direction.

It is interesting to compare the seismic responses of Cases 3 and 4, which use the modified Newton method of the initial and present stiffness and the time increment of 1.0×10^{-4} 5.0 and 5.0×10^{-4}, respectively. As shown in Figs. 8 and 9, almost identical results are computed for the acceleration response spectra and the crack distribution. Since Case 4 uses a longer time increment, we can expect that the modified Newton method with the present stiffness is more efficient than Newton's method with the initial stiffness, even though the use of the present stiffness needs additional numerical computation of updating the stiffness at each step.

In closing this section, we summarize the results of the trial simulation. Regarding the numerical analysis settings, the modified Newton method with the present stiffness seems better than that with the initial stiffness, and fast numerical computation is expected by setting a longer time increment and a smaller maximum iteration number, obtaining almost the same seismic structural responses. Regarding the constitutive relation implemented into the parallel FEM, we fail to fully examine its applicability for the extremely strong ground motions because Cases 5 and 6, with the amplification factor of 2.0 and 4.0, are not completed.

3. Improvement of Parallel Finite Element Method

In earthquake engineering, Newton's method and modified Newton's method with the initial stiffness are employed as an algorithm for solving Eq. (2). It is necessary to set a sufficiently large iteration number and a sufficiently small given increment for these algorithms. However, the results of the trial simulation suggest that either (or both) condition is not satisfied for the extremely strong ground motion. Also, the numerical computation is not completed when the amplification factors are set as 2.0 and 4.0. Therefore, it is necessary to improve an algorithm for solving Eq. (2), in addition to considering a proper setting of numerical computation.

Unlike metallic materials, concrete exhibits zero and negative stiffness as nonlinear behavior; see Fig. 10. For this zero and negative stiffness, modified Newton's method is employed in the seismic structural response analysis. The use of this algorithm is to save numerical computation costs of Newton's method that changes stiffness at each iteration. The numerical cost increases as the number of elements increases since the material nonlinearity is computed for each Gauss integration point. Hence, it is not realistic to employ Newton's method.

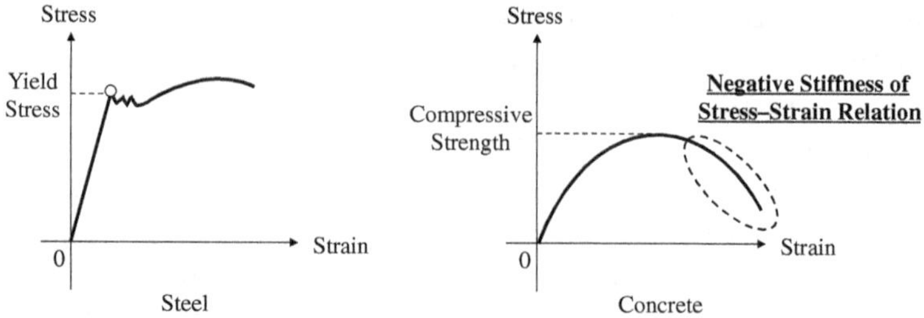

Fig. 10. Schematic view of nonlinearity of concrete.

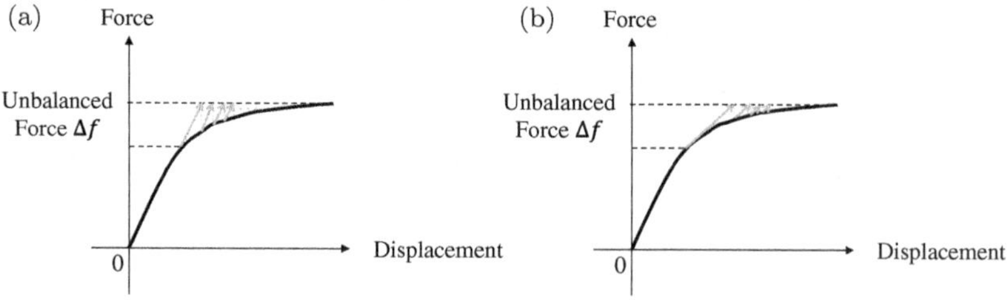

Fig. 11. Comparison of the modified Newton method with (a) initial, and (b) present stiffness.

The material nonlinearity shown in Fig. 10 reflects the nonlinear relation between the nodal displacement and force, which correspond to $[x]$ and $[y]$, respectively, in Eq. (2). For simplicity, we cast this relation to a displacement–force relation as shown in Fig. 11. The slope of the displacement–force relation becomes flat when it reaches a critical state, i.e., a structure reaches its capacity. At this state, the modified Newton method with the initial stiffness needs a greater number of iterations since the difference between the initial stiffness and the present stiffness becomes large. While extra computation cost is needed for recomputing the present stiffness, the modified Newton method can reach a converged solution in a fewer number of iterations. It should be noted that the modified Newton method with the present stiffness is a reasonable algorithm, compared with Newton's method which needs large numerical computation of updating stiffness at each iteration, or the modified Newton method of the initial stiffness which needs a large number of iterations.

While it can reach a converged solution for a sufficiently stiff relation, the modified Newton method with the present stiffness needs a larger iteration number and a smaller force increment when the slope of the displacement–force relation changes from positive to negative or changes greatly at the transition from soft plastic regime to a stiff elastic regime or vice versa. Such changes in the slope of the displacement–force relation often happen during the cyclic loading induced by input ground motion.

For a faster numerical computation of the seismic structural response analysis subjected to extremely strong ground motion, we improve the algorithm of solving the concrete constitutive relation of shearing; see Fig. 12. The improvement is aimed at removing the sudden change in the slope. As shown in Fig. 13, the stress–strain relation in shearing depends on the crack strain that is computed by the accumulation of plastic strain; see Appendix A. While the relation between stress and strain is smooth, the slope of the relation near the origin changes abruptly as the crack strain grows; for instance, the slope is almost zero for the crack strain of 0.05%, but it exceeds the slope of the initial elastic response for the crack strain of 0.02%.

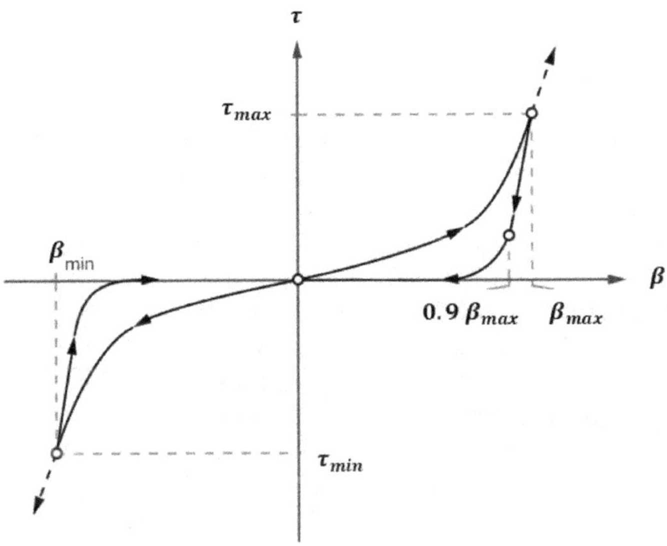

Fig. 12. Concrete stress–strain relation in shearing.

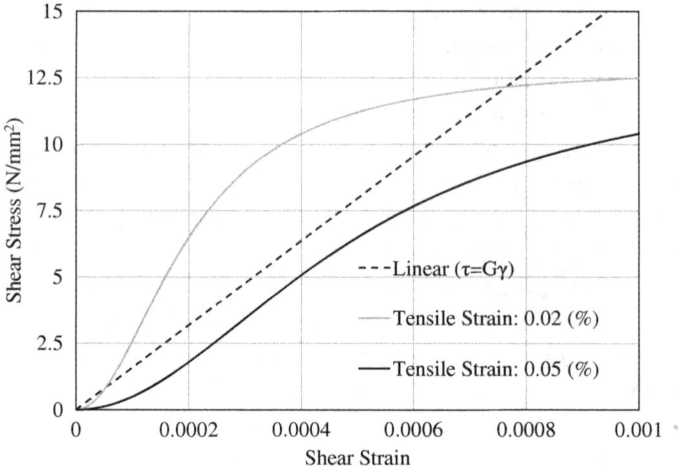

Fig. 13. Dependence of the stress–strain relationship in shearing on the crack strain.

To remove the abrupt change in the slope, we add the following condition to compute the slope of the stress–strain relation in shearing:

$$G^{ep} = \min\left[\frac{\partial\tau}{\partial\gamma}, G\right], \tag{4}$$

where $\frac{\partial\tau}{\partial\gamma}$ is an elastoplastic stiffness that is computed at the present state and G is the initial stiffness of linear elastic response. The algorithm for Eq. (4) is implemented into the parallel FEM. It should be emphasized that Eq. (4) contributes to the more stable computation of solving Eq. (2) as it results in a smoother change in $[A]$. Assigning this additional condition, the material constitutive relation is altered from its original form. The difference between this altered relation from the original relation should be minimum since this condition is applied only at very small strain with the initial state of cracking.

We examine the numerical computation performance of the improved parallel FEM. Table 3 shows the CPU time of the seismic structural response analysis using the original and improved algorithms. The amplification factor of the input ground motion is 1.5 and 2.0, and the numerical computation settings are the same as those of Case 4 shown in Table 2, i.e., the time increment is 5.0×10^{-4} s

Table 3. CPU time of seismic structural response analysis.

	Seismic motion multiplier	Algorithm	CPU time (h)	Ratio of CPU time
Old_1.5	1.5	Original	247.3	
Improved_1.5	1.5	Improved	120.6	2.05
Original_2.0	2.0	Original	319.3	
Improved_2.0	2.0	Improved	142.1	2.25

and the maximum iteration number is 10. As is seen, the CPU time is reduced more than half for the two cases if the improved algorithm is used.

Figure 14 compares the time series of the residual error and the iteration number for the amplification factor of 1.5 and 2.0, using the original and improved algorithms. As is seen, the iteration number does not reach the maximum number for the improved algorithm, while there are many steps at which the modified Newton method does not converge for the original algorithm. The residual error does not exceed the allowable value for the improved algorithm. It is clearly seen that the improved algorithm results in faster convergence of the modified Newton method.

It should be mentioned that in a numerical analysis model of low fidelity which uses coarse meshes, the dependence of the stress–strain relation slope on the crack strain is not critical because the crack strain grows sharply in large elements. However, the crack strain gradually increases in small elements of a numerical analysis model of high fidelity. Also, the time increment must be set to a smaller value for extremely strong ground motion. Thus, even though the crack strain gradually increases, it induces an abrupt decrease in the slope of the displacement–force relation when a high-fidelity model and extremely strong ground motion are used. This abrupt decrease is not noticed for a low-fidelity model or ordinarily strong ground motion.

In addition to the improvement of the algorithm of solving Equation (2), we need to improve the quality of the numerical analysis model of a nuclear power plant if a simulation of more accurate and higher resolution is needed. The quality of the model is determined

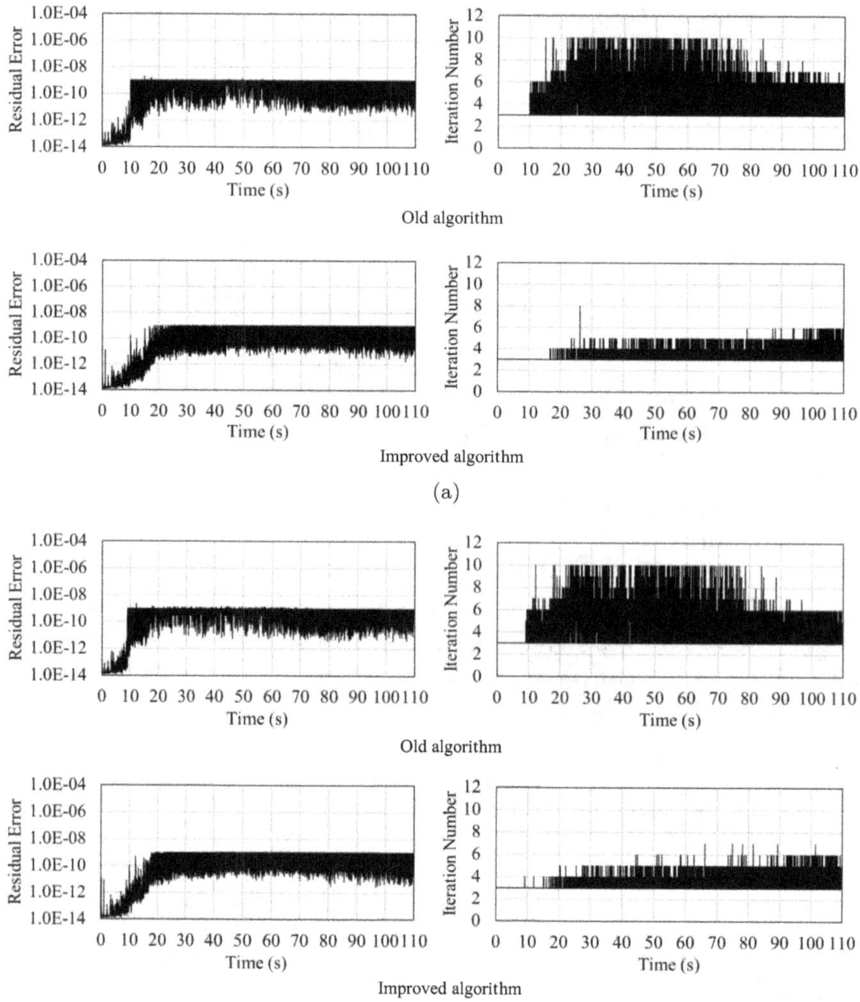

Fig. 14. Time series of residual error and iteration number. (a) Amplification factor of 1.5, and (b) amplification factor of 2.0.

by the configuration of elements; elements of twisted configuration result in a larger value of the condition number of $[A]$ of Eq. (2) and a larger number of iteration is needed for the conjugate gradient method; the dependence of the convergence on the model quality is a characteristic of the conjugate gradient method, which is often overlooked. It should be mentioned that the present numerical analysis model is of the largest scale compared with ordinary analysis models used in earthquake engineering, but its scale is not regarded as large in the standard of other fields of engineering.

4. Simulation of Extremely Strong Ground Motion

As explained in Section 2, the numerical computation is not completed for three cases of the trial simulation due to the improper setting in addition to poor performance of the nonlinear analysis algorithm implemented in the parallel FEM. We carry out a full-scale simulation for the seismic structural response analysis, improving the algorithm for the nonlinear analysis. The amplification factors of 2.0 and 4.0 are used in the full-scale simulation to synthesize an extremely strong ground motion; these two cases correspond to Cases 5 and 6, respectively, in the trial simulation. This section presents the results of the full-scale simulation, focusing on the computation performance and the computed seismic responses of the reactor building.

4.1. *Computation performance results*

First, we present the results of computation performance. The time series of the residual error and the iteration number for the amplification factor of 4.0 is presented in Fig. 15. The results of the amplification factor of 2.0 is the same as the examination test of the numerical computation performance for the improved algorithm shown in Section 3, Fig. 14(b). As is seen, the iteration number does not reach the maximum iteration number (10) in the whole period;

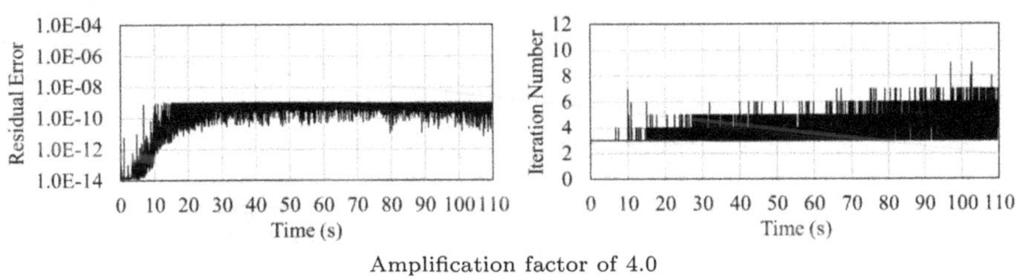

Fig. 15. The time series of the residual error and the iteration number.

for the amplification factor of 2.0, the iteration number gradually increases but stays around 4; and for the amplification factor of 4.0, the iteration number is generally greater than that for the amplification factor of 2.0 but remains less than 6 for most of the period. These numerical computation performances are much better than those of Case 4, as shown in Fig. 2; the iteration number is the smallest in the trial simulation but it often reaches the maximum iteration number. While the amplification factor of Case 4 is 1.5, the improved parallel FEM is capable of enabling better numerical computation performance for the amplification factor of 2.0 and 4.0.

As shown in Fig. 15, the residual errors are smaller for the amplification factor of 2.0 compared with the amplification factor of 4.0. The time series of the residual error of the amplification factor of 4.0 is similar to that of Case 4 of the trial simulation, shown in Fig. 2.

We summarize the CPU time and the total iteration number in Table 4; the case of the amplification factor of 1.5 is included, in addition to 2.0, and 4.0. The total iteration number is the sum of the iteration numbers of all time steps. As the amplitude of the input ground motion increases, the CPU time and the total iteration number increase, which is well expected. It should be noted that the CPU time increases 70% from the amplification factor of 2.0 to that of 4.0, while the total iteration number increases only 20%. This indicates that more CPU time is spent for each iteration to solve Eq. (2) for the stronger input ground motion. The numerical computation is used to update $[A]$ since the number of elements that enter elastoplastic regime increases for the stronger ground motion. It is also possible that the conjugate gradient method takes a longer CPU time since $[A]$'s condition number (that is defined as the ratio of the maximum and minimum eigenvalue) becomes larger and a greater number of iterations is needed.

Table 4. CPU time.

	CPU time (h)	Iteration number
Amplification factor 1.5	120.6	472626
Amplification factor 2.0	142.1	503271
Amplification factor 4.0	234.0	633570

4.2. *Seismic structural response results*

Since the numerical computation performance of the parallel FEM is improved, we study the results of the seismic structural responses subjected to extremely strong ground motion. It should be emphasized that the input ground motion is extremely strong, in particular, when the amplification factor of 4.0 is used. The results of the full-scale simulation should be regarded as an extrapolation since the seismic responses for such strong ground motion have not been experienced. From the viewpoint of material, however, the results are reliable as long as the stress and strain state induced by the ground motion remains in the range where the constitutive relations are applicable. It is thus important to examine the applicability of the constitutive relations implemented into the parallel FEM to the case of the extremely strong ground motion.

We first study the point-wise kinematic response, which is essential for seismic design. Figure 16 plots the acceleration response spectra in the x-direction for the observation points of OS and IC shown in Fig. 7. For OS, the responses of the two observation points are

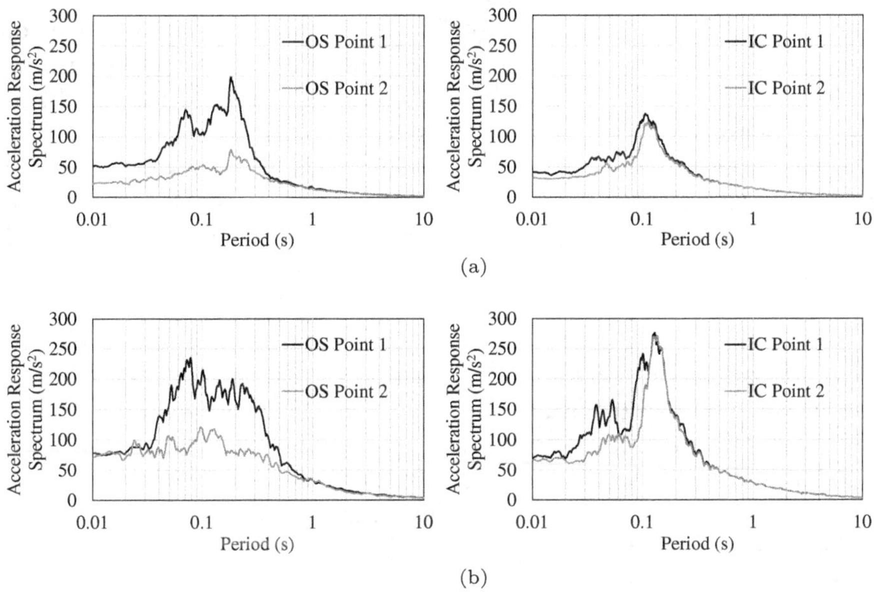

Fig. 16. The acceleration response spectra. (a) Amplification factor of 2.0, and (b) amplification factor of 4.0.

different as the point put on the surface normal to the input ground motion direction is shaken severely. Comparing the amplification factor of 4.0 with the amplification factor of 2.0, it is seen that frequency components around 0.08 s are larger, while frequency components longer than 1.0 s are almost the same. IC behaves like a rigid body since the difference between the two observation points is small. Comparing the amplification factor of 4.0 with the amplification factor of 2.0, it is seen that there are some differences in frequency components between 0.02 and 0.05 s. It should be noted that the amplitude of longer periods (0.3–0.8 s) increases for OS, but only the amplitude of shorter periods (0.02–0.1 s) increases for IC when the amplification factor changes from 2.0 to 4.0. It should also be noted that compared with OS, IC has a clear peak in the spectra, which corresponds to the natural vibration period. The period is slightly longer for the amplification factor of 4.0. This indicates that the material nonlinearity results in a longer natural period of the building.

Figure 17 plots the distribution of the maximum displacement in the reactor building; the maximum displacement refers to the maximum value of the displacement component in the x-direction, not the maximum value of the norm of the displacement vector. As is seen, the pattern of the distribution is similar for the two cases of the amplification factor of 2.0 and 4.0, although the amplitude is different more than ten times; the red color indicates 0.04 and 0.14 m for the factor of 2.0 and 4.0, respectively. It is interesting that while the amplitude of the maximum displacement becomes much larger than two times, the pattern of the distribution seems similar. The pattern is different locally, for instance, near the connecting part of OS and IC and the steel frame structure shown in the center sectional view No. 1.

It is seen from the kinematic structural responses shown in Figs. 16 and 17 that the reactor building remains intact even when it is subjected to the ground motion amplified four times. There can be some structure members and connecting parts which are locally damaged, but these parts cannot be identified in the kinematic responses.

Taking advantage of the numerical simulation, we study the stress responses to identify the location and degree of possible local damages. It should be noted that measuring stress is much more difficult than displacement or acceleration; using strain gauge measurement, we can measure a component of stress, but we cannot estimate the

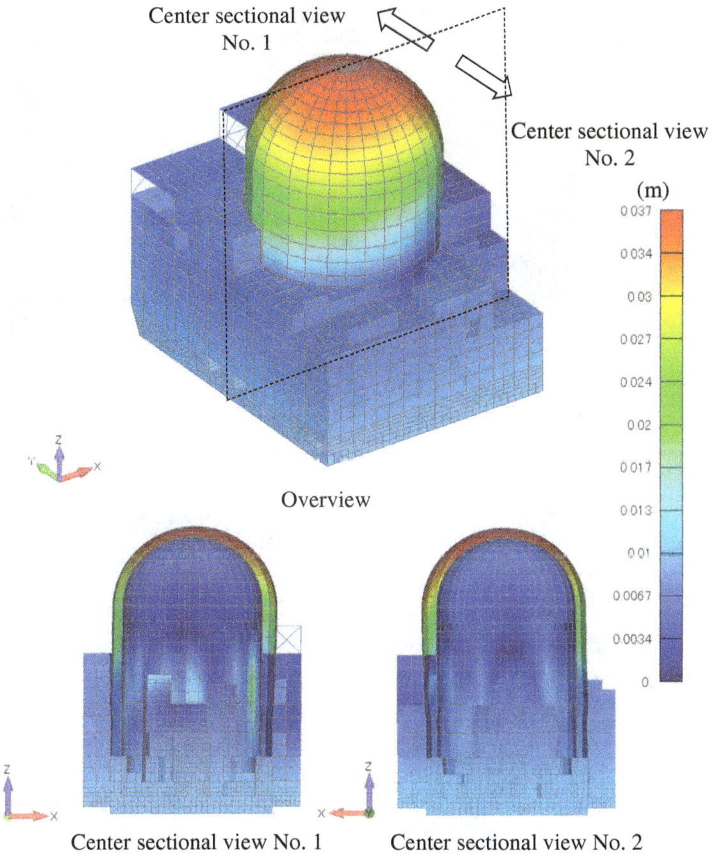

(a)

Fig. 17. The distribution of the maximum displacement. (a) Amplification factor of 2.0. (b) Amplification factor of 4.0.

stress state that is expressed in terms of the six components of stress. Even though we need to examine the accuracy of computing stress, it is useful to use the stress computed in the numerical simulation for the identification of local damage.

Figures 18 and 19 plot the time series of the compressive stress in concrete and tensile stress in steel reinforcement, respectively. We choose the element with the largest and smallest stress for the numerical analysis model of OS, and (a) and (b) of both figures are for the element with the largest and smallest stress, respectively;

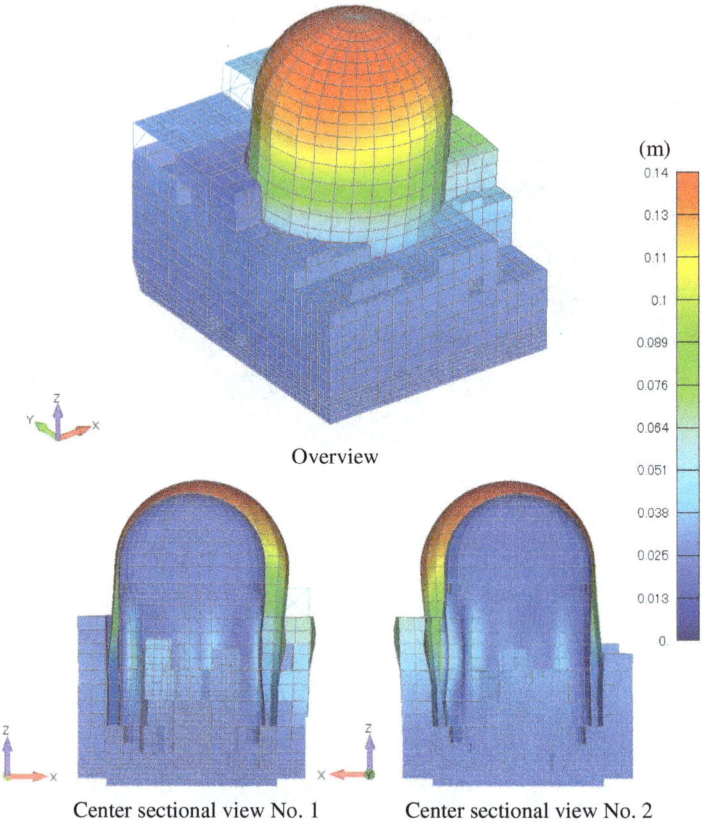

(b)

Fig. 17. (*Continued*)

the damage is measured in terms of the fracture parameter which will be explained later in this section. The stress is the principal value, i.e., the compressive or tensile stress refers to the minimum or maximum value of the principal value of stress, respectively.

For the element with largest stress, the compressive stress of concrete for the amplification factor of 4.0 becomes smaller than that of 2.0 after 40 s. This indicates that the concrete of this part is damaged and loses its compressive strength. For the element with smallest stress, the compressive stress is greater for the amplification factor of 4.0. But the compressive stress is much smaller than the compressive strength. For the steel reinforcement, the residual tensile stress

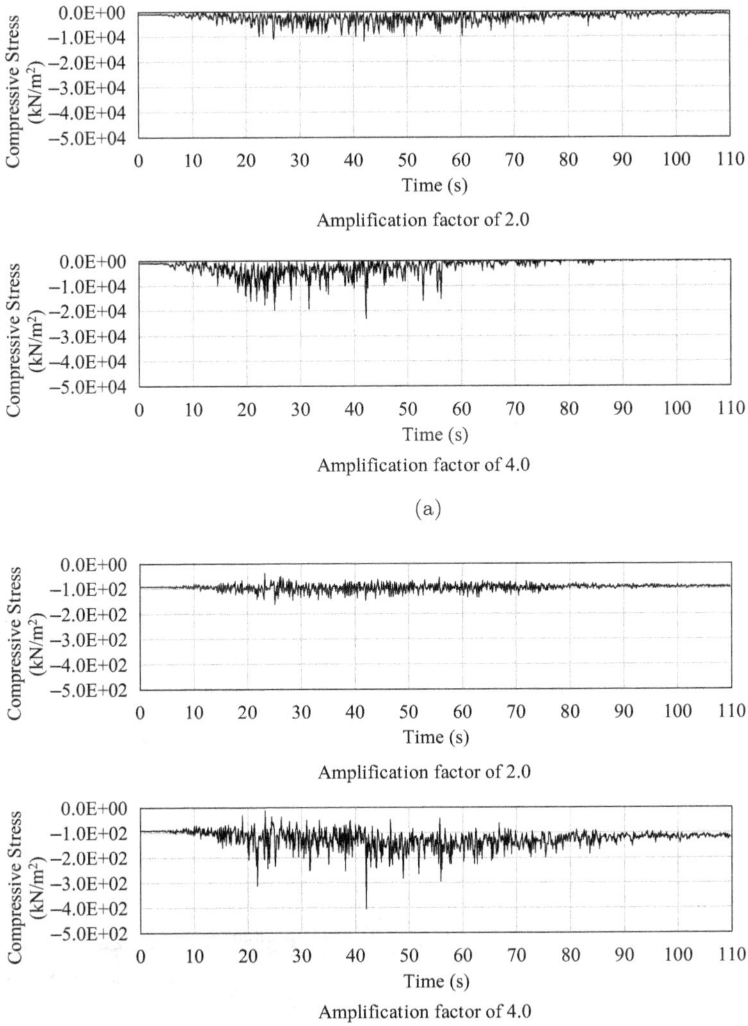

Fig. 18. The time series of compressive stress in concrete. (a) The element with largest stress, and (b) the element with smallest stress.

is generated for the element with largest stress for the amplification factor 4.0; residual tensile stress is generated in the element with smallest stress, too, but it is negligibly small compared with that generated in the element with largest stress. Due to local damage in the concrete, it cannot carry stress caused by the gravity force, and

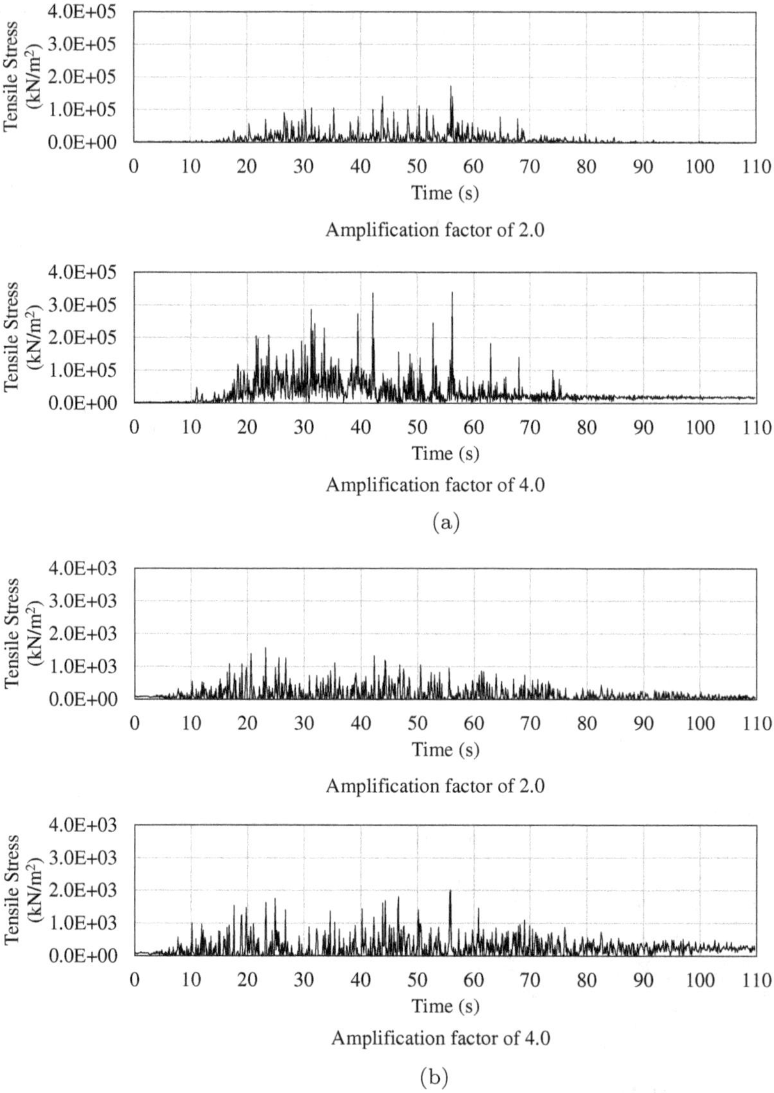

Fig. 19. The time series of tensile stress in steel reinforcement. (a) The element with largest stress, and (b) the element with smallest stress.

stress in reinforcement takes over this stress, which appears as residual stress after the input ground motion becomes zero. It should be noted that the yield stress of steel is 345,000 (kN/m^2) and the maximum value of the tensile stress in steel reinforcement of the element

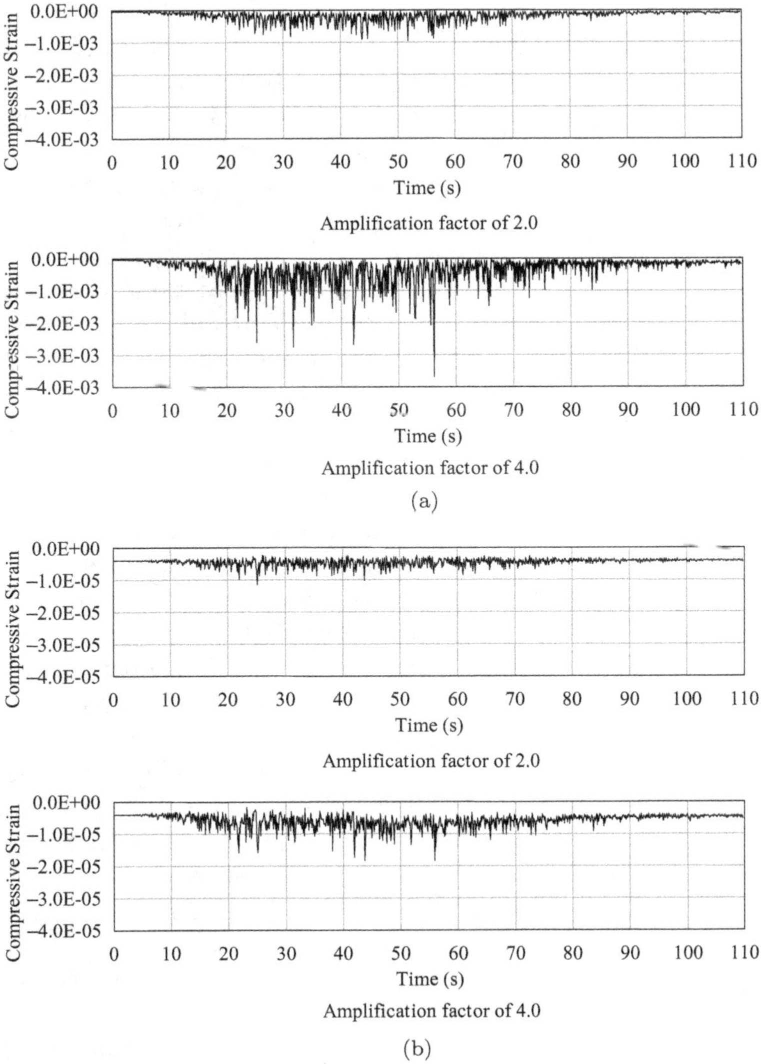

Fig. 20. The time series of compressive strain in concrete. (a) The element with largest stress, and (b) the element with smallest stress.

with largest stress is close to this value. We can expect an intact state of steel reinforcement even for the element with largest stress.

To further examine the state of concrete, Fig. 20 plots the time series of compressive strain at the element with largest and smallest stress; like compressive stress, compressive strain refers to the minimum of the principal strain. It is seen that residual compressive strain is generated for the amplification factor of 4.0. The amplitude of the residual strain remains less than 0.001, indicating that the concrete can carry some compressive stress even though its strength is considerably reduced. For the amplification factor of 2.0, the compressive residual strain is of the order of 0.0001 even for the element with largest stress. This confirms that the concrete is not severely damaged.

It is seen from Figs. 18 and 20 that considerable damages are induced in concrete for the amplification factor of 4.0. The damages can be critical as near-by reinforcement steel is about to yield; see Fig. 19. The full-scale simulation shows that while the reactor building is intact for the amplification factor of 2.0 and 4.0, the input ground motion of the amplification factor of 4.0 is critical as a slight increase of the amplification may result in the yielding of steel reinforcement at 60 s and local damages can spread to the surrounding parts. Another numerical simulation is needed to examine the occurrence of such overall failure induced by local steel reinforcement's yielding, by increasing the amplification factor of the input ground motion. Indeed, the ultimate safety of the nuclear power plant can be estimated by gradually increasing the amplification factor.

The soundness of steel reinforcement does not indicate the occurrence of overall failure induced by the local damages of concrete. Drawing such a conclusion is important for the full-scale simulation since it is aimed at extrapolating seismic responses subjected to experienced ground motion to those subjected to extremely strong ground motion which is not experienced. The conclusion is based on the soundness of steel reinforcement. Although the input ground motion is at the level of inexperience, the fact that structures with sound steel reinforcement are intact can be applied to the seismic structural response analysis for such strong ground motion. Stable numerical computation of the parallel FEM is essentially important to draw this conclusion. This is because the stress and strain states of steel and concrete computed for the amplification factor of 4.0 remain in the nonlinear material constitutive relation implemented into the parallel FEM.

We next examine the overall responses of the reactor building. Figures 21 and 22 show the distribution of the maximum compressive stress in concrete and the maximum tensile stress in steel reinforcement, respectively; as mentioned, the maximum stress is computed for the maximum value of principal values of stress, i.e., compressive or tensile stress for a negative or positive principal value, respectively.

As is seen in Fig. 21, there are some parts in OS of the reactor building where concrete is damaged for the amplification factor of 4.0; some parts of OS are damaged for the amplification factor of 2.0, but the degree of damage is much smaller than that for the amplification factor of 4.0. As shown in Fig. 22, there are some parts of OS where steel reinforcement is damaged for the amplification factor of 4.0; all steel reinforcements remain sound for the amplification factor of 2.0.

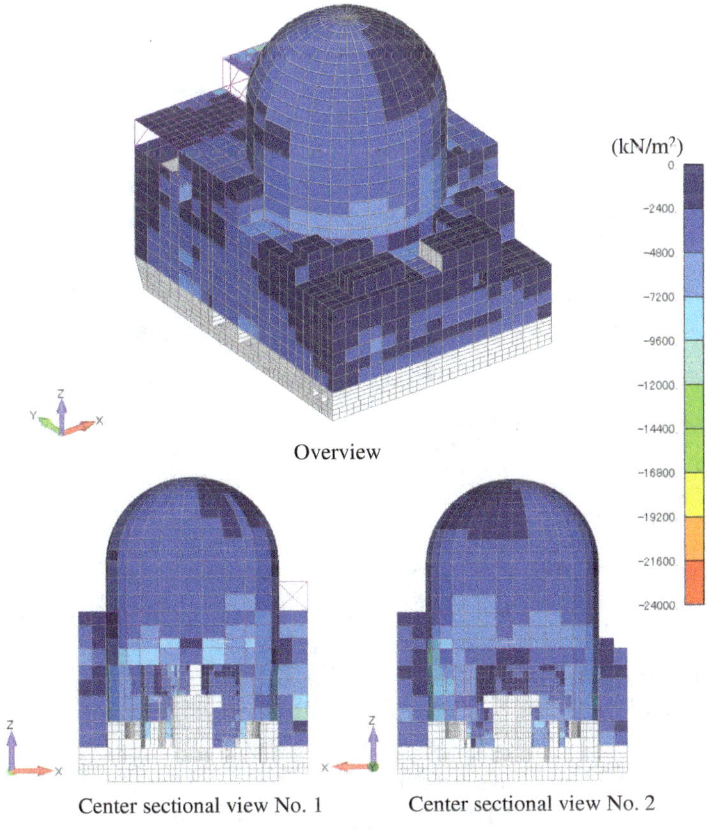

Fig. 21. The distribution of the maximum compressive stress in concrete. (a) Amplification factor of 2.0. (b) Amplification factor of 4.0.

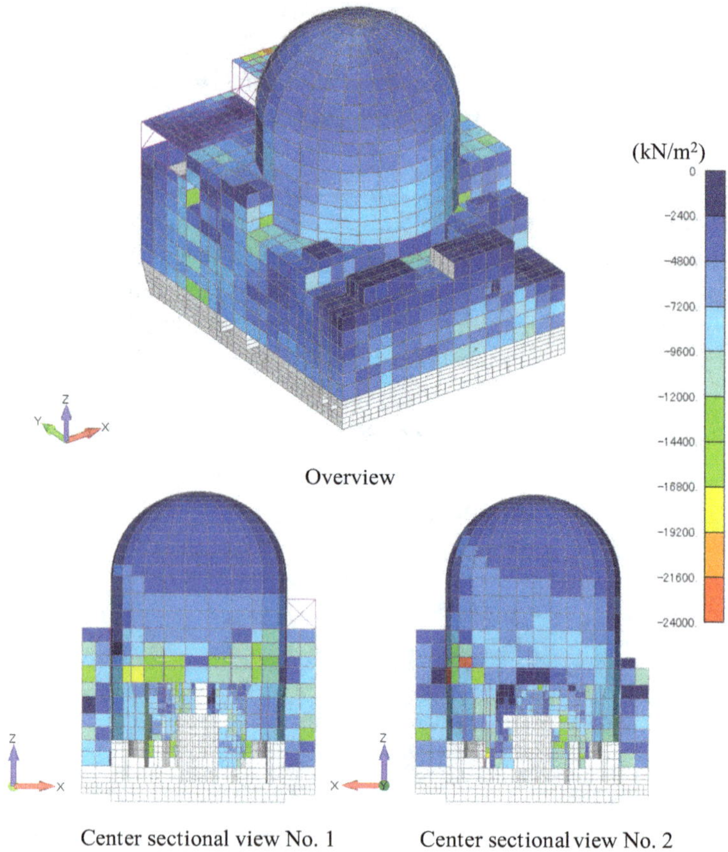

(b)

Fig. 21. (*Continued*)

The location of the damaged parts of concrete and steel reinforcement coincides with the crack distribution of the trial simulation of the amplification factor of 1.5; see Fig. 9.

It should be noted that IC remains intact even for the amplification factor of 4.0. The results of the full-scale simulation confirm the higher seismic safety of IC; see Figs. 21 and 22.

To examine the details of local concrete damages, Fig. 23 show the distribution of the fracture parameter of concrete. The fracture parameter defined in Appendix A is 1 for non-damaged concrete and 0 for fully damaged concrete. This parameter is used as a measure of the degree of concrete damage. If the fracture parameter is over, say, 0.95, concrete has certain strength for compression, tension, and shearing. As is seen, the fracture parameters are sufficiently large in the entire reactor building even for the multiplication factor of 4.0. The fracture parameter takes on the minimum value of 0.89 at the connecting part of OS and IC for the amplification factor of 4.0. The distribution of the fracture parameters supports that while local failure takes place at several parts, the reactor building remains intact even though it is subjected to the extremely strong ground motion of the multiplication factor of 4.0.

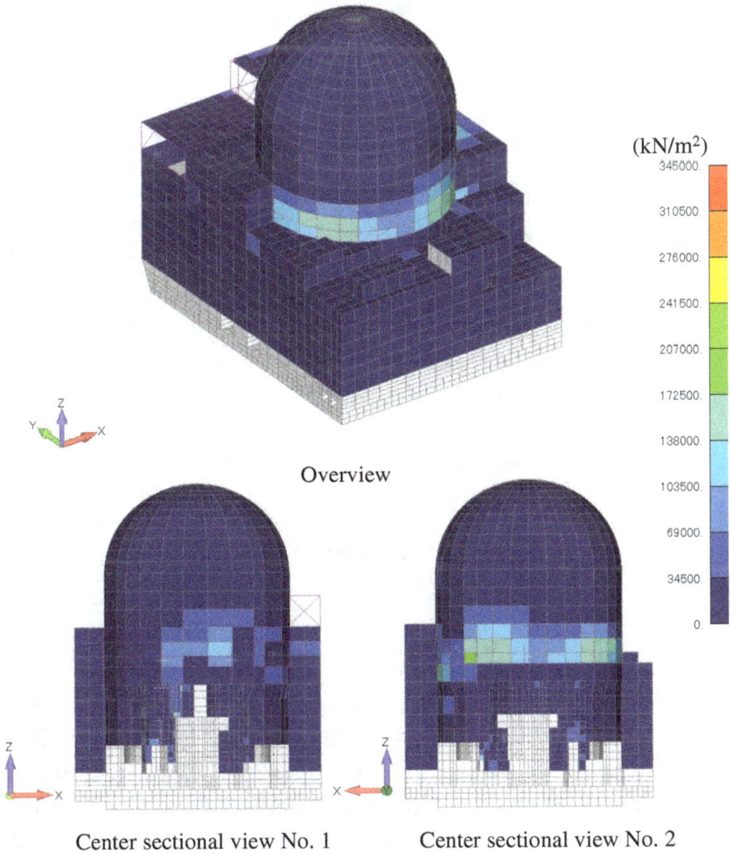

(a)

Fig. 22. The distribution of the maximum tensile stress in steel. (a) Amplification factor of 2.0. (b) Amplification factor of 4.0.

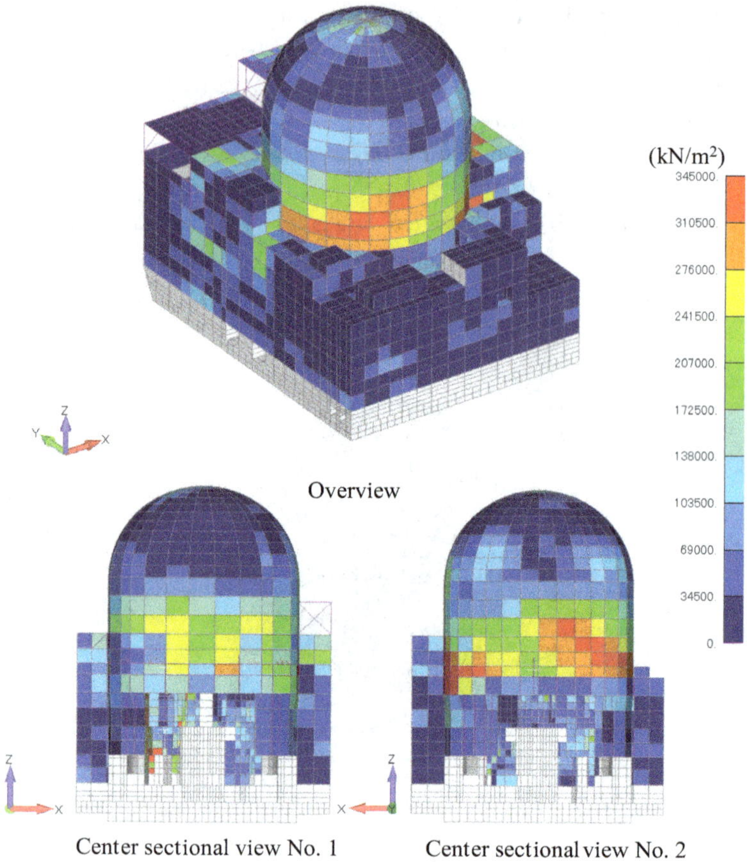

(b)

Fig. 22. (*Continued*)

To examine the stress and strain states of concrete at the element with largest stress, we plot the relationship between the compressive strain and stress during the seismic response in Fig. 24; the compressive strain and stress are the minimum value of the principal strain and stress. While it appears more or less linear for the amplification factor of 2.0, the relationship is complicated for the amplification factor of 2.0. This complicated relationship is due to the material nonlinearity that strongly depends on the history of loading, i.e., cycles of loading and unloading in various directions.

It is certainly true that the degree of nonlinearity for the relationship between the compressive strain and stress is large, in particular for the amplification factor of 4.0, as shown in Fig. 24(b). However, the peak value of the compressive stress is 2.48×10^4 kN/m^2 and does not reach the compressive strength, 4.92×10^4 kN/m^2; see Table 1. This result also confirms that while it is damaged, concrete of this element still possesses certain strength. Also, it is confirmed that the state of concrete stays in the regime of the material constitutive relation implemented into the parallel FEM.

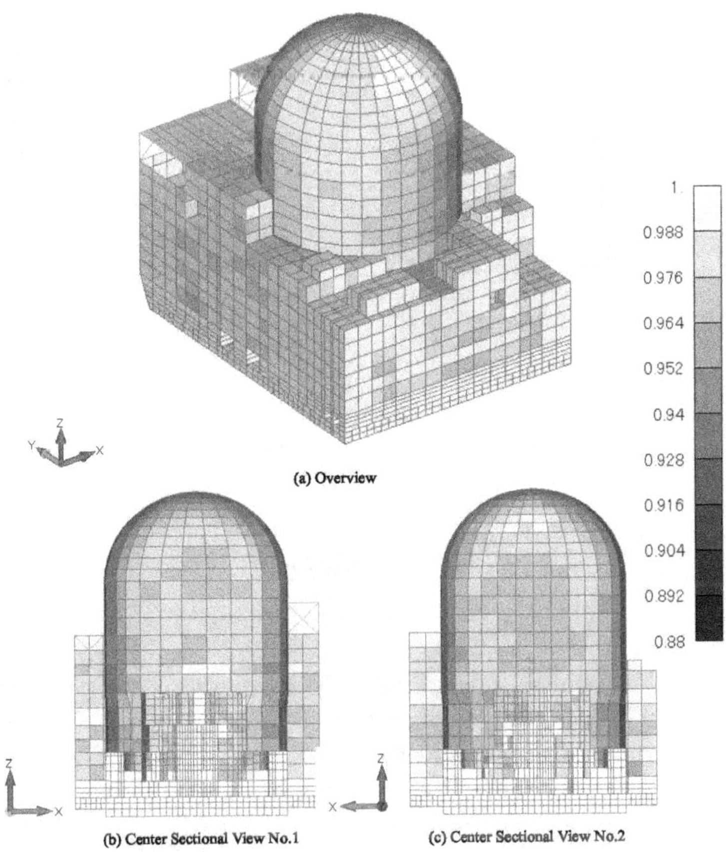

(a)

Fig. 23. The distribution of the fracture parameter of concrete. (a) Amplification factor of 2.0. (b) Amplification factor of 4.0.

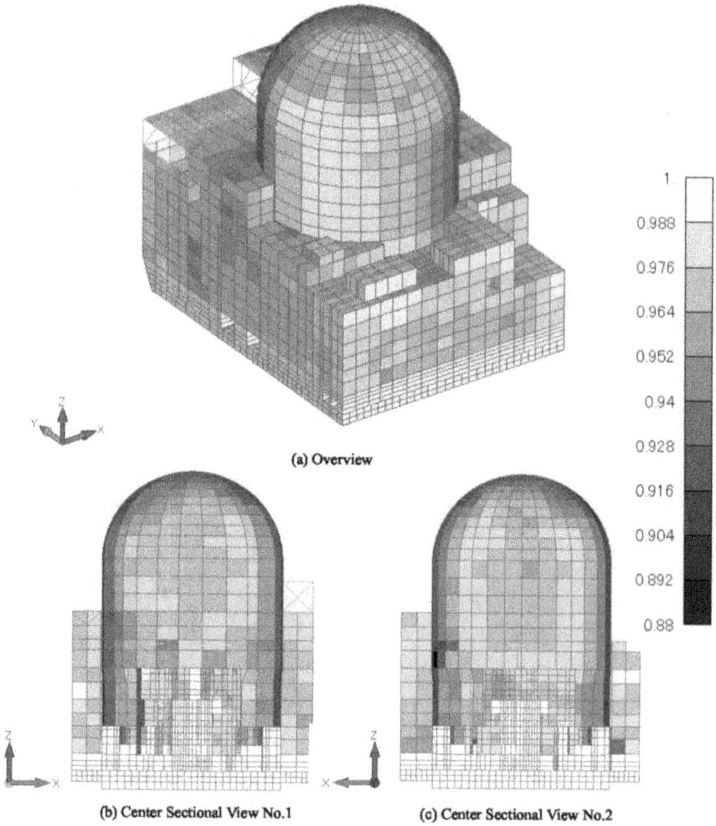

(b)

Fig. 23. (*Continued*)

Using the results of the full-scale simulation, we can make an element-wise evaluation of concrete soundness in terms of the fracture parameter shown in Fig. 23 or the stress–strain relationship shown in Fig. 24. As is seen, processing the simulation results properly, we can confirm the sound or damaged state of concrete at the element level using various physical quantities. However strong a ground motion is input, this element-wise evaluation is reliable because it is based on the evaluation of the material properties which are independent of the severity of the input ground motion. A numerical analysis

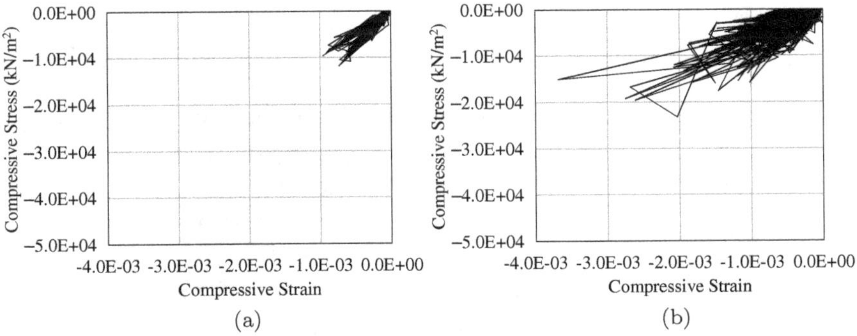

Fig. 24. The stress–strain relationship of the element with largest stress. (a) Amplification factor of 2.0, and (b) amplification factor of 4.0.

model of high fidelity enables us to examine such an element-wise (or point-wise) evaluation of seismic safety.

In closing this section, we discuss the usefulness of the parallel FEM. While the strong ground motion input is extremely strong and not experienced, it can complete the numerical analysis. The improvement of the nonlinear analysis algorithm succeeds in making stable numerical computations. The numerical results obtained seem reasonable, i.e., local failure occurs at some parts, but they do not lead to overall failure because local failure is only for concrete, and the steel reinforcement remains intact. It should be emphasized that by gradually increasing the amplification factor, we can observe the occurrence and progress of local damages, which eventually induce overall failure. Another merit of the full-scale simulation is that the gradual spread of the local damages due to the increase of input ground motion can be observed.

5. Concluding Remarks

Examining the results of the trial simulations for an extremely strong ground motion, we improve the algorithm for solving nonlinear matrix equations that is implemented in the parallel FEM. It is shown that due to this improvement, the computation of the seismic structural response analysis is completed even for an extremely strong ground motion multiplied by four times. There are some parts of the reactor building where concrete is damaged due to tensile cracking. However, the results showed that compressive strength remained in concrete, and the seismic safety of the reactor building for these extremely strong ground motions is confirmed in the full-scale simulation.

The input ground motion used in this chapter is much stronger than the ground motions which were observed. In this sense, the seismic response analysis presented here is regarded as an extrapolation of the numerical analysis that is made for experienced ground motions. The parallel FEM using the high-fidelity numerical analysis model is applicable to evaluate the seismic safety for such extremely strong ground motions since the evaluation is made element-wise and based on the material properties of concrete or steel which are independent of the input ground motion. The accurate numerical computation which is achieved by the parallel FEM is essential to achieve this element-wise evaluation of the reactor building.

It is necessary to increase the spatial and temporal resolution of the seismic structural response analysis for the safety evaluation of facilities that are placed in the reactor building; the natural frequency of the facilities is higher than that of the building. To meet this demand, we need higher computation performance of the parallel FEM and a numerical analysis model of higher fidelity. Further development is needed to improve the numerical computation performance of the parallel FEM and the method of constructing a high-fidelity numerical analysis model.

Appendix A. Constitutive Relation Implemented into Parallel FEM

To simplify expression, we use symbolic notation in this appendix. For instance, strain and stress tensors are denoted by $\boldsymbol{\epsilon}$ and $\boldsymbol{\sigma}$, elastic and plastic strain by $\boldsymbol{\epsilon}^e$ and $\boldsymbol{\epsilon}^p$, and strain increment by $\mathrm{d}\boldsymbol{\epsilon}$. Inner product, second-order contraction, and tensor product are denoted by $\cdot$, $:$, and $\otimes$, respectively.

A.1. *Concrete constitutive relation*

The 3D tensorial constitutive relation of concrete is based on the following three uniaxial experimental data: (1) compression; (2) tension;

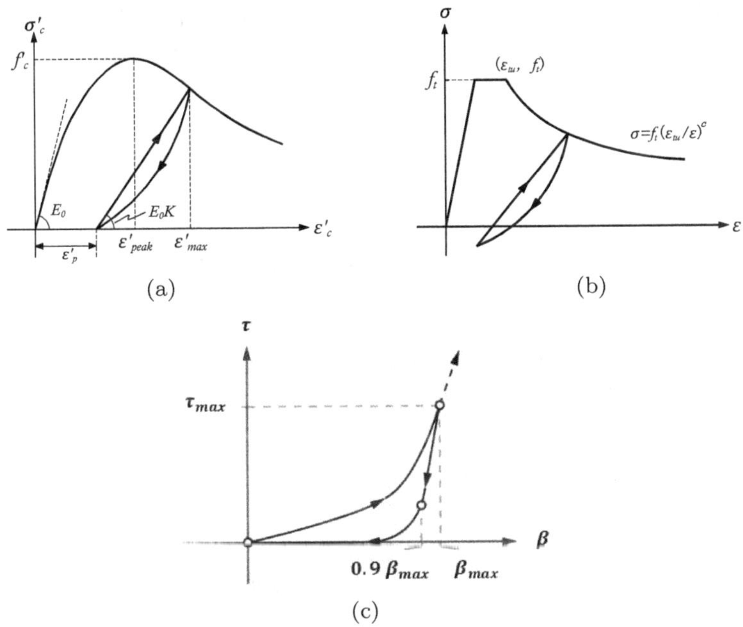

Fig. A.1. Uniaxial concrete constitutive relation. (a) Compression, (b) tension, and (c) shearing.

and (3) shearing. Figure A.1 shows the schematic view of the three uniaxial stress–strain relations; see Refs. 44 and 45. We formulate the constitutive relation following the standard plastic flow rule, by transforming the uniaxial experimental observation to a *yield function*, denoted by f, which describes plastic strain as its derivative and constraints stress and strain state so that $f = 0$ for elastoplastic deformation and $f < 0$ for elastic deformation.

In terms of a yield function f, the concrete constitutive relation is expressed as the following relation between $\mathrm{d}\boldsymbol{\sigma}$ and $\mathrm{d}\boldsymbol{\epsilon}$:

$$\mathrm{d}\boldsymbol{\sigma} = \boldsymbol{c}^{\mathrm{EP}} : \mathrm{d}\boldsymbol{\epsilon}, \tag{A.1}$$

where $\boldsymbol{c}^{\mathrm{EP}}$ is elasto-plasticity that is determined by f. For instance, if f is a function of $\boldsymbol{\sigma}$, it determines plastic strain increment as

$$\mathrm{d}\boldsymbol{\epsilon}^p = \mathrm{d}\lambda \frac{\partial f}{\partial \boldsymbol{\sigma}},$$

with $\mathrm{d}\lambda$ being a plastic loading parameter. Denoting elasticity tensor by $\boldsymbol{c}$, we can derive an explicit expression of $\boldsymbol{c}^{\mathrm{EP}}$ in terms of f as

$$\boldsymbol{c}^{\mathrm{EP}} = \boldsymbol{c} - \frac{(\boldsymbol{c}\!:\!\frac{\partial f}{\partial \boldsymbol{\sigma}}) \otimes (\boldsymbol{c}\!:\!\frac{\partial f}{\partial \boldsymbol{\sigma}})}{\frac{\partial f}{\partial \boldsymbol{\sigma}}\!:\!\boldsymbol{c}\!:\!\frac{\partial f}{\partial \boldsymbol{\sigma}}}. \tag{A.2}$$

If f depends on both $\boldsymbol{\sigma}$ and $\boldsymbol{\epsilon}$, an explicit expression of $\boldsymbol{c}^{\mathrm{EP}}$ can be derived; the expression of $\boldsymbol{c}^{\mathrm{EP}}$ becomes more complicated than Eq. (A.2).

A yield function for compression is a function of normal stress and strain component, denoted by σ and ϵ, and given as

$$f(\sigma, \epsilon) = \begin{cases} \sigma - \omega K_0 E_{co}(\epsilon - \epsilon^p) & \text{loading,} \\ \sigma - \omega \left(\sigma_{\max} - (\sigma_{\max} - \sigma_0) \dfrac{\epsilon_{\max} - \epsilon}{\epsilon_{\max} - \epsilon_o} \right) & \text{reloading,} \\ \sigma - \omega K_0 E_{co}(\epsilon - \epsilon_p))\alpha & \text{unloading,} \end{cases} \tag{A.3}$$

where ω and E_{co} are reduction constant and initial Young's modulus; $\sigma_{\max}$ and $\epsilon_{\max}$ are the maximum value of experienced stress and strain in compression; σ_0 and ϵ_0 are the present stress and strain; and ϵ_p, K_0, and α are given as

$$\epsilon_p = \left(\frac{\epsilon}{\epsilon_c} - \frac{20}{7} \left(1 - \exp\left(-0.35 \frac{\epsilon}{\epsilon_c} \right) \right) \right) \epsilon,$$

$$K_0 = \exp\left(-0.73 \frac{\epsilon}{\epsilon_c} \left(1 - \exp\left(-1.25 \frac{\epsilon}{\epsilon_c} \right) \right) \right),$$

$$\alpha = \text{slop} + \left(\frac{\sigma}{K_0 E_{co}(\epsilon_0 - \epsilon_p)} - \text{slop} \right) \left(\frac{\epsilon - \epsilon_p}{\epsilon_0 - \epsilon_p} \right)^{\mathrm{PN}},$$

with slop $= 2$ and PN $= 2$. This K_0 is the fracture parameter used in Section 3.

Like the one for compression, a yield function for tension is a function of normal stress and strain and is given as

$$f(\sigma,\epsilon)=\begin{cases}\sigma - f_t\left(\frac{\epsilon_{tu}}{\epsilon}\right)^c & \text{loading,}\\ \sigma - \left(\sigma_{t\max} - (\sigma_{t\max}-\sigma_{b0}-\sigma_{0cb})\dfrac{\epsilon_{t\max}-\epsilon}{\epsilon_{t\max}-\epsilon_o}\right) & \text{reloading,}\\ \sigma - E_{bo}(\epsilon-\epsilon_p))\alpha + \sigma_{b0} & \text{unloading,}\end{cases} \tag{A.4}$$

where f_t, c, and ϵ_{tu} are tensile strength, a coefficient for bonding, and critical strain for cracking, respectively; $\sigma_{t\max}$ and $\epsilon_{t\max}$ are the maximum value of experienced stress and strain in tension; ϵ_0 is the present strain; and σ_{b0} and E_{bo} are

$$\sigma_{b0} = -f_t\left(0.05 + 0.15\frac{\Delta\epsilon_{t\max}}{\epsilon_t}\right),$$

$$E_{bo} = \frac{\sigma_{t\max}-\sigma_{bo}}{\Delta\epsilon_{t\max}},$$

with $\Delta\epsilon_{t\max}$ being the maximum tensile strain component. Note that ϵ_p and α, which are shared by the yield function for compression (i.e., Eq. (A.3)), use slop = 2 and PN = 2 for tension.

A yield function for shearing is a function of shear stress and strain components, denoted by τ and γ, respectively, and normal strain component, denoted by ϵ, and given as

$$f(\tau,\gamma,\epsilon)=\begin{cases}\tau - f_{st}\dfrac{\beta^2}{1+\beta^2} & |\beta| > \beta_{\max},\\ \tau - \tau_{\max}\dfrac{\beta-\beta_{\max}}{0.15\beta_{\max}} & \beta_{\max} > |\beta| > 0.9\beta_{\max},\\ \tau - \dfrac{1}{3}\tau_{\max}\left(\dfrac{\beta}{0.9\beta_{\max}}\right)^9 & 0.9\beta_{\max} > |\beta|,\end{cases} \tag{A.5}$$

where $\beta = \frac{\gamma}{\epsilon}$; f_{st} is the shear strength on crack surfaces, given as $f_{st} = 3,8f_c^{1/3}$; and $\tau_{\max}$ and $\beta_{\max}$ are the maximum value of β and sheer stress, respectively.

A.2. *Soil constitutive relation*

Like concrete, the 3D tensorial constitutive relation of soil is based on the 1D shear experiment, which is given as the relation between shear stress and strain component, denoted by τ and γ, respectively. A multi-spring mechanism model regards this relation as a nonlinear shear spring and derives a 3D constitutive relation by distributing the springs in every direction; see, for instance, Iai *et al.*[46] and Iai and Kameoka.[47]

In a 2D setting, the distribution of shear springs is easily understood. Let the relation between τ and γ be expressed as

$$\tau = g(\gamma, \text{history of } \gamma), \tag{A.6}$$

and the increment of τ is given as $\mathrm{d}\tau = \frac{\partial g}{\partial \gamma}\mathrm{d}\gamma$. Denoting by $\boldsymbol{s}$ the direction rotated 90 degrees counterclockwise from $\boldsymbol{n}$, τ and γ are expressed in terms of $\boldsymbol{\sigma}$ and $\boldsymbol{\epsilon}$ as

$$\tau = \frac{1}{2}(\boldsymbol{n}\otimes\boldsymbol{s}+\boldsymbol{s}\otimes\boldsymbol{n}):\boldsymbol{\sigma}, \quad \gamma = \frac{1}{2}(\boldsymbol{n}\otimes\boldsymbol{s}+\boldsymbol{s}\otimes\boldsymbol{n}):\boldsymbol{\epsilon}.$$

Hence, we can determine stress increment, $\mathrm{d}\boldsymbol{\sigma}$, integrating Equation (A.6) as

$$\mathrm{d}\boldsymbol{\sigma} = \frac{1}{\pi}\int_0^{\pi} \frac{\partial g}{\partial \gamma}\mathrm{d}\gamma(\boldsymbol{n}\otimes\boldsymbol{s}+\boldsymbol{s}\otimes\boldsymbol{n})\,\mathrm{d}\theta, \tag{A.7}$$

where θ determines components of $\boldsymbol{n}$ in a certain Cartesian coordinate as $(\cos\theta, \sin\theta)$. Note that the component of $\boldsymbol{s}$ is $(-\sin\theta, \cos\theta)$.

A simple way to extend Eq. (A.7) to a 3D setting is to consider various planes and carry out integration on each of the planes. Since two parameters are needed to determine one plane, this results in a triple integral to compute $\mathrm{d}\boldsymbol{\sigma}$. In this expression, one shear spring, the direction of which is given as $\boldsymbol{n}$, is used on all planes which include $\boldsymbol{n}$ on it. To make the contribution of one shear spring one time and to reduce the triple integral to the double integral, we use the following equation:

$$\mathrm{d}\boldsymbol{\sigma} = \frac{3}{4\pi}\int_{\omega} \frac{\partial g}{\partial \gamma}\mathrm{d}\gamma(\boldsymbol{n}\otimes\boldsymbol{s}+\boldsymbol{s}\otimes\boldsymbol{n})\,\mathrm{d}s, \tag{A.8}$$

where ω stands for a spherical surface and a point on the surface determines $\boldsymbol{n}$, and $\boldsymbol{s}$ is the direction of the maximum shear strain component, i.e., $\boldsymbol{s}$ is parallel to the direction of a vector given as $d\epsilon \cdot \boldsymbol{n}$. As is seen, a shear spring in the $\boldsymbol{n}$ direction appears once in the integration, and the right-hand side of Eq. (A.8) is a double integral since two parameters are used to describe $\boldsymbol{n}$.

A.3. *Model for contact surface between building and ground*

For the lift-up analysis which considers detaching/attaching the reactor building on the ground beneath induced by strong vertical ground motion, we need to develop a mathematical model for the contact surface between the building and the ground. Physically, this model is simple, as the mechanism of detaching/attaching is friction. In terms of the traction on the contact surface and the displacement gap, denoted by $\boldsymbol{t}$ and $\Delta\boldsymbol{u}$, respectively, the model is expressed as

$$\begin{cases} \boldsymbol{t} = 0 & \text{detaching,} \\ \Delta u_n = 0, \Delta\boldsymbol{t}_s = \boldsymbol{0} & \text{attaching,} \end{cases} \tag{A.9}$$

where $\Delta u_n = \boldsymbol{n} \cdot \Delta\boldsymbol{n}$ and $\Delta\boldsymbol{t}_s = \Delta\boldsymbol{t} - (\boldsymbol{n} \cdot \Delta\boldsymbol{t})\boldsymbol{n}$, with $\boldsymbol{n}$ being the unit normal of the contact surface, are the normal component of the displacement gap and the traction vector parallel to the contact surface.

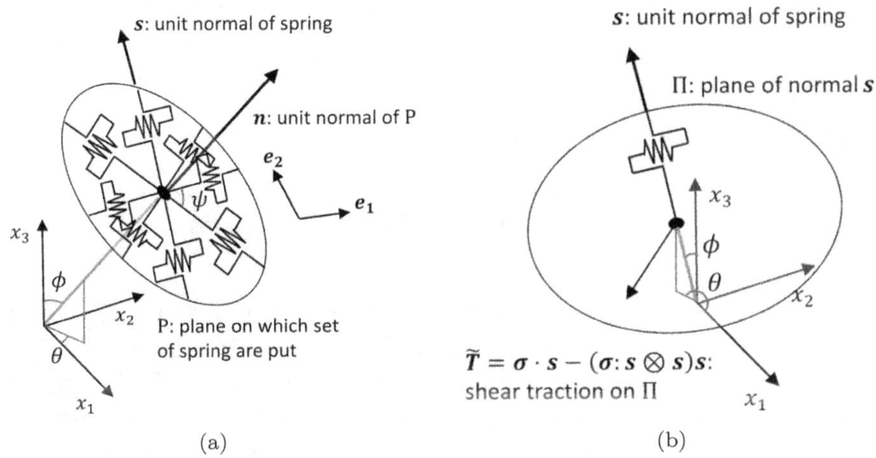

Fig. A.2. Multi-spring mechanism model of soil. (a) Ordinary model, and (b) reformulated model.

While Eq. (A.9) is simple, it is difficult to numerically compute this model due to the non-smooth change in $\boldsymbol{t}$ and $\Delta\boldsymbol{u}$. To explain the non-smooth change, we consider the relation between t_n and Δu_n, where $t_n = \boldsymbol{n}\cdot\boldsymbol{t}$ is the normal component of the traction. The slop of the t_n–Δu_n relation is 0 for $\Delta u_n > 0$ since $t_n = 0$. But the slope becomes infinite at $\Delta u_n = 0$ since no penetration of the building into the ground is allowed. A non-smooth change similar to this t_n–Δu_n relation happens for the $\boldsymbol{t}$–$\Delta\boldsymbol{u}$ relation.

To avoid the non-smooth change in the $\boldsymbol{t}$-$\Delta\boldsymbol{u}$ relation due to detaching/attaching, an artificial nonlinear spring is introduced as a mathematical model of the contact surface; see, for instance, Schellekens and De Bors.[48] We consider slip on the contact surface, which is given as the relation between t_s and Δu_s; denoting by $\boldsymbol{s}$ the direction of slip, we can express $\boldsymbol{t}$ and $\Delta\boldsymbol{u}$ as $\boldsymbol{t} = t_n\boldsymbol{n} + t_s\boldsymbol{s}$ and $\Delta\boldsymbol{u} = \Delta u_n\boldsymbol{n} + \Delta u_s\boldsymbol{s}$ with t_s and Δu_s being the components in the $\boldsymbol{s}$-direction. The mathematical model is thus formulated as

$$t_n = \begin{cases} k_n\Delta u_n & t_n < \sigma_t/k_n, \\ 0 & \Delta u_n > \sigma_t/k_n, \end{cases}$$

$$t_s = \begin{cases} k_s\Delta u_s & \Delta u_s < \tau_{ys}/k_s, \\ \tau_{yd} & \Delta u_s > \tau_{ys}/k_s, \end{cases} \tag{A.10}$$

where k_n and k_s are normal and shear spring constants, and σ_t, τ_{yd}, and τ_{ys} are the normal strength, dynamic shear strength, and static shear stress, respectively. Figure A.3 shows this mathematical model.

To implement Eq. (A.10) into the module for the contact surface model, we employ the following functional for the displacement gap:

$$U[\Delta\boldsymbol{u}] = \int_S \frac{1}{2}\Delta\boldsymbol{u}\cdot(k_n\boldsymbol{n}\otimes\boldsymbol{n} + k_s\boldsymbol{s}\otimes\boldsymbol{s})\cdot\Delta\boldsymbol{u}\,\mathrm{d}s, \tag{A.11}$$

where S stands for the contact surface. Note that $k_n\boldsymbol{n}\otimes\boldsymbol{n}$ and $k_s\boldsymbol{s}\otimes\boldsymbol{s}$ represent the normal and shear springs, respectively, and Eq. (A.11) is for the case when the normal and shear springs work. If the spring does not work, we replace $k_n\boldsymbol{n}\otimes\boldsymbol{n}$ and $k_s\boldsymbol{s}\otimes\boldsymbol{s}$ with a suitable function. As is seen, $U[\Delta\boldsymbol{u}]$ uses surface integral, which can be exactly computed by using Gauss integral when the discretization of FEM is applied.

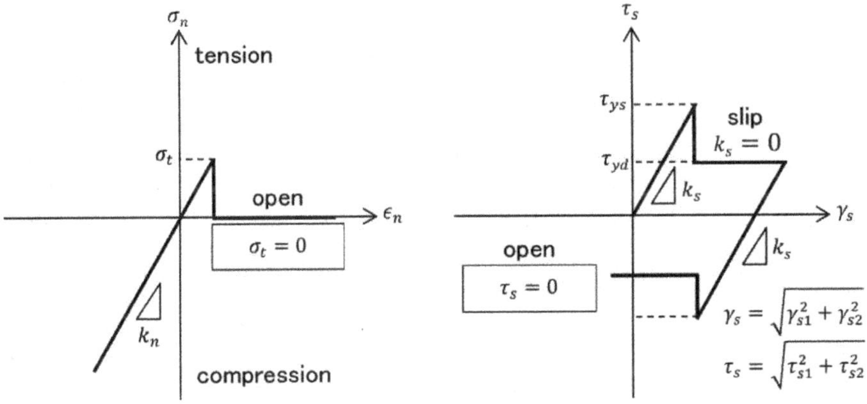

Fig. A.3. Contact surface model.

Appendix B. Capability of Liquefaction Analysis

In general, a liquefaction problem is divided into two groups, namely, a coupled problem of the soil deformation and the pore water flow and an uncoupled problem of the soil deformation accompanying the change in pore water pressure. Theoretically, the coupled problem can make a more accurate prediction for the occurrence and scale of liquefaction; the scale refers to the area and subsidence of liquefaction. The problem of soil deformation is used for the prediction of the occurrence (or non-occurrence) of liquefaction, making the safety-side setting of a numerical analysis model and material parameters.

Into the parallel FEM, the capability of solving the uncoupled problem of the soil deformation is implemented. In terms of soil displacement and stress, denoted by $\boldsymbol{u}$ and $\boldsymbol{\sigma}$, and pore water pressure, denoted by p, this problem is formulated as follows:

$$\begin{aligned} \rho\ddot{\boldsymbol{u}} - \boldsymbol{\nabla}\cdot\boldsymbol{\sigma} + \boldsymbol{\nabla}p = \boldsymbol{0}, \\ \dot{\epsilon} = f(p), \end{aligned} \tag{B.1}$$

where ρ is the soil density and ϵ is the soil volumetric strain; $\boldsymbol{\nabla}\cdot$ and $\boldsymbol{\nabla}$ stand for the divergence and gradient, respectively; and ϵ is computed from strain given as the symmetric part of the displacement gradient $(\boldsymbol{\nabla}\boldsymbol{u})$.

We present an example of the liquefaction analysis in this appendix; see, for instance, Chen *et al.*,[49] Hotta *et al.*,[50] Elgamal *et al.*,[51] Matsuo *et al.*,[52] Tamari *et al.*,[53] and Arango *et al.*[54] for the application of large-scale numerical computation to the liquefaction analysis. The target is a standard model of a nuclear power plant of US Department of Energy which consists of a reactor building and the ground. A 3D high-fidelity model of over 1,000,000 degrees-of-freedom is constructed for this plant; the number of nodes and elements is 482,880 and 595,727, respectively. The numerical analysis model is shown in Fig. B.1; a Cartesian coordinate system of (X, Y, Z) is employed. The input ground motion is shown in Fig. B.2; this motion is input simultaneously in the three directions, but the amplitude is set to half in the vertical direction.

To carry out a liquefaction analysis, we employ a two-layer ground model, consisting of a sand layer and bedrock; it is assumed that the sand layer is fully saturated, and liquefaction can take place there if sufficiently strong ground motion is input. Material properties and numerical setting are summarized in Tables B.1 and B.2, respectively. It should be noted that liquefaction characteristics' parameters such as the phase transformation angle, the overall dilatancy, the initial and final phases of dilatancy, the threshold limit of dilatancy, and the ultimate limit of dilatancy must be specified for the sand layer.

As explained in Appendix A, we reduce the triple integral of the original multi-spring mechanism model to the double integral. We numerically examine the validity of this reduction even though the expression of the double integral, Eq. (A.8), is mathematically derived from the expression which uses the triple integral. Figure B.3 shows the comparison of the shear stress–strain relation in the upper and lower part of the sand layer. Two components of horizontal shear stress and strain are used; (a) and (b) are the responses of the YZ- and ZX-components in the upper sand layer, and (c) and (d) are the responses of the YX- and ZX-components in the lower sand layer. As is seen, the relations computed by the double integral do not perfectly agree with those computed by the triple integral; the difference increases as the number of shear loading cycles

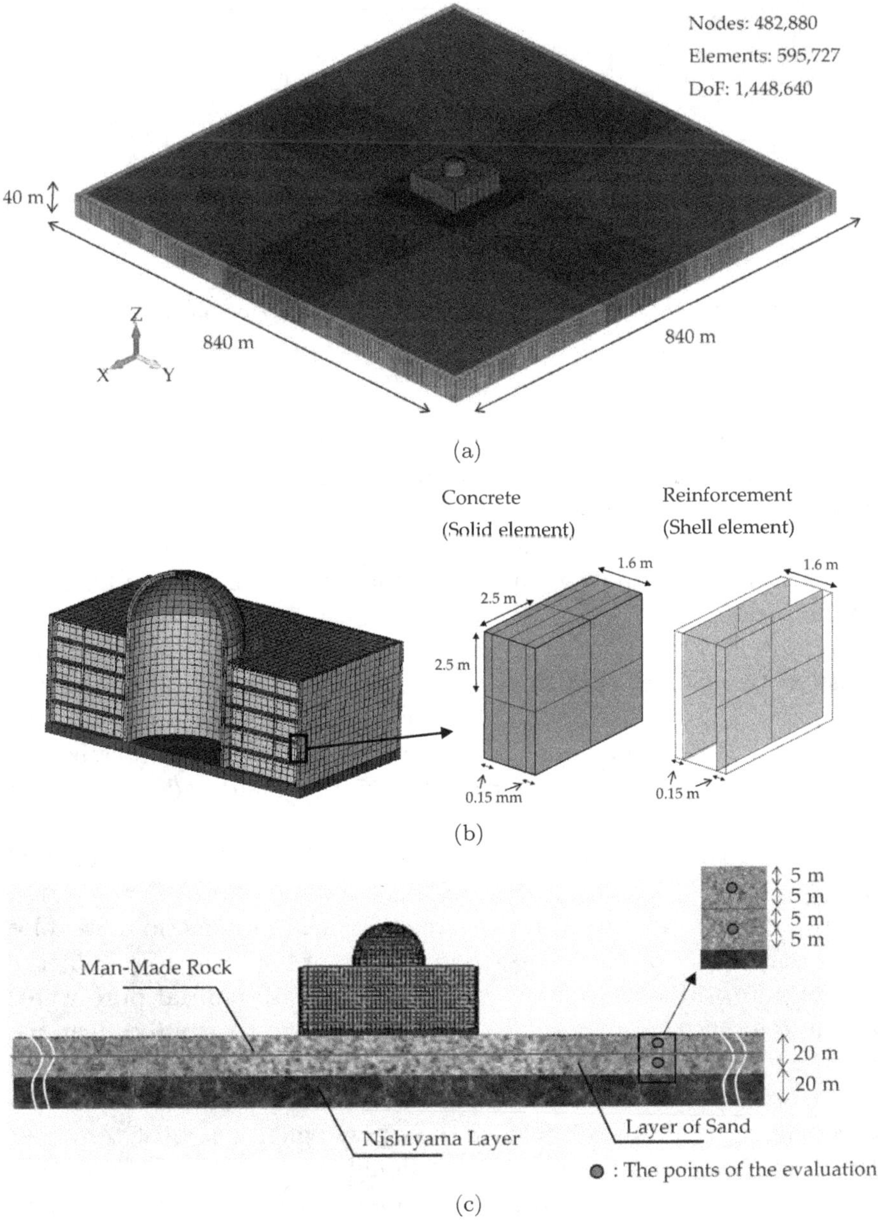

Fig. B.1. A numerical analysis model used for the liquefaction analysis. (a) Overview, (b) reactor building, and (c) soil and rock.

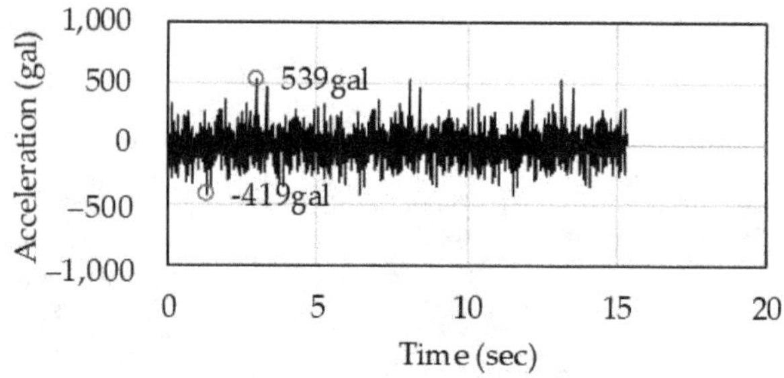

Fig. B.2. Input ground motion used for the liquefaction analysis.

increases. However, the difference is negligible; the maximum difference in shear strain is of the order of 0.0001 when the shear stress is 30 kN/m^2. The difference might be due to the numerical computation of the triple integral, as it limits the number of planes on which springs are two-dimensionally distributed for efficient numerical computation.

Figure B.4 shows the shear strain distribution on an artificial cross-section which cuts the reactor building and the ground when the numerical computation is completed; the shear strain is the horizontal XZ-component. It is seen that large shear strains are distributed along the interface between the man-made rock layer and the natural sand layer. The value of the maximum shear strain is of the order of 0.0001, suggesting a possible occurrence of liquefaction.

To further examine the occurrence of the liquefaction, we plot the distribution of the excess pore water pressure in Fig. B.5; the excess pore water ratio is computed by using the initial pore water as the reference. Higher excess pore waters are distributed near the interface between the sand layer and the bedrock layer, which is 20 m beneath the ground surface, and the maximum value of the excess pore water ratio is 0.52. Excess pore water is almost zero near the ground surface, suggesting that liquefaction does not occur near the ground surface, and larger shear deformation between the man-made rock and the sand layer is not caused by liquefaction in the sand layer.

Table B.1. Material properties.

(a) Deformation characteristics' parameters of the soil and rock

	Shear modulus	Bulk modulus	Reference confining pressure	Poisson's ratio	Reduction factor of shear and bulk modulus	Internal friction angle	Cohesion	Upper bound for hysteretic damping factor
Type	G_{ma} (kPa)	K_{ma} (kPa)	σ'_{ma} (kPa)	N	m	Φ_f (degree)	c (kPa)	$h_{\max}$
Layer of sand	207,000	540,000	190.0	0.33	0.5	45.0	0.0	0.155
Nishiyama layer	415,000	1,080,000	98.0	0.33	0.0	—	—	0.257

(b) Liquefaction characteristics' parameters of the soil

	Phase transformation angle	Overall dilatancy	Initial phase of dilatancy	Final phase of dilatancy	Threshold limit of dilatancy	Ultimate limit of dilatancy
Type	Φ_p (degree)	ω_1	p_1	p_2	c_1	S_1
Layer of sand	28.0	2.40	0.50	0.80	1.92	0.005

Table B.2. Setting of numerical analysis.

Category	Item	Numerical conditions
Nonlinear calculation	Iterative method	Newton Raphson method
	Convergence criterion	1.0×10^{-8}
	Maximum number of iterations	100
	Dynamic step	0.01 (s/step)
	Total step	1,536 (15.36 s)
Method to solve the inverse matrix	Solver	CG method
	Precondition	Slice Successive Over Relaxation
	Convergence criterion	1.0×10^{-8}
	Maximum number of iterations	100,000
	MPI processes	2-128

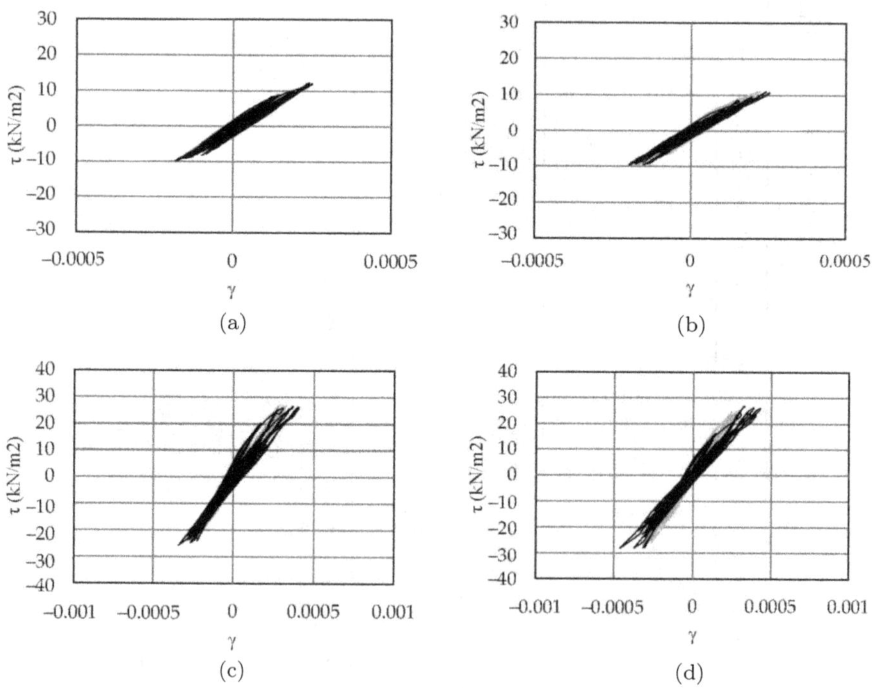

Fig. B.3. Comparison of shear stress–strain relation computed by the double and triple integrals.

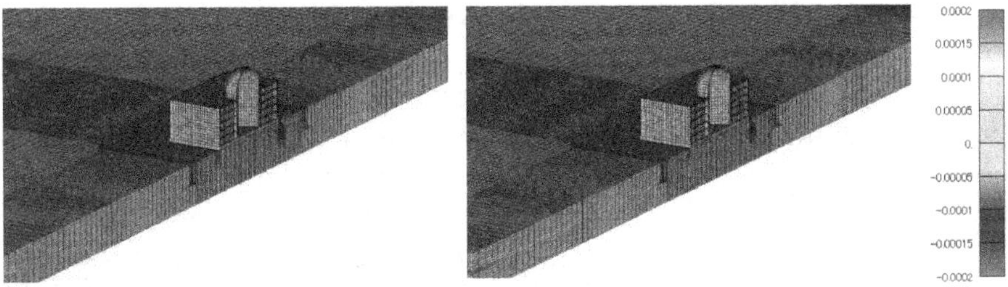

(a) (b)

Fig. B.4. Distribution of shear strain.

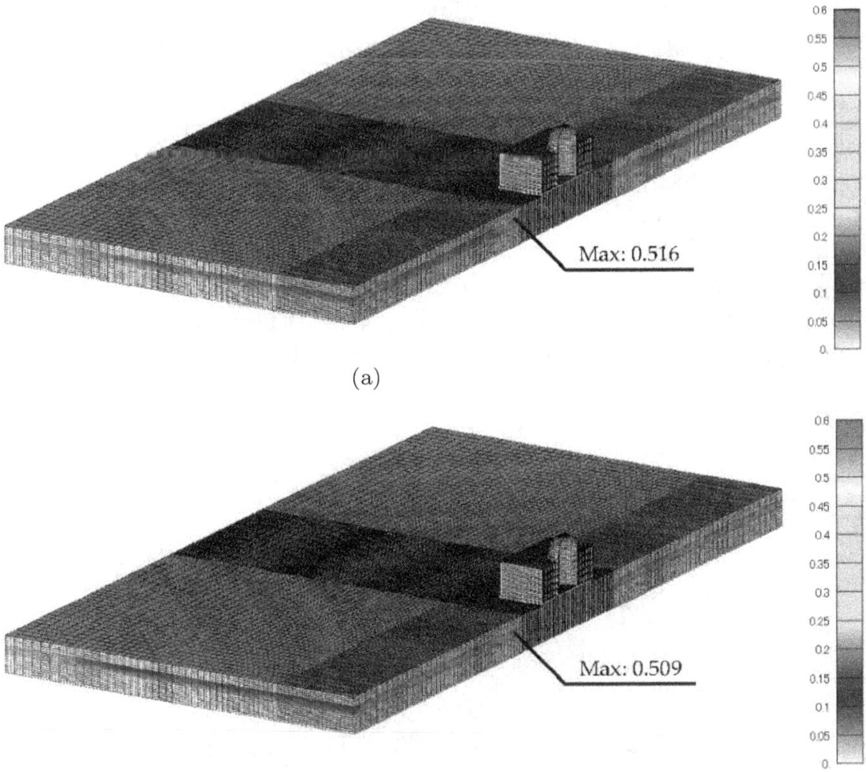

(b)

Fig. B.5. Distribution of the excess pore water pressure ratio. (a) Original model, and (b) proposed model.

Appendix C. Capability of Lift-Up Analysis

The lift-up analysis computes the structural responses induced by vertical ground motion that detaches the structure from the ground beneath and then attaches it to the ground. This detaching/attaching leads to an impulsive response of the structure, which could be a critical state for facilities located in it. The lift-up analysis uses the contact surface model explained in Appendix A, and the parallel FEM is capable of this analysis since the contact surface model is implemented into it; a simple joint element is also implemented.

We first verify the module for the contact surface model. An analysis model of a simple rectangular concrete slab put on the ground is used; see Fig. C.1. The material properties are summarized in Table C.1.

We first study the state of the contact surface by applying the gravity force. The number of Gauss integration points and the spring constant are changed to find a suitable setting for the contact surface; the number is 2×2, 4×4, and 6×6, and the normal and shear spring constants are $k_n = k_s = 1\times 10^9$ and 5×10^6 (kN/m/m^2), respectively.

The distribution of the vertical stress on the contact surface due to the gravity force is presented in Fig. C.2. When $k_n = k_s = 1\times 10^9$ (kN/m/m^2) is used, partial lift-up of the slab takes place for the

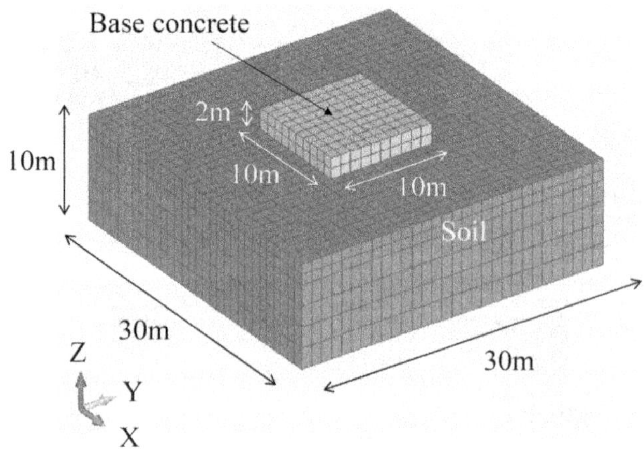

Fig. C.1. A simple slab model used to verify contact surface model.

Table C.1. Settings of numerical computation.

(a) Material properties of concrete and soil

	Young's modulus	Poisson's ratio	Density
	E (kPa)	ν	ρ (t/m^3)
Concrete	45,490,000	0.3000	2.4
Soil	1,330,000	0.3333	2.0

(b) Deformation characteristics' parameters of contact surface

Vertical spring coefficient	Shear spring coefficient	Cohesion	Friction coefficient	Tension stress
k_n ($kN/m/m^2$)	k_s ($kN/m/m^2$)	C (kN/m^2)	μ	σ_t (kN/m^2)
Parametric study	Parametric study	0.0	0.60	0.00

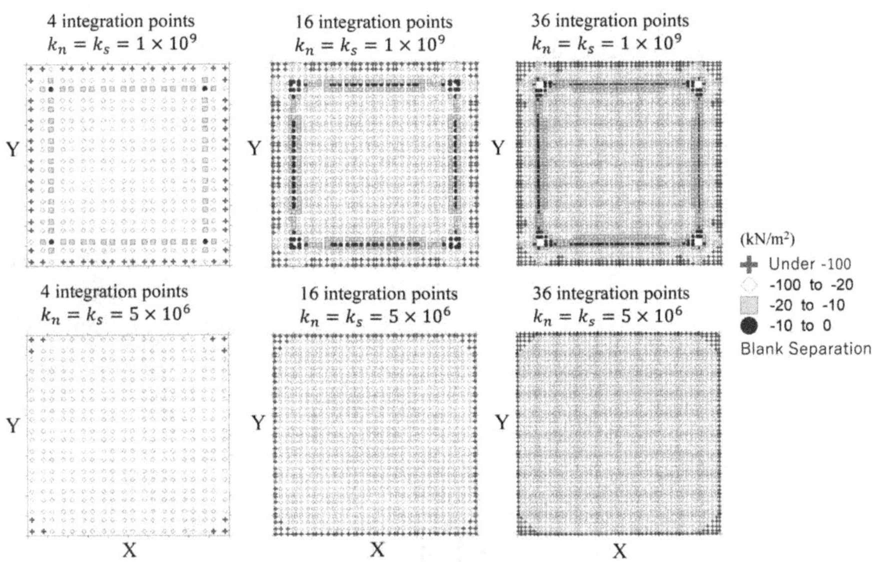

Fig. C.2. Distribution of the vertical stress due to gravity.

Gauss integration points of 4×4 and 6×6, and the vertical stress is not uniformly distributed on the surface; the non-uniformity of the stress distribution tends to increase as the number of the Gauss integration point increases. When $k_n = k_s = 5 \times 10^6$ (kN/m/m^2) is used, no lift-up takes place, and the non-smooth distribution of the vertical stresses is reduced; a large concentration of the vertical stress is induced at the slab's edge. While it seems odd, the occurrence of the stress concentration near the edge is mathematically correct as the solution of the continuum mechanics problem has a singularity at the edge.

We carry out seismic response analysis using a uniaxial spectrum wave shown in Fig. C.3 which is simultaneously input to the three directions and magnified 1.5 times. The time increment is 0.001 s, and the total number of time steps is 22,090. Two parameters of the Rayleigh damping are set as $(\alpha, \beta) = (1.473, 0.00127)$ for the slab and the ground and $(\alpha, \beta) = (0, 0.0001)$ for the contact surface.

Figure C.4 shows the distribution of normal stress on the contact surface at 5.42 s when the contact area between the foundation and the ground is minimized. As is seen, the distribution changes depending on the number of the Gauss integration points for $k_n = k_s = 1 \times 10^9$ (kN/m/m^2). However, the distribution is almost the same for $k_n = k_s = 5 \times 10^9$ (kN/m/m^2). The stress distribution during the seismic loading is affected by the initial stress distribution shown in Fig. C.2, which partially contributes less convergence for the case of $k_n = k_s = 1 \times 10^9$ (kN/m/m^2) compared with the case of $k_n = k_s = 5 \times 10^6$ (kN/m/m^2).

For $k_n = k_s = 5 \times 10^6$ (kN/m/m^2), the separation area becomes more stringent as the number of Gauss integration points increases. While more quantitative analysis is needed, it is seen that the numerical computation converges with respect to the number of Gauss integration points. At the slab's edge, there are no noticeable vertical

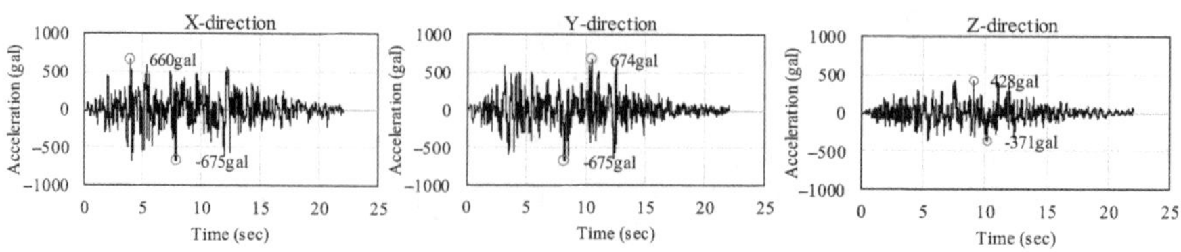

Fig. C.3. Input ground motion used for the lift-up analysis.

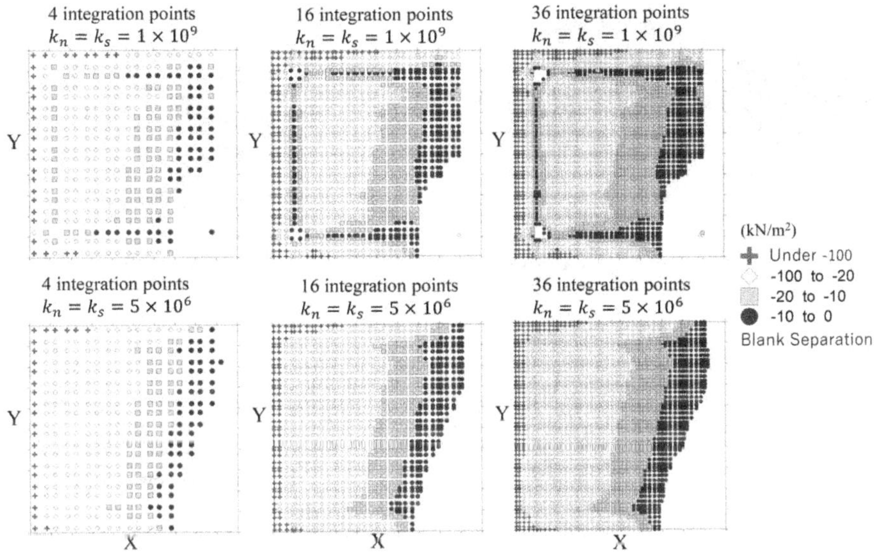

Fig. C.4. Distribution of vertical stress for the seismic response analysis.

stress oscillations. This is the effect of Rayleigh damping which serves as a numerical damping. While it is necessary to examine the most appropriate damping coefficients for the contact surface, the present setting of the damping coefficient together with the setting of k_n and k_s is sufficient to carry out the lift-up analysis.

Since the module for the contact surface model is verified, we carry out the lift-up analysis for a nuclear power plant building, using a standard model of the US Department of Energy. The numerical analysis model is presented in Fig. C.5. The reactor building is a hybrid model which uses solid and shell elements for concrete and steel reinforcement, respectively. The material properties of concrete and soil are the same as those used in the slab model; see Table C.1. The spring coefficients of the contact surface are set as $k_n = k_s = 5 \times 10^6$ (kN/m/m^2). The spectrum waves proposed by Kato are used as input ground motion; three acceleration components are given, and the maximum acceleration is set at 1,500 gal. The settings of numerical computation are summarized in Table C.2. Newmark-β method is employed for the time integration, with $(\beta, \gamma) = (0.25, 0.5)$. Based on the result of the eigenvalue analysis of the present numerical analysis model, we set the Rayleigh damping coefficients as

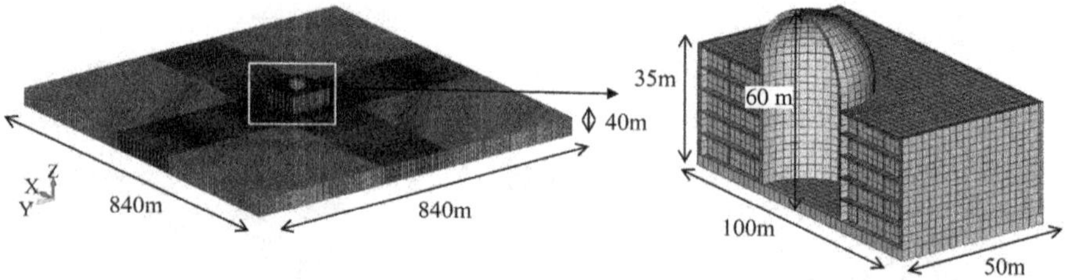

Fig. C.5. A numerical analysis model used for the lift-up analysis.

Table C.2. Setting of numerical analysis.

Category	Item	Numerical conditions
Nonlinear calculation	Iterative method	Newton Raphson method
	Convergence criterion	1.0×10^{-5}
	Maximum number of iterations	50
	Dynamic step	0.001 (s/step)
	Total step	22,090 (22.09 s)
Method to solve the inverse matrix	Solver	CG method
	Precondition	Slice successive over relaxation
	Convergence criterion	1.0×10^{-8}
	Maximum number of iterations	100,000
	MPI processes	64

$(\alpha, \beta) = (1.473, 0.00127)$ for the structure and the soil and $(\alpha, \beta) = (0, 0.0001)$ for the contact surface element.

We construct another numerical analysis model which employs standard joint elements instead of the contact surface elements developed. In a joint element, the average of the nodal force is used to compute the displacement gap, while an element using the contact surface model, which is called contact surface element hereinafter, uses the surface integral for more analysis of detaching/attaching. Partial separation is allowed for a contact element, while a joint element allows full or no separation.

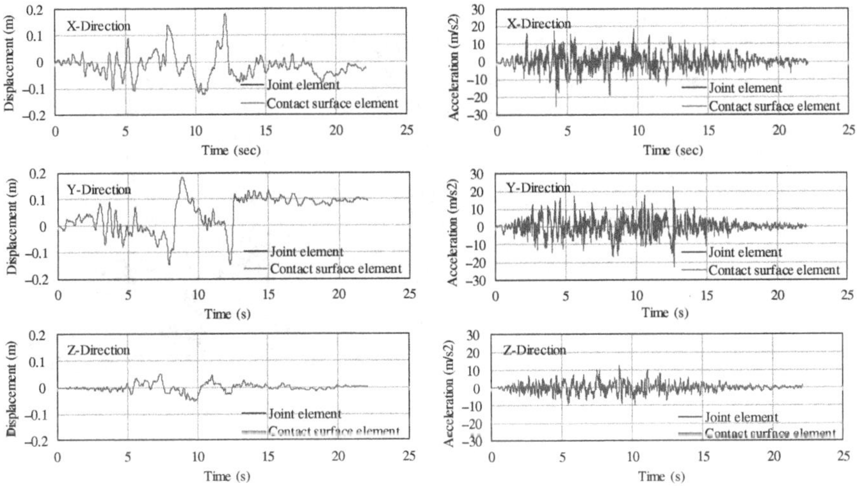

Fig. C.6. The time series of the displacement and the acceleration.

The time series of the displacements and the accelerations at the top of the reactor building is presented in Fig. C.6. The maximum horizontal displacement and acceleration are 0.2 m and 2,000 gal, respectively, and the horizontal acceleration becomes 1.3 times greater than the maximum input acceleration of 1,500 gal. The maximum vertical displacement and acceleration are 0.05 m and 1,000 gal.

The two numerical analysis models with joint elements and contact surface elements compute almost the same responses for the top of the reactor building. This agreement could be understood since the top of the reactor building represents the overall responses of the reactor building, and joint elements and contact surface elements do not cause significant differences in the overall responses even though local responses on the contact surface are different.

Figure C.7 shows the distribution of vertical normal stress on the contact surface at the time of 5.45 s when the contact area between the foundation and soil is minimized. It is seen that the distribution of vertical stress is different for the two numerical analysis models; the distribution of the numerical analysis model with joint elements seems slightly smoother. As mentioned, the stress distribution can be singular at the edge of the reactor building, and vertical stress does not have to be smooth near the edge. For the part of the separation

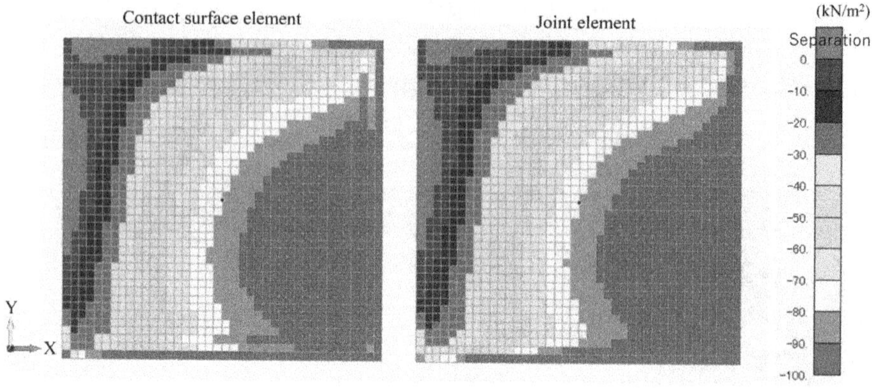

Fig. C.7. Distribution of the vertical stress.

Table C.3. Setting of numerical analysis.

Type	Total number of iterations for CG method	Computation time (s)
Contact surface element: 4 integration points	5,534,597	70,000
Contact surface element: 36 integration points	5,532,163	81,927
Joint element	5,598,364	71,118

(top left), however, the numerical analysis model with contact surface elements has a smoother distribution.

The iteration number of the conjugate gradient method and the CPU time are summarized in Table C.3; the two settings are used for the numerical analysis model with contact surface elements. As is seen, there is no significant difference in the iteration number and the CPU time. As the number of Gauss integration points increases, the CPU time increases, as expected. While the joint element uses point-wise relation between normal forces and displacement gap on the interface, the contact surface element uses surface integral for more accurate analysis. Similar numerical computation performance measured in terms of the CPU time suggests the potential usefulness of the surface element.

References

1. IAEA. Review of seismic evaluation methodologies for nuclear power plants based on a benchmark exercise (2013), https://inis.iaea.org/search/search.aspx?orig_q=RN:45042235.
2. A. Gurpinar, A review of seismic safety considerations in the life cycle of critical facilities, *J. Earthquake Eng.* **1**(1), 57–76 (1997).
3. G. L. Fenves, D. P. Poland, R. T. Eguchi, J. F. Hajjar, J. P. Lynch, and M. Nakashima, *Grand Challenges in Earthquake Engineering Research: A Community Workshop Report*. Washington DC: National Academies Press (2011).
4. M. Papadrakakis, P. Vagelis, and N. D. Lagaros, *Computational Methods in Earthquake Engineering*. Springer-Verlag, New York (2017).
5. B. Poursartip, A. Fathi, and J. L. Tassoulas, Large-scale simulation of seismic wave motion: A review, *Soil Dynamics and Earthquake Engineering*. **129**, 105909 (2020).
6. M. R. Riaz, H. Motoyama, and M. Hori, Review of soil-structure interaction based on continuum mechanics: Theory and use of high performance computing, *Geosciences*. **11**(2), 72 (2021).
7. N. Inoue, K. Yang, and A. Shibata, Dynamic non-linear analysis of reinforced concrete shear wall by finite element method with explicit analytical procedure, *Earthquake Eng. Struct. Dynam.* **26**(9), 967–986 (1997).
8. J. Tejchman, Fe-simulations of a direct wall shear box test, *Soils and Foundations*. **44**(4), 67–82 (2004).
9. M. Kawamura and J. Tanjung, 3D non-linear parallel FEM analysis for seismic earth pressures of a saturated soil layer, *J. Struct. Eng./Earthquake Eng. JSCE*. **19**(2), 119s–129s (2002).
10. Proceedings of Supercomputing '96, *Earthquake Ground Motion Modeling on Parallel Computers*. IEEE (1996).
11. T. Usami, Z. Lu, H. Ge, and T. Kono, Seismic performance evaluation of steel arch bridges against major earthquakes. Part 1: Dynamic analysis approach, *Earthquake Eng. Struc. Dynam.* **33**(14), 1337–1354 (2004).
12. D. Isobe and M. Tsuda, Seismic collapse analysis of reinforced concrete framed structures using the finite element method, *Earthquake Eng. Struct. Dynam.* **32**(13), 2027–2046 (2003).

13. W. Huang, P. L. Gould, R. Martinez, and G. S. Johnson, Non-linear analysis of a collapsed reinforced concrete chimney, *Earthquake Eng. Struct. Dynam.* **33**(4), 485–498 (2004).
14. S. Alali, L. Jianzhong, and G. Guanzhong, Double deck bridge behavior and failure mechanism under seismic motions using nonlinear analyzes, *Earthquake Eng. Eng. Vibr.* **12**(3), 447–461 (2013).
15. Proceedings of the International Conference on High Performance Computing, Networking, Storage and Analysis (SC'13), *IEEE Computer Society, Hysics-based Seismic Hazard Analysis on Petascale Heterogeneous Supercomputers.* New York: IEEE, (2013).
16. M. Hori, *Introduction to Computational Earthquake Engineering*, 3rd edition. London: World Scientific (2018).
17. R. Faria, N. V. Pouca, and R. Delgado, Seismic behaviour of a r/c wall: numerical simulation and experimental validation, *J. Earthquake Eng.* **6**(4), 473–498 (2002).
18. T. Miyamura, H. Akiba, and M. Hori, Large-scale seismic response analysis of super-high-rise steel building considering soil-structure interaction using k computer, *Int. J. High-Rise Build.* **4**(1), 75–83 (2015).
19. T. Miyamura, S. Tanaka, and M. Hori, Large-scale seismic response analysis of a super-high-rise-building fully considering the soil-structure interaction using a high-fidelity 3d solid element model, *J. Earthquake Tsunami.* **10**, 5 (2016).
20. The 14th World Conference on Earthquake Engineering, *Full 3D Seismic Response Analysis of Underground Ramp Tunnel Structure Using Large-scale Numerical Computation.* Beijing: IAEE, (2008).
21. Computational Methods in Structural Dynamics and Earthquake Engineering, *Seismic Performance Analysis of Underground Ramp Tunnel Structure Using 3D Massive Numerical Computation.* Corfu: ECCOMAS, (2011).
22. S. Yoshimura, M. Hori, and M. Ohsaki, *High-Performance Computing for Structural Mechanics and Earthquake/Tsunami Engineering.* New York: Springer International Publishing, (2015).
23. S. Tanaka, M. Hori, and T. Ichimura, Hybrid finite element modeling for seismic structural response analysis of a reinforced concrete structure, *J. Earthquake Tsunami.* **10**, 1640015 (2016).
24. H. Motoyama, W. Hotta, S. Suzuki, and M. Hori, Development and basic study on effectiveness of seismic response analysis method using a high-fidelity model for large-scale reinforced concrete structures, *JSCE, A2.* **75**(2), I_589–I_599 (2019), doi:10.2208/jscejam.75.2_I_589.
25. F. Forum, FrontISTR ver.4.5 (2022), http://www.multi.k.u-tokyo.ac.jp/FrontISTR/files/FISTRv4.5E.pdf.

26. Y. Saad, *Iterative Methods for Sparse Linear Systems.* New York: SIAM, (2003).
27. W. Hackbush, *Iterative Solution of Large Sparse Systems of Equations.* Applied Mathematical Sciences, Vol. 95. New York: Springer-Verlag, (1994).
28. H. Kawai, M. Ogino, R. Shioya, and S. Yoshimura, Large-scale elast-plastic analysis using domain decomposition method optimized for multi-core CPU architecture, *Key Eng. Mater.* **462–463**, 605–610 (2012).
29. O. C. Zienkiewicz and R. L. Taylor, *The Finite Element Method for Solid and Structural Mechanics.* New York: Elsevier, (2005).
30. T. Belytschko, W. K. Liu, and B. Moran, *Nonlinear Finite Elements for Continua and Structures.* New York: John Willy and Sons, (2000).
31. R. R. de Borst, *Computational Methods in Non-Linear Solid Mechanics*, Lecture Notes. Delft: Delft University of Technology, (1993).
32. J. Bonet and R. D. Wood, *Nonlinear Continuum Mechanics for Finite Element Analysis.* London: Cambridge University Press, (1997).
33. J. P. Wolf, *Soil-Structure-Interaction Analysis in Time Domain.* Prentice-Hall (1988).
34. K. D. Hjelmstad, *Fundamentals of Structural Mechanics*, 2nd edition. New York: Springer, (2005).
35. A. Pecker, J. J. Johnson, B. Jeremic, and N. Orbovic, *Methodologies for Seismic Soil-Structure Interaction Analysis in the Design and Assessment of Nuclear Installations.* International Atomic Energy Agency (2022).
36. K. Aki, Local site effect on ground motion, in J. V. T. Lawrence, (ed.), *Earthquake Engineering and Soil Dynamics II: Recent Advances in Ground-Motion Evaluation*, pp. 103–155. ASCE (1988).
37. B. Jeremic, N. Tafazzoli, T. Ancheta, O. N, and B. A, Seismic behavior of NPP structures subjected to realistic 3d, inclined seismic motions, in variable layered soil/rock, on surface or embedded foundations, *Nucl. Eng. Design.* **265**, 85–94 (2013).
38. H. Wang, H. Yang, Y. Feng, and B. Jeremic, Modeling and simulation of earthquake soil structure interaction excited by inclined seismic waves, *Soil Dynam. Earthquake Eng.* **146**, 106720 (2021).
39. R. E. Goodman, R. L. Taylor, and T. L. Brekke, A model for the mechanics of jointed rocks, *J. Soil Mech. Found. Div., ASCE.* **94**(3), 637–659 (1968).
40. J. Berenger, A perfectly matched layer for the absorption of electromagnetic waves, *J. Comput. Phys.* **114**, 185–200 (1994).
41. R. Clayton and B. Engquist, Absorbing boundary condition for acoustic and elastic wave equations, *Bullet. Seismol. Soc. America.* **67**, 1529–1540 (1977).

42. Y. S. Karinski, V. V. Shershnev, and D. Z. Yankelevsky, The effect of an interface boundary layer on the resonance properties of a buried structure, *Earthquake Eng. Struct. Dynam.* **33**, 227–247 (2004).
43. C. Cerjan, D. Kosloff, R. Kosloff, and M. Reshef, A nonreflecting boundary condition for discrete acoustic and elastic wave equations, *Geophys.* **50**, 705–708 (1985).
44. K. Maekawa, H. Okamura, and A. Pimanmas, *Non-Linear Mechanics of Reinforced Concrete.* New York: Taylor and Francis, (2003).
45. W. Hotta, S. Suzuki, and M. Hori, Implementation and verification of three-dimensional tensorial concrete constitutive relation on two-dimensional thick plate element, *JSCE, A1.* **77**(4), I_107–I_116 (2021), doi:10.2208/jscejseee.77.4_I_107.
46. S. Iai, Y. Matsunaga, and T. Kameoka, Strain space plasticity model for cyclic mobility, *Soils Found.* **29**, 27–56 (1990).
47. S. Iai and T. Kameoka, Finite element analysis of earthquake induced damage to anchored sheet pile quay walls, *Soils Found.* **33**, 71–91 (1993).
48. J. C. J. Schellekens and R. D. Borst, On the numerical integration of interface elements, *Int. J. Numer. Meth. Eng.* (1), 43–66 (1993).
49. J. Chen, T. Takeyama, H. O-Tani, K. Fujita, H. Motoyama, and M. Hori, Using high performance computing for liquefaction hazard assessment with statistical soil models, *Int. J. Comput. Meth.* **16**(05), 1840005 (2019).
50. W. Hotta, S. Suzuki, and M. Hori, On contraction of three-dimensional multiple shear mechanism model for evaluation of large scale liquefaction using high performance computing, *Geosciences.* **9** (2019).
51. A. Elgamal, E. Parra, Z. Yang, and K. Adalier, Numerical analysis of embankment foundation liquefaction countermeasures, *J. Earthquake Eng.* **6**(4), 447–471 (2002).
52. O. Matsuo, T. Shimazu, R. Uzuoka, M. Mihara, and K. Nishi, Numerical analysis of seismic behavior of embankments founded on liquefiable soils, *Soils Found.* **40**(2), 21–40 (2000).
53. Y. Tamari, I. Towhata, and M. Takahashi, Seismic soil-structure interaction of cross sections of flexible underground structures subjected to soil liquefaction, *Soils Found.* **43**(2), 69–88 (2003).
54. I. Arango, M. R. Lewis, and C. Kramer, Updated liquefaction potential analysis eliminates foundation retrofitting of two critical structures, *Soil Dynam. Earthquake Eng.* **20**(1–4), 17–25 (2000).

https://doi.org/10.1142/9781800614635_0005

Chapter 5

Earthquake Simulation on Unstructured Finite Elements Enhanced by High-Performance Computing

Kohei Fujita* and Tsuyoshi Ichimura†

Earthquake Research Institute and Department of Civil Engineering,
The University of Tokyo,
Bunkyo-ku, Tokyo, Japan
**fujita@eri.u-tokyo.ac.jp*
†ichimura@eri.u-tokyo.ac.jp

In this chapter, the finite element method using low-order unstructured elements, which is frequently used in the analysis of seismic phenomena, is examined from the viewpoint of high-performance computational science, and methods for solving very large-scale problems in a massively parallel manner at high speed are described. The governing equations and discretization for seismic and ground motion analysis and crustal deformation analysis are first presented, and then algorithms to realize these massively parallel computations on CPU- and GPU-based supercomputers are presented and their effective implementation methods are also shown. The performance evaluation of each algorithm on each supercomputer is also presented in detail.

1. Introduction

The process of a major earthquake is as follows: (i) the plate movement causes strain to accumulate in the plate and when it reaches a limit, a fault is generated; (ii) seismic waves generated by faulting propagate through the crust; (iii) seismic waves amplified in the soft layers near the surface cause damage to human life, property, and society (see Fig. 1). Advanced continuous observations have been conducted and various observation data have been accumulated for mitigating earthquake damage, and this accumulation is expected to be further accelerated with the development of 5G and post-5G networks as well as IoT technologies. The possibility of constructing a digital twin that can evaluate the past, present, and future by integrating data obtained from a large number of observation instruments with more detailed simulations is becoming apparent. Such a digital twin will enable us to conduct various scientific and engineering-based studies on earthquakes in more detail, such as conducting scientific research on various phenomena related to earthquakes and volcanoes, and to explore more deeply than before how to prevent and mitigate disasters directly or indirectly caused by earthquakes and volcanoes.

Seismic phenomena are complex and varied. However, despite differences in length scales and resolution, from the perspective of mathematical problems, many of the behaviors of seismic phenomena can be modeled as static/dynamic, nonlinear/linear responses, which

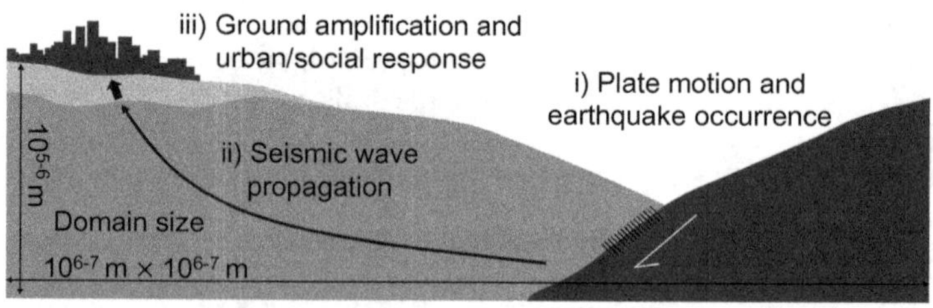

Fig. 1. Earthquake processes include (i) plate motion and earthquake occurrence, (ii) seismic wave propagation, and (iii) ground amplification and urban and social response. The analysis cost is enormous because of the large area covered and the very fine resolution required. Among them, the analysis cost of plate motion and earthquake generation by physics-based simulation is huge.

provides a good perspective on computational method development. Although the *constitutive laws* and *senses* differ depending on the phenomenon of interest, many phenomena can be reduced to problems of forward and inverse analysis. Since these are mathematical problems in which the (stress-free) boundary conditions and geometry have a strong influence on the solution, the finite element method based on unstructured elements can be used for highly detailed analysis. However, because the problem size is large and the required discretization resolution is high, to ensure the reliability of the solution (convergence), it is necessary to solve much larger problems (0.1 billion to 1 trillion DOF) than those handled by the conventional finite element method in fields such as the manufacturing industry (e.g., order of 1 million DOF). In order to solve ultra-large-scale problems using the finite element method, a method for generating ultra-large-scale finite element models and analysis methods based on these models are required, and there are strong expectations for contributions from high-performance computational science.

Although it is challenging to robustly construct large-scale finite-element models for complex 3D models, it is possible to construct finite-element models with more than one trillion degrees-of-freedom by using methods such as in Ref. 13 for stratified structures, which are frequently used in crustal deformation, seismic, and ground motion analysis. On the other hand, analysis using this super-large-scale finite element method is a more difficult problem. Although it may seem straightforward to solve super-large-scale problems by using supercomputers, the finite element method itself is not suitable for efficient computation on recent computer architecture, and solving very large-scale problems is not easy even when using very large-scale computer systems. Specifically, it is necessary to solve the bottleneck caused by the random memory access that constrains the performance of the core computation kernels of the finite element method, and to develop a method that can efficiently utilize a very large number of compute nodes on a very large computer system. This chapter presents some examples of such methodological developments for seismic wave propagation, ground motion, and crustal deformation analysis. In the following, we first describe the governing equations and discretization of the seismic and ground motion analysis and crustal deformation analysis discussed in this chapter. Next, algorithms and suitable implementations for (1) CPU-based

Finite Element Solver for Nonlinear Wave Propagation; (2) CPU-based Finite Element Solver on Specialized Mesh for Nonlinear Wave Propagation; (3) CPU-based Finite Element Solver for Crust Deformation; (4) CPU-based Finite Element Solver for Nonlinear Soil–Structure System Response; and (5) GPU-based Finite Element Solver for Nonlinear Soil–Structure System Response are described. Also, performance measurement results of each algorithm on each computer system is described in detail.

2. Background

First, seismic wave propagation and ground motion analysis are explained. At present, seismic wave propagation from a fault to the ground surface is often divided into two phases: seismic wave propagation, which deals with wave propagation from the fault to the engineering bedrock, and ground motion amplification, which deals with the amplification of seismic waves in surface soils above the engineering bedrock. Seismic wave propagation and ground motion amplification are regarded as wave propagation in solids, and are modeled with the governing equation

$$\frac{\partial}{\partial x_i}\left(c_{ijkl}(\mathbf{x})\frac{\partial}{\partial x_l}u_k(\mathbf{x},t)\right) - \rho(\mathbf{x})\ddot{u}_j(\mathbf{x},t) = f_j(\mathbf{x},t). \tag{1}$$

For simplicity, we use Cartesian coordinates for x_1, x_2, x_3. Here, $\mathbf{x}$ and t are spatial coordinates and time, $\frac{\partial}{\partial x_i}$ and $\ddot{}$ are partial derivative with respect to x_i and (second) order derivative with time. c_{ijkl}, u_i, f_i, and ρ indicate a heterogeneous elasticity tensor, displacement in ith direction, outer force in ith direction, and density, respectively. Seismic wave propagation is modeled as linear wave propagation where c_{ijkl} in Eq. (1) does not vary with time t and is often solved with fault rupture as input. Ground motion amplification is often modeled as nonlinear wave propagation where c_{ijkl} in Eq. (1) varies with the constitutive law of soils (e.g., nonlinear relationship between strain and stiffness) for the earthquake motion input from the engineering bedrock.

Numerical simulation of wave propagation is considered effective for these analyses, and many studies have been conducted. Since the numerical simulation of seismic wave propagation and ground motion amplification is strongly affected by the complex geometry of the model, the finite element method is also used in this chapter because it can handle the geometry of the model precisely and is often applied to numerical simulations that analytically satisfy stress-free boundary conditions at the surface. Implicit time integration is used because explicit time integration leads to severe Courant conditions due to the very small finite elements that are generated locally when modeling complex geometries. We will also use low-order unstructured tetrahedral finite elements, which are suitable for modeling complex geometries and are typically used in numerical analysis in mechanical engineering. Specifically, the Newmark-β method ($\delta = 1/2$, $\beta = 1/4$), which is an implicit time integration typically used in earthquake engineering, is used to discretize the time direction, and the finite element method based on the Galerkin method is used to discretize the space direction to obtain wave propagation and ground motion amplification.

The discretized governing equation in seismic wave propagation is

$$\left(\frac{1}{dt^2}\mathbf{M} + \frac{1}{dt}\mathbf{C} + \mathbf{K}\right)\mathbf{u}^n = \mathbf{f}^n + \mathbf{C}\mathbf{v}^{n-1} + \mathbf{M}\left(\mathbf{a}^{n-1} + \frac{4}{dt}\mathbf{v}^{n-1}\right), \tag{2}$$

with

$$\begin{cases} \mathbf{v}^n = -\mathbf{v}^{n-1} + \dfrac{2}{dt}(\mathbf{u}^n - \mathbf{u}^{n-1}), \\ \mathbf{a}^n = -\mathbf{a}^{n-1} - \dfrac{4}{dt}\mathbf{v}^{n-1} + \dfrac{4}{dt^2}(\mathbf{u}^n - \mathbf{u}^{n-1}), \end{cases} \tag{3}$$

where $\mathbf{u}$, $\mathbf{v}$, $\mathbf{a}$, $\mathbf{f}$ indicate displacement, velocity, acceleration, and outer force vectors, respectively. $\mathbf{M}$, $\mathbf{C}$, $\mathbf{K}$ are the mass, damping, and stiffness matrices, respectively. dt, n indicate time step width and number of time steps. Rayleigh damping is used for $\mathbf{C}$, where the element damping matrix $\mathbf{C}_e$ is indicated as

$$\mathbf{C}_e = \alpha\mathbf{M}_e + \beta\mathbf{K}_e,$$

using element mass matrix $\mathbf{M}_e$ and element stiffness matrix $\mathbf{K}_e$. Here, α and β are determined by solving the least squares problem

$$\text{minimize}\left[\int_{f_{\min}}^{f_{\max}}\left(h-\frac{1}{2}\left(\frac{\alpha}{2\pi f}+2\pi f\beta\right)\right)^2 df\right].$$

Here, $f_{\max}$, $f_{\min}$, h indicate maximum target frequency, minimum target frequency, and damping factor. First, outer force by the fault source is set to $\mathbf{f}^n$, and values up to the $n-1$th time step is used to solve Eq. (2) for displacement $\mathbf{u}^n$ at the nth time step. This value is used to obtain $\mathbf{v}^n$ and $\mathbf{a}^n$ using Eq. (3). The above is repeated to obtain displacement time history step by step. Here, the absorbing boundary conditions for absorbing the reflected waves at the bottom and sides of the domain is included in $\mathbf{C}$ together with the internal damping.

While many ways to discretize the nonlinear wave propagation equation are proposed, we use the discretized governing equation

$$\begin{aligned}&\left(\frac{4}{dt^2}\mathbf{M}+\frac{2}{dt}\mathbf{C}^n+\mathbf{K}^n\right)\delta\mathbf{u}^n\\&\quad=\mathbf{f}^n-\mathbf{q}^{n-1}+\mathbf{C}^n\mathbf{v}^{n-1}+\mathbf{M}\left(\mathbf{a}^{n-1}+\frac{4}{dt}\mathbf{v}^{n-1}\right),\end{aligned}\tag{4}$$

with

$$\begin{cases}\mathbf{q}^n=\mathbf{q}^{n-1}+\mathbf{K}^n\delta\mathbf{u}^n,\\ \mathbf{u}^n=\mathbf{u}^{n-1}+\delta\mathbf{u}^n,\\ \mathbf{v}^n=-\mathbf{v}^{n-1}+\dfrac{2}{dt}\delta\mathbf{u}^n,\\ \mathbf{a}^n=-\mathbf{a}^{n-1}-\dfrac{4}{dt}\mathbf{v}^{n-1}+\dfrac{4}{dt^2}\delta\mathbf{u}^n,\end{cases}\tag{5}$$

for modeling ground motion amplification, assuming that the time increment is sufficiently small. Here, $\delta\mathbf{u}$, $\mathbf{u}$, $\mathbf{v}$, $\mathbf{a}$, $\mathbf{f}$ indicate displacement increment, displacement, velocity, acceleration, and outer force vectors, respectively. $\mathbf{M}$, $\mathbf{C}^n$, $\mathbf{K}^n$ are the mass, damping, and stiffness matrices, respectively. dt, n indicate time step width and number of time steps. As the stiffness and damping changes every time step,

Rayleigh damping is used for $\mathbf{C}^n$, where the element damping matrix $\mathbf{C}_e^n$ is indicated as

$$\mathbf{C}_e^n = \alpha \mathbf{M}_e + \beta \mathbf{K}_e^n,$$

using element mass matrix $\mathbf{M}_e^n$ and element stiffness matrix $\mathbf{K}_e^n$. Here, α and β are determined by solving the least squares problem

$$\text{minimize} \left[\int_{f_{\min}}^{f_{\max}} \left(h^n - \frac{1}{2} \left(\frac{\alpha}{2\pi f} + 2\pi f \beta \right) \right)^2 \mathrm{d}f \right].$$

Here, $f_{\max}$, $f_{\min}$, h indicate maximum target frequency, minimum target frequency, and damping factor. The specific calculation procedure, taking into account the stiffness and damping changes due to the nonlinear constitutive law of the ground, is in conducting the following for each time step:

(1) Evaluate stiffness and damping based on the strain at the $n-1$th time step using the constitutive model.
(2) Evaluate $\mathbf{K}^n$ and $\mathbf{C}^n$ based on the stiffness and damping at the nth time step.
(3) Obtain $\delta \mathbf{u}^n$ by solving Eq. (4), and update values in Eq. (5).

Here, the absorbing boundary conditions for absorbing the reflected waves at the bottom and sides of the domain is included in $\mathbf{C}^n$ together with the internal damping. In addition to the input seismic motion from the bottom, forces due to ground continuity conditions are added to $\mathbf{f}^n$.

Equations (2) and (4) both require solving large-dimensional matrix equations, but they offer the great advantage of preventing instability in numerical analysis. Note that the Newmark-β method is a type of Taylor expansion, so even though the analysis is stable, the time step width must be small enough for the solution to converge. Empirically, the spatial discretization should be at least 8–10 elements per wavelength at the target frequency when using first-order elements, and at least 4–5 elements per wavelength at the target frequency when using second-order elements for obtaining a converged solution (note that more elements are needed when the wave is dominated by surface waves that are slower than body waves). As one of the procedures for quality assurance of the numerical solution, it is often necessary to check whether a converged solution is obtained by gradually decreasing the element size in a complex model based on the above guideline, which inevitably leads to the need to solve large-scale problems.

Next, crustal deformation analysis is described. Crustal deformation analysis analyzes the deformation of the Earth's crust due to fault movement using the following governing equation:

$$\frac{\partial}{\partial x_i}\left(c_{ijkl}(\mathbf{x})\frac{\partial}{\partial x_l}u_k(\mathbf{x},t)\right) = f_j(\mathbf{x},t). \tag{6}$$

Crustal deformation immediately after an earthquake is often modeled as a linear elasticity body, while post-earthquake relaxation behavior is often modeled as a viscoelastic body. This chapter focuses on the deformation of the crust immediately after an earthquake, because it is expected to contribute not only to the understanding of the earthquake mechanism by estimating fault behavior through inverse analysis, but also to the mitigation of tsunami damage by being used for tsunami wave source estimation.

Conventionally, analytical solutions such as that based on semi-infinite medium approximations have been used; however, with the increase in the number of observations, including those in the vicinity of the epicenter, and the growing reliability of crustal structure models, numerical simulations using 3D crustal structure models are now being developed. As in the seismic wave propagation analysis, the complex geometry of the model has a strong influence on the numerical solution; thus, numerical simulations using the finite element method are often used because they can handle the geometry of the model precisely and analytically satisfy the traction-free boundary conditions at the surface. By discretizing Eq. (6) in space by using the finite element method based on the Galerkin method, we obtain the following discretized governing equations for the analysis of crustal deformation (linear elasticity) immediately after an earthquake.

$$\mathbf{Ku} = \mathbf{f}, \tag{7}$$

where **K**, **u**, **f** are stiffness matrix, displacement, and outer force vectors, respectively.

Algorithm 1 Algorithm of conjugate gradient method. Here, numerical solution of $\mathbf{Ax} = \mathbf{f}$ with a positive definite symmetric matrix $\mathbf{A}$ is solved using initial solution $\mathbf{x}_0$, within relative error $\|\mathbf{f} - \mathbf{Ax}\|/\|\mathbf{f}\| < \epsilon$.

$\mathbf{r}_0 \Leftarrow \mathbf{f} - \mathbf{Ax}_0$
$\mathbf{p}_0 \Leftarrow \mathbf{r}_0$
$i \Leftarrow 0$
while $(\|\mathbf{r}_i\|/\|\mathbf{f}\| \geq \epsilon$ **do**
  $\alpha_i \Leftarrow \frac{(\mathbf{r}_i, \mathbf{p}_i)}{(\mathbf{p}_i, \mathbf{Ap}_i)}$
  $\mathbf{x}_{i+1} \Leftarrow \mathbf{x}_i + \alpha_i \mathbf{p}_i$
  $\mathbf{r}_{i+1} \Leftarrow \mathbf{r}_i - \alpha_i \mathbf{Ap}_i$
  $\beta_i \Leftarrow \frac{(\mathbf{r}_{i+1}, \mathbf{Ap}_i)}{(\mathbf{p}_i, \mathbf{Ap}_i)}$
  $\mathbf{p}_{i+1} \Leftarrow \mathbf{r}_{i+1} - \beta_i \mathbf{p}_i$
  $i \Leftarrow i + 1$
end while

As the DOF of the matrix equations in Eqs. (2), (4), and (7) become large, suitable solver development is required. For such large-degree-of-freedom finite element analyses, iterative solvers are often used, especially conjugate gradient methods, due to the sparsity of the target matrices, and this chapter also introduces studies along the lines of conjugate gradient methods. Specifically, in the CG method, starting at initial solution $\mathbf{x}_0$, the solution is refined iteratively following Algorithm 1 such that the relative error $(\|\mathbf{r}_i\|/\|\mathbf{f}\|$ becomes less than a small positive number $\epsilon \ll 1$.

The CG method for the finite element method comprises basic operations as shown in Algorithm 1, and it is necessary to devise ways to increase the speed and parallel performance based on the characteristics of the computer architecture and system. The main points are the preconditioner and the computation method of **Au**.

Preconditioner: In the CG method, the number of iterations to reach the desired approximate solution depends on the maximum/minimum eigenvalues and the distribution of eigenvalues of **A**. To reduce the number of iterations, **A** is multiplied by a matrix **M** such that **MA** has better convergence characteristics. Such processing is called preconditioning, and various methods have been

Algorithm 2 Computation of $\mathbf{f} = \mathbf{K}\mathbf{u}$ by the EBE method. cny_i^{ie} is the global node number of the ith local node number of element ie, $\mathbf{B}_i^{ie}$ and $\mathbf{D}_i^{ie}$ are strain–displacement transformation matrix and stiffness tensor of the ith Gauss point of element ie, respectively. $\mathbf{u_t}$, $\mathbf{f_t}$ are temporary vectors of length 30.

for $ie = 1, \ldots, ne$ **do**
  for $i = 1, \ldots, 10$ **do**
    for $j = 1, \ldots, 3$ **do**
      $u_t(3(i-1)+j) \Leftarrow u(3(cny_i^{ie}-1)+j)$
    end for
  end for
  $\mathbf{ft} = \sum (\mathbf{B}_i^{ie})^T \mathbf{D}_i^{ie} \mathbf{B}_i^{ie} \mathbf{u_t}$
  for $i = 1, \ldots, 10$ **do**
    for $j = 1, \ldots, 3$ **do**
      $f(3(cny_i^{ie}-1)+j) = f(3(cny_i^{ie}-1)+j) + f_t(3(i-1)+j)$
    end for
  end for
end for

proposed. In this chapter, we mainly discuss low-precision operations and methods that take advantage of the sparsity of the solution space.

Efficient Au computation method: When performing large-degree-of-freedom finite element analysis using a solver such as the CG method, it is sometimes difficult to keep $\mathbf{A}$ in memory, even for sparse matrices. In such cases, the element-by-element (EBE) method can be used to perform equivalent computation without storing $\mathbf{A}$ in memory. For simplicity, the calculation method of $\mathbf{f} = \mathbf{K}\mathbf{u}$ for static analysis ignoring boundary conditions is shown in Algorithm 2. Here, we consider a finite element model comprising ne 10-node tetrahedral quadratic elements with n nodes and 3 DOF per node; thus, $\mathbf{u}$ has $3n$ dimensions and $u(3(i-1)+j)$ is the displacement component of the i-node in the j direction, and $\mathbf{K}$ is the global stiffness matrix. The performance of EBE is highly dependent on the computer architecture and implementation method. This chapter details the actual implementation methods for several architectures.

3. CPU-Based Finite Element Solver for Nonlinear Wave Propagation

This section describes a general finite element solver for CPU-based massively parallel computing environments targeting nonlinear wave propagation in local ground structures.[1] During an earthquake, waves that enter from the engineering bedrock are amplified in the surface soft ground. Depending on the nonuniform nonlinear material properties of the 3D ground, waves may be locally amplified, leading to large ground motion or large strain levels at certain points, which is important when considering the seismic response of aboveground and underground structures. In this problem, Eq. (4) is the target equation. This problem is a large degree-of-freedom (DOF) problem that requires massively parallel computations since the target frequency is high and the target area is large. For solving such problems, we develop an adaptive CG type solver with mixed-precision and a multi-grid preconditioner. The solver is designed to have excellent convergence characteristics with good load balancing in massively parallel computing environments. It is also designed to be a most general method for solving finite element method with second-order unstructured tetrahedral mesh, capable of extension to a variety of target equations and for more sophisticated and specialized solvers in the later sections. In the following, we describe the details of the target problem and the developed adaptive conjugate gradient-type solver with mixed-precision and multi-grid preconditioner (hereafter called SC14 solver). We show the computation performance of the developed solver on the K computer, and show an example of a physics-based seismic wave amplification simulation.

3.1. *Algorithm design*

The amplification process of seismic waves is complex because it is affected by the 3D geometry of the surface and underground structure with heterogeneous nonlinear material properties. The computational cost required to solve the nonlinear wave equation in non-uniform grounds is much greater than the computational cost required for wave propagation simulations presented in previous literature. This is because (1) nonlinearity of soil reduces stiffness, decreasing the wavelength, and (2) deformation concentrates at parts of the ground with irregular geometry and material properties. These issues lead to a necessity to increase the resolution in spatial discretization; the spatial and temporal resolution required is 0.1 m and 0.001 s, respectively. For a target domain of 2,000 $\times$ 2,000 $\times$ 100 m with earthquakes of 30 s duration, the resulting DOF) of the discretized target equation (i.e., Eq. (4)) are on the order of 10^{10} DOF, and the resulting number of time steps is 3×10^4. Because most of the computation cost is involved in solving Eq. (4), the key point for achieving efficient analysis is the development of a suitable solver, considering the characteristics of the nonlinear dynamic finite element method and the target computer system. A finite element model (FE model) construction method that can generate models with a large number of unstructured low-order elements is also developed for modeling the complex ground structure.

3.1.1. *Development of solver algorithm*

In the targeted seismic wave amplification simulation, we compute $\delta\mathbf{u}^n$ with 30,000 time steps on a 10 billion DOF-order FEmodel that can ensure the convergence of $\mathbf{u}^n$ in terms of acceleration response and the convergence of structural response in an urban area. Thus, this problem results in solving a 10 billion DOF-order matrix equation in each of the 30,000-order time steps, where the matrix components are changed at every time step. In order to solve a large-DOF problem with relatively small computer memory, we develop a fast, iterative solver by considering the characteristics of the matrix equation. Although the nonlinear dynamic finite element method results in solving matrix equations repeatedly where its matrix components change every time step, there is a clue in that the characteristics of the matrix are not so bad due to the inertia terms involved. However, the convergence characteristics of the target matrix equation is worse than the convergence of a matrix equation of lower resolution problems, even if its DOF is the same. Thus, we need a fast iterative solver comprising a suitable preconditioner with lower computation cost and high scalability on massively parallel computer environments. Additionally, low memory footprint is required for the preconditioner because the DOF per compute node is large. As a result, a moderate preconditioner with less computation cost and high scalability in a many-compute-nodes environment is more suitable for this problem, when compared with a sophisticated preconditioner with higher computation cost and only moderate scalability. Although one candidate for such a solver is the preconditioned conjugate gradient (PCG) method[7] with a block Jacobi preconditioner, the number of iterations tend to increase when solving matrices with poor characteristics. Indeed, the number of iterations required for solving the performance measurement problems in this study required many iterations; thus, development of a much faster iterative solver for the target problem is desirable. Thus, we developed a solver for a huge DOF and a many time-step, physics-based seismic wave amplification simulation on massively parallel CPU-based systems based on a fast and scalable iterative solver using a moderate preconditioner with low computation cost. In the following, we describe the solver algorithm following Algorithms 3 and 4, which are designed based on the ideas (a)–(e).

Algorithm 3 Details of SC14 solver. Matrix vector product $(\frac{4}{dt^2}\mathbf{M}+\frac{2}{dt}\mathbf{C}^n+\mathbf{K}^n)(\ )$ is computed by EBE. diag[ ], $(\bar{\ })$ and ϵ indicates 3×3 block Jacobi of [ ], single-precision variable and tolerance for relative error. $(\ )_c$ indicates calculation related to FEmodel$_c$, and the other is the related calculation of FEmodel. $\bar{\mathbf{P}}$ is a mapping matrix, from FEmodel^i_c to FEmodel^i, which is defined by interpolating the displacement in each element of FEmodel^i_c.

---

read boundary condition of n-th time step

$\mathbf{b} \Leftarrow \mathbf{f}^n - \mathbf{Q}^{n-1} + \mathbf{C}^n\mathbf{v}^{n-1} + \mathbf{M}(\mathbf{a}^{n-1} + \frac{4}{dt}\mathbf{v}^{n-1})$

$\mathbf{x} \Leftarrow \frac{dt}{24}(-9\mathbf{v}^{n-4} + 37\mathbf{v}^{n-3} - 59\mathbf{v}^{n-2} + 55\mathbf{v}^{n-1})$

$\bar{\mathbf{B}} \Leftarrow diag[\frac{4}{dt^2}\mathbf{M} + \frac{2}{dt}\mathbf{C}^n + \mathbf{K}^n]$

$\bar{\mathbf{B}}_c \Leftarrow diag[\frac{4}{dt^2}\mathbf{M}_c + \frac{2}{dt}\mathbf{C}^n_c + \mathbf{K}^n_c]$

$\bar{\mathbf{A}}_c \Leftarrow \frac{4}{dt^2}\mathbf{M}_c + \frac{2}{dt}\mathbf{C}^n_c + \mathbf{K}^n_c$ (stored on memory with CRS format)

$\mathbf{r} \Leftarrow \mathbf{b} - \left(\frac{4}{dt^2}\mathbf{M} + \frac{2}{dt}\mathbf{C}^n + \mathbf{K}^n\right)\mathbf{x}$

$\beta \Leftarrow 0$

$i \Leftarrow 1$

(*outer loop start*)

while $\|\mathbf{r}\|_2/\|\mathbf{b}\|_2 \geq \epsilon$ **do**

  (*inner loop start*)

  $\bar{\mathbf{r}} \Leftarrow \mathbf{r}$

$\bar{\mathbf{z}} \Leftarrow \mathbf{B}^{-1}\mathbf{r}$
$\bar{\mathbf{r}}_c \Leftarrow \bar{\mathbf{P}}^T\bar{\mathbf{r}}$
$\bar{\mathbf{z}}_c \Leftarrow \bar{\mathbf{P}}^T\bar{\mathbf{z}}$
$\bar{\mathbf{z}}_c \Leftarrow \bar{\mathbf{A}}_c^{-1}\bar{\mathbf{r}}_c$ (*solved on FEmodel$_c$ by Algorithm 4 with ϵ_c^{in} and initial solution $\bar{\mathbf{z}}_c$*)
$\bar{\mathbf{z}} \Leftarrow \bar{\mathbf{P}}\bar{\mathbf{z}}_c$
$\bar{\mathbf{z}} \Leftarrow (\frac{4}{dt^2}\bar{\mathbf{M}} + \frac{2}{dt}\bar{\mathbf{C}}^n + \bar{\mathbf{K}}^n)^{-1}\bar{\mathbf{r}}$ (*solved on FEmodel by Algorithm 4 with ϵ^{in} and initial solution $\bar{\mathbf{z}}$*)
$\mathbf{z} \Leftarrow \bar{\mathbf{z}}$
(*inner loop end*)
if $i > 1$ **then**
  $\beta \Leftarrow (\mathbf{z}, \mathbf{q})/\rho$
end if
$\mathbf{p} \Leftarrow \mathbf{z} + \beta\mathbf{p}$
$\mathbf{q} \Leftarrow (\frac{4}{dt^2}\mathbf{M} + \frac{2}{dt}\mathbf{C}^n + \mathbf{K}^n)\mathbf{p}$
$\rho \Leftarrow (\mathbf{z}, \mathbf{r})$
$\alpha \Leftarrow \rho/(\mathbf{p}, \mathbf{q})$
$\mathbf{q} \Leftarrow -\alpha\mathbf{q}$
$\mathbf{r} \Leftarrow \mathbf{r} + \mathbf{q}$
$\mathbf{x} \Leftarrow \mathbf{x} + \alpha\mathbf{p}$
$i \Leftarrow i + 1$
end while
(*outer loop end*)
$\delta\mathbf{u}^n \Leftarrow \mathbf{x}$
$\mathbf{Q}^n \Leftarrow \mathbf{Q}^{n-1} + \mathbf{K}^n\delta\mathbf{u}^n$
$\mathbf{u}^n \Leftarrow \mathbf{u}^{n-1} + \delta\mathbf{u}^n$
$\mathbf{v}^n \Leftarrow -\mathbf{v}^{n-1} + \frac{2}{dt}\delta\mathbf{u}^n$
$\mathbf{a}^n \Leftarrow -\mathbf{a}^{n-1} - \frac{4}{dt}\mathbf{v}^{n-1} + \frac{4}{dt^2}\delta\mathbf{u}^n$
output results of n-th time step
modify material properties considering nonlinearity

Algorithm 4 PCG with a preconditioner. This is used for roughly solving $\mathbf{z} = (\frac{4}{dt^2}\mathbf{M} + \frac{2}{dt}\mathbf{C}^n + \mathbf{K}^n)^{-1}\mathbf{r}$ and $\mathbf{z}_c = \mathbf{A}_c^{-1}\mathbf{r}_c$ in Algorithm 3. $\mathbf{B}$ and ϵ^{in} are a block Jacobi matrix and the tolerance for the relative error. All the calculations are conducted using single-precision variables.

$\mathbf{x} \Leftarrow \mathbf{z}$
$\mathbf{e} \Leftarrow \mathbf{r} - (\frac{4}{dt^2}\mathbf{M} + \frac{2}{dt}\mathbf{C} + \mathbf{K})\mathbf{x} \quad (or \quad \mathbf{A}_c\mathbf{x})$
$\beta \Leftarrow 0$
$i \Leftarrow 1$
while $(\|\mathbf{e}\|_2/\|\mathbf{r}\|_2 \geq \epsilon^{in} \quad (or \quad \epsilon_c^{in})$ **do**
  $\mathbf{z} \Leftarrow \mathbf{B}^{-1}\mathbf{e} \quad (or \quad \mathbf{B}_c^{-1}\mathbf{e})$
  $\rho_a \Leftarrow (\mathbf{z}, \mathbf{e})$
  if $i > 1$ **then**
    $\beta \Leftarrow \rho_a/\rho_b$
  end if
  $\mathbf{p} \Leftarrow \mathbf{z} + \beta\mathbf{p}$
  $\mathbf{q} \Leftarrow (\frac{4}{dt^2}\mathbf{M} + \frac{2}{dt}\mathbf{C} + \mathbf{K})\mathbf{p} \quad (or \quad \mathbf{A}_c\mathbf{p})$
  $\alpha \Leftarrow \rho_a/(\mathbf{p}, \mathbf{q})$
  $\rho_b \Leftarrow \rho_a$
  $\mathbf{e} \Leftarrow \mathbf{e} - \alpha\mathbf{q}$
  $\mathbf{x} \Leftarrow \mathbf{x} + \alpha\mathbf{p}$
  $i \Leftarrow i + 1$
end while
$\mathbf{z} \Leftarrow \mathbf{x}$

(a) **Initial Solution Predictor:** For reduction in the number of iterations in the iterative solver, we predict the initial solution of the iterative solver as $\frac{dt}{24}(-9\mathbf{v}^{n-4} + 37\mathbf{v}^{n-3} - 59\mathbf{v}^{n-2} + 55\mathbf{v}^{n-1})$ based on the Adams–Bashforth method.[9] The use of this method with small time increments leads to initial solutions with relative error of $10^{-4\sim-3}$, leading to reduction in the number of conjugate gradient (CG) iterations.

(b) **Adaptive Conjugate Gradient Method[10]:** Instead of using a fixed matrix for the preconditioner to the CG method, the preconditioning matrix equation $(\frac{4}{dt^2}\mathbf{M} + \frac{2}{dt}\mathbf{C}^n + \mathbf{K}^n)\mathbf{z} = \mathbf{r}$ is solved roughly using another CG solver in the adaptive CG method. The conventional CG iteration and the CG iteration for the preconditioner are called the outer and inner loops, respectively. As the preconditioning matrix equation need only be solved approximately, this makes room for the following multi-grid and mixed-precision arithmetic methods.

(c) **Multi-Grid Method[7]:** In the inner loop, a rough solution is first computed on FEmodel_c, which is a finite-element model equivalent to the targeted FEmodel, but with coarser resolution. Using the solution on FEmodel_c as an initial solution, a rough solution is computed using the FEmodel with fewer iterations. Since the DOF and nodes to be communicated are much smaller on FEmodel_c than those for the FEmodel, the computation cost of the inner loop is reduced by the multi-grid method.

(d) **Mixed-Precision Arithmetic:** Although FP64 arithmetic is required in the outer loop, lower-precision arithmetic can be used for the inner loops as only an approximate solution is required for the preconditioner. Since the DOF of FEmodel_c is smaller than that of the FEmodel, and halving the data size by use of FP32, we can store the matrix of FEmodel_c in the compressed row storage (CRS) format. Also, the size of peer-to-peer communication is reduced by 2-fold for an FEmodel with FP32 and approximately 8-fold for FEmodel_c with FP32 when compared with that of FEmodel with FP64.

(e) **EBE Method[8]:** As the DOF computed by one compute node is large, the global $\mathbf{K}$ matrix of FEmodel required for solving Eq. (4) cannot be stored explicitly on memory. Thus, instead of storing $\mathbf{K}$, computation of $\mathbf{Kx}$ is computed by the EBE method on the second-order tetrahedral elements of the FEmodel. With the advantage of displacement–stress mixed finite-element method[12] and the orthogonal-basis function, we successfully transformed $\mathbf{Kx}$ into $\sum_i \mathbf{N_b}_i^T(\mathbf{D}(\mathbf{N_b}_i\mathbf{x}))$, where $\mathbf{D}$ and $\mathbf{N_b}_i$ are the elastic tensor and equivalent mode, respectively. Sophisticated reformulation and careful tuning can achieve a further reduction in the number of operations and higher peak performance ratios (e.g., 21.7% for EBE with single precision) on the K computer. Because the frequency of EBE with single precision is very high, this speedup significantly reduces computation time.

Although the computation cost of one outer loop is almost the same as that for one loop of PCG, the number of outer iterations is reduced significantly by the developed preconditioner. Moreover, the computation cost of the preconditioner itself is also reduced by the multi-grid and mixed-precision arithmetic methods, resulting in speedup of the total computation time. The solver algorithm is implemented by MPI-OpenMP hybrid parallelization, where FEmodel and FEmodel$_c$ are partitioned and assigned to MPI processes and each partition is computed on each MPI process with OpenMP. Point-to-point communication (`MPI_Isend, MPI_Irecv, MPI_Waitall`) after the matrix vector product and collective communication (`MPI_Allreduce`) after the inner product are conducted for synchronization.

3.1.2. *Development of finite element model construction method*

We describe the FE model construction method capable of constructing 10 billion DOF-order 3D FE models using 3D-layered ground structure data. A large number of unstructured second-order tetrahedral elements are required for evaluating stress and strain in complex ground structures at high resolution for large domains; thus, the computational cost of constructing such a large 3D FE model is expected to be considerable. A fully automated robust meshing algorithm that can generate elements of high quality is desirable. Since manual tuning to avoid elements with poor quality is extremely time-consuming for large models, various methods have been developed for generating FE models automatically. In this study, we extended the automated mesh generation method of Ref. 13 for construction of huge-DOF 3D FE models. In this method, a structured grid is used to decompose the simulation domain into background cells, and elements are constructed for each cell with small aspect ratios. This leads to robust generation of meshes with high-quality elements in full automation. Because the generation of elements in each cell is performed independently, the mesh generation can be conducted in parallel, leading to a reduction in the time for mesh generation. Here we generate linear-tetrahedral elements and cubic elements and then convert the cubic elements to linear tetrahedral elements. Next, the linear-tetrahedral elements are converted to second-order tetrahedral elements. The linear tetrahedral element model is used for the FEmodel$_c$ and the second-order tetrahedral element model is used for the FEmodel. Using METIS 5.1.0[11] with suitable options on a shared-memory computer, we partition FEmodel and FEmodel$_c$ such that the number of elements in each domain becomes balanced.

3.2. *Performance*

3.2.1. *Performance measurement environment*

We measure the performance of SC14 solver on the K computer, a massively parallel CPU-based supercomputer previously at RIKEN, Advanced Institute for Computational Science.[14] The K computer comprises 82,944 compute nodes, each with single SPARC64™ VIIIfx CPUs with 6 MB L2 shared cache. Each of the eight cores of the SPARC CPU has two sets of twin fused multiply-and-add (FMA) units, which lead to four multiplications and four additions in one clock cycle. The number of operations per clock cycle is the same for single and double precision floats. The frequency of the CPU is 2.0 GHz, leading to a peak performance of 8 FLOP $\times$ 2 GHz $\times$ 8 CPU cores = 128 GFLOPS per CPU. Each compute node has 16 GB of DDR3-SDRAM memory, with peak memory bandwidth of 64 GB/s. Tofu, a 6D interconnection network, is used for communication between nodes.[15] Each node can communicate in four directions simultaneously, with 5 GB/s throughput in each direction. An OpenMPI 1.4.3[16]-based MPI library optimized for the Tofu network was used, following the MPI 2.1 standard.[17]

3.2.2. *Performance measurement setting*

The effectiveness of the SC14 solver is shown by comparison of four solvers: the developed SC14 solver, SC14 (double precision, multi-grid) solver, SC14 (double precision, no multi-grid) solver, and PCGE. The SC14 (double precision, multi-grid) solver is a SC14 solver with the mixed-precision arithmetic disabled. SC14 (double precision, no multi-grid) solver is SC14 solver with both the mixed-precision arithmetic and the multi-grid preconditioner disabled. PCGE is the PCG solver where a 3×3 block Jacobi method is used for the preconditioner and the EBE method is used for matrix–vector products. The same tuning applied to SC14 solver is applied to all solvers. The modified Ramberg–Osgood model[2] and Masing rule[3] are used for the nonlinear constitutive models of the soil. Relative error of $\epsilon = 10^{-8}$, $\epsilon^{\text{in}} = 0.25$, $\epsilon_c^{\text{in}} = 0.5$ is used for the tolerance of the linear solvers. We use one compute node (one MPI process) for each FEmodel^i and FEmodel$_c^i$ and parallelized each MPI domain using OpenMP (i.e., $8m$ CPU cores are used for m K computer compute nodes).

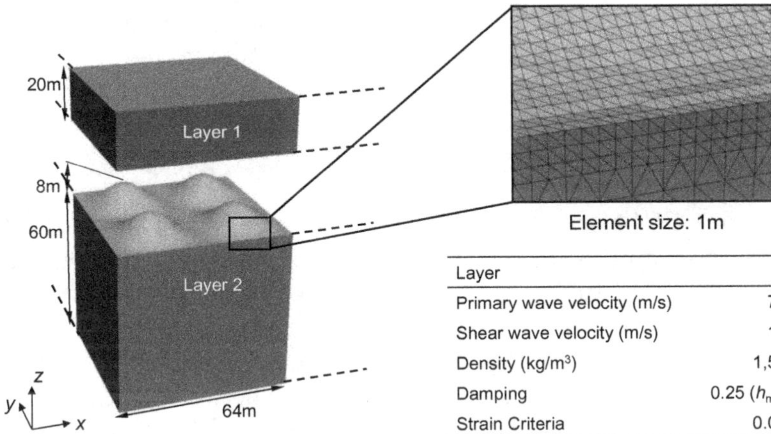

Layer	1	2
Primary wave velocity (m/s)	700	2,100
Shear wave velocity (m/s)	100	700
Density (kg/m^3)	1,500	2,100
Damping	0.25 (h_{max})	0.05
Strain Criteria	0.007	–

Fig. 2. 94,407,951-DOF ground model used for performance measurement. This model is duplicated in the x and y directions for making large-scale models for weak scaling measurements.

Table 1. Configuration of performance measurement models.

Model	# of MPI domains	# of CPU cores	Total DOF	DOF per MPI domain	# of PCGE iterations
1	4,608	36,864	1,505,755,135	326,767	355
2	9,216	73,728	3,010,699,983	326,681	355
3	18,432	147,456	6,019,818,087	326,596	355
4-A	9,216	73,728	12,038,067,615	1,306,213	355
4-B	18,432	147,456	12,038,067,615	653,107	355
4-C	36,864	294,912	12,038,067,615	326,553	355
5	82,944	663,552	27,082,129,647	326,511	355

We use a two-layered ground that mimicked actual ground structures for performance measurement (Fig. 2). Here, the ground structure is made periodical, so that it is suitable for measuring weak scaling. The first layer is a nonlinear soft layer and the second layer is a harder layer that mimics the bedrock layer, with bumps in the interface between the two layers that mimic local changes in the ground structure. This problem is duplicated in the x and y directions for making large problems for weak scaling measurements. The FE model of this problem was generated with 0.44 m-sized elements to obtain convergence in ground acceleration, even when the soils soften due to nonlinearity. We input a wave with 100 time steps of

$dt = 0.002$ s, which was the main part of the Kobe earthquake wave[18] from the bottom of the model. Table 1 shows the configuration of models used for measuring weak scaling and strong scaling.

3.2.3. *Performance measurement results*

Although the SC14 solver is suited for systems with fast FP32 floating-point computation capabilities, we measured performance using the K computer, as the time-to-solution for large-scale problems is important, and thus the scalability performance for solving large-scale problems becomes important. The ninth time step was used for performance measurement as it was not influenced by the fluctuations in the initial part of the wave and became a poor convergence problem for all of the models. Although the characteristics of the problem vary with the size of the problem, the weak scaling models used in this study have approximately the same convergence characteristics due to their periodic geometry. In fact, the number of loops required for convergence of the PCGE solver was found to be the same for all models (see Table 1), thus; we can see that this set of models is well suited for measuring weak scaling of the solvers.

We first measure weak scaling of the solver. Figure 3 shows the elapsed time and its breakdown. Here, inner coarse, inner fine, and outer indicate line 17 of Algorithm 3, line 19 of Algorithm 3, and the outer loop, respectively. The numbers in brackets indicate the number of loops required for the linear solver to converge. We can see that the required numbers of loops required for convergence were almost the same for all the models; thus, the model set is suitable for measuring weak scalability of the solver. We can see that the increase in elapsed time is small and that the solver scales well to a large number of K computer nodes. The smallest model with 1.5 billion DOF (model 1) was computed in 1.42 s using 36,864 CPU cores (4,608 compute nodes $\times$ 8 CPU cores, 8.69% of FP64 peak, 51.3 TFLOPS), and the largest model with 27 billion DOF (model 5) was computed in 1.63 s using the full K computer (82,944 compute nodes $\times$ 8 CPU cores = 663,552 CPU cores, 7.57% of FP64 peak, 0.804 PFLOPS), corresponding to 87.1% weak scaling efficiency. On large-scale problems, we can see larger increase in elapsed time of inner coarse loop when compared to the outer and inner fine loops. Although communication time was expected to increase with respect to model size as the number of hops in point-to-point communications and the number

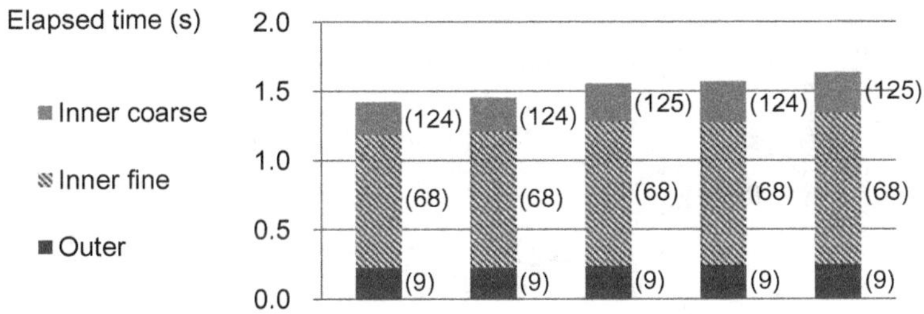

Model	1	2	3	4-C	5
Elapsed time (s)	1.42	1.45	1.55	1.56	1.63
FLOPS/Peak (%)	8.69	8.50	7.95	7.87	7.57
Mem. throughput/Peak (%)	13.43	13.14	12.31	12.18	11.72

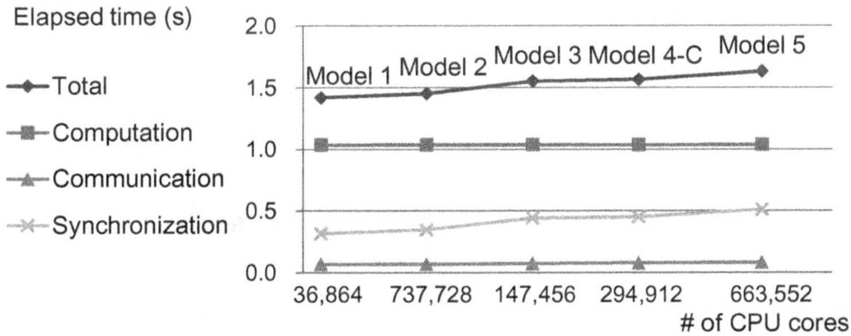

Fig. 3. Elapsed time and its breakdown (inner fine, inner coarse, outer loop) for the ninth time step of the weak scaling problem set. Numbers in brackets indicate the number of elapsed loops required for convergence of the linear solver. The breakdown of the elapsed time for computation, communication, and synchronization is also shown.

of processes involved in all-reduce operations would increase, the increase in communication time was only 0.081 s (model 5) − 0.067 s (model 1) = 0.014 s. Such a small increase in communication time is due to the reduction in communication size for point-to-point communication and interconnect hardware supported all-reduce operations.[15] On the other hand, the synchronization time has increased from 0.316 s (model 1) to 0.511 s (model 5). As the EBE and CRS cost is correlated with the number of elements and nodes, respectively, the larger imbalance in the number of nodes may be the cause of worse performance deterioration of the inner coarse solver. We can expect performance improvement by decreasing the imbalance in the number of nodes and elements by tuning the parameters of METIS.

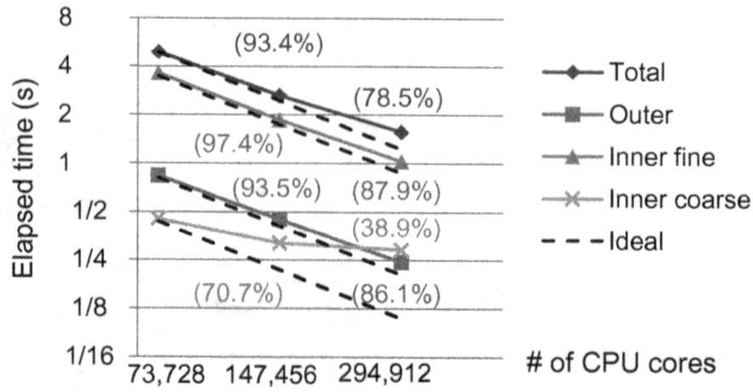

Model	4-A	4-B	4-C
Elapsed time (s)	4.92	2.63	1.56
FLOPS peak ratio (%)	9.99	9.36	7.87
Memory throughput peak ratio (%)	19.14	16.05	12.18

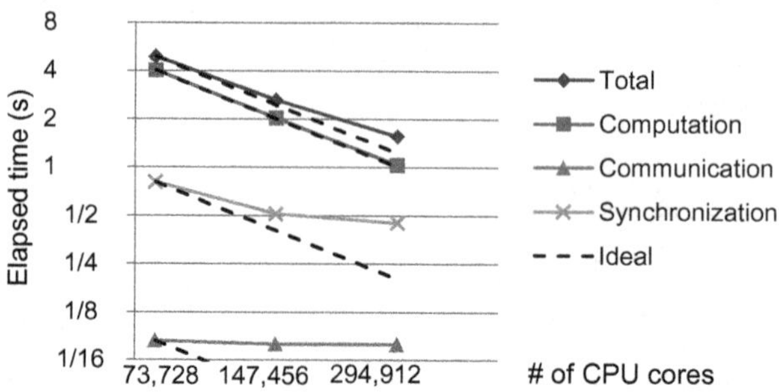

Fig. 4. Strong scaling measurement results. Numbers in brackets show the strong scaling efficiency with respect to Model 4-A (73,728 CPU cores). The breakdown of the elapsed time for computation, communication, and synchronization is also shown.

Figure 4 shows the strong scaling measurement results of the solver. We can see that the 12 billion DOF problem is solved in 4.92 s using 73,728 CPU cores (9,216 compute nodes × 8 CPU cores, 9.99% of FP64 peak, 117.8 TFLOPS) and that the same problem is solved in 1.56 s using 294,912 CPU cores (36,864 compute nodes × 8 CPU cores, 7.87% of peak, 371.4 TFLOPS), with strong scaling efficiency of 78.5%. From the breakdown of elapsed time, we can see that the outer and inner fine loops have high scalability; however, the increase in the time used for solving the inner coarse loop is prominent (with 38.9% strong scaling efficiency for the inner coarse loop part) and thus this is the main reason for the deterioration in the scalability of the whole solver. Next, a breakdown of the elapsed time is shown for

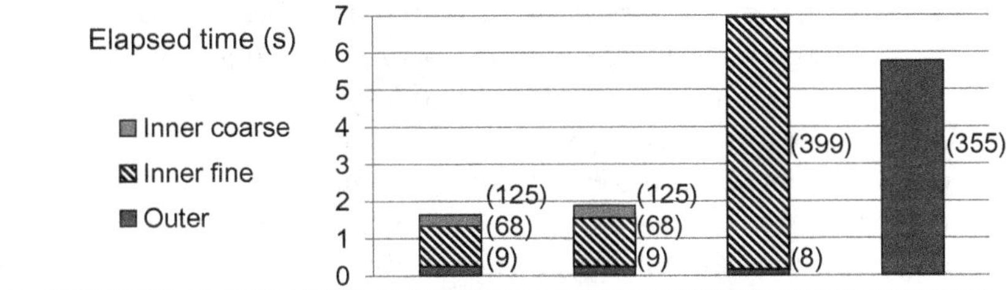

	SC14 (mixed-precision, multi-grid)	SC14 (double precision, multi-grid)	SC14 (double precision, no multi-grid)	PCGE
Elapsed time (s)	1.63	1.87	6.95	5.75
FLOPS peak ratio (%)	7.57	6.60	7.88	8.11
Memory throughput peak ratio (%)	11.72	19.40	20.40	21.66

Fig. 5. Performance comparison of SC14, SC14 (double precision, multi-grid), SC14 (double precision, no multi-grid), and PCGE solvers measured on Model 5 (27 billion DOF, 663,552 CPU cores). Numbers in brackets indicate the required iterations for convergence of the iterative solvers.

computation, communication, and synchronization. It can be seen that the communication and synchronization times have increased, but the communication time is still well contained. It is expected that a reduction in the load balancing of the inner coarse loops will lead to improved performance.

Figure 5 shows the effectiveness of mixed-precision arithmetic and the multi-grid preconditioner for model 5 using the whole K computer. While the number of floating-point operations per clock cycle is the same for single-precision and double-precision arithmetic on the K computer, the required data transfers and communication volume are halved when using single precision, which is expected to lead to performance improvement. First, we compare the SC14 solver with mixed-precision arithmetic and multi-grid preconditioner and the SC14 (double precision, multi-grid) solver. Since the required data transfer of the preconditioner doubles by use of double precision, the elapsed time increases by 1.15-fold of the SC14 solver. Next, we compare the SC14 solver and the SC14 (double precision, no multi-grid) solver. The computation of FEmodel_c^i (line 17 of Algorithm 3) uses linear elements with a CRS format, whereas the computation of FEmodel^i (line 19 of Algorithm 3) uses second-order elements with

EBE. Thus, the elimination of the inner coarse solver leads to higher peak performance; however, the number of loops required for convergence increases when using SC14 (double precision, no multi-grid) solver. As the required time per iteration also increases by use of the fine grids, the total time for solving the matrix equation increases to 4.26 times that of the SC14 solver. While most of the cost for the PCGE solver is involved in the double precision EBE computation of second-order elements, it can be computed efficiently by using the tuned EBE computation developed in this section; thus, the FLOPS and memory throughput are good. However, many iterations are required for convergence and thus the time required to solve the matrix equation is 3.53-fold of the SC14 solver. Although the PCGE solver is slightly better in terms of peak performance, we can see that the multi-grid method combined with mixed-precision arithmetic results in a speedup greater than 3-fold compared with the PCGE solver.

Finally, we show the total cost of the whole application on full K computer for solving 100 time steps in Fig. 6. In addition to the solver, the application involves preparing CRS of the coarse mesh in each time step, updating the material parameters using strain of the previous time step, and preparation of the block Jacobi matrix used in the preconditioner. Since the input is the same for all processes, process 0 reads the input wave from the file system and broadcasts

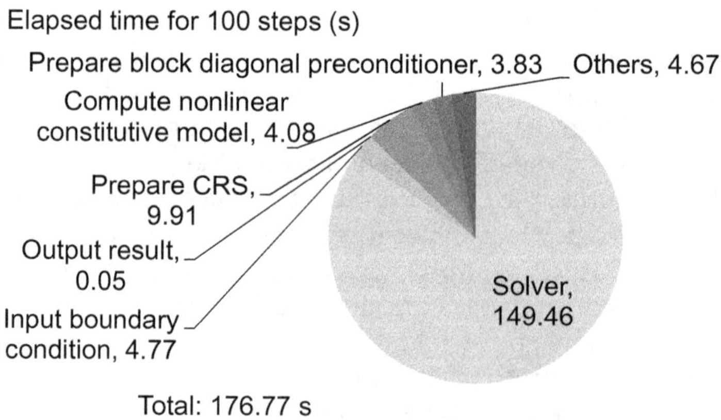

Fig. 6. Breakdown of the elapsed time of the whole application. Measured for computing 100 steps of Model 5 (27 billion DOF) on full K computer (663,552 CPU cores).

it to other processes. The output of the 100-step simulation is gathered per rank and outputted in binary to reduce the number of files and increase throughput. By such measures, the time spent for I/O is reduced to 2.7% of the whole application. Also, the other parts are suppressed to 12.7% with hybrid MPI-OpenMP implementations. Thus, we can see that most of the time is used for the solver (84.6%) and that the performance of the solver dominates the performance of the whole analysis, leading to a high performance of 0.736 PFLOPS (6.93% of peak) for the whole application. The SC14 solver algorithm is seen to be effective on these 100 steps, but it is also expected to be effective in solving poorer convergence problems with more complex grounds or input waves with phase differences.

3.3. *Example*

We show an application example of the SC14 solver in a physics-based urban earthquake simulation of a 2.0 × 2.0 km region of central Tokyo. The region comprises business, commercial, and densely built residential districts, with a total of 13,275 buildings. We model the ground structure of a 2.0 × 2.0 km domain with a depth of 100 m, comprising three soil layers based on the 5 m grid elevation map[19] and digital geotechnical database.[20] Figure 7 shows the geometry and the material properties of the ground. Here, the data at the nearest earthquake observation point[21] was used to set the material properties of the ground layers. The FE model was constructed with an element size small enough to obtain convergence in ground acceleration, even when the soils softened due to nonlinearity. This resulted in a model with 10.7 billion DOF and a minimum element size of 0.66 m. The Kobe earthquake wave[18] ($dt = 0.001$ s × 30,000 time steps) was inputted from the bottom of the model, and the simulation was conducted using 294,912 CPU cores (36,864 compute nodes × 8 CPU cores). $\epsilon = 10^{-8}$, $\epsilon^{\text{in}} = 0.25$, $\epsilon_c^{\text{in}} = 0.5$ were used for the tolerances of the preconditioning solvers.

Figure 8 shows the breakdown of the elapsed time on this problem. The required time to compute 30,000 time steps was 41,521 s including I/O, meaning that only 1.38 s was used for solving each time step. Most of the time was spent on the linear solver (35,245 s); thus, we can see that the fast solver enabled such a large-scale analysis. The performance of the whole application was also high due to

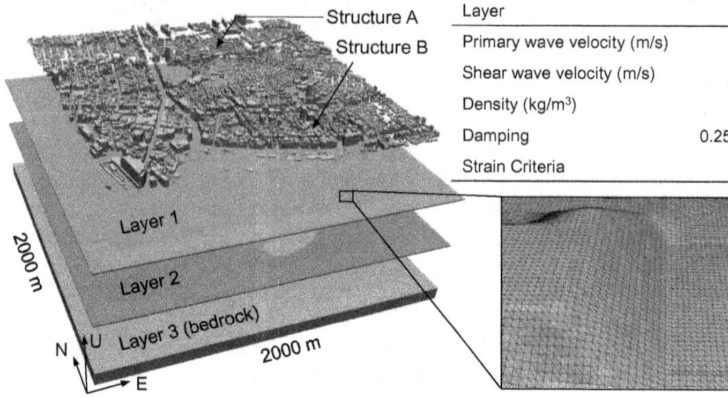

Layer	1	2	3
Primary wave velocity (m/s)	1,210	1,380	1,770
Shear wave velocity (m/s)	150	255	490
Density (kg/m^3)	1,500	1,800	1,900
Damping	0.25 (h_{max})	0.05	0.005
Strain Criteria	0.005	–	–

Fig. 7. Ground and structure model of the application example comprising a domain of 2.0 × 2.0 km with 13,275 structures. The three-layered ground was modeled with minimum element size of 0.66 m, leading to a 10,755,536,091 DOF model, and computed using 294,912 CPU cores (36,864 compute nodes) of the K computer.

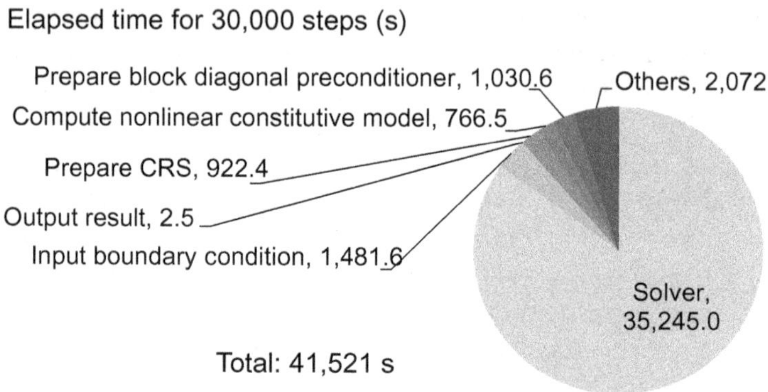

Fig. 8. Breakdown of elapsed time on practical application problem with 10.7 billion DOF. The performance of the solver and whole application was 0.421 PFLOPS (8.93% of peak) and 0.358 PFLOPS (7.59% of peak), respectively, for the initial 1,000 time steps.

the reduction in I/O cost (1,481.6 s for inputting data and 2.5 s for outputting the results). Displacements at the surface were outputted with 0.33-m spatial resolution every 10 time steps, leading to total output size of 1.32 TB. For the first 1,000 time steps, the performance of the linear solver and whole application was 0.421 PFLOPS (8.93% of FP64 peak) and 0.358 PFLOPS (7.59% of FP64 peak),

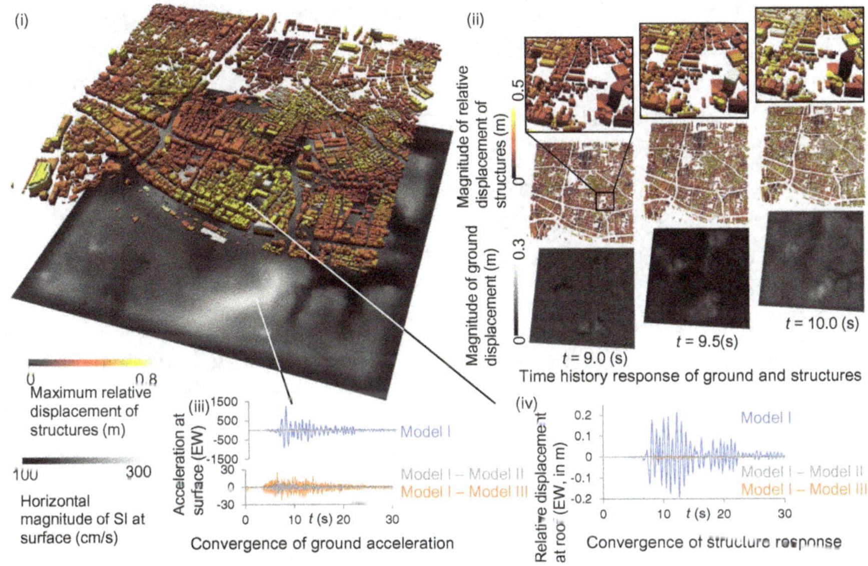

Fig. 9. Seismic response of ground and structures. (i) Horizontal magnitude of SI at surface and maximum relative displacement of structures, (ii) time history response snapshots, (iii) convergence of acceleration waveforms at surface (Point A) by increase in mesh resolutions (Model I (10.7 billion DOF model), Model II (with element sizes 1.5 times coarser than Model I), and Model III (with element sizes 3 times coarser than Model I)), and (iv) convergence of relative displacement response waveforms of Structure A (a two-story RC building located above Point A) obtained using acceleration waveforms computed using Models I, II, and III.

respectively; we can see that high performance was attained on an actual practical problem setting.

Figure 9(i) shows the ground surface response as well as the structural response computed using the computed ground motion as an input. From Fig. 9(ii), we can see that complex ground motion and structural response is obtained according to the ground motion and properties of each structure. In typical ground motion analyses, it is difficult to ensure the convergence of acceleration due to the huge computation costs; however, we can obtain highly reliable results that converge in terms of acceleration response in large domains by use of the SC14 solver. Here, we checked the numerical convergence of the response using three models with different resolutions: model I with the highest resolution (10.7 billion DOF model), model II with elements 1.5 times coarser than that of model I, and model III with elements 3 times coarser than that of model I. From Fig. 9(iii) showing the comparison of acceleration time history at the surface (Point A), we can see that the waveform converges with respect to the increase in the discretization resolution of the FE model (waveform of the finest model converges within 1% of the amplitude of the wave). As the responses of structures are dependent on the input acceleration at the ground surface, the responses of structures also converge with the increase in resolution; see Fig. 9(iv). We can see that high-resolution finite-element modeling of ground motion is essential for analyzing the response of structures. By comparing the ground and structure responses in Fig. 9(i), we can see that larger SI values (an index commonly used for evaluating the destructiveness of waves to buildings[22]) does not necessarily lead to large structural responses. Such a complex distribution of structure responses in an urban area can only be realized by combination of high-resolution 3D ground motion analysis and structural response analysis.

4. CPU-Based Finite Element Solver on Specialized Mesh for Nonlinear Wave Propagation

In this section, we describe a finite element solver on specialized meshes for CPU-based massively parallel computer environments targeting nonlinear wave propagation in local ground structures.[39] In recent computer architectures, performance differences between specific types of computation patterns and other computation patterns are increasing, and in particular, performance for random data access is becoming a bottleneck. By considering the characteristics of the target problem and computer algorithms, we can transform the computation patterns to ones that are suitable for the target computer architecture. Targeting the problem of nonlinear wave propagation in nonlinear soft soils (i.e., Eq. (4)) on a massively parallel CPU-based computer environment, we generate a mesh with mixed structured and unstructured tetrahedral elements, leading to a better computation pattern for a large part of the target domain. With algorithm development for SIMD-based CPU architectures, we can accelerate the computation. In the following, we describe the algorithm development, performance measurement results on the CPU-based K computer, and some application examples.

4.1. *Algorithm design*

To obtain numerical convergence of the acceleration waveform of the ground response and the building response in 100 km^2 city problems, nonlinear wave propagation on a 3D FE model with 100 billion DOF must be solved for each of the 10,000 time steps. This corresponds to the same target problem as in Section 3.2.2, but with a significantly larger problem size. The problem involves solving Eq. (4) with 100 billion DOF, with matrix component changing every time step due to nonlinearity, for each of the 10,000 time steps on a massively parallel computer environment. Due to the large DOF, development of an iterative solver with low memory footprint, suitable for the matrix property, is required for fast analysis. While many methods that can reduce the computation time by using a preconditioner with a high computation cost to improve the poor convergence of the target problem have been proposed for low-order unstructured finite element methods for solid problems (e.g., Refs. 4–6), our problem needed a much faster solver. The nonlinear wave propagation problem has the difficulty that the matrix components change with each time step and must be solved many times, but the matrix properties are not so bad because of the inertia term. Potentially, the PCG method using a block Jacobi preconditioner[7] is such a solver, but the number of iterations tends to increase when solving problems with poor convergence properties. The SC14 solver[1] was developed as a faster solver than PCG, but its performance is still insufficient for this problem. In this section, we improve the algorithm and implementation of the SC14 solver to solve a 1 trillion DOF problem.

We review the history of seismic linear wave propagation analysis, which deals with wave propagation phenomena similar to those targeted in our nonlinear wave propagation problem. The wave propagation problem is huge, requiring high-resolution wave propagation analysis over a large area of interest. The finite difference method was first used to study practical wave propagation in the crust because of its low computational cost and efficient memory access. However, the geometry of crust structure has significant effect on earthquake propagation; therefore, unstructured finite element methods that can accurately model the geometry were developed (e.g., Ref. 36). Although the unstructured finite element method involves a huge amount of computation due to random data access and cost of modeling complex finite element meshes, research based on the finite element family is increasing due to the advantages of precise geometry modeling and high accuracy numerical calculations (e.g., Refs. 27 and 31). More recently, a structured finite element method using an octree structure was developed to reduce the cost of the finite element method (e.g., Ref. 26). This method can be regarded as an extension of the structured finite difference method because of its ability to reduce the number of elements through multi-resolution octree modeling. Compared to the unstructured finite element method, the octree method reduces computational and memory access costs and facilitates robust generation of FE models. However, it still has shortcomings in terms of geometry modeling. Alternatively, finite element methods have been developed using hybrid elements of unstructured and structured elements (unstructured tetrahedral elements for regions of complex geometry and structured hexahedral elements for uniform regions) (e.g., Refs. 37 and 13). Based on these insights, we develop a method to reduce computation and memory access costs by octree-constraints in uniform regions, while using unstructured tetrahedral elements for regions of complex geometry for accurate numerical modeling.

Here, we enhance the SC14 solver based on these insights; we develop a fast solver capable of many-time-step nonlinear wave simulation on massively parallel computer systems by using preconditioning with reduced computational costs and low memory footprint, together with reduction in communication for high scalability. Algorithms 5 and 6 are used to solve $\delta \mathbf{u}^n$ of Eq. (4) for the nth time step, by combining methods (a) to (d) that follow. Hereafter, we refer to this solver as the SC15 solver.

Algorithm 5 Details of SC15 solver algorithm. Matrix vector product $\mathbf{A}^n\mathbf{y} = (\frac{4}{dt^2}\mathbf{M} + \frac{2}{dt}\mathbf{C}^n + \mathbf{K}^n)\mathbf{y}$ is computed with an EBE method with merged voxel elements: $\sum_i^{N^{\mathrm{V}}}(\frac{4}{dt^2}\mathbf{M}_i^{\mathrm{V}} + \frac{2}{dt}\mathbf{C}_i^{\mathrm{V}n} + \mathbf{K}_i^{\mathrm{V}n})\mathbf{y}_i^{\mathrm{V}} + \sum_i^{N^{\mathrm{T}}}(\frac{4}{dt^2}\mathbf{M}_i^{\mathrm{T}} + \frac{2}{dt}\mathbf{C}_i^{\mathrm{T}n} + \mathbf{K}_i^{\mathrm{T}n})\mathbf{y}_i^{\mathrm{T}}$, where $(\ )^{\mathrm{V}}$ indicates values related to merged voxel elements, $(\ )^{\mathrm{T}}$ indicates values related to tetrahedral elements, and subscript i indicates the ith element. $diag[\ ]$ and ϵ indicate the 3×3 block Jacobi of $[\ ]$ and tolerance for relative error. $(\ )_c$ indicates the calculation related to FEmodel$_c$, and the other is the related calculation of the FEmodel. $\bar{\mathbf{P}}$ is a mapping matrix, from FEmodel$_c^i$ to FEmodel^i, which is defined by interpolating the displacement in each element of FEmodel$_c^i$. $(\bar{\ })$ indicates FP32 variables, while other parts are conducted in FP64.

---

read boundary condition of n-th time step

$\mathbf{f} \Leftarrow \mathbf{F}^n - \mathbf{Q}^{n-1} + \mathbf{C}^n\mathbf{v}^{n-1} + \mathbf{M}(\mathbf{a}^{n-1} + \frac{4}{dt}\mathbf{v}^{n-1})$

$\mathbf{x} \Leftarrow \frac{dt}{24}(-9\mathbf{v}^{n-4} + 37\mathbf{v}^{n-3} - 59\mathbf{v}^{n-2} + 55\mathbf{v}^{n-1})$

$\bar{\mathbf{B}} \Leftarrow diag[\frac{4}{dt^2}\mathbf{M} + \frac{2}{dt}\mathbf{C}^n + \mathbf{K}^n]$

$\bar{\mathbf{B}}_c \Leftarrow diag[\frac{4}{dt^2}\mathbf{M}_c + \frac{2}{dt}\mathbf{C}_c^n + \mathbf{K}_c^n]$

$\mathbf{r} \Leftarrow \mathbf{f} - \mathbf{A}^n\mathbf{x}$ (* computed by EBE with voxel *)

$\beta \Leftarrow 0$

$i \Leftarrow 1$

(*outer loop start*)

while $\|\mathbf{r}\|_2/\|\mathbf{f}\|_2 \geq \epsilon$ **do**

  (*inner loop start*)

  $\bar{\mathbf{r}} \Leftarrow \mathbf{r}$

  $\bar{\mathbf{z}} \Leftarrow \mathbf{B}^{-1}\mathbf{r}$

  $\bar{\mathbf{r}}_c \Leftarrow \bar{\mathbf{P}}^T\bar{\mathbf{r}}$

  $\bar{\mathbf{z}}_c \Leftarrow \bar{\mathbf{P}}^T\bar{\mathbf{z}}$

  $\bar{\mathbf{z}}_c \Leftarrow (\frac{4}{dt^2}\bar{\mathbf{M}}_c + \frac{2}{dt}\bar{\mathbf{C}}_c^n + \bar{\mathbf{K}}_c^n)^{-1}\bar{\mathbf{r}}_c$ (*solved on FEmodel$_c$ by algorithm 6 with ϵ_c^{in} and initial solution $\bar{\mathbf{z}}_c$*)

  $\bar{\mathbf{z}} \Leftarrow \bar{\mathbf{P}}\bar{\mathbf{z}}_c$

  $\bar{\mathbf{z}} \Leftarrow (\frac{4}{dt^2}\bar{\mathbf{M}} + \frac{2}{dt}\bar{\mathbf{C}}^n + \bar{\mathbf{K}}^n)^{-1}\bar{\mathbf{r}}$ (*solved on FEmodel by algorithm 6 with ϵ^{in} and initial solution $\bar{\mathbf{z}}$*)

  $\mathbf{z} \Leftarrow \bar{\mathbf{z}}$

  (*inner loop end*)

  if $i > 1$ **then**

    $\beta \Leftarrow (\mathbf{z}, \mathbf{q})/\rho$

  end if

  $\mathbf{p} \Leftarrow \mathbf{z} + \beta\mathbf{p}$

  $\mathbf{q} \Leftarrow \mathbf{A}^n\mathbf{p}$ (* computed by EBE with voxel *)

  $\rho \Leftarrow (\mathbf{z}, \mathbf{r})$

  $\alpha \Leftarrow \rho/(\mathbf{p}, \mathbf{q})$

  $\mathbf{q} \Leftarrow -\alpha\mathbf{q}$

  $\mathbf{r} \Leftarrow \mathbf{r} + \mathbf{q}$

  $\mathbf{x} \Leftarrow \mathbf{x} + \alpha\mathbf{p}$

  $i \Leftarrow i + 1$

end while

(*outer loop end*)

$\delta\mathbf{u}^n \Leftarrow \mathbf{x}$

$\mathbf{Q}^n \Leftarrow \mathbf{Q}^{n-1} + \mathbf{K}^n\delta\mathbf{u}^n$

$\mathbf{u}^n \Leftarrow \mathbf{u}^{n-1} + \delta\mathbf{u}^n$

$\mathbf{v}^n \Leftarrow -\mathbf{v}^{n-1} + \frac{2}{dt}\delta\mathbf{u}^n$

$\mathbf{a}^n \Leftarrow -\mathbf{a}^{n-1} - \frac{4}{dt}\mathbf{v}^{n-1} + \frac{4}{dt^2}\delta\mathbf{u}^n$
output results of n-th time step
modify material properties considering nonlinearity

Algorithm 6 PCG method used to roughly solve $\mathbf{z} = (\frac{4}{dt^2}\mathbf{M} + \frac{2}{dt}\mathbf{C}^n + \mathbf{K}^n)^{-1}\mathbf{r}$ and $\mathbf{z}_c = (\frac{4}{dt^2}\mathbf{M}_c + \frac{2}{dt}\mathbf{C}_c^n + \mathbf{K}_c^n)^{-1}\mathbf{r}_c$ in Algorithm 5. The matrix vector product $\mathbf{A}^n\mathbf{y} = (\frac{4}{dt^2}\mathbf{M}+\frac{2}{dt}\mathbf{C}^n+\mathbf{K}^n)\mathbf{y}$ is computed using an EBE method with voxel: $\sum_i^{N^{\mathrm{V}}}(\frac{4}{dt^2}\mathbf{M}_i^{\mathrm{V}} + \frac{2}{dt}\mathbf{C}_i^{\mathrm{V}n} + \mathbf{K}_i^{\mathrm{V}n})\mathbf{y}_i^{\mathrm{V}} + \sum_i^{N^{\mathrm{T}}}(\frac{4}{dt^2}\mathbf{M}_i^{\mathrm{T}} + \frac{2}{dt}\mathbf{C}_i^{\mathrm{T}n} + \mathbf{K}_i^{\mathrm{T}n})\mathbf{y}_i^{\mathrm{T}}$, where $(\ )^{\mathrm{V}}$ indicates values related to merged voxel elements, $(\ )^{\mathrm{T}}$ indicates values related to tetrahedral elements, and the subscript i indicates the ith element. $\mathbf{B}$ and ϵ^{in} are a block Jacobi matrix and the tolerance for the relative error, respectively. All of the calculations are made using single-precision variables. The computation in loop 3 is merged into one do loop, which reduces the total memory access from 13 streams in the naive implementation to 8 streams per CG iteration.

$\mathbf{x} \Leftarrow \mathbf{z}$
$\mathbf{e} \Leftarrow \mathbf{r} - \mathbf{A}^n\mathbf{x}$ (* computed by EBE with voxel *)
$\beta \Leftarrow 0$
$i \Leftarrow 1$
$\mathbf{z\_or\_q} \Leftarrow \mathbf{B}^{-1}\mathbf{e}$
$\rho_a \Leftarrow (\mathbf{z\_or\_q}, \mathbf{e})$
$r2 \Leftarrow \|\mathbf{r}\|_2$
$err \Leftarrow \|\mathbf{e}\|_2/r2$
while $err \geq \epsilon^{in}$ **do**
  $\mathbf{p} \Leftarrow \mathbf{z\_or\_q} + \beta\mathbf{p}$ (* loop 1: two streams *)
  $\mathbf{z\_or\_q} \Leftarrow \mathbf{A}^n\mathbf{p}$ (* computed by EBE with voxel *)
  $\alpha \Leftarrow \rho_a/(\mathbf{p}, \mathbf{z\_or\_q})$ (* loop 2: two streams *)
  $\rho_b \Leftarrow \rho_a$
  (* loop 3 begin: four streams + one 3×3 stream ($\mathbf{B}^{-1}$) *)
    $\mathbf{e} \Leftarrow \mathbf{e} - \alpha\mathbf{z\_or\_q}$
    $\mathbf{x} \Leftarrow \mathbf{x} + \alpha\mathbf{p}$
    $\mathbf{z\_or\_q} \Leftarrow \mathbf{B}^{-1}\mathbf{e}$
    $\rho_a \Leftarrow (\mathbf{z\_or\_q}, \mathbf{e})$
    $err \Leftarrow \|\mathbf{e}\|_2/r2$
  (* loop 3 end *)
  $\beta \Leftarrow \rho_a/\rho_b$
  $i \Leftarrow i + 1$
end while
$\mathbf{z} \Leftarrow \mathbf{x}$

(a) **Adaptive conjugate gradient method[10] with mixed precision arithmetic and multi-grid method[7]**: The inverse of a matrix close to the matrix of the target equation is typically used as a preconditioner for the CG iteration. The adaptive CG method uses another CG solver to roughly solve this preconditioning equation instead of multiplying a fixed matrix. The original CG iteration is called the outer loop and the CG iteration in the preconditioner is called the inner loop. Because we only need to solve the matrix equation roughly in the inner loop, there is room for algorithm improvement. First, the computation of the inner loop is performed in FP32, since the precision of the preconditioning computation can be reduced. Faster computation can be expected due to improved peak performance on some hardware, reduced memory transfer size, increased apparent cache capacity, and reduced data size for MPI communication. Next, we use multi grid to reduce computation. Here, we use another model that is equivalent to the original problem, albeit at a coarser resolution. Hereafter, the original model is denoted as FEmodel and the coarsened model is denoted as FEmodel$_c$. In the inner loop, we first solve the problem roughly on FEmodel$_c$, and use this as an initial solution for the FEmodel, to reduce the computational cost of the preconditioner. Reducing the number of iterations in solving the FEmodel roughly and improving the convergence characteristics reduces the number of floating-point operations. Although additional computation on the FEmodel$_c$ is required, it has fewer DOF and fewer finite element nodes to be communicated between MPI processes than in the original problem. As a result, the computational cost and communication size were reduced compared to the FEmodel, significantly reducing the cost of the inner loop. When running on a massively parallel computer, the communication involved in mapping values between the fine and coarse grids becomes a bottleneck; thus, we use a first-order tetrahedral element model for the FEmodel$_c$, which is generated by removing the edge nodes of the second-order tetrahedral element model used for the FEmodel. The same parts of the sub-models are assigned to the same compute nodes, which enables mapping between FEmodel and FEmodel$_c$ with small communication cost.

(b) **EBE method**[8]: Due to the large DOF per compute node, the cost of updating matrices $\mathbf{K}$ and $\mathbf{M}$ can be high, and it sometimes becomes impossible to store the global matrix within the limited memory capacity. Therefore, matrix vector products $\mathbf{Ku}$ and $\mathbf{Mu}$ of the FEmodel and FEmodel$_c$ are calculated using the EBE method, which computes $\mathbf{Ku}$ and $\mathbf{Mu}$ element by element without storing the entire $\mathbf{K}$ and $\mathbf{M}$ matrices in memory. Using orthogonal functions for bases of stress (i.e., displacement–stress mixed formulation of finite element method[12]), we transform $\mathbf{Ku}$ into $\sum_i \mathbf{N_b}_i^T \mathbf{D} \mathbf{N_b}_i \mathbf{u}$, and compute the matrix vector product as $\sum_i \mathbf{N_b}_i^T (\mathbf{D}(\mathbf{N_b}_i \mathbf{u}))$. Here, $\mathbf{D}$ is the elasticity matrix and $\mathbf{N_b}_i$ is the equivalent mode. Since $\mathbf{N_b}_i$ and $\mathbf{D}$ comprise many zero components, we compute only non-zero components by explicitly expanding the computation in the finite element program. We also compute $\mathbf{Mu}$ element by element, and add the results to the global left-hand-side vector.

(c) **"Structuring" unstructured mesh**: When using highly efficient algorithms involving (a) and (b), the communication cost becomes negligible, and reducing computation time and improving load balancing becomes the key to speeding up the solver. Here, most of the time is spent on the EBE kernels of the inner fine, inner coarse, and outer loops. Although the computation cost is large, it involves indirect list accesses and data recurrence. EBE computation suitable for current CPUs with multi-core and single instruction multiple data (SIMD) units is important for speedup. In modern computers, structurally arranged computation leads to efficient computation, and structured geometry leads to reduced computation. Although unstructured meshes inherently involve unstructured calculations, we extract uniform regions inside the complex geometry and compute those regions using structured calculations. Here, a multi-resolution mesh of voxels and tetrahedra is generated, and each voxel is further decomposed with six tetrahedra; see Fig. 10. Although the resulting mesh is an unstructured mesh consisting of second-order tetrahedral elements, uniform regions inside the domain are structured; these are decomposed with six types of elements with fixed geometry. Since the geometry related part of the

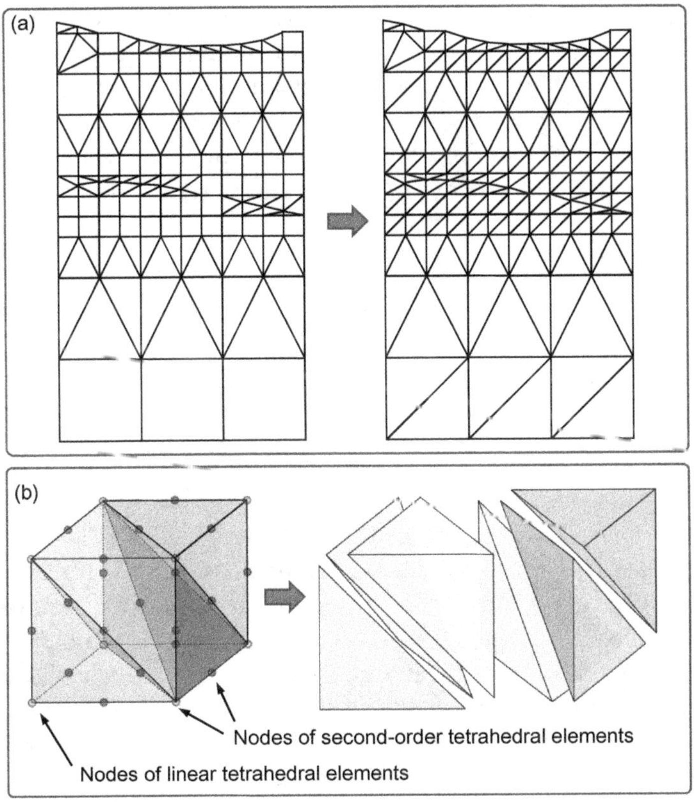

Fig. 10. "Structuring" an unstructured mesh. (a) A hybrid mesh comprising multi-resolution voxel elements and unstructured tetrahedral elements is converted to an unstructured tetrahedral mesh combined with a tetrahedral mesh with fixed geometry, and (b) decomposition of a voxel element into six tetrahedral elements with fixed geometry.

computation of $\mathbf{Ku}$ can be skipped, the number of floating point operations in EBE computation can be reduced by 2.12-fold when compared with using usual second-order tetrahedral elements. Also, when calculating six elements together in one voxel, the number of random accesses can be reduced from 60 (i.e., 10 nodes per second-order tetrahedral element $\times$ 6 elements) to 27. The same can be applied to first-order tetrahedral elements, with 3.55-fold reduction in floating point operations for the EBE computation, and reduction in random accesses from 24 (4 nodes $\times$ 6) to 8. Since computations and random accesses are reduced, we can expect shorter time-to-solution and improvement in computational efficiency. By storing the nodal data of elements in temporary buffers, we can transform the data accesses in EBE computation into contiguous access without data recurrence, which can be computed using SIMD units.

(d) **Steady contiguous memory access computation**: With the very fast matrix vector product computation enabled by (c), the time required for contiguous memory access computation (i.e., vector addition, inner product) becomes non-negligible in the total computation time. In such cases, the additional cost if cache thrashing occurs in these contiguous memory access computations can become large. Although the probability for cache thrashing to occur for a specific array length is very low, many partitions with different array lengths are generated when using graph partitioning libraries such as METIS.[11] Thus, the probability for cache thrashing to occur on at least one of the processes in a massively parallel environment becomes large, especially for the K computer where two-way L1 cache is equipped. In order to avoid cache thrashing on massively parallel computations, we allocate arrays with the same length for all processes, and use only their necessary parts. This leads to similar memory alignments for all processes, leading to a significant reduction in probability of cache thrashing. We also exchange the order of `do` loops inside the standard PCG, and merge these loops to reduce memory access from 13 streams to 8 (see Algorithm 6). Combination with use of FP32 data-types results in reduction of memory access by $(13 \times 2)/8 = 3.25$-fold of that of the standard PCG solver.

The SC15 solver uses the inner loop for preconditioning and the outer loop for the normal CG iteration using (a) and (b); efficient preconditioning is conducted with low memory footprint, fast convergence, and reduced computation, while the outer loop attains low memory footprint computation. Furthermore, (c) and (d) enable efficient memory access and reduction in computation by use of the constraints in the arithmetic space of (a) and (b). The computation cost of an outer loop iteration is similar to that of a PCG iteration, but the number of iterations becomes small due to the effective preconditioning by the inner loops. Since most computation cost is involved in the fast inner loops, the computation time of the entire solver is reduced. The SC15 solver is expected to be effective on systems with fast single-precision arithmetic (e.g., an Intel CPU or GPU) or systems with slow interconnects, but fast computation is also expected on the K computer, which has the same peak performance regardless of single or double precision arithmetic, due to reduced data access volumes. The Adams–Bashforth method[9] is used to predict the initial solution of the iterative solver when solving $\delta \mathbf{u}^n$. The FE models are generated as in Ref. 1, and partitioned using METIS.

4.2. *Performance*

4.2.1. *Performance measurement environment*

Although the SC15 solver is suitable for environments with fast FP32 arithmetic, performance was measured on the K computer at RIKEN, Advanced Institute for Computational Science,[14] as many computing nodes are available for solving large-scale problems. The K computer is a massively parallel CPU-based supercomputer comprising 82,944 compute nodes, each with a single SPARC64™ VIIIfx CPU. Each of the eight cores of the SPARC CPUs has two sets of fused multiply-add (FMA) SIMD units of length 2. This leads to FMA (two operations) $\times$ SIMD length 2 $\times$ 2 sets = 8 operations per clock. Consequently, the peak performance of each CPU is 8 operations $\times$ 8 cores $\times$ 2 GHz = 128 GFLOPS. The SIMD length is 2 for both FP32 and FP64 arithmetic, so that the peak performance of FP32 arithmetic is also 128 GFLOPS. Each core has two 32 KB two-way L1 caches, each for data and instructions, and the cores share a 6 MB 12-way L2 cache (shared with data/instruction). Each node has 16 GB of DDR3-SDRAM memory, with a peak memory bandwidth of 64 GB/s. Tofu, a 6D interconnection network, is used for communication between nodes.[15] Each node can communicate in four directions simultaneously with 5 GB/s throughput in each direction. An OpenMPI[16]-based MPI library optimized for the Tofu network is used.

4.2.2. *Performance measurement setting*

The effectiveness of the SC15 solver was compared to the SC14 and PCGE solvers. The SC14 solver is a solver that was developed in an SC14 paper[1] and described in Section 3. PCGE is a preconditioned conjugate gradient solver with EBE for matrix–vector products, which was referred to as a standard method in the SC14 paper.[1] The SC14 solver and PCGE were improved for solving large-scale problems. For the relative error tolerance of the outer loop, we use $\epsilon = 10^{-8}$. Relative error tolerance of $\epsilon^{\text{in}} = 0.25$ and $\epsilon_c^{\text{in}} = 0.5$ were used for the inner loops. Here, the partitioned models FEmodel^i and FEmodel_c^i are assigned to each compute node, and each partition is computed using OpenMP; thus, $8 \times m$ CPU cores were used when using m K computer compute nodes (m MPI processes).

A two-layer FE model was created mimicking actual ground, and the performance of nonlinear ground motion analysis was measured by uniformly inputting waves from the bottom of the model. The model has periodic geometry for use in weak scaling measurements; see Fig. 11 for the model and its material properties. The first layer is a soft layer that behaves nonlinearly for strong inputs, and the second layer is a hard layer that mimics engineering bedrock, with bumps between the layers. The model is copied in the x and y directions for generating larger models used for weak scaling measurements. The

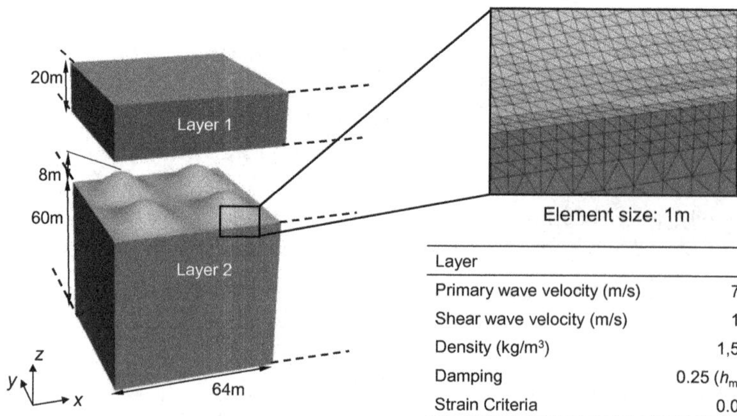

Layer	1	2
Primary wave velocity (m/s)	700	2,100
Shear wave velocity (m/s)	100	700
Density (kg/m³)	1,500	2,100
Damping	0.25 (h_{max})	0.05
Strain Criteria	0.007	-

Fig. 11. Ground model used for performance measurement. The model is duplicated in the x and y directions to make large-scale models with periodical configurations.

models are discretized with minimum element size of 0.444 m for convergence in the acceleration response even when the soil softens due to nonlinearity. The main part of the Kobe wave[18] ($dt = 0.002$s) observed during the 1995 Southern Hyogo Earthquake is used as an input. Table 2 shows the properties of the weak and strong scaling model sets. Although a periodical problem is used, the model is

Table 2. Configurations of performance measurement models.

Model	# of MPI processes	# of compute cores	Total DOF	Mean DOF per MPI domain
A-1	1,152	9,216	15,095,157,099	13,103,435
A-2	2,304	18,432	30,186,410,283	13,101,741
A-3	4,608	36,864	60,368,916,651	13,100,893
A-4	9,216	73,728	120,730,026,027	13,100,046
A-5	18,432	147,456	241,452,244,779	13,099,623
A-6	36,864	294,912	482,888,875,563	13,099,199
A-7	82,944	663,552	1,086,476,549,163	13,098,917
B-1	1,152	9,216	12,956,427,579	11,246,899
B-2	2,304	18,432		5,623,449
B-3	4,608	36,864		2,811,725
B-4	9,216	73,728	Same as above	1,405,862
B-5	18,432	147,456		702,931
B-6	36,864	294,912		351,466

Model	Total # of tetra elements	(# of voxel elements)	(# of tetra elements)	# of PCGE iterations
A-1	3,755,249,664	449,798,144	1,056,460,800	1,238
A-2	7,510,474,752	899,579,904	2,112,995,328	1,237
A-3	15,020,924,928	1,799,143,424	4,226,064,384	1,238
A-4	30,041,800,704	3,598,254,080	8,452,276,224	1,237
A-5	60,083,552,256	7,196,475,392	16,904,699,904	1,238
A-6	120,167,006,208	14,392,885,248	33,809,694,720	1,237
A-7	270,375,616,512	32,383,893,504	76,072,255,488	1,236
B-1	3,220,557,760	450,803,328	515,737,792	8,970
B-2				8,971
B-3				8,970
B-4	Same as above	Same as above	Same as above	8,970
B-5				8,969
B-6				8,969

Note: Model set A is a weak scaling model set, while model set B is a strong scaling model set.

partitioned by METIS and the model set is expected to have load balancing and communication characteristics similar to practical problems. Here, we solve 3 time steps for the weak scaling model set (model set A) and 25 time steps for the strong scaling model set (model set B). Although the convergence characteristics change with the model size, the model sets are made with periodical settings so that they have similar convergence characteristics regardless of the model sizes. In fact, the number of iterations required for convergence of the PCGE solver is approximately constant for all models in each model set; thus, we can see that the model set is suitable for measuring weak scaling.

4.2.3. *Weak scaling*

First, we measure weak scaling (see Fig. 12). The outer, inner fine, and inner coarse indicate the outer loop, line 18 of Algorithm 5, and line 16 of Algorithm 5, respectively. The dark gray/gray/light gray numbers indicate the number of iterations required for convergence. We can see that the number of iterations is similar across models; the model set is suitable for measuring weak scalability.

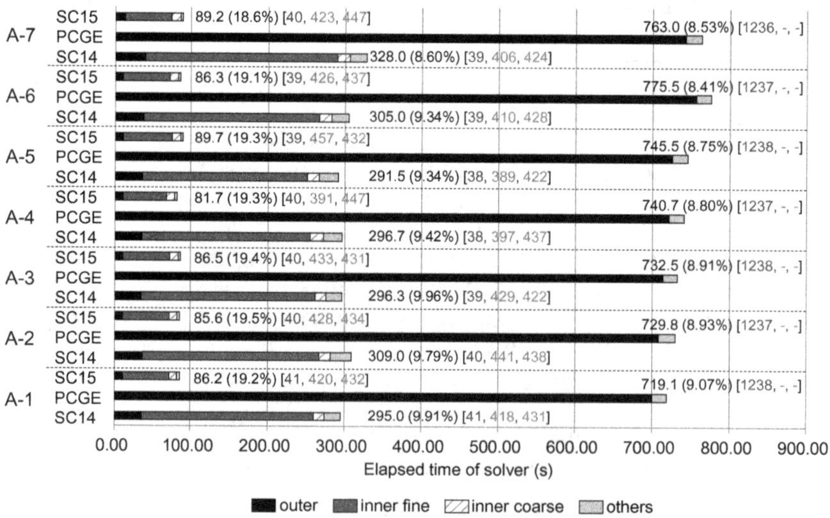

Fig. 12. Weak scaling results on model A. Numbers in brackets indicate floating point performance, while numbers in square brackets indicate iterations required for convergence of the solver (dark gray: outer, gray: inner_fine, light gray: inner_coarse).

The smallest model, A-1, with 15.1 billion DOF, is solved in 86.2 s with 28.3 TFLOPS (19.2% of FP64 peak) when using 9,216 CPU cores (1,152 nodes × 8 CPU cores), while the largest model, A-7, with 1.08 trillion DOF, is solved in 89.2 s with 1.97 PFLOPS (18.6% of FP64 peak) when using the full K computer system with 663,552 CPU cores (82,944 nodes × 8 CPU cores). The weak scaling efficiency up to the full system is 96.6%. In comparison, the elapsed time when using the SC14 solver is 328 s (8.60% of FP64 peak), and the elapsed time when using PCGE is 763 s (8.53% of FP64 peak). As expected, the FLOP count is reduced from 6.91E+17 (PCGE) → 3.00E+17 (SC14 solver) → 1.76E+17 (SC15 solver), and the time-to-solution is reduced by 8.55-fold and 3.67-fold from the PCGE and SC14 solvers, respectively. As the SC15 solver has better load balancing, the scalability when comparing 1,152 nodes and the full system improved from 89.9% (SC14 solver) and 94.2% (PCGE) to 96.6% (SC15 solver). Table 3 shows the elapsed time and imbalance of the matrix vector computation (mostly random access), other computation (mostly contiguous access), global communication, and point-to-point communication on the full system problem (model A-7). The maximum communication time

Table 3. FLOP count and breakdown of the elapsed time of SC15, SC14, and PCGE solvers on A-7 model (1.09 TDOF model running on full K computer system).

	SC15 solver	SC14 solver	PCGE solver
Total FLOP count	1.76E+17	3.00E+17	6.91E+17
Computation matrix vector product (s)	65.28 (72.21)	202.2 (232.3)	552.37 (607.5)
Computation others (s)	16.38 (19.51)	39.45 (85.43)	84.71 (188.0)
Communication all reduce (s)	0.19 (0.26)	1.12 (2.31)	0.91 (1.38)
Communication point-to-point (s)	1.36 (3.26)	1.31 (3.45)	4.17 (8.82)
Communication pack/unpack (s)	0.58 (1.64)	0.69 (2.50)	1.97 (4.19)
Total with barrier (s)	96.88	327.3	811.2
Total without barrier (s)	89.24	328.0	763.0

Note: Numbers indicate the mean elapsed time, while numbers in brackets indicate the maximum elapsed time among all MPI processes.

(global + point-to-point) of all processes is at most 3.52 s (3.9% of the total elapsed time) when using SC15 solver; we can see that the communication time is sufficiently small compared to the computation time. We can see that the imbalance of elapsed time in "Computation others" is large for SC14 solver and PCGE, which is the major reason for the imbalance of the full application. This imbalance ratio (i.e., maximum elapsed time among processes/average elapsed time of all processes) of 2.17 in the SC14 solver is greatly improved to 1.19 in the SC15 solver by avoiding L1 cache thrashing, which leads to an improvement in scalability of the full application. The combination of single-node performance and scalability improvements have lead to high peak performance of 18.9% of FP64 peak on full K computer (2.01 PFLOPS), which is very high for an unstructured low-order finite element method with implicit time integration.

4.2.4. *Strong scaling*

Next, we measure strong scaling. Figure 13 shows the elapsed time with the increase in the number of CPU cores. We can see that

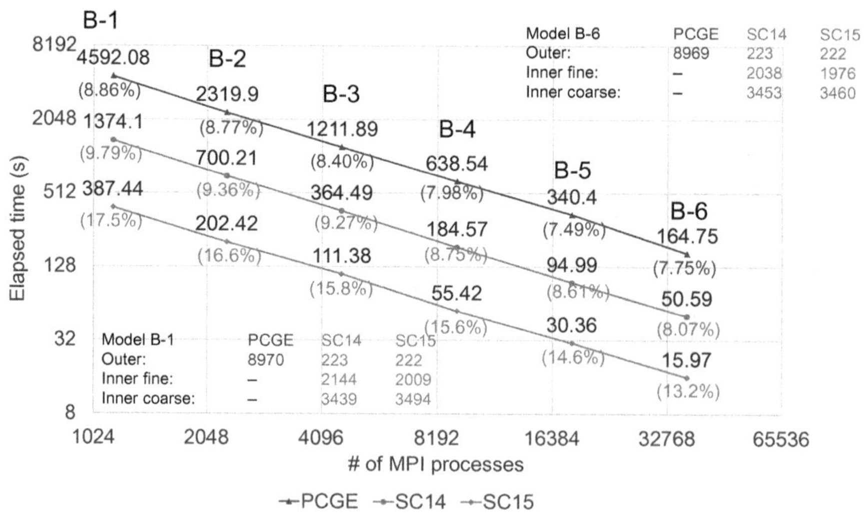

Fig. 13. Strong scaling results on model B. Numbers in brackets indicate floating-point performance. Black lines indicate ideal scaling. Insets in bottom left and top right indicate elapsed loops of outer, inner fine, and outer loops for model B-1 (1,152 processes), and model B-6 (36,864 processes), respectively.

good scalability is attained although there is some degradation in performance with an increase in CPU cores: the 13.0 billion DOF problem is solved in 387.4 s with 25.8 TFLOPS (17.52% of FP64 peak) with 9,216 CPU cores (1,152 nodes × 8 CPU cores, model B-1), in 55.4 s with 184.3 TFLOPS (15.62% of peak) with 73,728 CPU cores (9,216 nodes × 8 CPU cores, model B-4); and in 16.0 s with 623.8 TFLOPS (13.22% of FP64 peak) with 294,912 CPU cores (36,864 nodes × 8 CPU cores, model B-6). The strong scaling efficiency from 9,216 to 73,728 CPU cores is 87.4%, and the strong scaling efficiency from 9,216 to 294,912 CPU cores is 75.8%. The SC14 solver has better scalability (84.9% from 9,216 to 294,912 CPU cores), because the communication time is the same with SC15 solver, but computation takes longer. There is possibility of improvement of scalability by overlapping point-to-point communication with computation in the matrix vector products in CG iterations.

4.2.5. *Performance on practical problem*

Finally, we show the performance of the 0.133 TDOF application problem to be shown in Section 4.3. Figure 14 shows the time required to analyze the first 25 time steps. The elapsed time of the application (including I/O) is 50.6 s, with only 2.03 s used to solve each time step of the 0.133 TDOF dynamic nonlinear problem. Of the total elapsed time, 92.5% (46.8 s) is spent on the solver, indicating that large-scale analysis is enabled by sufficient speedup of the solver. All other

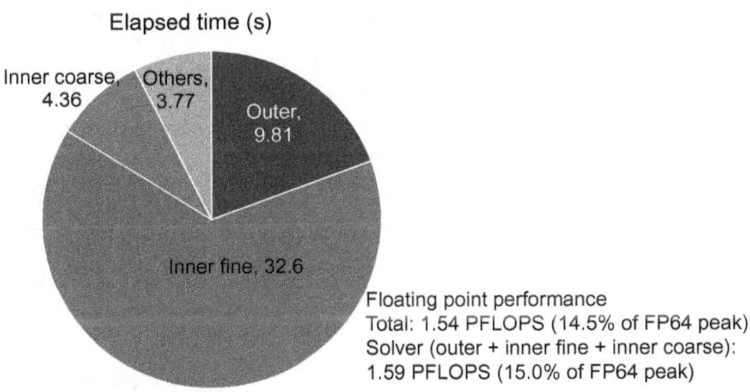

Fig. 14. Computation performance on the 0.133 TDOF application problem. Performance measured for the first 25 steps.

operations (I/O, updating the material property with the nonlinear constitutive model, and updating the block Jacobi matrix used for the block Jacobi preconditioner) required only 7.5% of the total time, leading to high application performance. The output data size for the 6,600 time step problem was 382 GB (surface displacements were output every $dt = 0.01$ s at 2 m resolution). Performance was 1.59 PFLOPS (15.0% of FP64 peak) for the solver and 1.54 PFLOPS (14.5% of FP64 peak) for the whole application, indicating high performance for practical problems. Compared to the SC14 application, the problem size increased by 0.133 trillion/10.7 billion = 12.4-fold and efficiency improved by 14.5%/7.59% = 1.91-fold. While the SC14 solver was only capable of executing applications on 36,864 nodes, the SC15 solver improves scalability and enables application runs over the full system. If the elapsed time required for computing one second in the simulation are compared, a 12.4-fold larger problem is computed in 0.63-fold the speed.[a] This corresponds to roughly a 7.87-fold speedup in the total elapsed time.

4.3. *Example*

In this section, we present a comprehensive earthquake simulation of an actual city performed on the K computer using the SC15 solver. Urban earthquake disasters involve the propagation of seismic waves in the crust due to fault rupture (seismic wave propagation), nonlinear wave propagation near the ground surface due to strong material nonlinearities (ground amplification), and building response due to ground shaking (structural response); see Fig. 15. A comprehensive physics-based simulation of these processes is expected to improve the reliability of urban earthquake hazard estimation, but this has not been possible due to the enormous computational cost. For this reason, statistical methods that relate seismic source intensity to ground motion intensity or ground motion intensity to building damage[23–25] are typically used, rather than physics-based evaluations. To achieve a breakthrough in earthquake disaster estimation, we present a comprehensive physics-based earthquake

[a](41,521 elapsed time/30.0 simulation time; SC14 solver)/(28,800 elapsed time/13.2 simulation time; SC15 solver) = 0.63.

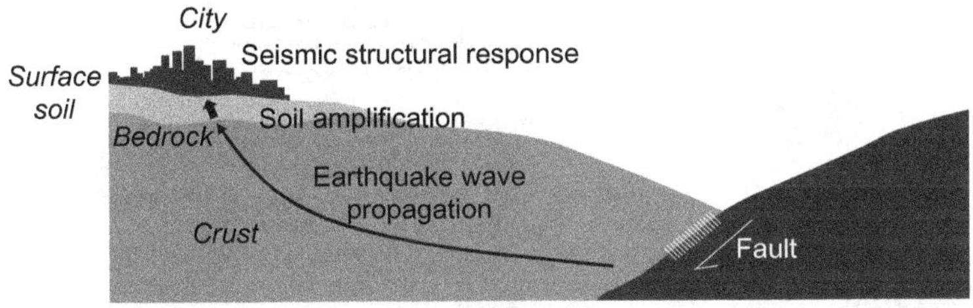

Fig. 15. Urban earthquake disasters consist of the processes of seismic wave propagation, soil amplification, and seismic structure response. Comprehensive physics-based earthquake simulations that include these processes are expected to be enormously computationally expensive, so supercomputing is expected to contribute to this effort.

simulation that covers all of these disaster processes using large-scale computation.

Here, we target the central area of Tokyo, an area measuring 10.25 × 9.25 km with 328,056 structures, which includes financial, business, and commercial centers, government offices, and densely populated residential districts, and simulate the seismic response of the city and evacuation of people in an assumed Tokyo earthquake. The earthquake wave propagation simulation and evacuation/recovery simulation shown as follows compare favorably to the previous largest-scale simulations, while the soil amplification simulation in this comprehensive simulation surpasses scales conducted in previous simulations by full utilization of the full K computer.

With the accumulation of crustal data and developments in high-performance computing, many studies have estimated ground motion based on 3D linear wave propagation. Computations for analyzing fault rupture and wave propagation at high resolution in a huge area are computationally expensive and have been studied extensively (e.g., Refs. 26, 27, 38 and 28). Here, we constructed a low-order unstructured/structured hybrid finite element model (an 18 billion first-order finite element, 56 billion DOF model, with a minimum element size of 5 m), targeting an area of 34.56 × 61.44 × 60 km, based on the crust model of Ref. 32. Using a source model of Ref. 33, we solved a 31.2 s linear earthquake propagation problem using implicit time integration. Because of the inherently large material heterogeneity of the crust, many elements are needed for discretization: we use 56 billion elements computed using 36,864

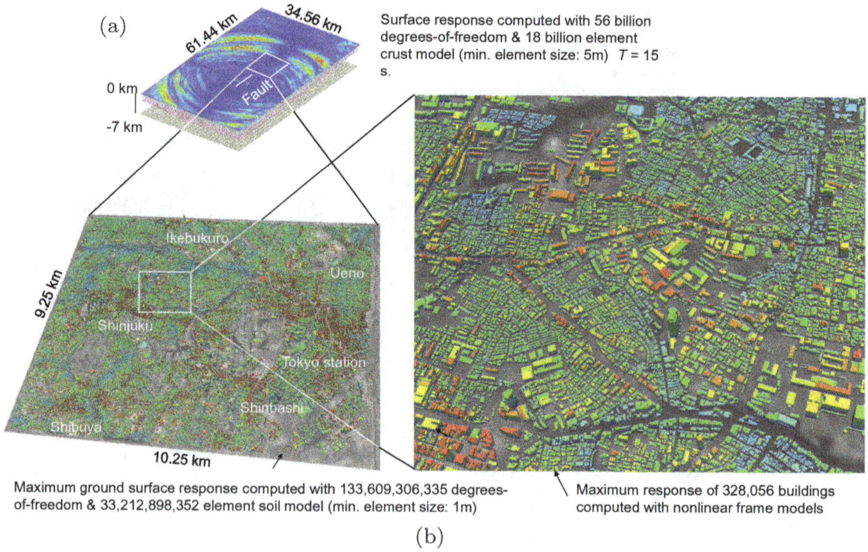

Fig. 16. Comprehensive simulation of an earthquake in Tokyo. (a) Snapshot of computed ground motion at surface. Due to the complex fault rupture process and crust structure, a complex distribution is observed (the mesh indicates interfaces of the crust structure, and color contour indicate velocity norm), and (b) snapshots of computed ground motions and seismic structural response; a complex distribution is observed due to the varying characteristics of structures and soil structure (black/white contours indicate maximum ground displacement of ground, while color contours indicate maximum relative displacement of structures).

compute nodes $\times$ 8 OpenMP threads = 294,912 CPU cores of the K computer for this problem. Figure 16(a) shows a snapshot of the computed ground motion for the assumed Tokyo earthquake. Here, we use enhanced code of Ref. 34. Although already a huge scale problem, its computation scale is still small compared to the nonlinear wave analysis involved in soil amplification shown next.

With the accumulation of urban spatial data, research on ground amplification and seismic structural response simulation using physics-based models of cities has begun. Although it is expected to be a frontier in supercomputing, it has been little studied (e.g., Refs. 29, 30 and 35) because of its huge computational cost. Ground amplification processes include local nonlinear responses due to the complex geometry and high material heterogeneity of the ground. Low-order unstructured 3D nonlinear wave simulations are used to model these phenomena, leading to huge computational costs. In the current state-of-the-art paper in nonlinear wave simulation (SC14 Gordon Bell prize finalist paper[1]) seismic response of a 2.0 × 2.0 km area of an actual city with 13,275 structures was computed using 36,864 compute nodes × 8 OpenMP threads = 294,912 CPU cores of the K computer using a 10.7 billion DOF and 2.67 billion finite-element model. In contrast, the seismic response analysis for an assumed Tokyo earthquake would require simulating a 10.25 × 9.25 × 0.24 km area with 0.133 tera-DOF and 33 billion elements; corresponding to 23.7-fold larger area and 12.4-fold larger DOF compared to the state-of-the-art computation conducted in the SC14 paper. Our model includes a complex three-layered ground structure constructed from the 5 m digital elevation map of the Geographic Information Authority of Japan, the Digital Soil Map,[20] and material properties obtained from the National Research Institute for Earth Science and Disaster Prevention,[21] and is discretized with an element size sufficiently small for convergence of the acceleration response, even when the ground is softened due to nonlinearity. Using this finite element model, we input the main part of the computed wave of the earthquake wave propagation simulation shown in the previous section, and compute the nonlinear wave equation (6,600 time steps with a 0.002 s time increment) using the full K computer (82,944 nodes × 8 CPU cores = 663,552 CPU cores). The calculated seismic motions are then used to calculate the response of the 328,056 structures located within the domain. Figure 16(b) shows snapshots of the computed ground motion and seismic structural response. It can be seen that not all structures exhibit significant response, even in areas of strong ground motion. Such complex responses in cities can only be computed by combining high-resolution 3D nonlinear wave simulations with response simulation of a large number of buildings in the city.

5. CPU-Based Finite Element Solver for Crust Deformation

This section describes the development of a finite element solver for crustal deformation analysis on massively parallel CPU-based computer systems.[41] Here, Eq. (7) is solved for computing the deformation of the crust due to plate motion and faulting. Such analysis is important for physics-based earthquake forecasting, which is based on assimilation of observation data and high-fidelity analysis.[42,43] In such forecasting methods, crustal movement observation data from GPS is used to estimate changes in the state (cohesive state) of the plate boundary, which is then compared with past earthquake occurrence patterns to estimate the probability of an earthquake. To be used for such purposes, the geometry of the 3D topography and crustal layer interface must be accurately modeled and the stresses and strains accurately calculated in order to update the nonlinear material constitutive models at plate boundaries. This leads to large-scale problems, requiring high-performance computing methods for analysis. In the following, we describe the details of the target problem and the developed adaptive CG type solver with mixed-precision and multi-grid preconditioner (hereafter referred to as the PASC17 solver). We show the performance on CPU-based K computer system, and show an application example of a crust deformation analysis.

5.1. *Algorithm design*

The major cost involved in the crust deformation problem is in solving Eq. (7) which governs the force equilibrium in the crust (i.e., solving). The time to solution for the implicit solver needs to be shortened for conducting multiple simulations during the data assimilation process; however, due to the difficulty of obtaining high peak performance and good load balance while simultaneously obtaining good convergence characteristics for iterative solvers, an all-purpose solver with short time to solution for a wide variety of problems has not yet been developed. One solution is to use a structured grid to obtain high performance and good load balancing. Although such structured grids are suitable for simple geometry problems, special care must be taken to obtain accurate stresses and strains in complex geometry areas when applied to complex geometry problems. Using an unstructured grid for explicit discretization of complex 3D geometry improves accuracy, but often results in poor load balancing and reduced peak performance. Although not optimal for static problems, the SC14 solver[1] has been applied to a crustal deformation problem with a domain of 2,944 km × 2,752 km × 850 km at 500 m resolution using 8,192 compute nodes of the K computer.[40] The problem size was 10 billion DOF, which is 10^{3-4} times larger than conventional crustal deformation analysis, but still only 38 times smaller than the 62.5 m resolution unstructured analysis that is possible with the development of memory and computationally efficient methods shown in this section. From this perspective, we have developed a fast and scalable solver suitable for the target static problem with general unstructured second-order tetrahedral finite element mesh. This solver enables multi-case crustal deformation analysis with stress/strain convergence using meshes generated by a wide variety of mesh generators, which are expected to contribute to practical inverse analyses of plate boundary state problems on high-performance computer systems.

In this section, the PASC17 solver is developed based on the SC14 solver developed for dynamic earthquake wave propagation analysis. The SC14 solver was developed for dynamic earthquake wave propagation problems; thus, there is room for improvement for static crust deformation problems. Therefore, we will develop an algorithm and implementation suitable for the target unstructured problem and the target compute architecture. The SC15 solver[39] was developed as a further accelerated dynamic seismic wave propagation analysis method based on the SC14 solver, but since the SC15 solver is only applicable to a dedicated mesh generator with a hybrid structured and unstructured mesh, the SC14 solver was used as the basis for development. The solver presented in this section is designed to be applicable to general meshes, so the SC14 solver is used as the base solver.

The main components leading to a fast unstructured low-order finite-element solver on recent massively parallel computers are as follows:

(1) a low cost preconditioner with high convergence improvement,
(2) fast matrix-vector products,
(3) both of the above with a low memory footprint and good scalability.

(1) is needed to reduce the total number of solver iterations, (2) is needed to reduce the cost per iteration, and (3) is needed to obtain sufficient performance on modern high-performance computers with relatively low memory and comprising many parallel components. Good scalability requires good load balancing and small and less frequent communications, which sometimes creates a trade-off between

(1) and (2). Thus, when designing a solver, it is necessary to comprehensively consider the target problem, the target computation system, and the available numerical algorithms.

First, we consider a suitable preconditioner for the target problem. Since a preconditioner essentially approximates the inverse of Eq. (7), considering Green's function of the target problem can provide insight into a good preconditioner. While Green's function of a dynamic problem has a range of influence proportional to the time step size, Green's function of a static problem has an infinite range of influence. Thus, for a static problem, we must collect information from distant points to obtain a solution at a given point. Since iterative solvers, which are widely used to solve large-scale problems, tend to eliminate local errors but not long-range errors with a small number of iterations, a preconditioner that can effectively reduce long-range errors would be suitable for this problem. An effective solution to this problem is to use multi-grids, in which coarse grids resolve long range errors and fine grids resolve local errors. Here, the target problem Eq. (7) is converted to solving the following:

$$(\mathbf{P}^T\mathbf{K}\mathbf{P})\mathbf{u}_c = (\mathbf{P}^T\mathbf{f}), \tag{8}$$

and its solution $\mathbf{u}_c$ is used to approximate $\mathbf{u}$ as follows:

$$\mathbf{u} \approx \mathbf{P}\mathbf{u}_c. \tag{9}$$

Here, $\mathbf{P}$ is a matrix that maps between the original grid $\mathbf{u}$ and the generated coarse grid $\mathbf{u}_c$. Equations (8) and (9) are used recursively to coarsen the problem to a given size. Since the total problem size is reduced on the coarse grid, fewer iterations are required to reduce long range errors.

In the case of structured grids, the use of geometric multi-grids (i.e., equally spaced coarsening for $\mathbf{P}$) often results in approximate solutions $\mathbf{P}\mathbf{u}_c$ that are very close to the exact solution $\mathbf{u}$. Thus, the use of multi-grid allows the overall solver to be significantly faster because the coarse grid effectively reduces long-range errors by a small number of iterations per computation and communication size, and local errors can be solved in a smaller number of iterations on the original mesh. In addition, the matrix–vector multiplication used to solve Eq. (8) can often be computed easily on the fly. For example, $\mathbf{P}^T\mathbf{K}\mathbf{P}$ of a structured grid is equivalent to $\mathbf{K}$ of a structured grid with twice the grid spacing. Thus, matrix-free multiplication can be used rather than precomputing, storing, and reading the global matrix from memory each time in an iterative solver. Since it is straightforward to obtain good load balance by allocating an equal number of elements per core, matrix-free multiplication achieves good scalability. This leads to a smaller memory footprint, less memory transfers, and good scalability for the coarse grid preconditioner.

However, geometric multi-grids cannot be used for unstructured grids (i.e., there is no concept of equally spaced coarsening); thus, $\mathbf{P}$ must be chosen in other ways (i.e., an algebraic multi-grid). While this approach can be used to approximate $\mathbf{u}$ for general unstructured problems, its approximation accuracy depends on the choice of $\mathbf{P}$ and is often less accurate than the geometric multi-grid. Also, matrix-free matrix–vector multiplication of $(\mathbf{P}^T\mathbf{KP})\mathbf{x}_c$ is often very costly; thus, $\mathbf{P}^T\mathbf{KP}$ is precomputed and stored in memory. This is a disadvantage when considering modern computers with limited memory bandwidth and relatively small memory capacity.

The key to an effective preconditioner for crust deformation problems is to combine the efficiency of geometric multi-grids and the flexibility of algebraic multi-grids to approximate long range errors with low computational cost, reduced communication, and a low memory footprint. Combining with a matrix-free matrix–vector multiplication method is expected to provide good scalability and fast time-to-solution with a low memory footprint. From this perspective, we developed the solver shown in Algorithms 7 and 8. The details of the solver are as follows.

Algorithm 7 Details of PASC17 solver. Matrix-vector products $\mathbf{K}(\ )$ is computed by the EBE method. diag[ ], $(^-)$, and ϵ indicate a 3×3 block Jacobi of [ ], a single-precision variable, and tolerance for relative error, respectively. $(\ )_1$ and $(\ )_2$ indicate calculation related to FEmodel_1 and FEmodel_2, respectively, and the other is the related calculation of the FEmodel. $\bar{\mathbf{P}}_1$ is a mapping matrix from FEmodel_1 to FEmodel, while $\bar{\mathbf{P}}_2$ is a mapping matrix from FEmodel_2 to FEmodel_1.

1: Set $\bar{\mathbf{A}}_2$ by coarsening $\bar{\mathbf{K}}_1$ (stored on memory with Compressed Row Storage format)
2: $\bar{\mathbf{B}} \Leftarrow diag[\bar{\mathbf{K}}]$, $\bar{\mathbf{B}}_1 \Leftarrow diag[\bar{\mathbf{K}}_1]$, $\bar{\mathbf{B}}_2 \Leftarrow diag[\bar{\mathbf{A}}_2]$
3: Set $\mathbf{f}$ according to slip state
4: $\mathbf{u} \Leftarrow 0$, $\mathbf{r} \Leftarrow \mathbf{f}$
5: $\beta \Leftarrow 0$, $i \Leftarrow 1$

(*outer loop start*)
while $\|\mathbf{r}\|_2/\|\mathbf{f}\|_2 \geq \epsilon$ **do**
(*inner loop start*)
$\mathbf{z} \Leftarrow \mathbf{B}^{-1}\mathbf{r}$
$\bar{\mathbf{r}} \Leftarrow \mathbf{r}$, $\bar{\mathbf{z}} \Leftarrow \mathbf{z}$
$\bar{\mathbf{r}}_1 \Leftarrow \bar{\mathbf{P}}_1^T\bar{\mathbf{r}}$, $\bar{\mathbf{z}}_1 \Leftarrow \bar{\mathbf{P}}_1^T\bar{\mathbf{z}}$
$\bar{\mathbf{r}}_2 \Leftarrow \bar{\mathbf{P}}_2^T\bar{\mathbf{r}}_1$, $\bar{\mathbf{z}}_2 \Leftarrow \bar{\mathbf{P}}_2^T\bar{\mathbf{z}}_1$
$\bar{\mathbf{z}}_2 \Leftarrow \bar{\mathbf{A}}_2^{-1}\bar{\mathbf{r}}_2$ (* inner loop 2: solved on FEmodel$_2$ by Algorithm 8 with tolerance ϵ_2, initial solution $\bar{\mathbf{z}}_2$, and maximum iteration n_2 *)
$\bar{\mathbf{z}}_1 \Leftarrow \bar{\mathbf{P}}_2\bar{\mathbf{z}}_2$
$\bar{\mathbf{z}}_1 \Leftarrow \bar{\mathbf{K}}_1^{-1}\bar{\mathbf{r}}_1$ (* inner loop 1: solved on FEmodel$_1$ by Algorithm 8 with tolerance ϵ_1, initial solution $\bar{\mathbf{z}}_1$, and maximum iteration n_1 *)
$\bar{\mathbf{z}} \Leftarrow \bar{\mathbf{P}}_1\bar{\mathbf{z}}_1$
$\bar{\mathbf{z}} \Leftarrow \bar{\mathbf{K}}^{-1}\bar{\mathbf{r}}$ (* inner loop 0: solved on FEmodel by Algorithm 8 with tolerance ϵ_0, initial solution $\bar{\mathbf{z}}$, and maximum iteration n_0 *)
$\mathbf{z} \Leftarrow \bar{\mathbf{z}}$
(*inner loop end*)
if $i > 1$ **then**
$\beta \Leftarrow (\mathbf{z}, \mathbf{q})/\rho$
end if
$\mathbf{p} \Leftarrow \mathbf{z} + \beta\mathbf{p}$
$\mathbf{q} \Leftarrow \mathbf{K}\mathbf{p}$
$\rho \Leftarrow (\mathbf{z}, \mathbf{r})$
$\alpha \Leftarrow \rho/(\mathbf{p}, \mathbf{q})$
$\mathbf{q} \Leftarrow -\alpha\mathbf{q}$
$\mathbf{r} \Leftarrow \mathbf{r} + \mathbf{q}$
$\mathbf{u} \Leftarrow \mathbf{u} + \alpha\mathbf{p}$
$i \Leftarrow i + 1$
end while(*outer loop end*)
output results $\mathbf{u}$

(a) **Adaptive Conjugate Gradient Method:** Instead of multiplying a fixed matrix approximating the inverse of the matrix $\mathbf{K}$ for preconditioning (i.e., $\mathbf{z} \Leftarrow \mathbf{C}\mathbf{r}$, where $\mathbf{C} \approx \mathbf{K}^{-1}$), the preconditioning matrix equation ($\mathbf{K}\mathbf{z} = \mathbf{r}$) is solved using another solver in the adaptive CG method.[10] We will refer to

Algorithm 8 PCG solver used to roughly solve $\bar{\mathbf{z}}_2 = \bar{\mathbf{A}}_2^{-1}\bar{\mathbf{r}}_2$, $\bar{\mathbf{z}}_1 = \bar{\mathbf{K}}_1^{-1}\bar{\mathbf{r}}_1$, and $\bar{\mathbf{z}} = \bar{\mathbf{K}}^{-1}\bar{\mathbf{r}}$ in Algorithm 7. $\mathbf{B}$ and ϵ are a block Jacobi matrix and the tolerance for the relative error, respectively. n indicates the maximum loop iterations. Matrix–vector multiplications $\mathbf{K}(\ )$ and $\mathbf{K}_1(\ )$ are computed using the EBE method for FEmodel and FEmodel$_1$, while matrix–vector multiplication $\mathbf{A}_2(\ )$ is computed by reading the global matrix stored in compressed row storage format for FEmodel$_2$. All calculations are performed using single-precision variables.

$\mathbf{x} \Leftarrow \mathbf{z}$
$\mathbf{e} \Leftarrow \mathbf{r} - \mathbf{K}\mathbf{x}$ (or $\mathbf{K}_1\mathbf{x}$ or $\mathbf{A}_2\mathbf{x}$)
$\beta \Leftarrow 0$
$i \Leftarrow 1$
while $\|\mathbf{e}\|_2/\|\mathbf{r}\|_2 \geq \epsilon$ and $i \leq n$ **do**
    $\mathbf{z} \Leftarrow \mathbf{B}^{-1}\mathbf{e}$ (or $\mathbf{B}_1^{-1}\mathbf{e}$ or $\mathbf{B}_2^{-1}\mathbf{e}$)
    $\rho_a \Leftarrow (\mathbf{z}, \mathbf{e})$
    if $i > 1$ **then**
        $\beta \Leftarrow \rho_a/\rho_b$
    end if
    $\mathbf{p} \Leftarrow \mathbf{z} + \beta\mathbf{p}$
    $\mathbf{q} \Leftarrow \mathbf{K}\mathbf{p}$ (or $\mathbf{K}_1\mathbf{p}$ or $\mathbf{A}_2\mathbf{p}$)
    $\alpha \Leftarrow \rho_a/(\mathbf{p}, \mathbf{q})$
    $\rho_b \Leftarrow \rho_a$
    $\mathbf{e} \Leftarrow \mathbf{e} - \alpha\mathbf{q}$
    $\mathbf{x} \Leftarrow \mathbf{x} + \alpha\mathbf{p}$
    $i \Leftarrow i + 1$
end while
$\mathbf{z} \Leftarrow \mathbf{x}$

the iterations of the original conjugate gradient loop as the outer loop (Algorithm 7, lines 20–30), and the iterations for solving the preconditioning matrix equation as the inner loop (Algorithm 7, lines 8–19). Since the preconditioning matrix equation need only be solved approximately, a fast method with limited accuracy can be used for the inner loop. Thus, although the inner loop incurs additional cost, an effective inner loop solver can reduce the total number of iterations of the costly outer loop, thus increasing the overall speed of the solver.

(b) **Integrated Geometric/Algebraic Multi-Grid Preconditioner:** To take advantage of the efficiency of geometric multi-grid and the flexibility of algebraic multi-grid, these methods are combined in an inner-loop solver. The idea here is that a second-order tetrahedral element mesh is mapped to a linear tetrahedral mesh, even if the mesh itself is unstructured. Thus, we apply geometric coarsening from the target second-order tetrahedral mesh to a linear tetrahedral mesh. Next, an algebraic multi-grid is applied to the linear tetrahedral mesh. Compared to pure algebraic multi-grids, this approach has three advantages. The first is that the number of iterations of the fine grid solver is reduced because of the good approximation accuracy of the geometric coarsening. The second is the applicability of matrix-free matrix–vector multiplication in the linear tetrahedral solver (i.e., $\mathbf{P}^T\mathbf{KP}$ of the second-order tetrahedral mesh is equivalent to $\mathbf{K}$ of the linear tetrahedral mesh), leading to reduced memory footprint and memory access costs. The third is the reduction of the large memory footprint required for algebraic multi-grids; after geometric coarsening, the grid is significantly smaller than the original grid. For example, the DOF of a typical second-order tetrahedral mesh are reduced by 8-fold by geometric coarsening and further reduced by an additional 8-fold after additional algebraic coarsening. This makes algebraic multi-grid practical for solving large problems on computer systems with limited memory capacity. Multi-grids are often used in a V-cycle or W-cycle preconditioner. However, when frequent mappings occur between grids, the limited approximation accuracy of algebraic multi-grids becomes a bottleneck. Thus, we use a simple but robust approach for the inner loop solver. Using an iterative solver, we first solve the preconditioning matrix equation roughly on the coarsest grid (FEmodel$_2$; Algorithm 7, line 13). The obtained solution is then used as the initial solution for the linear tetrahedral model (FEmodel$_1$; Algorithm 7, line 15). Finally, this solution is used as an initial solution to solve the preconditioning matrix equation on the original mesh (FEmodel). This approach reduces the frequency of mapping between grids and reduces the overall cost of the inner loop solver. In other words, long-range errors are first solved on the coarsest grid and the solution is refined as one progresses to finer grids. Hereafter, the solver for the coarsest loop will be referred to as inner loop level 2, the solver for the linear mesh as inner loop level 1, and the solver for the second-order mesh as inner loop level 0. A standard 3×3 block Jacobi PCG solver (Algorithm 8) is used to solve all inner loops. Implementing inner loop levels 0 and 1 is straightforward, but implementing algebraic multi-grid (inner loop level 2) in a distributed memory parallel environment requires careful consideration. Here, C nodes (nodes that remain in FEmodel_2) and F nodes (nodes eliminated in FEmodel_2) are assigned such that adjacent MPI partitions have consistent C and F nodes across the MPI boundaries. Here, MPI boundary nodes are first colored synchronously among MPI processes, and then internal nodes are colored within each process. This enables coarsening of internal and MPI boundary nodes, which leads to smaller matrix and vector sizes and less MPI communication in the level 2 solver.

(c) **Element-by-Element Method:** The EBE method[8] computes matrix–vector multiplication $\mathbf{y} \Leftarrow \mathbf{Kx}$ element-wise as:

$$\mathbf{y} \Leftarrow \sum_{e} (\mathbf{Q}_e^T(\mathbf{K}_e(\mathbf{Q}_e\mathbf{x}))). \tag{10}$$

$\mathbf{K}_e$ is generated from a small amount of information (i.e., node coordinates and material property); thus, the EBE method can be thought of a type of matrix–vector multiplication that transforms part of the memory access costs of standard matrix–vector multiplication (i.e., reading a global matrix from memory) to in-cache compute costs. Compared to standard matrix-vector multiplication, where non-uniform memory access is involved and the load balance among parallel processing elements is easily unbalanced, the EBE method provides a good load balance by equalizing the number of elements per core, thus improving the parallel efficiency. Thus, the EBE method is expected to reduce the time to solution when used on modern massively parallel computers with relatively low B/F performance. In the computation of Eq. (10), we use multi-core coloring and SIMD buffering techniques that utilize multi-core and SIMD parallel compute elements, and hide memory access latency with software prefetching. This highly tuned EBE kernel is used for matrix–vector multiplication at the outer loop (Algorithm 7, line 24) and inner loop levels 0 and 1 (Algorithm 8, lines 2 and 12). Since the EBE method cannot be used for inner loop level 2, the global matrix is stored in CRS format and read from memory for matrix-vector products in inner loop level 2 (Algorithm 8, lines 2 and 12).

(d) **Mixed-Precision Arithmetic:** While crustal deformation analysis requires double precision in the final solution, the preconditioner only needs to be calculated roughly. Thus, the inner loops are computed in single-precision, while the outer loop is computed in double precision. By setting appropriate thresholds for the inner loop solvers, almost all calculations can be transferred to the inner loop. This halved the memory footprint, memory transfer size, and communication size, and doubled the apparent cache size. Recent compute architectures have high arithmetic capability for single-precision floating-point operations, so improvements in peak performance is also expected.

In summary, we developed a method that combines the advantages of geometric and algebraic multi-grids for unstructured finite element crustal deformation problems at low total computational cost. Furthermore, the solver components are computed using a fast, scalable, and memory-efficient kernel, resulting in high performance on modern multi-core and SIMD-based massively parallel computer systems. In the following, we evaluate the performance of the developed solver on the K computer and an Intel Xeon cluster system.

5.2. *Performance*

5.2.1. *Performance measurement environment*

Targeting large-scale problems, we measure performance on the K computer system at RIKEN, Advanced Institute for Computational Science.[14] The K computer is a massively-parallel supercomputer comprising 82,944 compute nodes, each with a single SPARC64 VIIIfx CPU. Each of the eight cores of the SPARC CPU has two sets of fused multiply-add (FMA) SIMD units of length 2. This leads to FMA (two operations) $\times$ SIMD length 2 $\times$ 2 sets = 8 operations per clock. Consequently, the peak performance of each CPU is 8 operations $\times$ 8 cores $\times$ 2 GHz = 128 GFLOPS. The SIMD length is 2 for both single- and double-precision arithmetic; thus the peak performance of single-precision arithmetic is also 128 GFLOPS. Each core has two 32-KB two-way L1 caches, one for data and one for instructions, and the the cores share a 6-MB 12-way L2 cache (shared with data/instruction). Each node has 16 GB of DDR3 SDRAM memory with a peak memory bandwidth of 64 GB/s. Tofu, a 6D interconnection network, is used for communication between nodes.[15] Each node can communicate in four directions simultaneously with 5-GB/s throughput in each direction. An OpenMPI[16]-based MPI library optimized for the Tofu network is used.

5.2.2. *Performance measurement setting*

Using this system, the performance of the developed PASC17 solver is compared with the PCGE and SC14 solvers. PCGE is a CG solver with a 3×3 block Jacobi preconditioner using the EBE method and was considered a standard solver in SC14 and SC15 papers.[1,39] This is equivalent to skipping lines 10–18 in Algorithm 7. The SC14 solver is the SC14 Gordon Bell Prize Finalist solver[1] and uses two-step geometric multi-grid and mixed-precision arithmetic. This is equivalent to skipping lines 12–14 in Algorithm 7. Both PCGE and SC14 solver implementations were generated by skipping the appropriate lines in the PASC17 solver. Thus, all solver implementations use a highly tuned EBE kernel for matrix–vector multiplication. We solve all problems up to relative error tolerance of $\epsilon = 1.0 \times 10^{-8}$. Table 4 summarizes the tolerance and maximum number of iterations for the PASC17 and SC14 solvers. All of the following weak and strong scaling problems were generated for a horizontally layered two-layer crust structure with material properties summarized in Table 5. Periodic fault input and boundary conditions were used to make the problem suitable for measuring weak scaling. Although a periodical setting, METIS 5.1.0[11] was used to partition the mesh for distributed memory computation; thus, load balancing and communication characteristics similar to the actual problem can be expected. MPI was used for inter-node communication and eight OpenMP threads per node were used for intra-node computation.

5.2.3. *Weak scaling*

We first measured weak scaling using model set A, fixing the problem size to 13.7 million DOF per MPI process (Table 6). These models are intended to measure the performance of the largest problems within

Table 4. Error tolerance ϵ_i and maximum iteration n_i used for solving measurement models and application problem.

	ϵ_0	ϵ_1	ϵ_2	n_0	n_1	n_2
PASC17 solver	0.1	0.05	0.025	30	300	3,000
SC14 solver	0.1	0.05	—	30	3,000	—

Table 5. Material properties of performance measurement models.

Layer #	V_p (m/s)	V_s (m/s)	ρ (kg/m^3)
1	1,600	400	1,850
2	5,800	3,350	2,700

Note: Here, V_p, V_s, and ρ indicate primary wave velocity, secondary wave velocity, and density, respectively.

Table 6. Model configurations.

Model	MPI procs.	CPU cores	DOF	DOF/ proc.	Elements	PCGE iterations
A-1	1,152	9,216	15,782,835,951	13,700,378	3,932,160,000	8,152
A-2	2,304	18,432	31,553,601,711	13,695,140	7,864,320,000	8,966
A-3	4,608	36,864	63,062,312,751	13,685,398	15,728,640,000	9,107
A-4	9,216	73,728	126,100,445,871	13,682,774	31,457,280,000	9,858
A-5	18,432	147,456	252,111,071,151	13,677,901	62,914,560,000	9,026
A-6	36,864	294,912	504,173,743,791	13,676,588	125,829,120,000	9,154
B-1	1,152	9,216	15,782,835,951	13,700,378	3,932,160,000	8,152
B-2	2,304	18,432	15,782,835,951	6,850,189	3,932,160,000	9,064
B-3	4,608	36,864	15,782,835,951	3,425,095	3,932,160,000	8,153
B-4	9,216	73,728	15,782,835,951	1,712,547	3,932,160,000	9,719
B-5	18,432	147,456	15,782,835,951	856,274	3,932,160,000	9,143
C-1	16	128	125,177,217	7,823,576	30,720,000	4,925
C-2	16	96	125,177,217	7,823,576	30,720,000	4,926
D	36,864	294,912	381,383,242,611	10,345,682	95,259,986,496	37,373

Note: Model set A is weak-scaling model set; model set B is a strong-scaling model set. Model D is application problem.

the available memory. The number of PCGE iterations required for convergence for these models are almost constant (8,152 to 9,858 iterations); thus, this model set is suitable for measuring weak scaling. Figure 17 shows the performance of the developed solver.

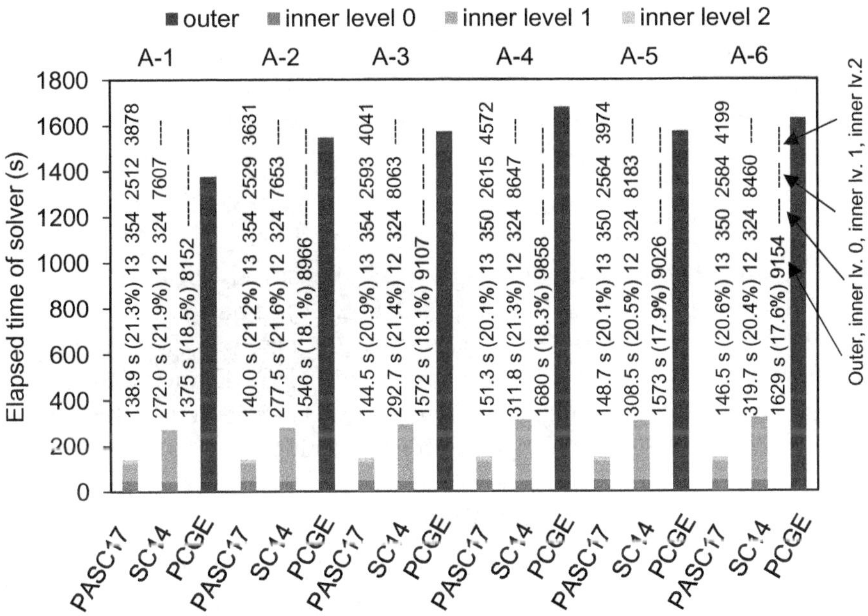

Fig. 17. Weak scaling results on model set A measured on the K computer. Here, numbers indicate elapsed time, floating-point arithmetic efficiency to peak, and elapsed iterations for the outer loop, and inner level 0, 1, and 2 loops.

The number of iterations required for convergence was nearly constant from model A-1 (9,216 CPU cores) to A-6 (294,912 CPU cores) with weak scaling efficiency of 94.8%. This high scalability enabled us to achieve high computational efficiency (20.6%; 0.970 PFLOPS) on the largest model using 294,912 CPU cores.

Next, we compare the performance of the PCGE and SC14 solvers with the largest model A-6. For the total number of iterations required for convergence, the PCGE solver required 9,154 iterations, the SC14 solver required 12 (outer) + 324 (inner level 0) + 8,460 (inner level 1) = 8,796 iterations, and the developed solver required 13 (outer) + 350 (inner level 0) + 2,584 (inner level 1) + 4,199 (inner level 2) = 7,146 iterations. These results indicate that the use of multi-grid is effective in reducing long-range errors with fewer total iterations. With mixed-precision arithmetic and smaller grid size (DOF for MPI rank #0 was reduced from FEmodel: 13,911,087 $\rightarrow$ FEmodel_1: 1,772,985 $\rightarrow$ FEmodel_2: 270,795), the elapsed time per iteration was reduced from 0.18 s for the outer loop to 0.13 s (inner level 0), 0.032 s (inner level 1), and 0.0046 s (inner level 2). This further increased the overall speed of the solver, i.e., the solver was 11.1-fold faster than PCGE and 2.18-fold faster than the SC14 solver.

5.2.4. *Strong scaling*

Next, using the same model as A-1, we measured strong scaling by varying the number of CPU cores (see Fig. 18). The solver scales well up to 147,456 CPU cores with strong scaling efficiency of 65.7%, and obtains nearly constant speed-up ratios with PCGE and SC14 solver for all problem sizes. Such good scalability is expected to be due to the solver algorithms with good load-balancing.

5.2.5. *Performance on Intel Xeon system*

Clusters equipped with Intel CPUs are widely used on a daily basis by researchers in solid Earth sciences; thus speedup on such systems is also of importance. As an example of such a system, performance

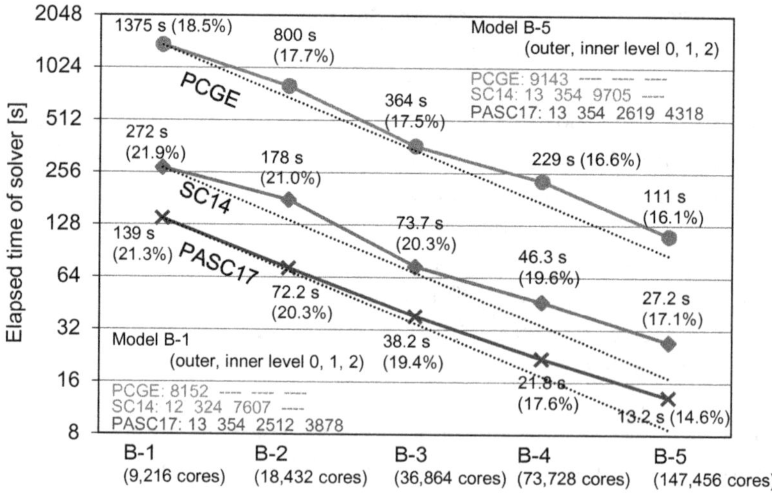

Fig. 18. Strong scaling measurement results on model set B measured on the K computer. Here, insets indicate elapsed time and floating-point arithmetic efficiency to peak, and the number of iterations for models B-1 and model B-5 is indicated in the bottom left and top right corners. Dashed lines indicate ideal scaling.

was measured on a compute cluster with Intel Xeon Haswell-EP CPUs. The main differences from the K computer nodes are the longer SIMD width (increased from 128 bits to 256 bits) and lower memory bandwidth capability (bytes/FP64_FLOPS is reduced from 0.5 of the K computer to 0.15 in the Xeon cluster). Here, we solved the same problem on 16 nodes of the K computer (with total hardware capability of 2048 FP64_TFLOPS, 2048 FP32_TFLOPS, and 1024-GB/s memory bandwidth) and four nodes of dual Intel Xeon E5-2670 v3 cluster (with total hardware capability of 3532 FP64_TFLOPS, 7064 FP32_TFLOPS, and 546-GB/s memory bandwidth). The same code was used for both systems, with additional SIMD directives for the Xeon cluster to facilitate use of AVX2 SIMD instructions on the Intel compiler (version 17.0.0). Figure 19 shows the performance results. The compute-intensive solvers using the EBE method (inner

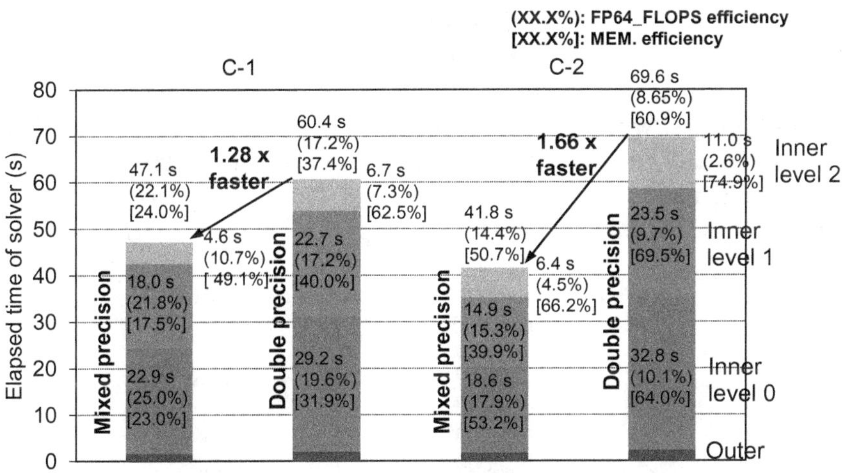

Hardware capability	16 nodes of K computer (8 core SPARC64 VIIIfx @ 2.0GHz per node)	4 nodes of Intel Xeon cluster (dual 12 core Xeon E5-2670 v3 @ 2.30GHz per node)
FP64 peak	2,048 GFLOPS	3,532 GFLOPS
Peak memory bandwidth	1,024 GB/s	546 GB/s

Fig. 19. Performance comparison with Intel Xeon cluster and K computer using model set C. Plain numbers indicate elapsed time, numbers in brackets indicate efficiency to FP64 peak, and numbers in square brackets indicate memory bandwidth utilization.

loops 0 and 1) were faster on the Xeon cluster, while the memory-access-intensive solver (inner loop 2) was faster on the K computer. Thus, the solver achieves performance improvement according to the hardware capability of the target compute system. In total, the solver obtained 14.4% of FP64_peak with the Xeon cluster for the total solver. To evaluate the effectiveness of mixed-precision arithmetic, we compared performance with a solver without mixed-precision arithmetic (i.e., all calculations are performed in FP64). The speedup using mixed-precision arithmetic was 1.28-fold on the K computer and 1.66-fold on the Xeon cluster. The mixed-precision arithmetic algorithm effectively reduced memory access costs for both systems. Also, wider SIMD lanes available on the Xeon cluster were utilized. Such effective speedup can be expected on other computer systems with high FP32 compute capabilities.

5.3. *Example*

As an example application of the developed solver, a crustal deformation analysis of eastern Japan was conducted. Here, we modeled a 576 km × 800 km × 400 km region consisting of eight crustal layers with the material properties listed in Table 7 and calculated the response to a given fault slip. Figure 20(a) shows the generated unstructured FE model that reflects the 3D geometry of the Earth's crust. To check the convergence of the numerical solution,

Table 7. Material properties of the Eastern Japan crust model (from J-SHIS[32]).

Layer #	V_p (m/s)	V_s (m/s)	ρ (kg/m^3)
1	1,600	400	1,850
2	2,000	650	1,950
3	2,600	1,100	2,150
4	3,600	1,800	2,300
5	4,500	2,400	2,450
6	4,800	2,800	2,500
7	5,500	3,150	2,600
8	5,800	3,350	2,700

Note: V_p, V_s, and ρ indicate primary wave velocity, secondary wave velocity, and density, respectively.

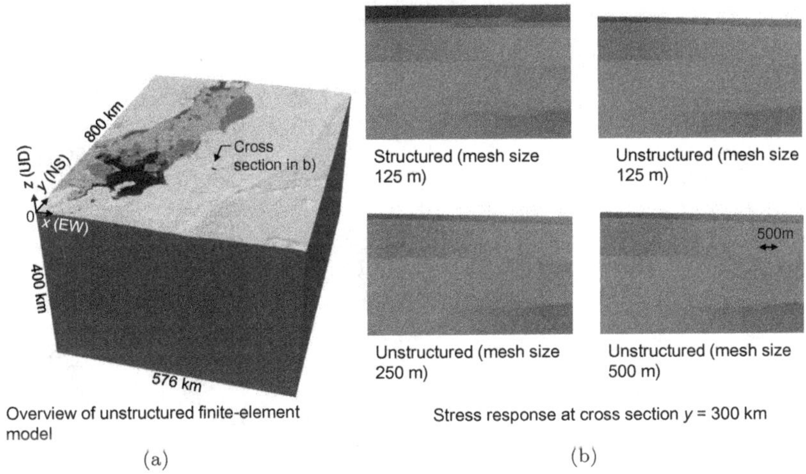

Fig. 20. Crust deformation analysis of Eastern Japan using structured and unstructured finite-element analysis. (a) Generated unstructured FE model, with 3D crust layer geometry and topography included in the unstructured finite element model. The surface is flattened for the structured FE model, and (b) stress response for structured and unstructured analysis at $y = 300$ km plane. The developed second-order tetrahedral finite element solver is used for solving all models (each voxel in structured grid is discretized with six tetrahedral elements).

we generate a second-order tetrahedral mesh with an element size of 125 m (381 billion DOF; computed using 294,912 CPU cores), element size of 250 m (DOF becomes approximately $1/2^3 = 1/8$; computed using 36,864 CPU cores), and element size of 500 m (DOF becomes approximately $1/4^3 = 1/64$). The resolutions of the analyses are 62.5 m, 125 m, and 250 m, respectively, because second-order tetrahedral elements are used. Figure 20(b) shows the the stress distribution calculated by the developed solver; we can see that the stress distribution is smooth in all models and that stress converges in the 250-m element model. In this problem, the use of second-order tetrahedral elements was sufficient for stress convergence. For comparison of accuracy, a structured grid model with flattened surface and material interface geometry was also calculated. Figure 20(b) shows the response of a 125-m structured grid model with 388 billion DOF. This is computed using 294,912 CPU cores of the K computer[14] with measured memory usage of 0.34 PB; thus, the scale of this structured grid analysis could be considered a large-scale problem in terms of memory usage. Nevertheless, jagged stress patterns

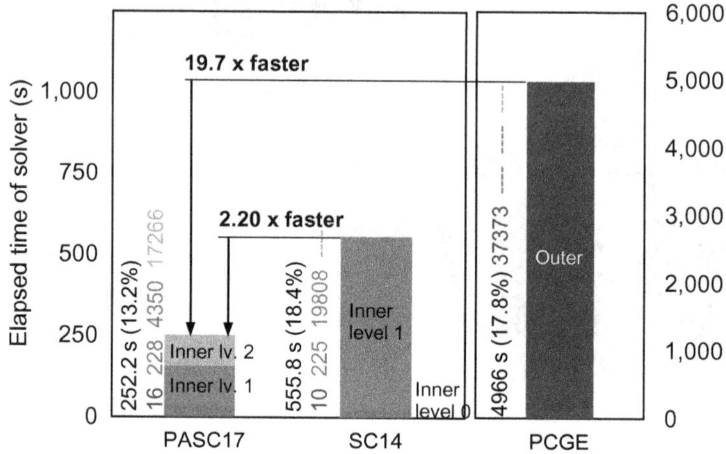

Fig. 21. Performance of application problem on the K computer. Insets indicate elapsed time, efficiency to FP64 peak, and number of outer loops, and inner level 0, 1, and 2 loops, respectively.

are seen at the crust layer boundaries, indicating that the resolution is insufficient for stress convergence. It can be seen that the low-order unstructured analysis computed with 36,864 CPU cores is more accurate than the structured analysis using 294,912 CPU cores.

Figure 21 shows the performance results for the developed solver when running on 294,412 CPU cores of the K computer for this Eastern Japan problem (model D in Table 6). This problem has higher resolution than the performance measurement models (resolution is 62.5 m) and thus is expected to have worse convergence characteristics. In this case, the performance efficiency decreased to 13.2% of FP64 peak; however, the speedup from PCGE increased to 19.7-fold due to the increase in the ratio of time used for the memory bandwidth intensive inner level 2 solver, which is more effective with actual high-resolution problems that are more difficult to converge with fine-grid solvers. In this case, the speedup from the SC14 solver was 2.20-fold. The time to prepare the preconditioner and output the results took 57.9 s (developed solver), 20.7 s (SC14 solver) and 20.3 s (PCGE), resulting in total elapsed times of 310.1 s, 576.5 s, and 4986 s, respectively. Thus, the developed solver obtains speedup on both hypothetical performance measurement models and actual crust structure models, even when including additional costs required for preparing the level 2 matrix (Algorithm 7, line 1).

6. CPU-Based Finite Element Solver for Nonlinear Soil–Structure System Response

Even with the significant improvement in memory bandwidth in recent computer architectures, the B (bytes/s; memory transfer capability)/F (FLOPS; floating point computation capability) ratio has not improved significantly due to the increasing floating point computation capabilities. Therefore, the recent trend in computer architecture is toward low B/F, making it difficult to find a suitable environment for applications that assume a relatively high B/F. Low-order unstructured finite element method is an example of such methods, which are becoming increasingly challenging to perform on modern computer architectures. In this section, we describe the development of a solver that utilizes the available computation capabilities of recent computer architectures to improve performance of the finite element method. Here, we target nonlinear soil-structure system response, which requires considerably larger cost than that for computing response of only soils. By considering the characteristics of the solution in the time domain, the IPDPS18 solver[44] described in this section improves peak performance and time to solution compared to previous solvers such as the SC14 solver. In the developed solver, several time steps of a nonlinear time-evolution problem are solved simultaneously to concentrate computation. This significantly improves the efficiency of sparse matrix-vector multiplication in low-order unstructured finite element methods. Concentrating computation using time-parallel computation is also applicable to general nonlinear time-evolution problems; thus, this algorithm can be applied to a wide range of nonlinear time-evolution problems. The methods presented in this section are also expected to improve the performance of low-order finite element simulations used to solve nonlinear time evolution problems in various fields that have not yet benefited from recent low B/F computer architectures. We show performance measurements on massively parallel CPU-based K computer and Intel Xeon Phi CPU-based Oakforest-PACS systems, and show an application example.

6.1. *Algorithm design*

The nonlinear dynamic low-order unstructured finite element method with implicit solvers, which is widely used in various fields as it is suitable for accurate modeling of nonlinear time evolution problems in highly heterogeneous complex geometry media, is also suffering from low B/F. The development of implicit solvers for huge problems is an important topic in HPC community. For example, a highly efficient implicit solver using semi-structured grids has been proposed for quasi-static problems, achieving efficient computation,[45] and a highly efficient implicit solver using structured grids has been proposed for time-dependent problems.[46] It can be seen that using structured (or semi-structured) grids for discretization yields very high computational efficiency. In fact, the implementation of structural discretization on unstructured elements has shown high performance in low-order unstructured finite element implicit solvers for nonlinear dynamics problems.[39] However, it is difficult to assume a structured grid inside an unstructured element for general problems due to the complexity of the target geometry and material heterogeneity. Therefore, it is difficult to avoid random accesses on unstructured grids, which significantly reduce computational efficiency. Similar bottlenecks occur in the finite element method using unstructured low-order elements, a standard tool in manufacturing, due to the difficulty of structural discretization in the target problem. The SC14 solver,[1] which has demonstrated high performance of 7.57% peak FLOPS on the K computer, is a state-of-the-art implicit solver for nonlinear dynamic problems with a pure unstructured finite element mesh (i.e., structured grids are not mixed). However, it is difficult to improve the performance of finite element applications (including the SC14 solver) on recent low B/F architectures because its main computation kernel are sparse matrix vector multiplications that require significant random access. With the development of new computer architectures with smaller B/F, the importance of developing new algorithms that perform better than the SC14 solver on low B/F supercomputers is increasing.

We target a dynamic nonlinear wave propagation problem using a dynamic nonlinear 3D finite element method with low-order solid elements, which is expressed as Eqs. (4) and (5). The earthquake problem involves a large-domain nonlinear time evolution problem with a locally complex structure; the 3D finite-element method is suitable for modeling complex geometries and analytically satisfying the traction-free boundary conditions on surfaces. For example, the soil–structure analysis used as an example in this section is $10^3 \times 10^3 \times 10^{1\sim 2}$m in size with structures with very complex geometry that varies at a resolution of $10^{-2\sim 1}$m. Since the bulk of the computation cost is incurred in solving Eq. (4), a fast, scalable solver suitable for solving this equation on modern low B/F computers is the target problem of this section.

6.1.1. *Details of solver algorithm*

For calculations where random access dominates, such as finite element analysis with sparse matrices, it is difficult to take advantage of the high F capability of modern low B/F computers. Modifying the solver algorithm to increase the computational density is one way to take advantage of the high F capability. Thus, we develop the IPDPS18 solver, which can efficiently solve equations by utilizing (1) a time-parallel computation algorithm, (2) solver algorithm utilizing density in space and arithmetic, and (3) concentrated computation kernels. The developed algorithm comprises Algorithms 9 to 11. We explain the ideas as follows:

Algorithm 9 Algorithm to obtain $\{\delta\mathbf{u}^i\}_{i=1}^{n_t}$ in Eq. (4) in IPDPS18 solver. $\mathbf{B}$ is a 3×3 block Jacobi matrix of $\mathbf{A}$. $(\ )_c$ indicates computation of FEmodel_c, while others indicate computation of FEmodel. ϵ denotes the tolerance of the solver. The loop from lines 7 to 12 is terminated based on the error of the current step i; thus, solutions $\{\delta\mathbf{u}^j\}_{j=i+1}^{i+3}$ in future time steps $j = i+1, ..., i+3$ are not solved accurately; however, they are used as good initial solutions in the next time steps. We denote the DOF of FEmodel and FEmodel_c as N and N_c, respectively; thus, $\mathbf{A}^i$ and $\mathbf{B}^i$ are sparse matrices of size $N \times N$; $\mathbf{b}^i, \delta\mathbf{u}^i$, and $\mathbf{f}^i$ are vectors of size N; and $\mathbf{B}_c^i$ are sparse matrices of size $N_c \times N_c$. Note that all computations are performed in FP64 unless otherwise indicated.

1: $\delta\mathbf{u}^0 \leftarrow 0, \delta\mathbf{u}^1 \leftarrow 0, \delta\mathbf{u}^2 \leftarrow 0, \delta\mathbf{u}^3 \leftarrow 0$
2: **for** $i = 1$ to n_t **do**
3:     update material properties for $\mathbf{A}^i$ using $\delta\mathbf{u}^{i-1}$

4: set $\mathbf{B}^i$ and $\mathbf{B}_c^i$ according to material property
5: set $\mathbf{b}^i$ using $\mathbf{f}^i$ and $\delta\mathbf{u}^{i-1}$
6: set $\delta\mathbf{u}^{i+3}$ using $\delta\mathbf{u}^{i-1}$ and guesses $\{\delta\mathbf{u}^j\}_{j=i}^{i+2}$ with Adams-Bashforth predictor
7: **while** $|\mathbf{A}^i\delta\mathbf{u}^i - \mathbf{b}^i|/|\mathbf{b}^i| > \epsilon$ **do**
8: update material properties for $\{\mathbf{A}^j\}_{j=i+1}^{i+3}$ using $\{\delta\mathbf{u}^j\}_{j=i}^{i+2}$
9: set $\{\mathbf{B}^j\}_{j=i+1}^{i+3}$ and $\{\mathbf{B}_c^j\}_{j=i+1}^{i+3}$ according to guessed material properties
10: set $\{\mathbf{b}^j\}_{j=i+1}^{i+3}$ using $\{\mathbf{f}^j\}_{j=i+1}^{i+3}$ and guesses $\{\delta\mathbf{u}^j\}_{j=i}^{i+2}$
11: solve $\{\mathbf{A}^j\delta\mathbf{u}^j = \mathbf{b}^j\}_{j=i}^{i+3}$ (* outer loop: solved by Algorithm 10 with initial guess $\{\delta\mathbf{u}^j\}_{j=i}^{i+3}$ using mixed precision solver *)
12: **end while**
13: output results of i-th time step $\delta\mathbf{u}^i$
14: **end for**

Algorithm 10 Algorithm to solve $\{\mathbf{A}^j\delta\mathbf{u}^j = \mathbf{b}^j\}_{j=i}^{i+3}$ (Algorithm 9, line 11) using an adaptive CG method, mixed-precision arithmetic, multi-grid, and the EBE method with initial solution $\{\delta\mathbf{u}^j\}_{j=i}^{i+3}$ in IPDPS18 solver. Matrix–vector multiplication $\mathbf{A}^j(\ ) = (\frac{4}{dt^2}\mathbf{M} + \frac{2}{dt}\mathbf{C}^j + \mathbf{K}^j)(\ )$ is computed by concentrated EBE kernels. $\mathbf{B}$ indicates a 3×3 block Jacobi matrix of $\mathbf{A}$, and $(\bar{\ })$ indicates computation in single precision (all other computations are double precision). ϵ is the tolerance of the solver. $(\ )_c$ indicates computation of FEmodel$_c$, while others are computation of FEmodel. Here, N and N_c denote the DOF of FEmodel and FEmodel$_c$, respectively; thus, $\mathbf{A}^i$ and $\mathbf{B}^i$ are sparse matrices of size $N \times N$ and $\delta\mathbf{u}^i$ are vectors of size N. $\mathbf{r}, \mathbf{f}, \mathbf{z}, \mathbf{p}, \mathbf{q}$, and $\mathbf{x}$ are a set of four vectors corresponding to $j = i, \ldots, i+3$ (size $4 \times N$), and $\mathbf{r}_c, \mathbf{z}_c$ are a set of four vectors corresponding to $j = i, \ldots, i+3$ (size $4 \times N_c$). α, β, ρ are a set of four scalars corresponding to $j = i, \ldots, i+3$.

$\mathbf{r} \leftarrow \mathbf{f} - \{\mathbf{A}^j\ \delta\mathbf{u}^j\}_{j=i}^{i+3}$
$\beta \leftarrow 0$
for $k = 1$ to 2 **do**
(* inner loop start *)
$\bar{\mathbf{r}} \leftarrow \mathbf{r}$
$\bar{\mathbf{z}} \leftarrow \{(\mathbf{B}^j)^{-1}\}_{j=i}^{i+3}\mathbf{r}$

7: $\bar{\mathbf{r}}_c \leftarrow \bar{\mathbf{P}}^T \bar{\mathbf{r}}$
8: $\bar{\mathbf{z}}_c \leftarrow \bar{\mathbf{P}}^T \bar{\mathbf{z}}$
9: $\bar{\mathbf{z}}_c \leftarrow \{(\bar{\mathbf{A}}_c^j)^{-1}\}_{j=i}^{i+3} \bar{\mathbf{r}}_c$ (* inner coarse loop: solved by Algorithm 11 with ϵ_c^{in} and initial solution $\bar{\mathbf{z}}_c$ *)
10: $\bar{\mathbf{z}} \leftarrow \bar{\mathbf{P}} \bar{\mathbf{z}}_c$
11: $\bar{\mathbf{z}} \leftarrow \{(\bar{\mathbf{A}}^j)^{-1}\}_{j=i}^{i+3} \bar{\mathbf{r}}$ (* inner fine loop: solved by Algorithm 11 with ϵ^{in} and initial solution $\bar{\mathbf{z}}$ *)
12: $\mathbf{z} \leftarrow \bar{\mathbf{z}}$
13: (* inner loop end *)
14: **if** $k > 1$ **then**
15: $\beta \leftarrow (\mathbf{z}, \mathbf{q})/\rho$
16: **end if**
17: $\mathbf{p} \leftarrow \mathbf{z} + \beta \mathbf{p}$
18: $\mathbf{q} \leftarrow \{\mathbf{A}^j\}_{j=i}^{i+3} \mathbf{p}$
19: $\rho \leftarrow (\mathbf{z}, \mathbf{r})$
20: $\alpha \leftarrow \rho/(\mathbf{p}, \mathbf{q})$
21: $\mathbf{q} \leftarrow -\alpha \mathbf{q}$
22: $\mathbf{r} \leftarrow \mathbf{r} + \mathbf{q}$
23: $\{\delta \mathbf{u}^j\}_{j=i}^{i+3} \leftarrow \{\delta \mathbf{u}^j\}_{j=i}^{i+3} + \alpha \mathbf{p}$
24: **end for**

Algorithm 11 CG solver algorithm with a 3×3 block Jacobi preconditioner used in IPDPS18 solver. Used to roughly solve $\mathbf{z}_c = \{(\mathbf{A}_c^j)^{-1}\}_{j=i}^{i+3} \mathbf{r}_c$ (line 9 of Algorithm 10) and $\mathbf{z} = \{(\mathbf{A}^j)^{-1}\}_{j=i}^{i+3} \mathbf{r}$ (line 11 of Algorithm 10). ϵ^{in} is the tolerance of the solver. $\mathbf{B}$ indicates a 3×3 block Jacobi matrix of $\mathbf{A}$. All computations are performed in single precision. We denote the DOF as N; thus, $\mathbf{A}^i, \mathbf{B}^i$ are sparse matrices of size $N \times N$, $\mathbf{x}, \mathbf{z}, \mathbf{r}, \mathbf{p}, \mathbf{q}$, and $\mathbf{e}$ are sets of four vectors corresponding to $j = i, \ldots, i+3$ (size $4 \times N$), and $\alpha, \beta, \rho_a, \rho_b$ are sets of four scalars corresponding to $j = i, \ldots, i+3$.

$\mathbf{x} \leftarrow \mathbf{z}$
$\mathbf{e} \leftarrow \mathbf{r} - \{\mathbf{A}^j\}_{j=i}^{i+3} \mathbf{x}$
$\beta \leftarrow 0$
$i \leftarrow 1$
while $\|\mathbf{e}\|_2 / \|\mathbf{r}\|_2 \geq \epsilon^{\text{in}}$ (or ϵ_c^{in}) **do**
$\mathbf{z} \leftarrow \{(\mathbf{B}^j)^{-1}\}_{j=i}^{i+3} \mathbf{e}$
$\rho_a \leftarrow (\mathbf{z}, \mathbf{e})$

if $i > 1$ **then**
  $\beta \leftarrow \rho_a/\rho_b$
end if
$\mathbf{p} \leftarrow \mathbf{z} + \beta\mathbf{p}$
$\mathbf{q} \leftarrow \{\mathbf{A}^j\}_{j=i}^{i+3}\mathbf{p}$
$\alpha \leftarrow \rho_a/(\mathbf{p}, \mathbf{q})$
$\rho_b \leftarrow \rho_a$
$\mathbf{e} \leftarrow \mathbf{e} - \alpha\mathbf{q}$
$\mathbf{x} \leftarrow \mathbf{x} + \alpha\mathbf{p}$
$i \leftarrow i + 1$
end while
$\mathbf{z} \leftarrow \mathbf{x}$

Time-parallel computation algorithm: Time-parallel computation algorithms have been developed to speed up nonlinear time-history analysis (e.g., Ref. 52). Here we develop a time-parallel computation algorithm that significantly improves computational performance and reduces time to solution when compared to these studies. As shown in Algorithm 9, several time steps are solved simultaneously, where the current time step is solved accurately, while future time steps are solved with limited accuracy. In the developed method, four time steps are solved simultaneously (i.e., time steps $i+1$, $i+2$, and $i+3$ are solved while solving time step i). An iterative solver is used to solve the four time steps simultaneously such that the solution for time step i is calculated with adequate accuracy. The solutions for the other three time steps are not exact, but are used as initial solutions for the solver for the next time step, reducing the number of solver iterations. As a result, the number of solver iterations in line 11 of Algorithm 9 was reduced to about 1/4. However, as the FLOP count per iteration to solve $\{\mathbf{A}^j\delta\mathbf{u}^j = \mathbf{b}^j\}_{j=i}^{i+3}$ increases by four times, the total number of arithmetic counts of the solver is mostly unchanged. The time-parallel algorithm becomes effective when combined with the concentrated computation kernel; however, we first describe an efficient solver algorithm based on the density of solutions in space and arithmetic space that reduces the total cost of the solver.

Solver algorithm utilizing density in space and arithmetic: We use Algorithm 10 to solve $\{\mathbf{A}^j\delta\mathbf{u}^j = \mathbf{b}^j\}_{j=i}^{i+3}$ efficiently by utilizing solution density and arithmetic space. In the standard CG method, a matrix similar to the inverse of the target matrix is multiplied as a preconditioner. In contrast, the adaptive CG method,[10] which roughly solves the preconditioning equations rather than multiplying a fixed preconditioning matrix, is used to make active use of solution density and arithmetic space. Here, another CG solver is used to solve the preconditioning equation. Therefore, we refer to the iteration of the original CG solver as the outer loop and the iteration of the preconditioning equation solver as the inner loop. Since the target problem requires high precision, the outer loop requires double-precision calculations, but the preconditioning equation only needs a rough solution, so it is possible to introduce methods to reduce the computational complexity. First, by employing single-precision (FP32) arithmetic in the inner loop, the amount of memory transfer, memory footprint, and communication size are reduced. This also increases the apparent cache size, and allows more SIMD lanes to be used in recent architectures. Second, the number of computations are reduced using the multi-grid method.[7] Here we use a spatially coarsened model (FEmodel$_c$) with equivalent properties to the original target model (FEmodel). In the inner loop, a rough solution is obtained using FEmodel$_c$, and this solution is used as an initial solution for the FEmodel solver. Although additional computations on FEmodel$_c$ are required, the total cost of the inner loop is reduced because the number of FEmodel iterations is reduced. Note that the computation and communication cost of FEmodel$_c$ is much smaller than that of FEmodel. Because the communication involved in mapping between fine and coarse grids can be a bottleneck on massively parallel computers, we use a second-order tetrahedral element model for FEmodel and a first-order tetrahedral element model with the same mesh for FEmodel$_c$ (i.e., FEmodel$_c$ is generated by removing the edge nodes of each tetrahedron in FEmodel); thus, mapping between grids can be performed without communication. The preconditioning equations (Algorithm 10, lines 9 and 11) are solved roughly using a conjugate gradient solver with the 3×3 block Jacobi preconditioner in Algorithm 11. Hereafter, the inner loops of FEmodel and FEmodel$_c$ will be called inner fine and inner coarse loops, respectively. Compared to the FEmodel of the double-precision outer loop, the message size of the point-to-point communication of the FEmodel of the inner fine loop is reduced to one-half, and that of the FEmodel$_c$ of the inner coarse loop is reduced to one-eighth.

Concentrated computation kernels: Matrix–vector multiplication $\mathbf{Ku}$ of FEmodel in the inner fine loop is the most costly kernel in this algorithm. Although it is possible to store the matrix $\mathbf{K}$ on memory (e.g., using compressed row storage), generating and storing four matrices at each time step is inefficient in terms of time and memory footprint. Therefore, instead of storing matrices in memory and multiplying as $\mathbf{K}^j\mathbf{u}^j$, we adopted the EBE method,[8] which computes $\mathbf{K}_e^j\mathbf{u}_e^j$ for each element and sums it to a global vector as follows:

$$\mathbf{f}^j \leftarrow \sum_e \mathbf{Q}_e \mathbf{K}_e^j \mathbf{Q}_e^T \mathbf{u}^j, \quad (j = i, \ldots, i+3). \tag{11}$$

Here, $\mathbf{Q}_e$ is the matrix mapping local nodal values to global nodal values and $\mathbf{K}_e^j$ is the element stiffness matrix of each element. Although written in a matrix form, $\mathbf{Q}_e$ correspond to random accesses in the actual code, where $\mathbf{u}_e^j \leftarrow \mathbf{Q}_e^T\mathbf{u}^j$ are random loads and $\mathbf{f}^j \leftarrow \sum_e \mathbf{Q}_e\mathbf{f}_e^j$ (where, $\mathbf{f}_e^j \leftarrow \mathbf{K}_e^j\mathbf{u}_e^j$) are random loads and stores. Thus, EBE computation consists of a random-access part and a computation-dominated part (i.e., $\mathbf{f}_e^j \leftarrow \mathbf{K}_e^j\mathbf{u}_e^j$). To reduce the total kernel time, it is necessary to run both parts of the kernel with high efficiency. First, we describe efficient computation of computationally expensive $\mathbf{f}_e^j \leftarrow \mathbf{K}_e^j\mathbf{u}_e^j$. Rather than considering $\mathbf{K}_e^j\mathbf{u}_e^j$ $(j = i, \ldots, i+3)$ as the [multiplication of a matrix and one vector] × four, we consider it a single [multiplication of four similar matrices and four vectors]. This leads to concentrated computation of the kernel. Specifically, $\mathbf{K}_e^j\mathbf{u}_e^j$ is transformed analytically to $\sum_k \mathbf{N_b}_k^T\mathbf{D}^j\mathbf{N_b}_k\mathbf{u}_e^j$ $(j = i, \ldots, i+3)$ using orthogonal functions for the bases of stress in displacement–stress mixed formulation of the finite element method.[12] We compute this multiplication in the order $\sum_k \mathbf{N_b}_k^T(\mathbf{D}^j(\mathbf{N_b}_k\mathbf{u}_e^j))$. Here, $\mathbf{D}^j$ are the elastic matrices for $j = i, \ldots, i+3$ in time step i and $\mathbf{N_b}_k$ are equivalent modes. The data size required for this computation is not large (i.e., $\mathbf{N_b}_k$ are independent of j), but the amount of computation increases, allowing use of the high floating-point computation capability leading to a shorter elapsed time. Since $\mathbf{N_b}_k$ and $\mathbf{D}^j$ consist of many zero components, the calculation is expanded into a scalar expression and only the necessary components are computed to eliminate unnecessary calculations. For the random access part of the kernel, summing element-wise results into $\mathbf{f}^j$ in Eq. (11) involves data recurrence in nodes shared by several elements. Therefore, it is important to compute this part with an algorithm suitable for many-core and wide SIMD architectures. Since the direct implementation of Eq. (11) cannot be parallelized element wise by SIMD due to data recurrence, we follow[39] and buffer the nodal data of each element as follows:

$$\mathbf{f}_e^j \leftarrow \mathbf{K}_e^j \mathbf{Q}_e^T \mathbf{u}^j, \tag{12}$$

and sum them as follows:

$$\mathbf{f}^j \leftarrow \sum_e \mathbf{Q}_e \mathbf{f}_e^j. \tag{13}$$

Then, the computationally expensive element loop of Eq. (12) can be computed in parallel using SIMD instructions. Element coloring is also used for multi-core computation of random load and store for Eq. (13). This concentrated and parallelized computation is very effective in recent architectures with low B/F. Table 8 shows the performance of the second-order tetrahedral element matrix–vector multiplication kernel at single precision on several computational environments. When calculated separately, the elapsed time for four matrices multiplied by four vectors is four times longer than for one matrix multiplied by one vector. On the other hand, by concentrating computation and utilizing the high F capabilities, this can be computed in 2.82-fold time on K computer with a relatively high B/F ratio of 0.5, and 2.05-fold time on Intel Xeon Phi (Knights Landing)-based Oakforest-PACS (OFP) with a B/F ratio of 0.16. In general, the random access capability of the SIMD architecture is not high; thus, the efficiency of unstructured computation drops sharply as the SIMD width is increased. Using the EBE, the single vector multiplication on OFP is high (18.6% of FP64 peak) for a 512-bit SIMD-based architecture for unstructured computation. From such high baseline performance, we can attain an additional two-fold performance increase by concentrating the computation, which leads to 35.7% efficiency. High performance gains were also observed on the 256-bit SIMD multi-core based FX100, which is a prototype system for the post-K computer. Thus, the effectiveness of concentrated computation and efficient SIMD/thread parallelization are very high. This concentrated computation is implemented for all matrix-vector multiplications in the solver (i.e., second-order tetrahedral elements

Table 8. Single-precision EBE kernel performance for second-order tetrahedral elements (measured for matrix–vector multiplication 529 times for a problem with 6,534,144 elements and 9,044,560 nodes).

	K computer	FX100	Oakforest-PACS	Reedbush-H
Nodes	16	4	2	1
Processes/node	1	4	8	1 GPU only
Threads/process	8	8	8	—
SIMD width (FP64)	2	4	8	—
SIMD width (FP32)	2	8	16	—
Memory bandwidth (TB/s)	1.02	1.92	0.98	0.73
Peak FP64 (TFLOPS)	2.05	4.05	6.09	5.30
Peak FP32 (TFLOPS)	2.05	8.09	12.2	10.6
EBE for 1 vector				
Elapsed time	10.29 s	4.89 s	5.11 s	4.55 s
Performance (GFLOPS)	503	1029	1068	1222
Ratio to peak FP64	24.5%	25.4%	18.6%	22.5%
Ratio to peak FP32	24.5%	12.7%	9.31%	11.3%
EBE for 4 vector				
Elapsed time	29.00 s	10.06 s	10.46 s	10.56 s
Performance (GFLOPS)	638	2121	2044	2096
Ratio to peak FP64	31.1%	52.4%	35.7%	38.6%
Ratio to peak FP32	31.1%	26.2%	17.8%	19.3%
Elapsed time ratio (4 vector/1 vector)	2.81	2.05	2.05	2.32

Note: Here the K computer equips SPARC64 VIIIfx CPU $\times 1$ per node, FX100 (FUJITSU Supercomputer PRIMEHPC FX100[47]) equips SPARC64 XIfx CPU $\times 1$ per node, Oakforest-PACS equips Intel Xeon Phi 7250 CPU $\times 1$ per node, Reedbush-H[48] equips NVIDIA P100 GPU $\times 2$ per node.

in FP64 and first-order tetrahedral elements in FP32). In addition, the time-parallel algorithm leads to concentration of other parts of the solver, leading to efficiency of the whole application on low B/F systems.

The core idea of this algorithm design is to utilize the unused F of modern supercomputers. One of the surest ways to leverage F without increasing data transfer costs is to preserve the data access and compute structure (for example, the data access patterns of unstructured elements and the computation structure of the **K** matrices are retained in our method). In supercomputers with a large number of computation cores, local imbalances can lead to major degradation in computational performance. Thus, compared to general algorithm designs, it is important to design algorithms that do not disrupt uniformity in time and space when targeting supercomputers. Based on this idea, multiple computations are performed with the same data access and the same computation structure using the time-parallel algorithm of Algorithm 9. Although Algorithm 10 requires more data access and computation per iteration than the standard CG solver, the reduction in the number of iterations by Algorithm 9 reduces the time to solution. The reduced number of iterations reduces the total time required for thread synchronization and communication latency, thereby improving scalability. Thus, it can be seen that this is an example of an application-oriented algorithm development that takes into account the architectural characteristics of modern supercomputers and the target problem.

6.2. *Performance*

In this section, we compare the developed IPDPS18 solver to SC14 solver. The SC14 solver, designed for the K computer, employs multigrid methods and mixed-precision arithmetic for nonlinear dynamic problems. Since the SC14 solver uses the EBE method for sparse matrix–vector multiplication, it is expected to be faster than standard sparse-matrix solvers, even on computers with low B/F. As both IPDPS18 and SC14 solvers are sophisticated in their algorithms and implementations, it is difficult to explain their differences; however, if we consider IPDPS18 solver as a multi-grid solver in the temporal-spatial-arithmetic domain, we can consider the SC14 solver to be the same as IPDPS18 solver without time parallelism. Due to concentrated computation by time parallelization, the IPDPS18 solver is expected to be faster than the SC14 solver, especially on computers with low B/F.

6.2.1. *Performance measurement environment*

The IPDPS18 solver algorithm is faster when the speed of the concentrated computation kernel is increased. In particular, larger speedup is expected when the efficiency of the four-vector matrix–vector multiplication kernel is significantly improved from that of the single-vector version (Table 8). Therefore, we measure performance on Oakforest-PACS, which has a high relative speed for matrix–vector products and is available for many nodes. Performance measurements were also performed on the K computer, which has a relatively large B/F, to compare the performance characteristics of IPDPS18 and SC14 solvers.

Oakforest-PACS (OFP): OFP[50] is a supercomputer system introduced by the Joint Center for Advanced HPC,[51] which was established by The University of Tokyo and the University of Tsukuba. The system comprises 8,208 nodes with a 68-core Intel Xeon Phi 7250 (Knights Landing) CPU, 96 GB of DDR4 RAM, and 16 GB of stacked 3D MCDRAM. The Omni-Path Architecture provides a 100-Gbps interconnection between nodes with full bisection bandwidth fat tree. Each core has two 512-bit vector units and supports AVX-512 SIMD instructions. Therefore, the peak performance of a single node at single precision is 1.4 (GHz) $\times$ 2 (FMA) $\times$ 16 (SIMD width) $\times$ 2 (sets per core) $\times$ 68 (cores) = 6,092.8 GFLOPS, while that at double precision is 3,046.4 GFLOPS. The measured performance of the STREAM Triad benchmark is 80.1 GB/s for DDR4 RAM and 490 GB/s for MCDRAM.

K computer (K): K[14] consists of 82,944 compute nodes, each with a single eight-core SPARC64 VIIIfx CPU. Each core has two sets of SIMD FMA arithmetic units of width 2. Thus, the FP64 peak performance of the CPU is 2 (GHz) $\times$ 2 (FMA) $\times$ 2 (SIMD width) $\times$ 2 (sets per core) $\times$ 8 (cores) = 128 GFLOPS. The SIMD width does not change regardless of precision; thus, the peak performance of the CPU for FP32 is also 128 GFLOPS. Each node has 16 GB of DDR3 SDRAM, with a peak bandwidth of 64 GB/s. Each node can communicate simultaneously in four directions at 5 GB/s throughput in each direction on a 6D interconnection network.

6.2.2. *Problem measurement setting*

In earthquake simulations, soft ground subjected to strong shaking is expensive to analyze because of the large domain, high resolution required, and poor convergence due to nonlinearities in material properties. Therefore, speeding up this problem is expected to lead

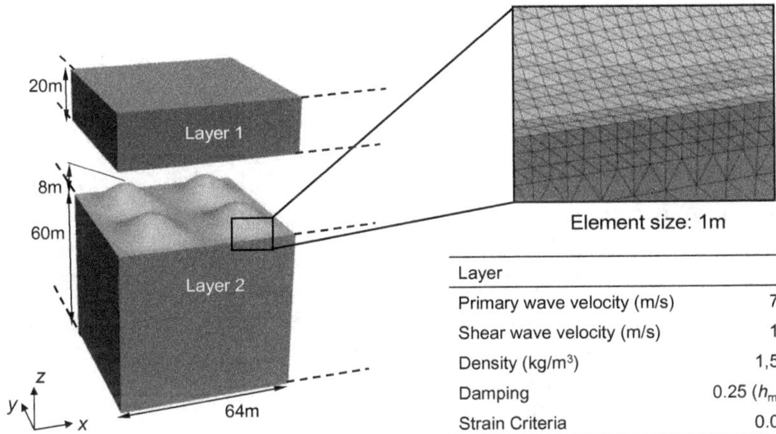

Layer	1	2
Primary wave velocity (m/s)	700	2,100
Shear wave velocity (m/s)	100	700
Density (kg/m^3)	1,500	2,100
Damping	0.25 (h_{max})	0.05
Strain Criteria	0.007	–

Fig. 22. Performance measurement model. The model is duplicated in the x and y directions to make a periodic problem suitable for measuring weak scaling.

to speeding up other earthquake simulations as well. Thus, a two-layered ground that mimics the actual ground is used for performance measurement (Fig. 22). The first layer is a soft layer that behaves nonlinearly with respect to strong shaking, while the second layer is a hard layer that mimics bedrock. The domain is replicated in the x and y directions to generate periodical problems for measuring weak scaling. Models are partitioned using the METIS graph partitioning method so that the load balancing characteristics of the performance measurement problems are close to those of actual problems with complex geometries. We input the earthquake wave observed at Kobe in 1995 with a time step increment of $dt = 0.05$ s, and measured the time required to solve 25 time steps. Although this problem setting may appear to be similar to the one used in the SC14 paper,[1] it has poorer convergence characteristics that are similar to the convergence characteristics of the oil-refinery plant problem discussed in Section 6.3. The relative error tolerance of the solvers are set to $\epsilon = 1.0 \times 10^{-8}$ for all problems. Although the sensitivity of the tolerances of the multi-grid preconditioner (i.e., ϵ_c^{in} and ϵ^{in}) to the solver's performance is not high, we set optimal values by considering the architecture of each system (i.e., $\epsilon_c^{\text{in}} = 0.5$, $\epsilon^{\text{in}} = 0.25$ for OFP and $\epsilon_c^{\text{in}} = 0.5$, $\epsilon^{\text{in}} = 0.4$ for the K computer). We use eight OpenMP threads per MPI process for both systems, and flat memory mode and quadrant cluster mode with one thread per physical core is used for OFP.

6.2.3. *Time-to-solution*

First, the portability of the state-of-the-art SC14 solver on OFP is evaluated. The problem size per compute node was set such that the same problem could be solved using all compute nodes in both systems. As shown in Table 9, the number of iterations of PCGE[b] is stable among all models. Thus, these problem settings are suitable for measuring weak scaling. We compared the performance with WO-1 (128 nodes, corresponding to 1/64 of OFP) and WK-1 (1,152 nodes, corresponding to 1/72 of the K computer). As shown in Fig. 23, we can see that both the SC14 solver and IPDPS18 solver algorithms are not memory bound due to the use of EBE for matrix–vector products; only 9% to 28% of memory bandwidth was used for the whole application. For memory bandwidth kernels such as vector addition operations, near peak memory bandwidth was obtained on both OFP and K computer. Thus, it can be seen that the algorithm design avoids memory bandwidth bottlenecks in the majority of the solver components, while memory bandwidth bound kernels are also well tuned for both the SC14 and IPDPS18 solvers. Although the SC14 solver achieved 11.1% of peak floating-point performance on the K computer, its performance is 2.26% of the FP64 peak on OFP. Thus, it can be seen that this state-of-the-art K computer application is not necessarily performing well on the newer generation of supercomputers like OFP. Nevertheless, this performance is better than the 1.5% performance obtained in the HPCG benchmark,[49] and the SC14 solver can be used as a suitable performance benchmark for low-order unstructured finite element methods. In contrast, the IPDPS18 solver obtained a high efficiency of 15.6% in the low-order unstructured finite element method on a 512-bit SIMD many-core system. The improvement in peak performance led to reduction in time-to-solution; IPDPS18@OFP is 5.02-fold faster

[b]PCGE is a conjugate gradient solver referred for comparison with the SC14 solver in Ref. 1. It uses a 3 × 3 block Jacobi matrix for preconditioning with the same EBE kernel of the SC14 solver. Thus, we can regard PCGE as a well-tuned solver based on a standard algorithm.

Table 9. Model configurations (WO and SO are weak scaling and strong scaling model sets for OFP.

Model	Compute nodes	MPI procs.	CPU cores	DOF (billion)	DOF /proc.	Elements (billion)	PCGE iterations
WO-1	128	1,024	8,192	1.800	1,758,232	0.447	78,781
WO-2	256	2,048	16,384	3.599	1,757,700	0.894	—
WO-3	512	4,096	32,768	7.198	1,757,331	1.788	—
WO-4	1,024	8,192	65,536	14.393	1,757,075	3.577	—
WO-5	2,048	16,384	131,072	28.784	1,756,891	7.155	—
WO-6	4,096	32,768	262,144	57.565	1,756,760	14.310	—
WO-7	8,192	65,536	524,288	115.125	1,784,555	28.621	—
WK-1	1,152	1,152	9,216	1.600	1,389,304	0.397	78,763
WK-2	2,304	2,304	18,432	3.200	1,388,920	0.795	78,790
WK-3	4,608	4,608	36,864	6.398	1,388,551	1.590	78,655
WK-4	9,216	9,216	73,728	12.795	1,388,364	3.180	78,467
WK-5	18,432	18,432	147,456	25.586	1,388,181	6.360	78,193
WK-6	36,864	36,864	294,912	51.170	1,388,088	12.720	77,965
WK-7	82,944	82,944	663,552	115.125	1,387,987	28.621	77,629
SO-1	128	1,024	8,192	7.198	7,029,325	1.788	—
SO-2	256	2,048	16,384	7.198	3,514,662	1.788	—
SO-3	512	4,096	32,768	7.198	1,757,331	1.788	—
SO-4	1,024	8,192	65,536	7.198	878,665	1.788	—
SO-5	2,048	16,384	131,072	7.198	439,332	1.788	—
SK-1	1,152	1,152	9,216	6.398	5,554,207	1.590	78,642
SK-2	2,304	2,304	18,432	6.398	2,777,103	1.590	78,648
SK-3	4,608	4,608	36,864	6.398	1,388,551	1.590	78,655
SK-4	9,216	9,216	73,728	6.398	694,275	1.590	78,630
SK-5	18,432	18,432	147,456	6.398	347,137	1.590	78,652

Note: WK and SK are weak scaling and strong scaling model sets for the K computer. WO-7 and WK-7 are identical problems computed on full OFP and full K computer.

than SC14@OFP. An algorithm that improved the peak performance of the core kernel with a similar total number of FLOP count reduced the time to solution and improved the overall peak performance of the application. Another reason for the short time to solution is that the implementation is suitable for the OFP architecture. For example, in the EBE calculation of a single vector in Eq. (13), SIMD could not

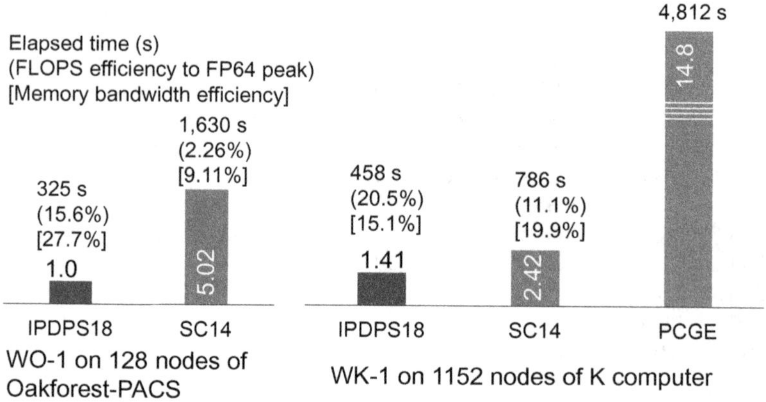

Fig. 23. Time-to-solution of IPDPS18@OFP. Time-to-solution is normalized with IPDPS@OFP. Performance to peak hardware FP64 FLOPS is indicated in brackets, while memory bandwidth is indicated in square brackets. Memory bandwidth is normalized by STREAM Triad MCDRAM bandwidth (490 GB/s) for OFP and DDR hardware peak (64 GB/s) for the K computer.

be applied due to random store accesses, but by computing four vectors as a set, the load/store operation can be conducted using SIMD operations. Thus, SIMD is used in all parts of the EBE calculation in the IPDPS18 solver. In addition to the solver, the four block Jacobi matrices $\mathbf{B}^j$ are computed as a set; thus, computation of the whole application becomes concentrated. Another example is the effective use of the fast memory with limited capacity. While the memory footprint of the SC14 solver was 11.5 GB per node (i.e., the whole application fits on the 16 GB MCDRAM), the IPDPS18 solver does not fit in the fast memory due to the additional three vectors. Thus, for a given problem size and available memory, we automatically assign arrays to MCDRAM in order of usage priority.[c] For example, arrays used in the multi-grid preconditioner are allocated first in fast memory automatically. Such use of SIMD units and hierarchical memory for unstructured mesh computations is expected to be useful

[c]Arrays are allocated in fast memory automatically for a given problem size and available memory according to its priority using the first touch policy of NUMA. Here, static arrays are zero-filled in order of priority in the application program and run using the FLAT memory mode of the second-generation Xeon Phi with Intel MPI option I_MPI_HBW_POLICY=hbw_preferred.

for other architectures in the near future. To observe the effectiveness of the development in terms of algorithm and implementation, we compared the elapsed time between WO-1 on OFP and WK-1 on the K computer (Fig. 23). Since SIMD is not wide (width 2 for FP32), B/F is high (0.5), and memory is not hierarchical, it is difficult to obtain substantial performance improvements by changing the implementation on the K computer. Thus, the speedup obtained by IPDPS18 solver compared to SC14 solver on the K computer (786 s/458 s = 1.7) can be considered to represent the effectiveness of the algorithm. The concentrated core kernel contributes to about 2.5-fold speedup on OFP; thus, we consider that the remaining two-fold speedup is due to the implementations suitable for many-core, wide SIMD, limited B/F, and hierarchical memory architecture. Based on the above, the developed solver was able to obtain high performance due to its ability to perform concentrated computation and reduce the number of iterations. In addition, the implementation of the solver is suitable for OFP's architecture. When compared with PCGE, a well-tuned conventional solver on the K computer, IPDPS18@OFP demonstrates a 14.8-fold speedup. The combination of architectural development, application algorithms, and effective implementation has resulted in a significant improvement in time-to-solution.

6.2.4. *Weak scaling*

We measure weak scaling on OFP and the K computer (Fig. 24). Ideally, we would measure scaling with respect to the best serial algorithm, but we are unable to set up an appropriate serial problem that reflects the convergence characteristics of large-scale problems. Therefore, we started scaling measurements with 128 nodes for OFP and 1,152 nodes for the K computer. The weak scaling efficiency of IPDPS18@OFP from WO-1 (128 nodes) to WO-7 (8,192 nodes; full system) was 83.9%, while the weak scaling efficiency of IPDPS18@K from WK-1 (1,152 nodes) to WK-7 (82,944 nodes; full system) was 91.4%. We can see that the scalability of IPDPS18 is very good, for example, IPDPS18@K was better than the 88.8% obtained by SC14@K. This is due to the relative reduction in thread synchronization and communication latency cost obtained by the four-fold increase in computation per iteration. In addition, measures taken

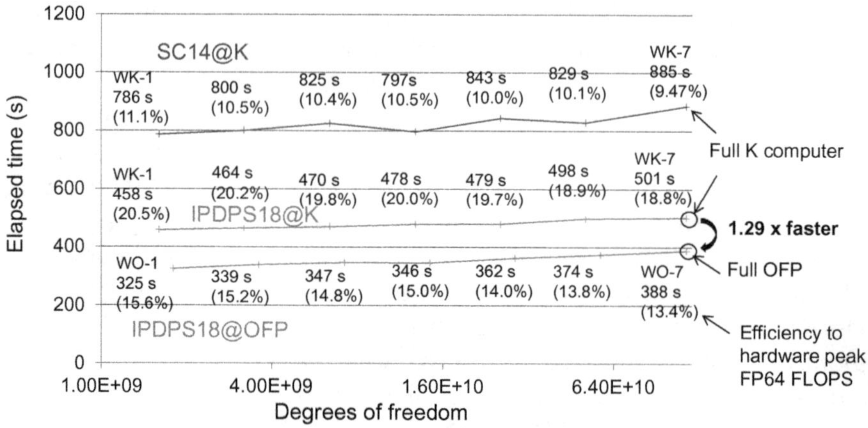

Fig. 24. Weak scaling results. Elapsed time and performance efficiency to hardware FP64 peak are shown.

to avoid cache thrashing that can occur in unstructured applications on massively parallel systems are expected to be contributing to the high scalability (i.e., by reduction in the number of streams and memory allocation alignment). The time-to-solution of IPDPS18@(full OFP) is 388 s, which is 2.28-fold faster than the state-of-the-art SC14@(full K).

6.2.5. *Strong scaling*

On the K computer, the IPDPS18 solver is about 1.8-fold faster than the SC14 solver, with high strong scaling efficiency of 78.9% from SK-1 to SK-5 (Fig. 25). On OFP, the IPDPS18 solver exhibits a high strong scaling efficiency of 72.6% from SO-1 to SO-5. Super-linear speedup was observed on SO-2 due to the suitable use of MCDRAM, i.e., arrays used for the preconditioner could not fit in MCDRAM for SO-1 but were allocated on MCDRAM for SO-2. These results indicate that the reduction of problem size by the multi-grid method and the reduction of memory footprint by mixed-precision arithmetic are effective for obtaining high performance in recent hierarchical memory systems.

In summary, by developing an algorithm and a suitable implementation that obtain high peak performance with nearly constant FLOP count, the IPDPS18 solver obtains fast time-to-solution on OFP, which has lower B/F than the K computer.

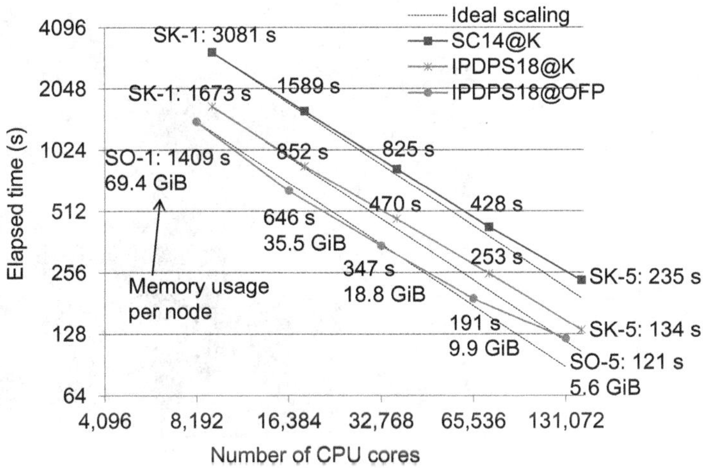

Fig. 25. Strong scaling results. Elapsed time and memory footprint for IPDPS18@OFP are shown.

6.3. *Example*

As an application example of the developed method, nonlinear seismic response analysis of industrial plants of high social value (e.g., oil refinery plants and nuclear power plants) is calculated with an ultra-high fidelity model. Here, a coupled ground-plant nonlinear dynamic finite element problem was simulated for an oil refinery plant on a two-layer ground using the seismic input used in Section 6.2. In addition to the solver development, FE model generation is another bottleneck in finite element analyses. Here, we developed a parallel unstructured FE model generation method based on an octree-based multiple-material method.[53] A coupled soil–plant model with 11,321,249,889 second-order tetrahedral elements and 49,064,764,344 DOF was generated in 17 h 22 min (62,551 s) using 64 cores of an SGI UV300, which is a 512 core, 24.5 TB cache-coherent shared memory system consisting of 32 sockets of 16-core Intel Xeon E7-8867 v3 CPUs and 768 slots of 32GB DDR4 DRAM connected with SGI NUMAlink ASIC technology. This model is automatically and robustly generated from CAD data of an oil refinery plant.[54] Figures 26(a)–26(c) show the generated model for a 262.1 × 262.1 × 50.7 m domain discretized with 16 mm second-order tetrahedral elements. In conventional structural simulations,

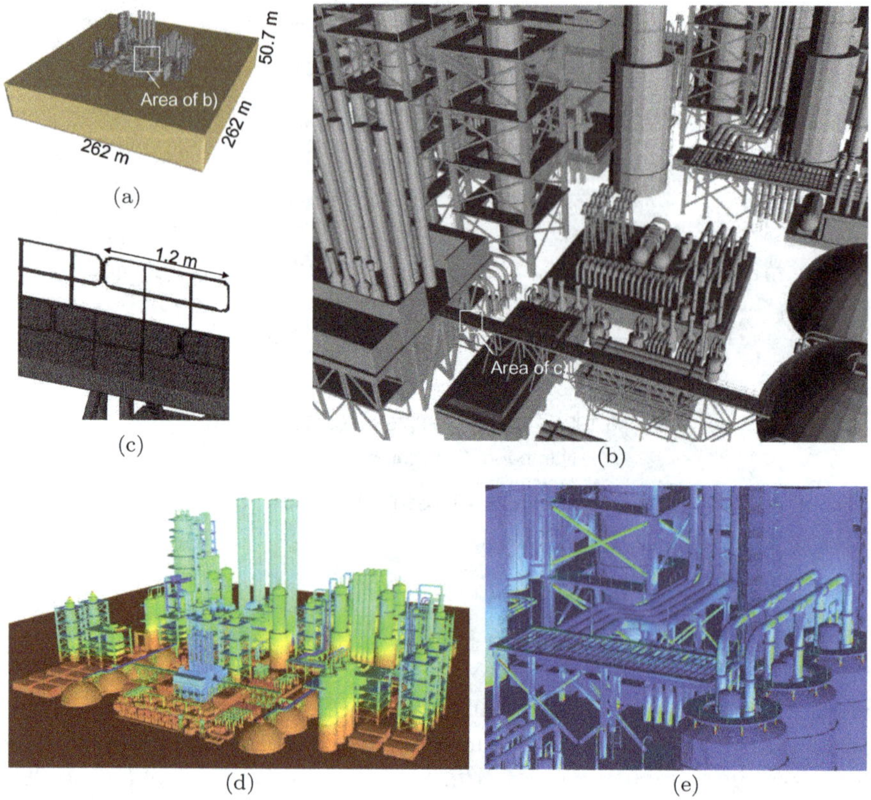

Fig. 26. Application example using the IPDPS18 solver. (a) overall view of the finite element plant model surrounded by a two-layered ground modeled with 11,321,249,889 second-order tetrahedral elements and 49,064,764,344 DOF, (b) close up view of the finite element model, (c) complex geometries modeled with minimum element size of 16 mm, (d) displacement response to ground shaking, and (e) stress concentration, which can be evaluated at high-resolution for all parts of the model for engineering purposes.

only structural components (e.g., beams and columns) are modeled. In contrast, structural and nonstructural components (e.g., pipes and chillers), which are essential for evaluating the plant operability, are modeled at ultra-high resolution. We used the IPDPS18 solver on the full K computer to analyze this model. Figs. 26(d) and 26(e) show the displacement and stress response at the 255th time step. Because nonstructural members are modeled in 3D, we can evaluate stress concentration on complex geometries and member joints, which are of engineering interest for seismic design considerations. The performance of the first 25 steps of the plant problem was 1.72 PFLOPS (16.2% of FP64 peak), which is close to the 18.8% of FP64 peak obtained from the performance measurements in Section 6.2. Therefore, we expect high performance on other practical nonlinear dynamic seismic problems with poor convergence and complex geometry.

7. GPU-Based Finite Element Solver for Nonlinear Soil–Structure System Response

In recent years, the development of artificial intelligence, such as machine learning, has gained momentum and has been shown to provide significant performance when combined with appropriate computing hardware, such as GPUs. This section describes a solver for seismic nonlinear soil structure system response enhanced by artificial intelligence (AI) and transprecision computing, based on the idea of reducing data transfer by localization and homogenization in preconditioned conjugate gradient solvers. AI-based methods and equation-based solver methods have different objectives and therefore have different accuracy, robustness, and computational cost. Therefore, the key to this development was to develop a cost-effective and robust algorithm that incorporates AI methods into an equation-based solver. In the following, the development of a nonlinear wave propagation solver (hereafter referred to as the SC18 solver[59]) is described in detail, with performance measurements and application examples on the GPU-based supercomputers Summit and Piz Daint.

7.1. *Algorithm design*

Here, we summarize the design of the SC18 solver that is developed by analyzing the characteristics of the target problem, and explain the algorithm and implementation details. Our target is to solve $\mathbf{A}^n\ \delta\mathbf{u}^n = \mathbf{b}^n$ in Eq. (4), derived from the 3D nonlinear finite element method using unstructured second-order tetrahedral elements in FP64 for each of the $10^{3\sim4}$ time steps within relative error tolerance of 10^{-8}. Here, $\mathbf{A}^n = (\frac{4}{dt^2}\mathbf{M} + \frac{2}{dt}\mathbf{C}^n + \mathbf{K}^n)$ and $\mathbf{b}^n = \mathbf{f}^n - \mathbf{q}^{n-1} + \mathbf{C}^n\mathbf{v}^{n-1} + \mathbf{M}(\mathbf{a}^{n-1} + \frac{4}{dt}\mathbf{v}^{n-1})$, and both change

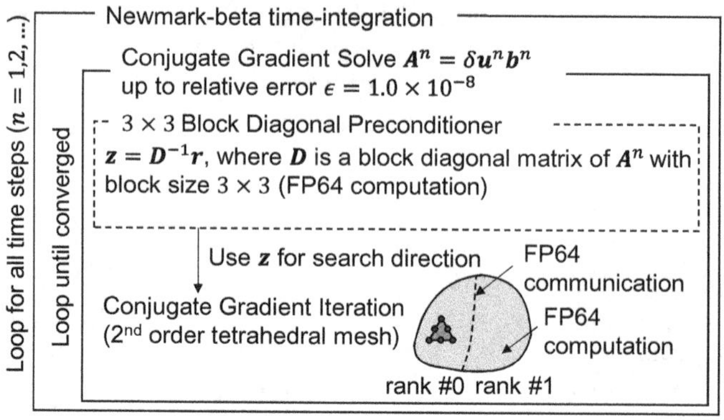

Fig. 27. Algorithm of PCGE. The matrix–vector product with FP64 computation and FP64 communication for ghost layer updates is performed once per CG iteration. Preconditioning is conducted using a 3 × 3 block diagonal matrix of $\mathbf{A}^n$. The light gray area represents the target domain, while the dashed line represents the MPI partition boundary.

at each time step n. This problem is characterized by the fact that the DOF of $\mathbf{u}^n$ is greater than 10^{10}, and since $\mathbf{A}$ is sparse, access to $\mathbf{u}^n$ is random. As $\mathbf{M}$ is diagonally dominant, small dt allows the property of $\mathbf{A}^n$ to be solvable for even very complex urban models. In addition, the problem size is large compared to the memory capacity of current computer systems. Therefore, it is impractical to allocate $\mathbf{A}^n$ on memory, even when using a compressed format such as compressed row storage. One of the standard methods for solving this problem is the PCGE solver, which is a CG solver using a 3×3 block diagonal matrix of $\mathbf{A}^n$ as a preconditioner[7] (Fig. 27), and the EBE method which computes a matrix–vector product on the fly without storing the global matrix $\mathbf{A}^n$ in memory.[8] However, since PCGE is computed in FP64 and the ghost layer is updated by point-to-point communication in FP64, the computational performance of PCGE is insufficient for modern supercomputers, which have relatively fast computing performance compared to memory and interconnect bandwidth. In addition, although the PCGE converges, its convergence is slow; thus, there is room for significant improvement in time-to-solution by development of more sophisticated preconditioners. The SC14 solver[1] can solve similar problems, but it is designed for previous generation computers and does not perform optimally on recent supercomputers. In addition, our target problem has significantly poor convergence, which could be improved with a more aggressive algorithm. The SC18 solver developed is based on the idea of reducing data transfer through localization and homogenization in CG solvers by use of AI and transprecision computing. Although the SC18 solver uses advanced algorithms, including AI, these are used only for preconditioning. Thus, the numerical results of the developed solver agree with PCGE with sufficient accuracy (relative error less than the error tolerance of 10^{-8}). In the following, we describe the details of the developed solver following Fig. 28.

7.1.1. *Enhancement of preconditioner by artificial intelligence*

As poor convergence is due to the nonuniform distribution of the strength of graph connectivity (i.e., matrix components) in the target matrix, we train an AI (in particular, an artificial neural network) to estimate the relationship between localization of graph connectivity strength and the reduction of convergence, and use this AI to estimate the parts of the matrix that degrade convergence performance. The idea here is to improve the convergence of the solver by suppressing the concentration of connectivity strength in the AI-estimated graph using local operations. In other words, the AI is trained on a small problem with graph characteristics similar to the larger target problem, using a data set of graph connectivity strengths and distributions of poor convergence. AI is then used to improve convergence by estimating areas of the larger target problem where convergence is poor, which is later solved extensively in the solver by performing local operations. Note that AI is used only to efficiently detect parts of the problem that result in poor convergence; thus, there is no deterioration in the accuracy of the simulation results and only the time to solution is improved. A graph connectivity data set of small problems may be small from a supercomputing perspective, but when considered as training data for AI, it turns out to be a large and high-quality data set. Thus, these small data sets have the potential to improve large-scale supercomputing.

Although the concept above is straightforward and general-purposed, careful consideration is needed to develop an AI that effectively estimates areas of poor convergence. The implementation

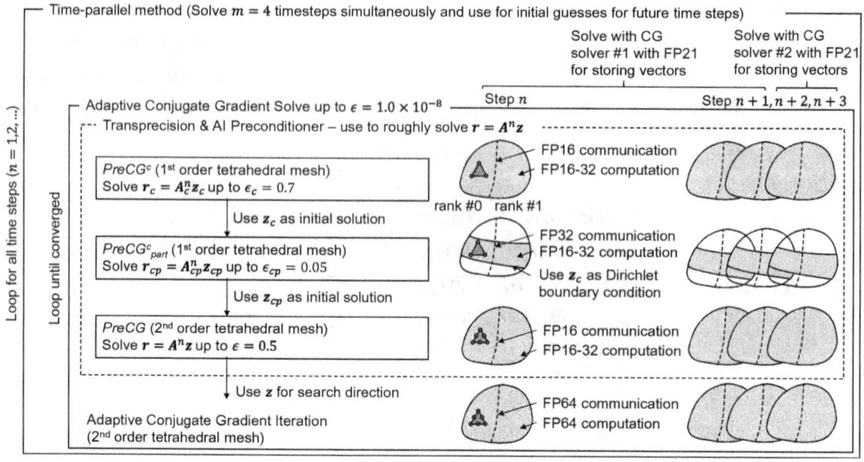

Fig. 28. Implicit solver algorithm using AI and transprecision computing for nonlinear time history unstructured low-order finite element simulations. The solver employs an adaptive CG method to efficiently solve the preconditioning equation $\mathbf{r} = \mathbf{A}^n\mathbf{z}$ with AI and transprecision computing. Here, the linear tetrahedral element CG solver $PreCG^c$ first roughly solves the preconditioning equation $\mathbf{r} = \mathbf{A}^n\mathbf{z}$ and its solution is used as an initial solution for the linear tetrahedral element CG solver $PreCG^c_{\text{part}}$. The light gray area represents the target area and the dashed line represents the MPI partition boundary. AI is used to estimate which parts of the problem has poor convergence characteristics. These parts are included in the $PreCG^c_{\text{part}}$ part and are solved extensively, leading to reduction of the computation cost when compared to solving the whole domain using $PreCG^c$. Next, the results of $PreCG^c_{\text{part}}$ are used as an initial solution for the second-order tetrahedral element CG solver $PreCG$ to obtain a highly accurate estimate of the preconditioning equation $\mathbf{r} = \mathbf{A}^n\mathbf{z}$. In these preconditioners, transprecision computing is used to reduce computational and communication costs (transprecision is used in EBE kernels as FP16-FP32, storing vectors in CG solvers as FP21, and in point-to-point communication as FP16 communication). Because the search direction of the FP64 CG solver points to a very accurate solution, the number of iterations of the FP64 CG loop is greatly reduced, allowing almost all calculations to be performed with low-precision arithmetic. Furthermore, the time-parallel algorithm allows solving $m = 4$ time steps concurrently; the four time steps are split to two, and solved in parallel. This leads to more concurrency in each CG solver that enables overlapping point-to-point communication with memory bandwidth bound kernels. Since all components of the PCGE are conducted in FP64, the SC18 solver is expected to be significantly faster on recent computers, which have higher computational performance on lower precision arithmetic.

is described as follows. In preconditioned CG methods, a preconditioning matrix $\mathbf{D}$ is used as $\mathbf{z} = \mathbf{D}^{-1}\mathbf{r}$ to improve convergence of the solver (Fig. 27). In contrast, we use the adaptive CG method, which solves the preconditioning equation $\mathbf{r} = \mathbf{A}^n\mathbf{z}$ roughly with another CG solver for efficient implementation of a more sophisticated preconditioner.[10] However, training an AI to estimate the relationship between graph connectivity strength and poor convergence in this manner would result in poor generalization ability. We first homogenize this problem because second-order tetrahedral elements have two types of nodes (i.e., edge and vertex nodes), which have completely different graph connectivity properties. Instead of using the target second-order tetrahedral element problem $\mathbf{r} = \mathbf{A}^n\mathbf{z}$, we use a coarse but equivalent problem $\mathbf{r}_c = \mathbf{A}_c^n\mathbf{z}_c$ with linear tetrahedral elements. This homogenizes the graph connectivity properties of the target matrix $\mathbf{A}_c^n$. We will refer to solving $\mathbf{r} = \mathbf{A}^n\mathbf{z}$ and $\mathbf{r}_c = \mathbf{A}_c^n\mathbf{z}_c$ in the preconditioner by the CG method as *PreCG* and *PreCG*^{c}, respectively. Next, we use the fact that the solution $\mathbf{u}$ can be expressed as $\mathbf{u}(\mathbf{x}) = \int \mathbf{G}(\mathbf{x},\mathbf{y})\mathbf{b}(\mathbf{y})dV$ using Green's function $\mathbf{G}$ and the force distribution $\mathbf{b}$, which reflects the differential equation and model properties. Considering the properties of $\mathbf{G}$ (its main part is given as a perturbation of $e^{|\mathbf{y}-\mathbf{x}|/c_2/dt}$ with time increment dt, shear wave velocity c_2 near point $\mathbf{x}$, and the distance d_{ij} between node i and its nearest node j), we expect that the connectivity strength of graph vertex i of $\mathbf{A}_c^n$ is governed with the frequency of strength of $d_{ij}/c_2/dt$. As $e^{|\mathbf{y}-\mathbf{x}|/c_2/dt}$ decays rapidly with distance, the interaction between nodes are localized; thus, the independence between parameters is enhanced, leading to an AI with higher accuracy. A deep learning-based AI was trained to relate these characteristic values to the residuals of small-scale problems. Although it is very difficult to develop an AI with high accuracy for heterogeneous problems, localization and homogenization to characterize the target problem enables the development of an AI with high generalization ability (SC17 Best Poster Award[55] is one example of constructing an AI with high generalization ability using such characteristics of the system). The part of $\mathbf{A}_c^n$ estimated to have poor convergence by AI is denoted as $\mathbf{A}_{cp}^n$, and we refer to solving $\mathbf{r}_{cp} = \mathbf{A}_{cp}^n\mathbf{z}_{cp}$ using the CG method as *PreCG*$^c_{\text{part}}$. The equations are solved in the order of *PreCG*^{c}, *PreCG*$^c_{\text{part}}$, and *PreCG*; thus, we expect improvement in convergence compared to when only *PreCG* is used. We use an AI

Table 10. Effectiveness of AI-based preconditioner.

	Without AI	With AI
CG iterations	132,665	88
$PreCG^c$ iterations	—	5,803
$PreCG^c_{\text{part}}$ iterations	—	26,826
$PreCG$ iterations	—	3,103
Number of floating point operations	184.7 PFLOP	33.2 PFLOP

Note: The number of iterations and number of floating point operations (FLOP count) required for solving 25 time steps for model W-1 described in Section 7.2.2 are shown.

trained using a small part of the measurement problem described in Section 7.2.2 as a training data set, and use it to extract parts of $\mathbf{A}^n_c$ in the problem used in Section 7.2.3. The resulting number of finite element nodes in $\mathbf{A}^n_{cp}$ was 17,162,689, which corresponds to 10.9% of the total number of nodes in $\mathbf{A}^n_c$ (i.e., 157,339,343 nodes). It took 88 CG iterations, 5,803 $PreCG^c$ iterations, 26,826 $PreCG^c_{\text{part}}$ iterations, and 3,103 $PreCG$ iterations to solve 25 time steps of this problem. On the other hand, PCGE required 132,665 CG iterations; thus, the convergence was improved by using the AI preconditioner (Table 10). Furthermore, as the floating point arithmetic for each iteration of $PreCG^c$ and $PreCG^c_{\text{part}}$ is smaller than that of a CG iteration, this leads to reduction of arithmetic count by 184.7 PFLOP/33.2 PFLOP = 5.56 times, which leads to significant speedups. As will be shown in Section 7.2.3, applying the same AI to larger problems to extract $\mathbf{A}^n_{cp}$ leads to similar speedup; thus, AI-based preconditioners can be effective for larger problems.

7.1.2. *Enhancement of preconditioner by transprecision computing*

As solving the preconditioning equation $\mathbf{r} = \mathbf{A}^n\mathbf{z}$ does not require high accuracy, we use FP32 arithmetic for $PreCG^c$, $PreCG^c_{\text{part}}$, and $PreCG$, and aim to improve computational efficiency by further use of the arithmetic space. Support for a variety of arithmetic data types has become a recent hardware trend, aiming for flexible use of the arithmetic space and improved system performance. For example, IEEE754 FP16[60] arithmetic units have been implemented in recent GPUs in addition to FP32 and FP64 arithmetic units.[61,62]

In addition, the trend toward systems with relatively slow memory and interconnect bandwidth relative to high-speed arithmetic units continues, and the reduction of data transfer size by use of low-precision data types is also important for shortening time-to-solution. However, since the exponents and fractions of FP16 are limited (1 sign bit, 5 exponent bits, and 10 fractional bits), it is not straightforward to use FP16 for general scientific calculations. Using FP16 for preconditioning of the implicit solver is an option; however, the range of values is large for general matrices. Therefore, when global matrices are stored in FP16, only very limited types of equations can be solved with meaningful accuracy. Indeed, the wide range of values of the matrix components in the target city problem makes it difficult to use FP16 directly, even for preconditioning purposes. On the other hand, from the perspective of the finite element method formulation, each element has a small value range because the problem is discretized by the local basis functions of each element. The large range of components in the global matrix is caused by assembling elements with different characteristics. Thus, we use FP16 for element-wise computation, which is used for local expansion in the finite element method. Here, FP16 is used in local matrix vector products in the EBE method, which is a matrix-free, matrix-vector product method. The following is a detailed description of the transprecision EBE kernel.

In the EBE method, matrix–vector product $\mathbf{y} \leftarrow \mathbf{Ax}$ is computed as follows:

$$\mathbf{y} \leftarrow \sum_{e} (\mathbf{Q}^{eT}(\mathbf{A}^{e}(\mathbf{Q}^{e}\mathbf{x}))). \tag{14}$$

Here, $\mathbf{A} = \sum \mathbf{Q}^{eT}\mathbf{A}^{e}\mathbf{Q}^{e}$, and $\mathbf{Q}^{e}$ is a mapping matrix between the local node number of element e and the global node number. As direct computation of Eq. (14) in FP16 leads to frequent overflow/underflow of variables, we use $\mathbf{x}_h^e \leftarrow \mathbf{f}(\mathbf{Q}^e\mathbf{x}_s)$, $\alpha_s^e \leftarrow g(\mathbf{A}_s^e)$, and $\beta_s^e \leftarrow h(\mathbf{Q}^e\mathbf{x}_s)$, and we compute $\mathbf{y}_s \leftarrow \sum_e \mathbf{Q}^{eT}\alpha_s^e\beta_s^e\mathbf{B}_h^e(\mathbf{x}_h^e)$. Here, subscripts s and h indicate values and functions in FP32 and FP16, respectively, and $\mathbf{f}$ and g represent functions that make the values of vector $\mathbf{x}_h^e$ and the values used in function $\mathbf{B}_h^e$ close to 1 (under infinite numerical accuracy, these functions satisfy $\mathbf{A}^e\mathbf{x}^e = g(\mathbf{A}^e)h(\mathbf{x}^e)\mathbf{B}^e(\mathbf{f}(\mathbf{x}^e))$). Although random addition to $\mathbf{y}_s$ and computation of scalars α_s^e, β_s^e are computed in FP32, the most costly $\mathbf{B}_h^e(\mathbf{x}_h^e)$ can be computed in

FP16. Thus, use of this transprecision kernel on systems with high FP16 performance is expected to lead to faster computation.

For EBE accelerated by FP16, the time involved in random load of $\mathbf{x}_s$ and random addition to $\mathbf{y}_s$ dominates the total elapsed time. For reduction in random access costs, we use a time-parallel algorithm developed in Refs. 44 and 63. Based on the idea that the connectivity between nodes and elements is independent of time, several time steps are solved in parallel to reduce random data access. When solving m time steps in parallel, the arithmetic count required per iteration for the iterative solver is m times larger than the standard solver. However, the obtained solution for future time steps can be used for highly accurate initial solutions, reducing the total number of solver iterations by approximately m-fold. Thus, a time-parallel algorithm can reduce the number of random accesses while maintaining the same number of operations, thus reducing the time to solution in recent computers. Furthermore, by dividing the m time steps into two groups, two independent solvers can be run simultaneously, allowing overlapping computation and communication between the two solvers. We have also implemented methods to reduce atomic add to L2 cache by utilizing the GPU's shared memory.

Table 11 shows the EBE kernel performance. Here, FP64 (one vector) and FP32 (one vector) compute Eq. (14) directly in FP64/FP32 and are used in the PCGE and SC14[1] solvers, respectively. FP16-32 (two vectors) is the developed transprecision EBE kernel with the time-parallel algorithm solved with $m = 4$ parallel time steps grouped into groups of two time steps. Note that elapsed time is normalized per vector. For Piz Daint and Summit with FP16 cores, FP16-32 (two vectors) improved FLOPS efficiency by 2.63- and 3.26-fold, respectively, compared to FP64. Since the K computer does not support FP16 operations and thus uniform precision of FP32 is used, we can see that the 2.35-fold performance improvement is obtained by reducing data transfer size by changing from FP64 to FP32 and reducing the number of random accesses using the time-parallel algorithm.

In GPU-accelerated systems, interconnect bandwidth and GPU device memory bandwidth are relatively low compared to the high arithmetic performance per node, so point-to-point communication and GPU device memory transfer must be reduced to obtain scalability and performance. Therefore, transprecision communication

Table 11. Performance of EBE matrix-vector product kernel (elapsed time per vector is shown).

System	Elapsed time	FLOPS efficiency to FP64 peak (%)
One K computer node		
FP64 (one vector)	56.71 ms	10.9
FP32 (one vector)	46.43 ms	13.2
FP32 (two vectors)	18.53 ms	25.7
One P100 GPU on Piz Daint		
FP64 (one vector)	785.3 us	19.1
FP32 (one vector)	623.4 us	24.0
FP16-32 (two vectors)	392.7 us	50.3
One V100 GPU on Summit		
FP64 (one vector)	413.5 us	21.9
FP32 (one vector)	400.3 us	22.8
FP16-32 (two vectors)	178.2 us	71.4

was used for updating the neighbor ghost layer of the EBE computation results, and transprecision data storage was used for the vectors stored in CG solvers. When vector data $\mathbf{y}_s$ are sent/received between two processes, we send vector $\mathbf{y}_h \leftarrow \mathbf{y}_s/|\mathbf{y}_s|$ and scalar $|\mathbf{y}_s|$, and we unpack the data by $\mathbf{y}'_s \leftarrow |\mathbf{y}_s|\mathbf{y}_h$ at the recipient (L infinity norm is used for $|.|$). Although $\mathbf{y}'_s$ includes small errors, the errors are localized at only MPI partition boundaries; thus, it does not change the convergence characteristics on most problems. The CG vector quantities are stored in a custom 21-bit data type with 1 sign bit, 8 exponent bits, and 12 fractional bits (hereafter called FP21). Three FP21 values are stored in 64-bit memory and converted to three FP32 values for computation. After computation on FP32 arithmetic units, results are converted to FP21 values and stored in 64-bit memory. Although additional data conversion is required, FP21 data types are only used for memory bandwidth bound kernels; thus, reduction in data size by using FP21 leads to speedup even when this conversion is added. Furthermore, the FP16 communication and FP21 data types are used only for preconditioning and do not alter the final result in double precision. As will be shown in Section 7.2.3, the number of iterations does not change significantly even if FP16 communication and FP21 data types are used. Thus, the inter-node communication and device memory transfer can be reduced to about 1/4 and 1/3, respectively, compared to the case of using FP64.

7.2. *Performance*

7.2.1. *Performance measurement environment*

The SC18 solver algorithm is accelerated on hardware with high transprecision kernel performance. In addition, improved scalability is expected in large systems where communication is a bottleneck. Therefore, we measured performance on Piz Daint and Summit, which were also used to compare kernel performance. We also performed performance measurements on the K computer to check the differences in the performance characteristics compared to the SC14 solver.

Summit[56]: Summit consists of 4,608 IBM Power System AC922 nodes, each with two IBM POWER9 (22 SIMD multi-core) processors and six NVIDIA Volta V100 GPUs. Each node has 512 GB of DDR4 memory used by the POWER9 processors, and each of the six accelerators has 16 GB of 900 GB/s high bandwidth memory (HBM2). The compute nodes are connected by a dual-rail InfiniBand EDR network with node injection bandwidth of 25 GB/s in both send/receive directions. The nodes are interconnected in a non-blocking fat-tree topology, with SHArP technology designed to offload collective operation processing to the network.[57] The peak FP64 performance of a node is 7.8 TFLOPS $\times$ 6 = 46.8 TFLOPS, which gives 215 PFLOPS for the entire system. The peak performance of FP32 and FP16 is 2 and 4 times that of FP64. Tensor Cores with mixed FP16 and FP32 arithmetic are also available, but not used in this study.

Piz Daint[58]: Piz Daint is a hybrid system consisting of 1,431 Cray XC40 (CPU only) nodes and 5,320 Cray XC50 (CPU + GPU) nodes. This study used the XC50 nodes each with an Intel Xeon E5-2690 v3 2.60 GHz CPU (12 cores) and an NVIDIA Tesla P100 GPU. Each XC50 node has 64 GB of memory for the Xeon processor and 16 GB of 732 GB/s bandwidth HBM2 for the GPU. The compute nodes are connected via Cray Aries routing and communication ASICs with the Dragonfly network topology, which provides a node injection bandwidth of 10.2 GB/s in both the send/receive directions. The peak FP64 performance is 4.7 TFLOPS per node and 25 PFLOPS for the entire system. The peak performance of FP32 and FP16 is 2 and 4 times that of FP64.

K computer[14]: The K computer comprises 82,944 compute nodes, each with a single eight-core SPARC64 VIIIfx CPU. The FP64 peak performance of the CPU is 2 (GHz) $\times$ 2 (FMA) $\times$ 2 (SIMD width) $\times$ 2 (sets per core) $\times$ 8 (cores) = 128 GFLOPS, which yields 10.6 PFLOPS for the entire system. Note that the SIMD width does not change regardless of precision; thus, the peak performance for FP32 is the same. Each node has 16 GB of DDR3 SDRAM, with a peak bandwidth of 64 GB/s. Tofu, a six-dimensional interconnection network, is used for communication between nodes.[15] Each node can communicate simultaneously in four directions at 5 GB/s throughput in each direction.

7.2.2. *Performance measurement settings*

Urban problems involving underground structures are large in scale and nonlinear with respect to earthquake shaking. In addition, the coefficient contrasts among materials are significant, resulting in poor convergence. Thus, these problems are among the most costly in urban earthquake simulations. Therefore, speeding up the computation of these problems is expected to lead to speeding up other urban seismic problems as well. Therefore, we will target a problem that mimics an underground structure with a hard layer in a two-layered ground (Fig. 29). The hard layer is a linear material with properties similar to those of reinforced concrete underground structures, the first soil layer is a nonlinear material that softens with strain, and the second soil layer is bedrock with linear material properties. To measure weak scaling, we replicated this problem in the x, y directions. Although the problem settings are periodic, METIS graph partitioning[11] was used to partition the problem for parallel computation so that the load balancing characteristics are similar to those of an real urban problems. A wave of $dt = 0.01$ s (1995 JMA Kobe wave[18]) was input and the time required to solve 25 time steps was measured. Semi-infinite absorbing boundary conditions were used for the sides and bottom of the model. An arbitrary nonlinear constitutive model can be used, but we used the modified

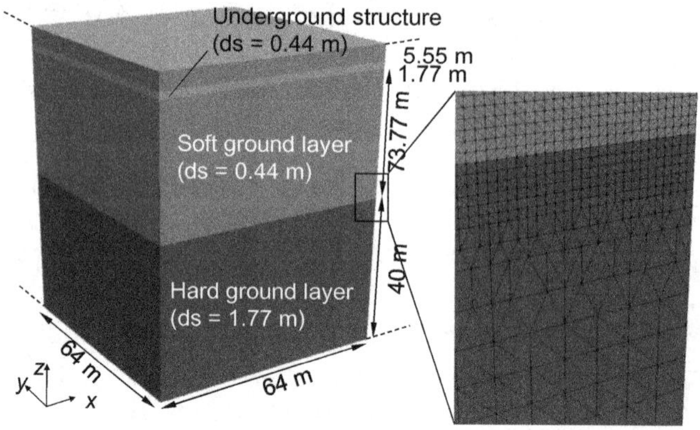

Fig. 29. Measurement model generated using Ref. 13. This model is replicated in the x and y directions to create a periodic problem suitable for measuring weak scaling.

Ramberg Osgood model[2] and Masing rule[3] for nonlinear modeling of the soil.

Using this problem, we compared the performance of the developed SC18 solver, the standard PCGE solver, and the latest SC14 solver. Because of the algorithmic complexity of the SC14 and SC18 solvers, it is difficult to make rigorous cost comparisons on a per-component basis. However, since both methods are based on adaptive CG methods, the SC14 solver can be considered an SC18 solver without AI-based $PreCG^c_{\text{part}}$, FP16/FP21 computation and communication, and time-parallel algorithms. PCGE can be thought of as an SC14 solver with FP64 computation and communication and $PreCG^c$ disabled. Note that the relative error tolerance for all solvers was set to $\epsilon = 1.0 \times 10^{-8}$. For SC18 solver, relative error tolerances of 0.7, 0.05, and 0.25 were used for the $PreCG^c$, $PreCG^c_{\text{part}}$, and $PreCG$ preconditioning solvers, respectively. For the preconditioning solvers in the SC14 solver, we used the error tolerances documented in Ref. 1. One GPU was used per MPI process for Piz Daint, one GPU per MPI process (6 MPI processes per node) for Summit and eight OpenMP threads were used per MPI process for the K computer. Hardware counters were used to measure FLOPS on the K computer, and nvprof was used to measure FLOPS on Piz Daint and Summit. PCGE was computed in FP64, while the SC14 solver was computed in mixed precision of FP32 and FP64.

The implementation of the transprecision computing algorithm following Fig. 28 is called the SC18 solver, and the same algorithm with FP16 communication disabled (FP16 communication in Fig. 28 is performed in FP32) is called the SC18-a solver, which will be used to analyze communication performance. Since the K computer does not support FP16 operations and has relatively high memory bandwidth, the computation and communication shown as FP16 and FP21 in Fig. 28 were performed using FP32 (this implementation is called the SC18-b solver). All measurements include outputting results to files.

7.2.3. *Performance measurement results*

In the urban simulation targeted in this study, a nonlinear response problem with 10^{10} DOF and $10^{3\sim4}$ time steps must be solved within the $10^{0\sim1}$ h job time limit on recent supercomputers. Because the memory size is small compared to the computational capability, the full memory size is used to analyze large-scale problems. Therefore, a problem with 12.3 million DOF per MPI process is used for measurement on Piz Daint and Summit (Table 12). Since this problem size does not fit in each node of the K computer, problem size of 6.15 million DOF per MPI process was used for measurement on the K computer.

First we compare the performance of the PCGE, SC14, and SC18 solvers using model W-1 in Table 12. As shown in Figs. 30–32, the

Table 12. Model configurations for Piz Daint/Summit (without brackets) and K computer (with brackets). 1 GPU is used per MPI process for Summit/Piz Daint, while 8 OpenMP threads are used per MPI process for K computer. 1 GPU is used for 2 K computer nodes.

Model	# of MPI processes	DOF	DOF per process	# of elements
W-1	288 (576)	3,545,198,451	12,309,716 (6,154,858)	883,104,768
W-2	576 (1,152)	7,088,615,271	12,306,623 (6,153,311)	1,766,209,536
W-3	1,152 (2,304)	14,174,736,543	12,304,458 (6,152,229)	3,532,419,072
W-4	2,304 (4,608)	28,345,910,535	12,302,912 (6,151,456)	7,064,838,144
W-5	4,608 (9,216)	56,687,546,151	12,301,984 (6,150,992)	9,419,784,192
W-6	6,144 (12,288)	75,580,545,159	12,301,521 (6,150,760)	18,839,568,384
W-7	12,288 (24,576)	151,152,541,191	12,300,825 (6,150,412)	37,679,136,768
W-8	24,576 (49,152)	302,293,683,783	12,300,361 (6,150,180)	75,358,273,536

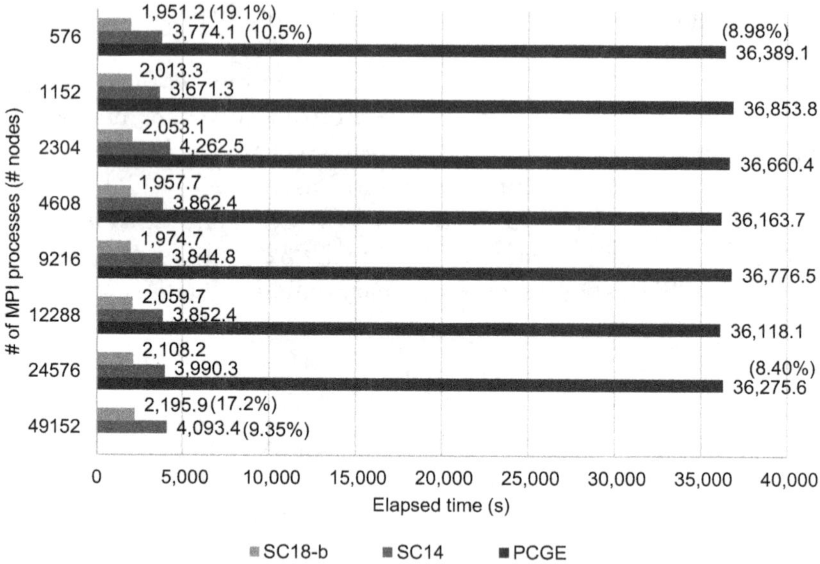

Fig. 30. Weak scaling results on the K computer, showing elapsed time and performance efficiency to FP64 peak.

SC18 solver was 18.6-fold, 24.9-fold, and 25.3-fold faster than PCGE on the K computer, Piz Daint, and Summit, respectively. This is due to the reduced FLOP count and the improved peak performance; the arithmetic count of the SC18 solver (33.2 PFLOP) was reduced by 5.56-fold from PCGE (184.7 PFLOP) on Summit. The SC14 solver, which also uses advanced preconditioning for reduction of arithmetic counts, still required 44.0 PFLOP; we can see that the arithmetic count is reduced significantly by use of the SC18 solver algorithm. Furthermore, the SC18 solver's highly efficient, transprecision EBE kernel improved the FLOPS efficiency on Summit from 4.27% of PCGE to 19.5% in SC18 solver, leading to a 25.3-fold speedup. This is also 3.99-fold faster than the state-of-the-art SC14 solver. The ratio of elapsed time between Piz Daint and Summit is 110.7 s/75.8 s = 1.46, which is close to the 1.66-fold hardware peak FLOPS capability difference between the P100 and V100 GPUs. Thus, the developed solver can improve its performance as hardware performance improves.

We next see the weak scaling performance on the K computer (Fig. 30). Even with FP16 communication disabled, high scalability of 88.8% (from W-1 (576 processes) to W-8 (49,512 processes)) was

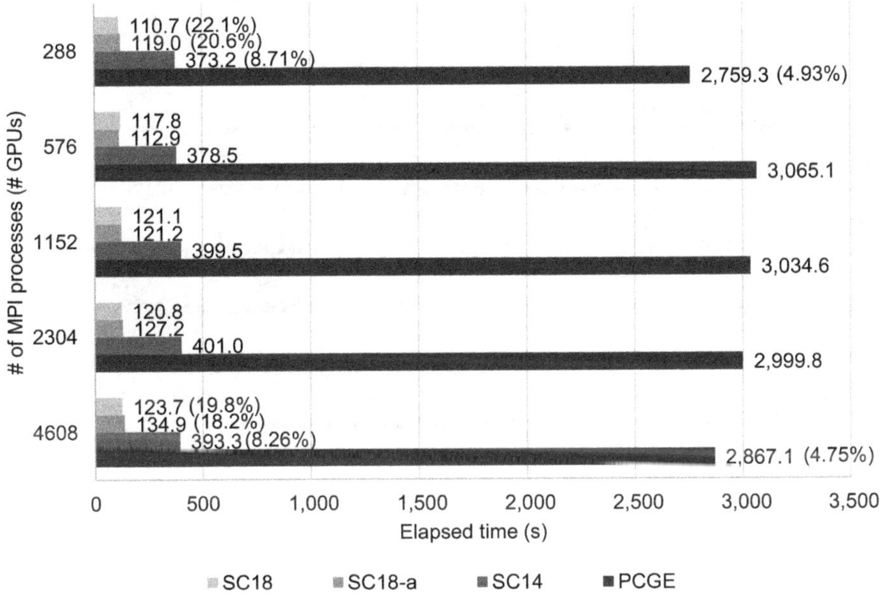

Fig. 31. Weak scaling results on Piz Daint, showing elapsed time and performance efficiency of FP64 peak.

obtained on the K computer. The high performance on a single node and high scalability to many nodes leads to high peak performance of 17.2% of peak FP64 FLOPS on the largest problem (W-8), which is 1.86-fold faster than the SC14 solver.

Next we see the effectiveness of FP16 communication on Piz Daint (Fig. 31). The floating point arithmetic capability to interconnect bandwidth ratio of Piz Daint is 4.7 TFLOPS : 10.2 GB/s, which is relatively low compared to the 128 GFLOPS : 5 GB/s of the K computer; thus, performance and scalability are expected to improve using FP16 communication. The same solution was obtained with less than a 5% difference in the number of iterations required to solve 25 time steps using FP32 communication (SC18-a solver) and FP16 communication (SC18 solver). That is, the iterations required for [CG, $PreCG^c$, $PreCG^c_{\text{part}}$, $PreCG$] were [119, 4516, 26593, 2448] and [117, 4453, 26553, 2422] for SC18-a and SC18 solvers, respectively. This has resulted in an approximately 50% reduction in communication size and a 1.07-fold speedup on model W-1 with 288 GPUs. Scalability is improved by reduction in communication; the weak scaling efficiency was improved from 88.2% of the SC18-a solver

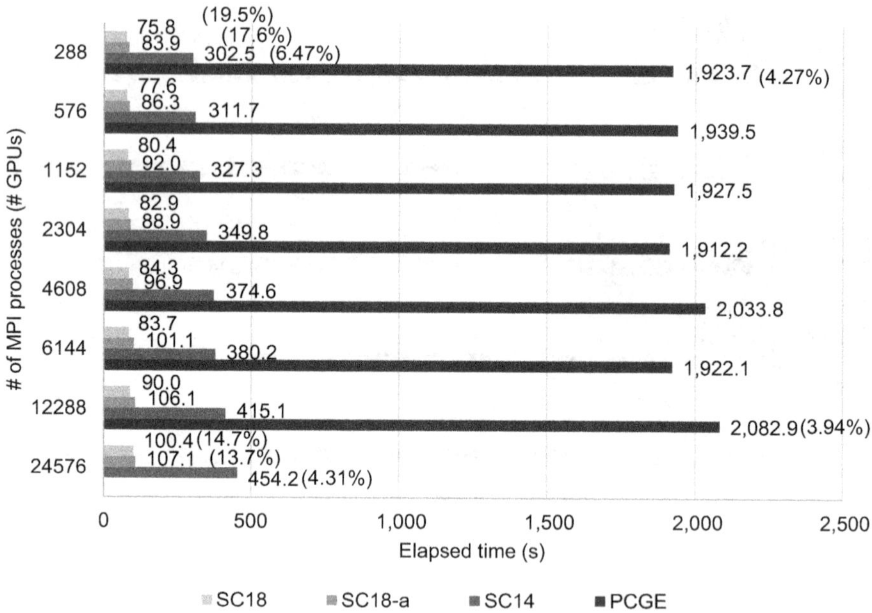

Fig. 32. Weak scaling results on Summit, showing elapsed time and performance efficiency to FP64 peak.

to 89.5% of the SC18 solver (measured from W-1 (288 nodes) to W-5 (4,608 nodes, near full system)). This resulted in a 23.2-fold speedup over PCGE and 3.18-fold speedup over the SC14 solver when measured on the full Piz Daint system. The SC18 solver achieved 19.8% of FP64 peak on full Piz Daint, which is very high for low-order unstructured implicit finite element simulation on a GPU supercomputer.

Finally, we measure weak scaling on Summit (Fig. 32). Summit's FLOPS/network capability ratio is 46.8 TFLOPS : 25 GB/s, which is even lower than Piz Daint; therefore, the SC18 solver with reduced communication is expected to be further effective. As expected, the performance of the SC18 solver with FP16 communication was 1.10-fold faster than the version using FP32 communication (SC18-a solver) on model W-1. Combined with the high weak scalability efficiency of 75.5%, a 4.52-fold speedup was attained from the SC14 solver on W-8 (4,096 nodes with 24576 GPUs, near full system). The performance was 28.2 PFLOPS (14.7% of FP64 peak) on 4,096 nodes of Summit.

In summary, the developed SC18 solver achieved very high performance on small problems: 19.5% of FP64 peak on Summit and 22.1% of FP64 peak on Piz Daint. For the K computer with fast interconnects, high scalability was attained even without the use of FP16 communication. High scalability was attained on Piz Daint and Summit by using the transprecision algorithm to reduce point-to-point communication. This high scalability is due to the increased number of arithmetic per iteration due to the time-parallel scheme, which reduced the number of synchronizations between parallel computing elements in addition to the reduced communication.

7.3. *Example*

As an application example of the developed SC18 solver, an earthquake simulation of a city with complex geometry comprising ground, underground structures, and high-rise buildings was performed on Summit (see Fig. 33). Here, the finite element mesh

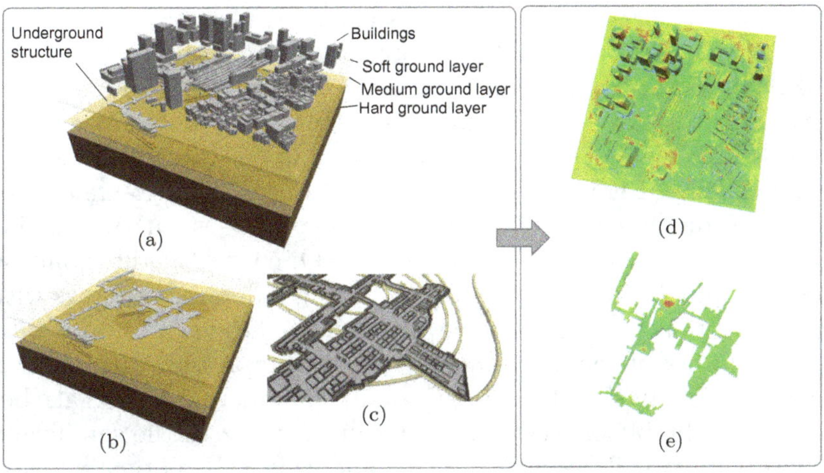

Fig. 33. Application example of SC18 solver computed on 384 Summit nodes (2,304 V100 GPUs): (a) 1,024 × 1,024 × 370 m city model based on Refs. 19 and 20 with underground and building structures surrounded by two-layered ground modeled with 3,961,851,160 elements and 16,291,917,564 DOF, (b) location of underground structure, (c) closeup of underground structure modeled with 25 cm second-order tetrahedral elements (brown part indicates $PreCG^c_{\text{part}}$ part extracted by AI), (d) displacement response to ground shaking, and (e) displacement response of underground structures.

was generated using SGI UV300 and finite element analysis was computed using 2,304 Summit GPUs, corresponding to problem size of W-4. As large uncertainty is involved in data of urban problems, and many stakeholders are involved in decision-making and evaluation, it is becoming increasingly important to achieve high-quality computing that accounts for such uncertainty. On the other hand, because of the large nonlinearity of urban responses, handling these uncertainties requires computationally expensive methods such as Monte Carlo simulations. As a result, urban simulations often require many cases of huge computations. The development of an accelerated solver, such as the one presented in this section, is expected to be effective in addressing these problems.

8. Conclusion and Future Tasks

As described previously, the development of algorithms and implementations based on high-performance computational science has realized a very large-scale, massively parallel, high-speed finite element analysis capable of handling tens of billions to trillions of DOF. This is expected to enable analysis of static/dynamic and non-linear/linear behaviors of scientific to engineering problems with different length scales, required resolutions, and constitutive laws using highly detailed FE models (e.g., analysis of archipelago-scale stress accumulation processes, M9-class source processes, urban shaking, and coupled ground/structure response). Of course, at this point, we have only just begun to be able to analyze the data, and it goes without saying that research and development is needed to improve the reproducibility of the phenomena through comparison with observation data. If analysis using a highly detailed model that can make better use of valuable observation data can be realized, the significance of observation will become clearer, and this will motivate further advanced observation. It is also expected that such development will make the path toward the construction of a next-generation digital twin of the city, society, and the Earth based on Big Data & Extreme Computing in the Society 5.0 era, and demonstrate how such digital twins are effective for disaster mitigation.

In addition to the realization of the above-mentioned ultra-large-scale physical simulations by high-performance computational science, new possibilities through the fusion of computational science and information science should not be overlooked. For example, methods that predict damage based on an artificial intelligence model (surrogate model) constructed from virtual large-scale dense observation data of inexperienced large earthquakes using very large-scale physical simulations are being developed. Also, methods that speed up large-scale physical simulations by use of artificial intelligence models that have learned the target differential equations, as well as methods that accelerate simulations using data generated during simulations, are also being explored. We can see that the possibility of development of algorithms that are different from conventional ones are increasing in potential.

High-performance computational science is not only expected to contribute to the improvement of national and regional disaster mitigation capabilities by improving the ability to analyze seismic problems, but has potential to evolve into a new field of science by integration with information science.

Acknowledgment

Part of this work was supported by Research and Application Development in Post-K Project for Social and Scientific Important Tasks, Task 3 "System for Integrated Simulation of Earthquake and Tsunami Hazard and Disaster". Part of this work was supported by JSPS KAKENHI Grant Numbers JP18H05239, JP18K18873.

This chapter shows the overview of research reported in papers.[1,39,41,44,59] The details of each method and results are available in each paper.

References

1. T. Ichimura, K. Fujita, S. Tanaka, M. Hori, M. Lalith, Y. Shizawa, and H. Kobayashi, Physics-based urban earthquake simulation enhanced by 10.7 BlnDOF x 30 K time-step unstructured FE non-linear seismic wave simulation, in: *Proceedings of the International Conference on High Performance Computing, Networking, Storage and Analysis* (SC'14), (2014), pp 15–26.

2. I. M. Idriss, R. D. Singh, and R. Dobry, Nonlinear behavior of soft clays during cyclic loading, *J. Geotech. Eng. Div.*, **104** (1978), pp. 1427–1447.
3. G. Masing, Eigenspannungen und verfestigung beim messing, in: *Proceedings of the 2nd International Congress of Applied Mechanics* (1926), pp. 332–335 (in German).
4. M. Ogino, R. Shioya, and H. Kanayama, An inexact balancing preconditioner for large-scale structural analysis, *J. Conput. Sci. Technol.* **2(1)** (2008), 150–161.
5. H. Kawai, M. Ogino, R. Shioya, and S. Yoshimura, Large-scale elast-plastic analysis using domain decomposition method optimized for multi-core CPU architecture, *Key Eng. Mater.* **462–463** (2011), 605–610.
6. H. Akiba, T. Ohyama, Y. Shibata, K. Yuyama, Y. Katai, R. Takeuchi, T. Hoshino, S. Yoshimura, H. Noguchi, M. Gupta, J. Gunnels, V. Austel, Y. Sabharwal, R. Garg, S. Kato, T. Kawakami, S. Todokoro, and J. Ikeda, Large scale drop impact analysis of mobile phone using ADVC on Blue Gene/L, in: *Proceedings of the 2006 ACM/IEEE conference on Supercomputing (SC'06)*. New York, NY, USA: ACM, (2006), Article 46. DOI=10.1145/1188455.1188503
7. Y. Saad, 2003, *Iterative Methods for Sparse Linear Systems*, 2nd edition. SIAM.
8. J. M. Winget and T. J. R. Hughes Solution algorithms for nonlinear transient heat conduction analysis employing element-by-element iterative strategies, *Comput. Meth. Appl. Mech. Eng.* **52** (1985), 711–815.
9. E. Hairer, S. P. Norsett, and G. Wanner, *Solving Ordinary Differential equations I: Non Stiff Problems*, 2nd edition. Springer (1993).
10. G. H. Golub and Q. Ye Inexact conjugate gradient method with inner-outer iteration. *SIAM, J. Sci. Comput.* **21**(4) (1997), 1305–1320.
11. METIS 5.1.0, http://glaros.dtc.umn.edu/gkhome/metis/metis/overview.
12. O. C. Zienkiewicz and R. L. Taylor. *The Finite Element Method for Solid and Structural Mechanics*, 6th edition. Elsevier (2005).
13. T. Ichimura, M. Hori, and J. Bielak. A Hybrid multiresolution meshing technique for finite element three-dimensional earthquake ground motion modeling in basins including topography, *Geophys. J. Int.*, **177** (2009), 1221–1232.
14. H. Miyazaki, Y. Kusano, N. Shinjou, F. Shoji, M. Yokokawa, and T. Watanabe. Overview of the K computer system, *FUJITSU Sci. Tech. J.* **48**(3) (2012), 302–309.

15. Y. Ajima, T. Inoue, S. Hiramoto, and T. Shimizu. Tofu: Interconnect for the K computer, *FUJITSU Sci. Tech. J.* **48**(3) (2012), 280–285.
16. OpenMPI, http://www.open-mpi.org/.
17. MPI: A Message-Passing Interface Standard, Version 2.1, http://www.mpi-forum.org/docs/mpi21-report.pdf.
18. Strong ground motion of The Southern Hyogo prefecture earthquake in 1995 observed at Kobe JMA observatory, Japan Meteorological Agency, http://www.data.jma.go.jp/svd/eqev/data/kyoshin/jishin/hyogo_nanbu/dat/H1171931.csv.
19. 5m mesh digital elevation map, Tokyo ward area, Geospatial Information Authority of Japan, http://www.gsi.go.jp/MAP/CD-ROM/dem5m/index.htm.
20. National Digital Soil Map, The Japanese Geotechincal Society, http://www.denshi-jiban.jp/.
21. Strong-motion seismograph networks (K-NET, KiK-net), National Research Institute for Earth Science and Disaster Prevention, http://www.kyoshin.bosai.go.jp/.
22. G.W. Housner. Spectrum intensities of strong-motion earthquakes, in: *Symposium on Earthquakes and Blast Effects on Structures*. Los Angeles, CA (1952).
23. P. Somerville, N. Collins, N. Abrahamson, R. Graves, and C. Saikia, Ground motion attenuation relations for the central and eastern united states, Final Report (June 30, 2001), http://earthquake.usgs.gov/hazards/products/conterminous/2008/99HQGR0098.pdf.
24. Second report of the Nankai Trough Large Earthquake Model Committee, Cabinet Office, Government of Japan (August 28, 2012), http://www.bousai.go.jp/jishin/nankai/model/index.html.
25. Disaster assessment of Tokyo due to large earthquakes such as the Nankai Trough Earthquake, Tokyo Metropolitan Government (May 14, 2013), http://www.bousai.metro.tokyo.jp/taisaku/1000902/1000402.html.
26. T. Tiankai, Y. Hongfeng, L. Ramirez-Guzman, J. Bielak, O. Ghattas, M. Kwan-Liu, and D. R. O'Hallaron, From mesh generation to scientific visualization: an end-to-end approach to parallel supercomputing, in: *Proceedings of the 2006 ACM/IEEE Conference on Supercomputing (SC '06)*. New York, NY, USA: ACM (2006), Article 91. doi:10.1145/1188455.1188551.
27. M. Rietmann, P. Messmer, T. Nissen-Meyer, D. Peter, P. Basini, D. Komatitsch, O. Schenk, J. Tromp, L. Boschi, and D. Giardini, Forward and adjoint simulations of seismic wave propagation on emerging large-scale GPU architectures, in: *Proceedings of the International Conference on High Performance Computing, Networking,*

Storage and Analysis (SC '12), Article 38, 11 pages. Los Alamitos, CA, USA: IEEE Computer Society Press, (2012).

28. Y. Cui, E. Poyraz, K. B. Olsen, J. Zhou, K. Withers, S. Callaghan, J. Larkin, C. Guest, D. Choi, A. Chourasia, Z. Shi, S. M. Day, P. J. Maechling, and T. H. Jordan, Physics-based seismic hazard analysis on petascale heterogeneous supercomputers, in: *Proceedings of the International Conference on High Performance Computing, Networking, Storage and Analysis*, (SC'13), Article 70, 12 pages. New York, NY, USA: IEEE Computer Society Press, (2013).
29. R. Taborda and J. Bielak, Large-scale earthquake simulation: Computational seismology and complex engineering systems, *Comput. Sci. Eng.* **13** (2011), 14–27.
30. T. Ichimura, K. Fujita, M. Hori, T. Sakanoue, and R. Hamanaka, Three-dimensional nonlinear seismic ground response analysis of local site effects for estimating seismic behavior of buried pipelines, *J. Press. Vessel Technol., Am. Soc. Mech. Eng.*, **136** (2014), 041702.
31. A. Heinecke, A. Breuer, S. Rettenberger, M. Bader, A-A. Gabriel, C. Pelties, A. Bode, W. Barth, X-K. Liao, K. Vaidyanathan, M. Smelyanskiy, and P. Dubey, Petascale high order dynamic rupture earthquake simulations on heterogeneous supercomputers, in: *Proceedings of the International Conference on High Performance Computing, Networking, Storage and Analysis*, (SC'14), pp 3–14. (2014).
32. Japan Seismic Hazard Information Station, National Research Institute for Earth Science and Disaster Prevention, http://www.j-shis.bosai.go.jp/en/.
33. Report of the Assumed Tokyo Large Earthquake Model Committee, Cabinet Office, Government of Japan (December 28, 2013), http://www.bousai.go.jp/kaigirep/chuobou/senmon/shutochokkajishinmodel/.
34. P. E. B. Quinay, T. Ichimura, M. Hori, A. Nishida, and S. Yoshimura, Seismic structural response estimates of a high-fidelity fault-structure system model using multiscale analysis with parallel simulation of seismic wave propagation, *Bullet. Seismol. Soc. Am.* **103**, 2094–2110, (2013).
35. K. Fujita, T. Ichimura, M. Hori, L. Maddegedara, and S. Tanaka, Scalable multicase urban earthquake simulation method for stochastic earthquake disaster estimation, *Procedia Comput. Sci.* (2015).
36. H. Bao, J. Bielak, O. Ghattas, D. R. O'Hallaron, L. F. Kallivokas, J. R. Shewchuk, and J. Xu, Earthquake ground motion modeling on parallel computers, in *Proceedings of the 1996 ACM/IEEE Conference on Supercomputing* (1996).

37. T. Ichimura, M. Hori, and H. Kuwamoto, Earthquake motion simulation with multi-scale finite element analysis on hybrid grid, *Bull. Seismol. Soc. Am.* **97**, 1133–1143, (2007).
38. C. Yifeng Cui, K. B. Olsen, T. H. Jordan, K. Lee, J. Zhou, P. Small, D. Roten, G. Ely, D. K. Panda, A. Chourasia, J. Levesque, S. M. Day, and P. Maechling, Scalable earthquake simulation on petascale supercomputers, in: *Proceedings of the 2010 ACM/IEEE International Conference for High Performance Computing, Networking, Storage and Analysis* (SC '10), pp. 1–20. IEEE Computer Society, Washington, DC, USA, (2010), doi:10.1109/SC.2010.45.
39. T. Ichimura, K. Fujita, P.E.B. Quinay, L. Maddegedara, M. Hori, S. Tanaka, Y. Shizawa, H. Kobayashi, and K. Minami, Implicit nonlinear wave simulation with 1.08T DOF and 0.270T unstructured finite elements to enhance comprehensive earthquake simulation, in: *Proceedings of the International Conference on High Performance Computing, Networking, Storage and Analysis* (SC'15), (2015).
40. T. Ichimura, R. Agata, T. Hori, K. Hirahara, C. Hashimoto, M. Hori, and Y. Fukahata, An elastic/viscoelastic finite element analysis method for crustal deformation using a 3D island-scale high-fidelity model, *Geophys. J. Int.* (2016), doi:10.1093/gji/ggw123.
41. K. Fujita, T. Ichimura, K. Koyama, H. Inoue, M. Hori, and L. Maddegedara, Fast and scalable low-order implicit unstructured finite-element solver for Earth's crust deformation problem, in: *Proceedings of the Platform for Advanced Scientific Computing (PASC) Conference* (2017), Article No. 11, doi:10.1145/3093172.3093236.
42. T. Hori, M. Hyodo, R. Nakata, S. Miyazaki, and Y. Kaneda, A forecasting procedure for plate boundary earthquakes based on sequential data assimilation, *Oceanography* **27**(2), 94–102, (2014).
43. T. Hori, M. Hyodo, S. Miyazaki, and Y. Kaneda, Numerical forecasting of the time interval between successive M8 earthquakes along the Nankai Trough, southwest Japan, using ocean bottom cable network data, *Marine Geophys. Res.*, **35**, 285–294, (2014).
44. T. Ichimura, K. Fujita, M. Horikoshi, L. Meadows, K. Nakajima, T. Yamaguchi, K. Koyama, H. Inoue, A. Naruse, K. Katsushima, M. Hori, and M. Lalith, A fast scalable implicit solver with concentrated computation for nonlinear time-evolution problems on low-order unstructured finite elements, in: *32nd IEEE International Parallel and Distributed Processing Symposium*, (2018).
45. J. Rudi, A. C. I. Malossi, T. Isaac, G. Stadler, M. Gurnis, P. W. J. Staar, Y. Ineichen, C. Bekas, A. Curioni, and O. Ghattas, An

extreme-scale implicit solver for complex PDEs: Highly heterogeneous flow in earth's mantle, in: *Proceedings of the International Conference on High Performance Computing, Networking, Storage and Analysis* (SC'15). p. 5, ACM, (2015).

46. C. Yang, W. Xue, H. Fu, H. You, X. Wang, Y. Ao, F. Liu, L. Gan, P. Xu, L. Wang, G. Yang, and W. Zheng. 10M-Core scalable fully-implicit solver for nonhydrostatic atmospheric dynamics, in: *Proceedings of the International Conference on High Performance Computing, Networking, Storage and Analysis (SC'16)*, p. 6, IEEE Press (2016).
47. FUJITSU Supercomputer PRIMEHPC FX100, http://www.fujitsu.com/global/products/computing/servers/supercomputer/primehpc-fx100/.
48. Reedbush-H, http://www.cc.u-tokyo.ac.jp/system/reedbush/index-e.html.
49. HPCG benchmark (Nov. 2016), http://www.hpcg-benchmark.org/.
50. Oakforest-PACS, http://www.cc.u-tokyo.ac.jp/system/ofp/index-e.html.
51. JCAHPC, http://jcahpc.jp/eng/index.html.
52. X. Liu, E. Chow, K. Vaidyanathan, and M. Smelyanskiy Improving the performance of dynamical simulations via multiple right-hand sides, in: *IEEE 26th International Parallel and Distributed Processing Symposium*, (2012).
53. K. Fujita, K. Katsushima, T. Ichimura, M. Hori, and M. Lalith. Octree-based multiple-material parallel unstructured mesh generation method for seismic response analysis of soil-structure systems, *Procedia Comput. Sci.* **80**, 1624–1634, (2016).
54. TurboSquid, Oil refinery (ID: 791179), http://www.turbosquid.com/.
55. T. Ichimura, K. Fujita, T. Yamaguchi, M. Hori, M. Lalith, and N. Ueda, AI with super-computed data for monte carlo earthquake hazard classification, in: *Research Poster for SC17: International Conference for High Performance Computing, Networking, Storage and Analysis*, (2017).
56. Summit, https://www.olcf.ornl.gov/olcf-resources/compute-systems/summit/.
57. R. L. Graham, D. Bureddy, and P. Lui, Scalable hierarchical aggregation protocol (SHArP): A hardware architecture for efficient data reduction, in: *2016 First International Workshop on Communication Optimizations in HPC (COMHPC)*, pp. 1–10, Salt Lake City, UT (2016).
58. Piz Daint, https://www.cscs.ch/computers/piz-daint/.

59. T. Ichimura, K. Fujita, T. Yamaguchi, A. Naruse, J. C. Wells, T. C. Schulthess, T. P. Straatsma, C. J. Zimmer, M. Martinasso, K. Nakajima, M. Hori, and L. Maddegedara, A fast scalable implicit solver for nonlinear time-evolution earthquake city problem on low-ordered unstructured finite elements with artificial intelligence and transprecision computing, in: *SC '18 Proceedings of the International Conference for High Performance Computing, Networking, Storage, and Analysis, vol. 49*, pp. 1–49, (2018).
60. D. Zuras, *et al.*, *IEEE Standard for Floating-Point Arithmetic*, pp. 1–70, IEEE Std 754-2008 (2008).
61. NVIDIA Tesla P100 white paper, https://images.nvidia.com/content/pdf/tesla/whitepaper/pascal-architecture-whitepaper.pdf.
62. NVIDIA Tesla V100 GPU architecture, https://images.nvidia.com/content/volta-architecture/pdf/volta-architecture-whitepaper.pdf.
63. K. Fujita, K. Katsushima, T. Ichimura, M. Horikoshi, K. Nakajima, M. Hori, and L. Maddegedara, Wave propagation simulation of complex multi-material problems with fast low-order unstructured finite-element meshing and analysis, in: *Proceedings of the International Conference on High Performance Computing in Asia-Pacific Region (HPC Asia 2018)*, pp. 24–35, ACM, (2018).

https://doi.org/10.1142/9781800614635_0006

Chapter 6

Large-Scale Finite Element Simulation of Surface Faulting

Masataka Sawada*,§, Kazumoto Haba†,¶, and Muneo Hori‡,‖

**Central Research Institute of Electric Power Industry, Sustainable System Research Laboratory, Abiko, Japan*

†Taisei Co., Nuclear Facility Division, Shinjuku-ku, Tokyo, Japan

‡Japan Agency for Marine-Earth Science and Technology, Research Institute for Value-Added-Information Generation, Kanazawa-ku, Yokohama, Japan

§sawada@criepi.denken.or.jp

¶hb-kzm00@pub.taisei.co.jp

‖horimune@jamstec.go.jp

Reliable estimation of surface fault displacements is crucial to the safety of nuclear power plant facilities. It is necessary to develop a numerical method for the estimation. In this study, we developed a finite element method in which the following two functions were implemented: (1) a symplectic time integration of an explicit scheme to properly conserve the energy of the fault; and (2) rigorously formulated joint elements of high order. The finite element method (FEM) was enhanced with parallel computing. We applied the developed method to simulate two recent earthquakes which induced surface earthquake faults in Japan, the 2014 Nagano-ken–hokubu earthquake (reverse faulting) and the 2016 Kumamoto earthquake (right lateral strike-slip faulting). As a result, we succeeded in reproducing the occurrence and fault displacements observed in the two earthquakes. We then proposed a predictive simulation by inputting the slip distributions of the primary fault, which were determined using a strong ground motion prediction method for an earthquake, with a specified source fault and presented simulations to consider various uncertainties in the fault rupture process for surface fault analysis.

1. Introduction

When a large-scale inland earthquake occurs, infrastructures and buildings may be damaged not only by seismic motion but also by fault displacement on the ground surface. The displacement and deformation by surface earthquake faults have received increasing attention since the 1995 Hyogo-ken Nanbu earthquake, and many infrastructures were damaged by fault displacement occurring on the ground surface during the 1999 Kocaeli earthquake in Turkey and the Chi-Chi earthquake in Taiwan. Therefore, the prediction of surface ground deformation caused by fault displacement in the base rock and measures to mitigate damage of infrastructures have been paid attention to.

In Japan, since the 2011 Tohoku earthquake, surface earthquake faults have occurred in the 2011 Fukushima-ken Hamadori earthquake, the 2014 Nagano-ken-hokubu earthquake, and the 2016 Kumamoto earthquake. In particular, damage to buildings directly above or very close to surface earthquake faults was reported in the 2016 Kumamoto earthquake.

The new regulatory standard for nuclear power plants (NPPs) in Japan, which came into effect after the 2011 off the Pacific coast of Tohoku earthquake, requires that future displacement on not only faults considered as seismic sources but also small faults and landslide surfaces distributed on the site must be investigated for the foundation ground of important facilities.

When designing and assessing the safety of structures against fault displacement, it is necessary to quantitatively estimate the fault displacement in the vicinity of the structures (hazard assessment). In NPPs, important structures such as reactor buildings are constructed based on detailed topographical, geological, and fault investigations, avoiding primary faults (or main faults) which are thought to be direct extensions from seismic source faults. On the other hand, there are relatively small secondary faults (sub-faults), which are not directly related to the primary fault but were formed as a result of the primary fault's activity. In addition, faults that branch off from the primary fault (branch faults) and are small in size may be difficult to distinguish from secondary faults. The above fault classification is shown in Fig. 1. In order to evaluate the safety of an NPP facility against fault displacement, it is necessary to evaluate the occurrence and displacement of sub-faults in the vicinity of the facility when the primary fault moves.

The purpose of this study is to develop a numerical analysis method to evaluate the occurrence of displacement in sub-faults distributed near structures during a large crustal earthquake, which is associated with the activity of the primary fault.

A study of the propagation of fault slip generally requires three-dimensional analysis, and the computational resources required are very large. For example, the number of elements is on the order of 1,000,000 even when a 1-km-scale domain of interest is discretized with a spatial resolution of 10 m. For a dynamic analysis of this model, a seismic wave velocity of 1,000 m/s would require time increments of 0.01 s or less. Therefore, high-performance computing is essential for reliable prediction of fault displacement.

Numerical analysis of fault displacement is a friction problem involving rupture on a fault plane. In such problems, it is necessary to pay attention to mathematical considerations such as existence, uniqueness, and stability/unstability of a solution, as well

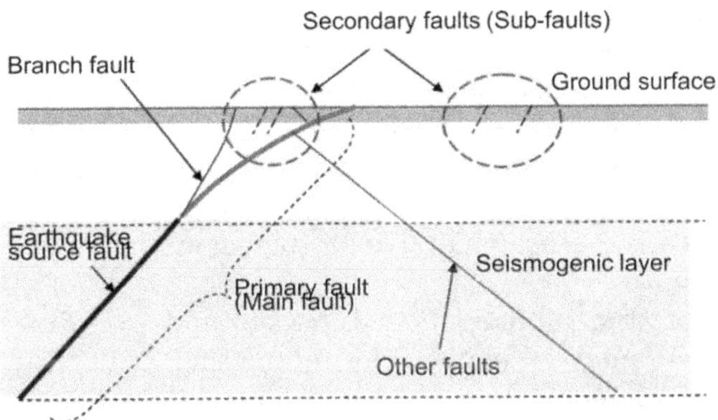

Fig. 1. Fault classification.

as to numerical methods that satisfy conservation laws. From the viewpoint of numerical analysis, it is important to solve mathematical problems for which solutions exist and stable solutions can be obtained.

Numerical simulations based on continuum mechanics are promising methods for the estimation of surface fault displacements. However, there are major difficulties in simulating fault rupture processes. One challenge is the requirement of significant numerical computations to simulate such processes for a target area that is only a few hundred meters in size. In practice, the following two-step simulation method is adopted[1]: (1) evaluating the boundary displacement of an area containing a target facility by determining the crustal deformation based on the elastic theory of dislocation[2] and (2) evaluating the deformation of the target area using a detailed two-dimensional model with high resolution and fidelity. A few researchers have conducted dynamic fault rupture simulations to evaluate ground motions or rupture propagation using the finite difference method, boundary element method, or finite element method (FEM).[3–7] It was difficult to conduct dynamic rupture simulations at a sufficiently high resolution because it requires a large amount of computation resources. In recent years, however, fault displacement analyses using dynamic rupture simulation with high-resolution detailed analytical models have begun to be reported.[8–11]

The amount of fault displacement on the ground surface is considered to be affected by the topography, strata structure, rock properties, and other factors near the ground surface. In order to evaluate these effects, it is necessary to construct analytical models with complex geometries of strata boundaries and ground surfaces. The FEM is a numerical analysis method suitable for analytical models with such complex geometries. In FEM, joint elements are used to construct analytical models of ground with discontinuous surfaces such as faults, but the joint elements commonly used are not rigorously derived and are not suitable for solving mathematical problems involving rupture.

Based on the aforementioned background, we first proposed a numerical analysis method for evaluating fault displacement. The analytical model is a rock mass including a fault plane, and the friction on the fault plane is represented using joint elements.

The analysis is dynamic and energy is conserved. Therefore, the numerical method developed includes three-dimensional joint elements and symplectic time integration with excellent energy conservation. In particular, the governing equations are described by the Hamiltonian canonical equations in order to apply higher-order symplectic time integration.

Next, we developed an FEM program for fault displacement evaluation. This was accomplished by implementing the aforementioned three-dimensional joint elements, the Hamiltonian canonical equation, and symplectic time integration into an open-source parallel FEM program.

The developed program was applied to problems of ground-fault for basic verification. The applicability of quasi-static analysis to multi-case analysis with different conditions to evaluate uncertainties was also discussed. As numerical analysis is utilized as a useful tool for manufacturing, assurance of the quality of numerical analysis has become an essential requirement. Verification and validation are the main means of quality assurance. These are also important in fault displacement analysis. In this study, we confirmed the applicability of the developed analysis method by applying it to real earthquakes with surface earthquake faults.

Fault displacement analysis involves uncertainties in the point stress–strain relationship and input displacement. Observations of surface earthquake faults also contain uncertainties due to measurement errors and coarse resolution. In this study, we evaluated fault displacement considering the uncertainty of input conditions by taking the variation of physical properties as an example, confirmed the applicability of the evaluation method, and summarized the effects of physical properties of the ground and fault on the results of fault displacement evaluation.

The structure of this chapter is as follows: in Section 2, we describe the analysis of physical phenomena and mathematical problem setting, the rigorous formulation of three-dimensional joint elements, symplectic time integration with review of its basic characteristics and their implementation in a parallel FEM program; in Section 3, we apply the developed programs to a relatively simple ground-fault problem and perform basic verification; in Section 4, we discuss the validity of the proposed analysis method and model by analyzing two real earthquakes with surface earthquake faults using the developed program and comparing it with the observed results; in Section 5, we propose a method to set slip distribution on primary fault for predictive simulation; in Section 6, multi-case analysis for the variation of physical properties is presented as an example of uncertainty quantification; and the conclusions are summarized in Section 7.

2. Numerical Analysis Method

2.1. *High-performance computing for fault displacement evaluation*

We propose a numerical method using high-performance computing to evaluate fault displacement with a certain degree of reliability. From a continuum mechanics perspective, a fault can be regarded as a displacement discontinuity that has slipped in the past. Fault behavior is a rupture phenomenon in which a new slip occurs on the fault plane. By setting up a stress–strain relationship in the ground and a stress–slip relationship on the fault plane, and by providing appropriate initial and boundary conditions, a continuum mathematical problem for fault behavior can be established.

As shown in Fig. 2, a continuum model of the ground and the fault is constructed, and the spread and dispersion processes of the slip generated at the depth of the fault Δ are evaluated to predict the fault displacement that appears at the ground surface δ. Two factors dominate the spread and dispersion process of slip: friction on the fault plane, which transitions from static to dynamic friction,

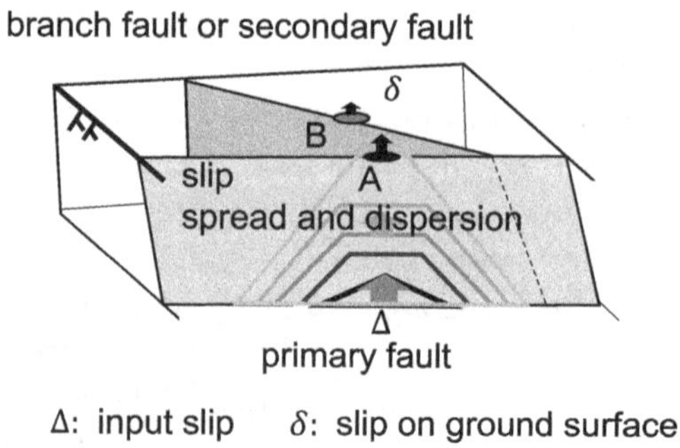

Fig. 2. Spreading and dispersing of a slip on faults.

and the rock mass deformation, which reduces the concentration of strain associated with rigid body displacement.

The slip–shear stress relationship of a fault plane should be treated with careful attention. Considering slip–shear stress relationship as friction, dynamic friction is generally smaller than static friction. As a slip–shear stiffness relationship, we simply consider a relationship in which the shear stiffness κ decreases as the slip u increases and becomes a constant value when the limit value (critical slip) $u_{\rm cr}$ is exceeded, as shown in the left side of Fig. 3. The slip–shear stress relationship shown on the right side of Fig. 3 is obtained. In this relationship, there are three solutions of u for the traction τ. These three solutions are stable, unstable, and stable in increasing order. In such a slip–shear stress relationship, a "jump" occurs, in which the slip increases significantly when the shear stress increases slightly. This suggests that the fault behavior may not change smoothly and gradually, but may slip abruptly.

It should be noted that in failure phenomena, the effect of variations in input conditions is significant. This is related to the stability and instability of the solution described above. If the solution is mathematically stable, it does not change much even if the input conditions are slightly disturbed, but if it is unstable, it varies greatly due to disturbance of the input conditions. When the aforementioned slip–shear stress relationship is applied to a fault plane, the solution is initially stable, but as the stress increases, the stability of the solution is rapidly lost and it becomes unstable. Then, a "jump" of slip to the next stable solution occurs. Due to this instability, as shown in

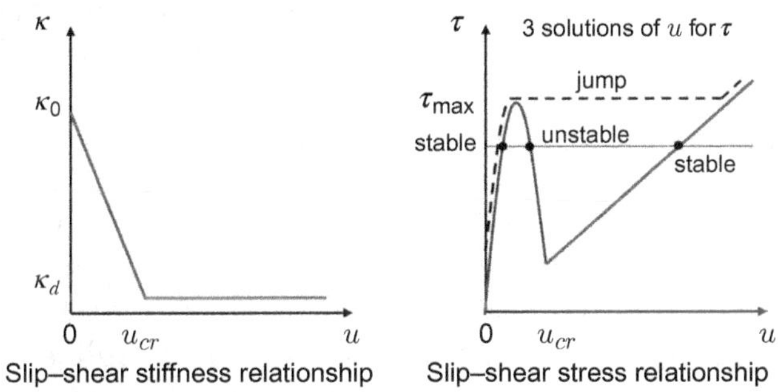

Fig. 3. Slip–spring constant relationship and slip–traction relationship.

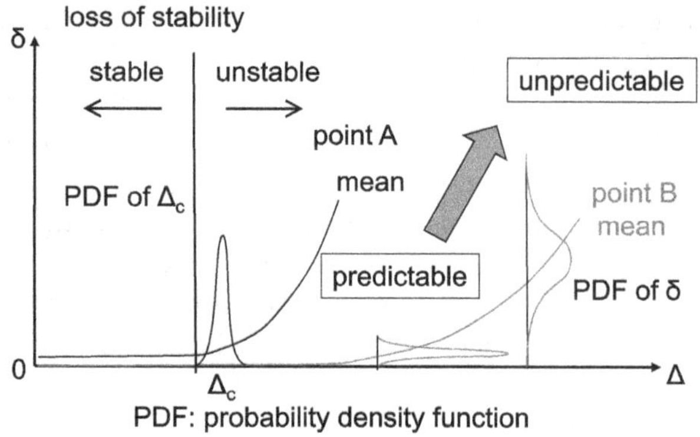

Fig. 4. Evaluation of surface fault displacement.

Fig. 4, the variation σ_δ of the solution δ with respect to the variation of input conditions is small before failure, but as failure progresses, σ_δ is expected to become larger. Therefore, a single solution will not correctly predict the actual fault behavior. In order to predict fault behavior, it is essential to vary the stress–strain relationship of the ground, the slip–shear stress relationship of the fault plane, and the initial and boundary conditions in order to obtain various solutions by numerical analysis.

If the variation σ_δ is too large, it means that it is unpredictable. On the other hand, the new regulatory standard for NPPs requires that the presence of surface displacement be evaluated. In the phase where the slip reaches the ground surface, the instability is less advanced than in the phase where the ground surface displacement becomes larger. Therefore, the critical input slip Δ_c, which is the minimum input slip that causes the ground surface displacement, can be evaluated with relatively small variation compared to δ. Therefore, it is important to evaluate Δ_c prior to evaluating δ.

As mentioned previously, three-dimensional analytical models of ground containing fault planes generally have a large number of degrees of freedom. Therefore, numerical analysis programs are required to solve problems with large degrees of freedom (capability computing). In addition, since fault behavior is greatly affected by variations in input conditions, the prediction of fault displacement requires quantification of the variation in the solution by performing many calculations under different conditions (capacity computing).

From these perspectives, high-performance computing is essential for the prediction of fault displacement. In addition, rigorous formulations and algorithms must be used to accurately solve the unstable problem of fault rupture.

The analytical model is a rock mass ground including a fault plane, and the slip–shear stress relationship of the fault plane is considered by using joint elements. For example, the discontinuous surfaces such as faults distributed in the foundation rock of important facilities of NPPs are grasped through detailed geological surveys. Therefore, it is reasonable to represent these discontinuities explicitly using joint elements. This joint element must be formulated rigorously to solve the fault rupture problem with instability with high accuracy.

Whether the slip propagates unstably or stably in the fault plane, the energy input in the fault plane is converted to frictional work of the slip in the fault plane or to strain energy of the surrounding ground. To guarantee the accuracy of the solution, energy conservation should be satisfied. The energy conservation required here is not only that of the entire domain, but also that of the local domain including the fault. When analyzing continuum mechanics problems numerically with FEM, energy conservation is related to the time integration algorithm. FEM is usually formulated using the Lagrangian, which can be converted to an equivalent Hamiltonian. Numerical analysis, in which the canonical equation of the Hamiltonian is solved by symplectic integration, is effective for energy conservation.

Therefore, the numerical methods proposed in this study for fault displacement evaluation are rigorous joint element and symplectic time integration. In this section, we present the formulation of these methods.

2.2. *Three-dimensional joint element*

In joint elements that are widely used in the field of rock mechanics, nodal forces are directly calculated from relative nodal displacements on a discontinuity.[12] By using this method, element stiffness matrices can be easily calculated. However, they differ from the standard method for constructing finite elements. Joint elements for fault displacement simulations should be defined to ensure proper convergence for the chosen element sizes and applicability to curved faults. We developed a rigorous joint element in which element stiffness matrices are derived from the Lagrangian on a discontinuous surface using an isoparametric formulation.[13] The following is a formulation for a second-order triangular joint element.

The variant of the Lagrangian δL_J is given by the following area integral on the discontinuous surface S:

$$\delta L_J = -\int \delta \boldsymbol{u}\,(2k_n \boldsymbol{n} \otimes \boldsymbol{n} + 2k_t \boldsymbol{t} \otimes \boldsymbol{t})\,\boldsymbol{u} dS, \tag{1}$$

where $\boldsymbol{u}\,(= \boldsymbol{u}_+ - \boldsymbol{u}_-)$ is the relative displacement; $\boldsymbol{u}_+$ and $\boldsymbol{u}_-$ are the displacements on the top and bottom surfaces of S, respectively; $\boldsymbol{n}$ is the unit normal vector of S; $\boldsymbol{t}$ is the unit tangential vector of S; and k_n and k_t are the normal and tangential spring constants, respectively. $\boldsymbol{A} \otimes \boldsymbol{B}$ denotes the tensor product of vectors $\boldsymbol{A}$ and $\boldsymbol{B}$.

Considering the case that S is a curved surface, an isoparametric formulation is adopted. The three-dimensional global coordinate system $\boldsymbol{x}$ of S is given by the following function of the two-dimensional local coordinate system $\boldsymbol{\xi}$ (see Fig. 5):

$$\boldsymbol{x}\,(\boldsymbol{\xi}) = \sum_{a=1}^{6} N^a\,(\boldsymbol{\xi})\,x^a, \tag{2}$$

where $\boldsymbol{x}^a$ $(a = 1, 2, \ldots, 6)$ is the nodal coordinate of the joint element and N^a is the shape function. The area integral on S is transformed into the area integral using $\boldsymbol{\xi}$, and the Gaussian integral is applied.

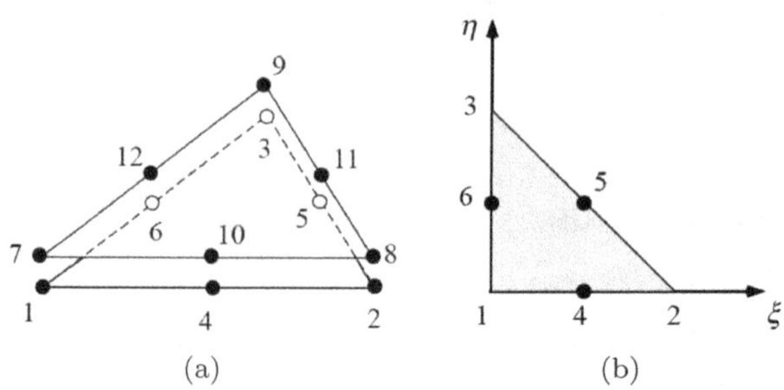

Fig. 5. Three-dimensional second-order triangular joint element and its local coordinate system. (a) Global coordinate system, and (b) local coordinate system.

The unit normal vector $\boldsymbol{n}$ is formulated as follows:

$$\boldsymbol{n} = s_\xi \times s_\eta / \left| s_\xi \times s_\eta \right|, \tag{3}$$

where $\boldsymbol{A} \times \boldsymbol{B}$ denotes the vector product of vectors $\boldsymbol{A}$ and $\boldsymbol{B}$. The differential global coordinate $\boldsymbol{s}$ is defined as follows:

$$s_\xi = \partial \boldsymbol{x} / \partial \boldsymbol{\xi}, \quad s_\eta = \partial \boldsymbol{x} / \partial \boldsymbol{\eta}. \tag{4}$$

The unit vectors have the following relationship:

$$\boldsymbol{t} \otimes \boldsymbol{t} = \boldsymbol{e} - \boldsymbol{n} \otimes \boldsymbol{n}, \tag{5}$$

where $\boldsymbol{e}$ is the identity matrix. The element stiffness matrix $\boldsymbol{E}_J^{ab}$ is obtained as follows:

$$\boldsymbol{E}_J^{ab} = \int 2N^a \left(k_n \boldsymbol{n} \otimes \boldsymbol{n} + k_t \boldsymbol{t} \otimes \boldsymbol{t} \right) N^b \left| s_\xi \times s_\eta \right| d\xi d\eta. \tag{6}$$

2.3. *Hamiltonian and symplectic time integration*

2.3.1. *Lagrangian and Hamiltonian*

Procedures for deriving a Hamiltonian from a Lagrangian are well established. In Fig. 6, we provide a schematic view of the procedure

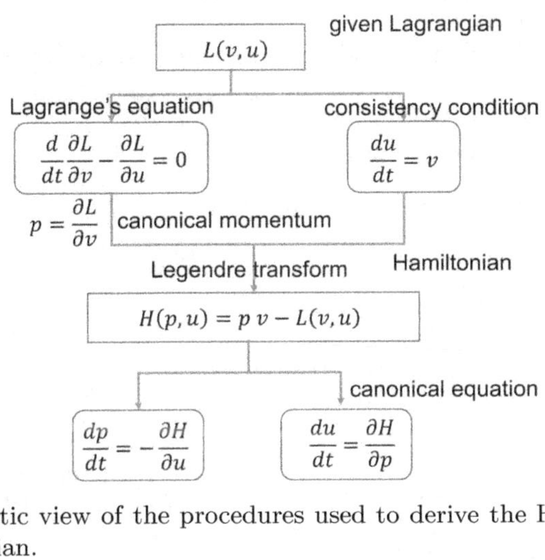

Fig. 6. Schematic view of the procedures used to derive the Hamiltonian from a given Lagrangian.

used for deriving a Hamiltonian from a Lagrangian of two variables $\mathcal{L}(v, u)$ with $v = \dot{u}$. The derivation of the Hamiltonian $\mathcal{H}$ requires the use of two equations, namely Lagrange's equation and a consistency condition. Lagrange's equation is derived from the stationary condition, and the consistency condition is assigned for the two arguments of the Lagrangian as an argument corresponding to the time derivative of the other (see Fig. 6). The Hamiltonian is calculated by using the Legendre transformation of $\mathcal{L}$, and the time evolution of the system is uniquely defined by the canonical equation (Hamilton's equation). The Hamiltonian represents the total energy of the system, which is the sum of kinetic and potential energy.

2.3.2. *Symplectic time integration*

Symplectic integrations are designed for numerical solutions of canonical equations.[14–16] A widely used class of symplectic integration is formed by using splitting methods. If the Hamiltonian $\mathcal{H}$ is separable, it is expressed in the following form:

$$\mathcal{H}(u, p) = \mathcal{T}(p) + \mathcal{V}(u), \tag{7}$$

where $p(= \frac{\partial \mathcal{L}}{\partial v})$ is the canonical momentum, $\mathcal{T}$ is the kinetic energy of the system, and $\mathcal{V}$ is the potential. The canonical equations are expressed as follows:

$$\frac{du}{dt} = \nabla_p \mathcal{T}(p) \quad \text{and} \quad \frac{dp}{dt} = -\nabla_u \mathcal{V}(u). \tag{8}$$

The symplectic Euler method follows the first-order integration:

$$\begin{cases} u_{n+1} = u_n + h\nabla_p \mathcal{T}(p_n), \\ p_{n+1} = p_n - h\nabla_u \mathcal{V}(u_{n+1}), \end{cases} \tag{9}$$

where h is a time increment, and n is the value of the nth step. We obtain an approximate solution of u_{n+1} and p_{n+1} (in this order) only by substitution.

We define the mapping S_U^h, S_P^h as follows:

$$S_U^h : \begin{bmatrix} u \\ p \end{bmatrix} \mapsto \begin{bmatrix} u + h\nabla_p \mathcal{T}(p) \\ p \end{bmatrix}, \tag{10}$$

$$S_P^h : \begin{bmatrix} u \\ p \end{bmatrix} \mapsto \begin{bmatrix} u \\ p - h\nabla_u \mathcal{V}(u) \end{bmatrix}. \tag{11}$$

For x_n defined as $x_n = [u_n^T, p_n^T]^T$, we rewrite Eq. 9 more simply by using composite mapping $S_E^h = S_P^h S_U^h$ as follows:

$$x_{n+1} = S_E^h(x_n). \tag{12}$$

Symplectic integrations of higher orders are obtained as follows. We prepare real numbers $b_1, b_2, \ldots, b_s, \hat{b}_1, \hat{b}_2, \ldots, \hat{b}_s$ (s is a natural number) and define a map S_i by substituting b_i or $\hat{b}_i$ times h in the S_E^h mapping as follows:

$$S_i = S_P^{\hat{b}_i h} S_U^{b_i h}. \tag{13}$$

We define the scheme from x_n to x_{n+1} by using S_i as a component as follows:

$$x_{n+1} = S^h(x_n) \quad (S^h = S_s \cdots S_2 S_1). \tag{14}$$

We implement the following two methods into a parallel FEM program:

(1) The Verlet method[17] is a second-order integration of Eq. (8) with $s = 2$ as follows:

$$\begin{array}{l} b_i : 0 \;\; 1 \\ \hat{b}_i : \frac{1}{2} \;\; \frac{1}{2}. \end{array} \tag{15}$$

(2) The third-order symplectic integration ($s = 3$) discovered by Ruth (1983)[18] was given as follows:

$$\begin{array}{l} b_i : \frac{7}{24} \;\; \frac{3}{4} \;\; -\frac{1}{24} \\ \hat{b}_i : \frac{2}{3} \;\; -\frac{2}{3} \;\; 1. \end{array} \tag{16}$$

These methods are not only symplectic, but also simple and stable, requiring only assignment calculations (explicit). However, the disadvantages are that symplecticism is generally broken when the time increment is changed, efficiency is reduced at higher orders due to the large number of steps, and local errors are large. Therefore, it is necessary to select an appropriate order and time interval.

2.3.3. *Simple spring problem*

We investigated the efficiency of the symplectic time integration when compared to that of the Newmark β method.[19] The Newmark β method is a widely used implicit time-integration method that is often used with the Newton–Raphson method. We solved the simple spring model shown in Fig. 7. The spring coefficient k exhibits nonlinearity similar to that shown in Fig. 3. For the condition $u \leq u_{\mathrm{cr}}$, the canonical equation was derived as follows:

$$\begin{aligned} \dot{u} &= \frac{p}{m}, \\ \dot{p} &= -2k_0 u + 2C_k u^2 + 2f, \end{aligned} \tag{17}$$

where u is the relative displacement, p is the relative momentum, m is the mass, $C_k(= k_0 - k_d)/u_{\mathrm{cr}}$ is the reduction rate of the spring coefficient, k_0 is the initial spring coefficient, k_d is the final spring coefficient, u_{cr} is the critical relative displacement, and f is the external force.

The parameter values for the proposed model were $m = 1.0$ kg, $f = 0.4$ N, $k_0 = 0.75$ N/m, $C_k = 0.25$ N/m^2. The maximum relative displacement was obtained from the following equation:

$$u_{\max} = \frac{3k_0}{4C_k}\left(1 - \sqrt{1 - \frac{16}{3}\frac{TC_k}{k_0^2}}\right). \tag{18}$$

In the graphs in Fig. 8, we normalized the relative displacement u by $u_{\max} = 1.737652$ m and the elapsed time by the period $\bar{t} = 8.265$ s, which was obtained by the following equation in *Mathematica*:

$$m\ddot{u} = -2k_0 u + 2C_k u^2 + 2f. \tag{19}$$

Equation (19) is equivalent to the canonical equation (17).

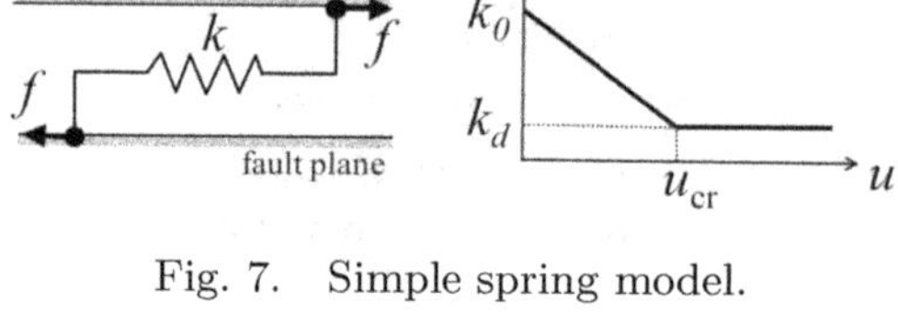

Fig. 7. Simple spring model.

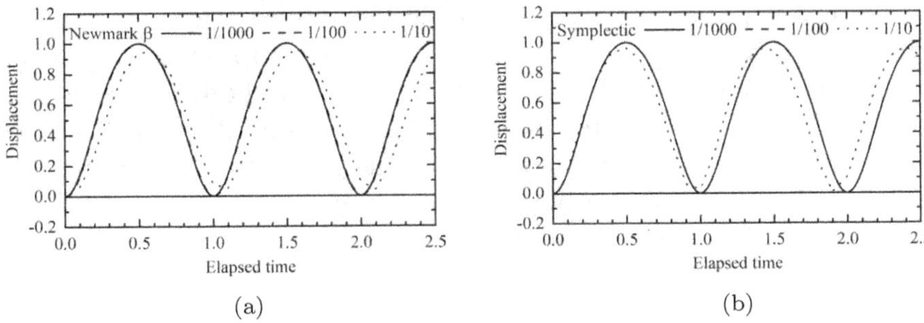

Fig. 8. Progression of the relative displacement. (a) Newmark β, and (b) symplectic.

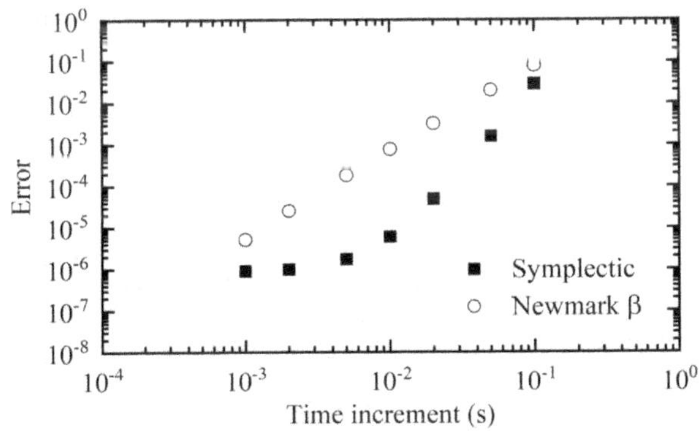

Fig. 9. Evaluation of error.

The parameters of the Newmark β method were $\beta = 0.25$ and $\gamma = 0.5$. We adopted a second-order symplectic integration ($s = 2$) such that the order of convergence was the same as that in the Newmark β method. Several cases were calculated with different time steps of 1/10, 1/20, 1/50, 1/100, 1/200, 1/500, and 1/1000 of the period $\bar{t} = 8.265$ s.

Figure 8 presents the progression of the relative displacement from the cases of $h = \bar{t}/1000$, $h = \bar{t}/100$, and $h = \bar{t}/10$. In both the Newmark β method and symplectic integration, the calculated amplitude and period were extremely good with the exception of the case with the maximum time step ($h = \bar{t}/10$).

Figure 9 presents the relationship between the error of the solution and the size of the time step. We defined the error as the difference between the normalized relative displacement and the analytical solution (zero) after a period. Second-order convergence was obtained from the Newmark β method. The error in the symplectic integration method was lower than that in the Newmark β method. Furthermore, the convergence in the symplectic integration method was faster than that in the Newmark β method. Thus, the result from the proposed symplectic integration method was better than that of the Newmark β method, which is a widely used implicit method.

2.4. *Implementation to parallel finite element program*

We implemented a rigorous joint element and symplectic time integration method with Hamiltonian formulation into an open-source parallel FEM program FrontISTR[20] to develop a numerical tool for fault displacement problems.[13]

2.4.1. *Nonlinear spring model on fault*

We implemented the joint element with a nonlinear spring model to represent the fault movement.[21] The following is the constitutive equation for shear movement of the joint element:

$$\tau = \kappa u, \tag{20}$$

where τ is the shear stress and u is the slip on the fault plane. The spring coefficient per unit area (shear stiffness) κ is described by the following function of the slip u:

$$\kappa(u) = \begin{cases} \kappa_0 - \frac{\kappa_0 - \kappa_d}{u_{\mathrm{cr}}} u & (u \leq u_{\mathrm{cr}}), \\ \kappa_d & (u > u_{\mathrm{cr}}), \end{cases} \tag{21}$$

where κ_0 is the initial shear stiffness, κ_d is the final shear stiffness, and u_{cr} is the critical slip (see Fig. 10). In dynamic simulations, the following slip trajectory u_{traj} is used instead of slip u since the slip direction might vary with the passage of time.

$$u_{\mathrm{traj}} = \int_0^t |\dot{u}| d\tau. \tag{22}$$

It should be noted that we do not use any sliders or dashpots in κ.

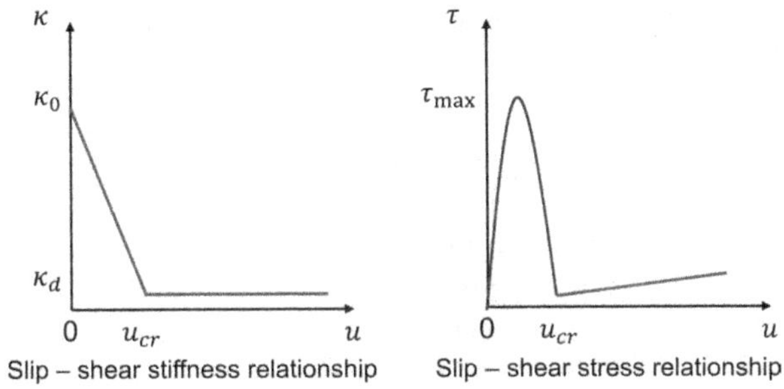

Fig. 10. Nonlinear slip–shear stress relationship on fault.

We consider the effect of normal stress on the shear stiffness of faults, and we define the following equation:

$$\kappa_0 = \widetilde{\kappa}_0 \, (\sigma_n + b), \tag{23}$$

where $\widetilde{\kappa}_0$ is the shear stiffness per unit of normal stress, σ_n is the normal stress on the fault plane, and b is a constant. We assume that the first peak of the slip–shear stress relationship shown in Fig. 10 is identical to the shear strength. The parameters κ_0 and b in Eq. 23 can be determined such that the shear strength satisfies Coulomb's law of friction as follows:

$$\widetilde{\kappa}_0 = \frac{4\,(1 - \kappa_d/\kappa_0)}{u_{\mathrm{cr}}} \tan\phi, \tag{24}$$

$$b = \frac{4c}{\widetilde{\kappa}_0 u_{\mathrm{cr}}} \, (1 - \kappa_d/\kappa_0), \tag{25}$$

where c is the cohesion and ϕ is the friction angle.

2.4.2. *Hamiltonian for finite element method*

The Lagrangian is typically used to formulate FEM although the Hamiltonian is expected to also work in the formulation. In this section, we describe a continuum Hamiltonian and a canonical equation for FEM. We provided a rigorous derivation of a Hamiltonian from a given Lagrangian of solid continuum.[13] We commence from the case in which the density of a continuum Lagrangian, denoted by l, is given for a continuum V. Specifically, l is assumed as a function of the velocity vector and strain tensor as denoted by $\boldsymbol{v}$ and $\boldsymbol{\epsilon}$, respectively. Using displacement vector $\boldsymbol{u}(\boldsymbol{x}, t)$ with $\boldsymbol{x}$ and

t denoting a point and time, respectively, $\boldsymbol{v}$ and $\boldsymbol{\epsilon}$ are expressed as follows:

$$\boldsymbol{v}(\boldsymbol{x},t) = \dot{\boldsymbol{u}}(\boldsymbol{x},t) \quad \text{and} \quad \boldsymbol{\epsilon}(\boldsymbol{x},t) = \text{sym}\{\nabla \boldsymbol{u}\}(\boldsymbol{x},t), \tag{26}$$

where $\dot{\boldsymbol{A}}$ and $\nabla \boldsymbol{A}$ are the temporal and spatial derivative of $\boldsymbol{A}$, respectively, and sym denotes the symmetric part of the second-order tensor, and infinitesimally low deformation is assumed throughout the study. The continuum Lagrangian is defined as the volume integral of l as follows:

$$\mathcal{L}[\boldsymbol{v},\boldsymbol{\epsilon}] = \int_V l(\boldsymbol{v},\boldsymbol{\epsilon})dv. \tag{27}$$

Specifically, a square bracket implies that $\mathcal{L}$ is a functional for a vector function and a tensor function, while a parenthesis (or a round bracket) implies that l is a function of a vector and a tensor, and the same symbols $\boldsymbol{v}$ and $\boldsymbol{\epsilon}$ are used in $\mathcal{L}$ and l for simplicity.

We consider the simplest case in which V is linearly elastic. In this case, the continuum Lagrangian density is explicitly expressed with a Cartesian coordinate system and index notation as follows:

$$l(\boldsymbol{v},\boldsymbol{\epsilon}) = \frac{1}{2}\rho v_i v_i - \frac{1}{2}c_{ijkl}\epsilon_{ij}\epsilon_{kl} + b_i u_i, \tag{28}$$

where ρ is the density, c is the elasticity tensor, and b is the body force.

In order to derive the FEM for Hamiltonian, we use a discretized displacement function as follows:

$$\boldsymbol{u}^*(\boldsymbol{x},t) = \sum \boldsymbol{u}^\alpha(t)\xi^\alpha, \tag{29}$$

where u^α is a displacement vector of the αth node, and ξ^α is an associated shape (or interpolation) function. For a given continuum Lagrangian density, we obtain the following expression:

$$\mathcal{L}^*[\{\boldsymbol{v}^\alpha\},\{\boldsymbol{u}^\alpha\}] = \int_V l^*(\{\boldsymbol{v}^\alpha\},\{\boldsymbol{u}^\alpha\})dv, \tag{30}$$

with $\boldsymbol{v}^\alpha = \dot{\boldsymbol{u}}^\alpha$ and $l^* = l(\sum \boldsymbol{v}^\alpha\xi^\alpha, \sum \text{sym}\{\boldsymbol{u}^\alpha \otimes \nabla\xi^\alpha\})$. We define a new variable as $\boldsymbol{p}^\alpha = \partial\mathcal{L}^*/\partial\boldsymbol{v}^\alpha$, which is the momentum of the

αth node, and we derive a continuum Hamiltonian from $\mathcal{L}^*$ as follows:

$$\mathcal{H}^*[\{\boldsymbol{p}^\alpha\},\{\boldsymbol{u}^\alpha\}] = \sum \boldsymbol{v}^\alpha \cdot \boldsymbol{p}^\alpha - \mathcal{L}^*[\{\boldsymbol{v}^\alpha\},\{\boldsymbol{u}^\alpha\}]. \tag{31}$$

The canonical equation of the $\mathcal{H}^*$ is as follows:

$$\begin{bmatrix} \dot{\boldsymbol{p}}^\alpha \\ \dot{\boldsymbol{u}}^\alpha \end{bmatrix} = \begin{bmatrix} -\frac{\partial \mathcal{H}^*}{\partial \boldsymbol{u}^\alpha} \\ \frac{\partial \mathcal{H}^*}{\partial \boldsymbol{p}^\alpha} \end{bmatrix}. \tag{32}$$

For the purpose of simplicity, we assumed that the derivatives of $\mathcal{H}^*$ were given as a linear combination of its arguments, u_j^β and p_j^β, i.e.,

$$\frac{\partial \mathcal{H}^*}{\partial u_i^\alpha} = \sum_{\beta,j} K_{ij}^{\alpha\beta} u_j^\beta - b_i^\alpha \quad \text{and} \quad \frac{\partial \mathcal{H}^*}{\partial p_i^\alpha} = \sum_{\beta,j} R_{ij}^{\alpha\beta} p_j^\beta, \tag{33}$$

where $K_{ij}^{\alpha\beta} = \partial \mathcal{H}^{*2}/\partial u_i^\alpha \partial u_j^\beta$ and $R_{ij}^{\alpha\beta} = \partial \mathcal{H}^{*2}/\partial p_i^\alpha \partial p_j^\beta$. We obtained two vectors, $[u]$ and $[p]$, by arranging u_i^α and p_i^α, respectively, and we similarly obtain two matrices and one vector, $[K]$, $[R]$, and $[b]$, by arranging $K_{ij}^{\alpha\beta}$, $R_{ij}^{\alpha\beta}$, and b_i^α, respectively. In terms of these vectors and matrices, the continuum Hamiltonian of FEM is expressed as follows:

$$\mathcal{H}^* = \frac{1}{2}[p]^T[R][p] + \frac{1}{2}[u]^T[K][u] - [u]^T[b], \tag{34}$$

where superscript T denotes the transpose. The canonical equation is as follows:

$$[\dot{p}] = -[K][u] + [b] \quad \text{and} \quad [\dot{u}] = [R][p]. \tag{35}$$

It should be noted that $[K]$ is a stiffness matrix and that $[R]$ is an inverse of mass matrix $[M]$; $[K]$ and $[M]$ are constructed as follows:

$$[K] = \int_V [B]^T[C][B]dv \quad \text{and} \quad [M] = \int_V \rho[N]^T[N]dv, \tag{36}$$

where $[B]$ is a B matrix, $[C]$ is a matrix of elastic constants, and $[N]$ is a matrix of the shape function. We used the lumped mass matrix in the following simulations. The matrices are identical to those used in ordinary FEM.

The continuum Hamiltonian with solid and joint elements is defined as follows:

$$\mathcal{H}^* = \frac{1}{2}[p]^T[R][p] + \frac{1}{2}[u]^T[K_s][u] + \int_t [\dot{u}][K_j][u]d\tau - [u]^T[b], \quad (37)$$

where $[K_s]$ is a stiffness matrix of solid elements, and $[K_j]$ is a stiffness matrix of joint elements. While we considered nonlinearity of the joint elements, we applied only linear elastic materials to the solid elements in the study. When we derived Eq. 37, we used the Lagrangian for the rigorous joint element described in Section 2.2. The canonical equation is as follows:

$$[\dot{p}] = -([K_s] + [K_j])[u] + [b] \quad \text{and} \quad [\dot{u}] = [R][p]. \quad (38)$$

3. Verification of Numerical Analysis Program

3.1. *One-dimensional problem*

Having developed a rigorous formulation for a nonlinear joint element, we examined its convergence with respect to mesh size of 10, 5, 2.5, 2, and 1 m, with Fig. 11 showing the numerical analysis model for the 10-m mesh. Second-order triangular joint elements were used to construct the fault, while other components were constructed using second-order tetrahedral elements. An external force, denoted by f_0, was uniformly loaded on one side of the model. We set $\kappa_0 = 1.5 \times 10^2$ MPa/m, $\kappa_d = \kappa_0/20$, and $u_{\text{cr}} = 0.1$ m for Eq. 21. The rock mass

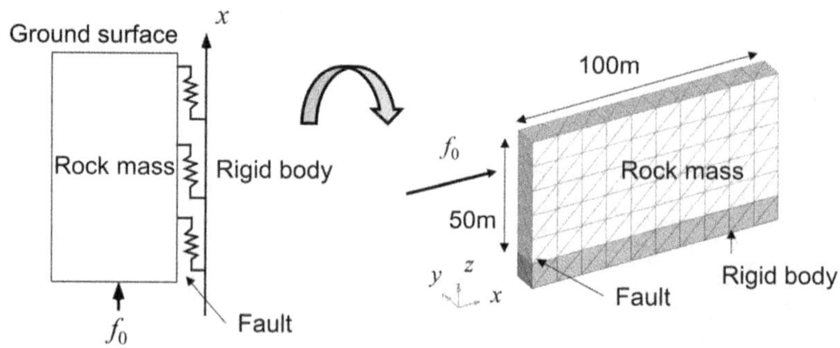

Fig. 11. Analysis model with mesh size of 10 m for numerical experiment of joint elements.

exhibits linear elasticity with Young's modulus of 1.0×10^4 MPa and Poisson's ratio of 0.0.

Figure 12 presents deformation at $f_0 = 10$ MPa and evaluation points. Slip on the fault is represented by the joint elements.

Figure 13 presents the relationship between f_0 and the displacement at the evaluation points. Displacement jump is observed just after $f_0 = 8$ MPa. Figure 14 presents the relationship between the mesh size and the relative error of the displacement at $f_0 = 10.0$ MPa, where the relative error is defined as the difference relative to the solution of the analysis model with a mesh size of 1 m. As expected, a second-order solution convergence rate was obtained with respect to the mesh size.

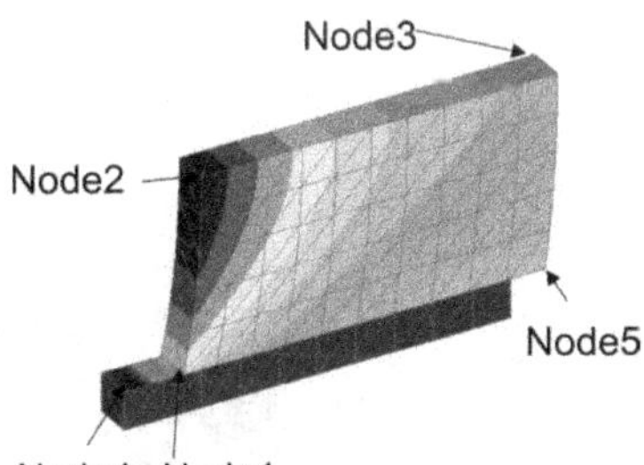

Fig. 12. Deformation and evaluation points. Deformation is emphasized by a factor of 20. The contour presents displacement distribution. Nodes 1, 2, 3, 4, and 5 are evaluation points.

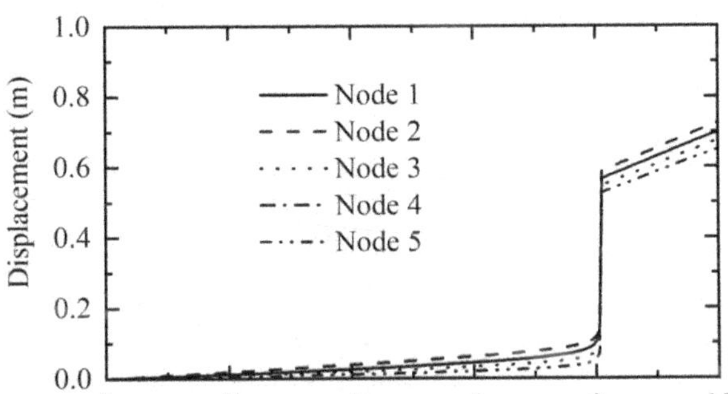

Fig. 13. Relationship between f_0 and the displacement at five evaluation points.

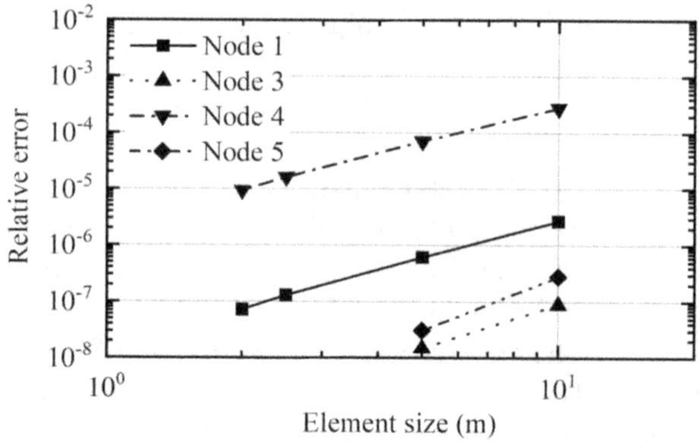

Fig. 14. Relationship between mesh size and relative displacement error at $f_0 = 10.0$ MPa. The relative error is defined by the difference relative to the solution of the analysis model with a mesh size of 1 m. The relative error at Node 2 is not plotted since relative error is calculated zero in the case of 5-m mesh.

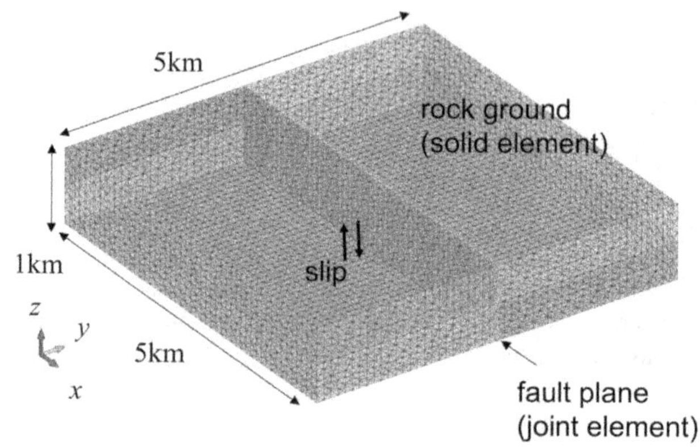

Fig. 15. Model for the test simulation 1.

3.2. *Three-dimensional problem*

3.2.1. *Test one: Single-fault problem*

We applied the developed parallel FEM program to the test simulations for three-dimensional problems.[19] With respect to the test simulation, we modeled a $5 \times 5 \times 1$ km domain of the ground that included a vertical fault plane. Figure 15 presents the FE mesh of the model. In the first step, our target number of degrees-of-freedom was less than 1,000,000. The model consisted of 1,000 second-order triangle elements for the fault, 150,000 second-order tetrahedral elements, 216,342 nodes, and 649,026 degrees-of-freedom.

The triangle slip distribution that was forcibly applied on the central 2.5 km of the fault bottom was treated as an external force. The maximum slip was 2 m at the midpoint of the fault bottom where a side of the fault was displaced 1 m upwards and the other side was displaced 1 m downwards (see Fig. 16). The slip was forced in 2 s, and thus the maximum slip rate was 1 m/s at the midpoint of the fault bottom. The other outer boundaries were traction-free.

Table 1 presents the material parameters for the test simulation. The rock mass was assumed to be a linear elastic body. We tried two values for the final shear stiffness κ_d, which were (1) 1/20 and (2) 1/1000 of initial shear stiffness κ_0. In the test simulation, we neglected the dependency of the shear stiffness of the fault on the confining pressure.

Figure 17 presents the slip–shear stress relationship in the problem. In Case (2), recovery of shear stress was not observed, and the stable solution of the slip for the final shear stiffness was extremely high. We determined these parameter values by considering the rock and fault at depths ranging from 0 to 1 km.

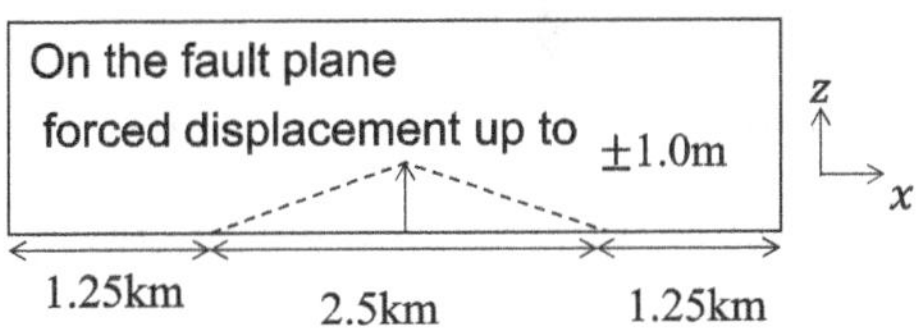

Fig. 16. Slip distribution forcibly applied on the bottom of the fault.

Table 1. Material parameters for test simulation.

	Property (unit)	Symbol	Value
Rock	Young's modulus (MPa)	E	1.0×10^4
	Poisson's ratio	ν	0.3
	Density (kg/m^3)	ρ	2.0×10^3
Fault	Initial shear stiffness (MPa/m)	κ_0	1.5×10^2
	Final shear stiffness (MPa/m)	κ_d	(1) $\kappa_0/20$, (2) $\kappa_0/1000$
	Critical slip (m)	u_{cr}	0.1

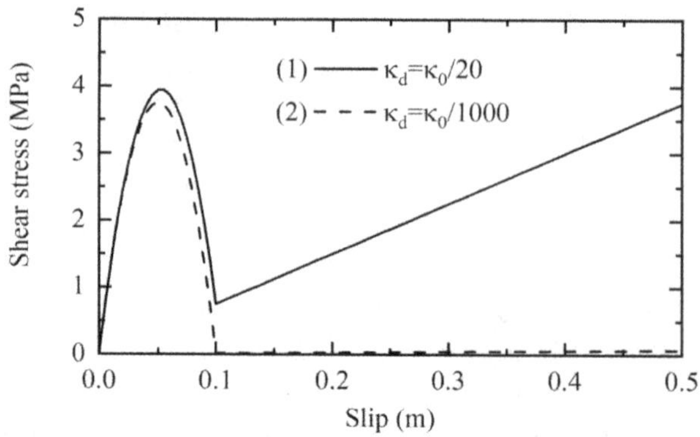

Fig. 17. Slip–shear stress relationship on the fault in the test problem.

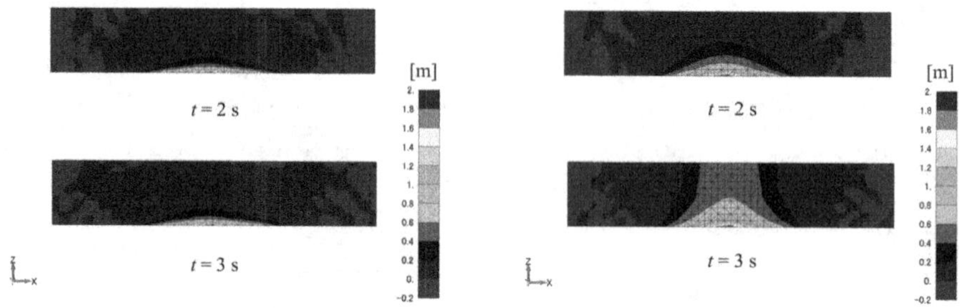

(1) $\kappa_d = \kappa_0/20$ (2) $\kappa_d = \kappa_0/1000$

Fig. 18. Slip distribution on the fault ($t = 2$–3 s).

Figure 18 presents the contour of the slip distribution at $t = 2$ s and $t = 3$ s on the fault plane. In Case (1), the slip did not propagate after the input slip was given until $t = 2$ s. Conversely, in Case (2), the slip propagated and reached the ground surface, and the surface slip was up to 0.6 m.

Figure 19 presents the slip progression at three depths at the mid-point of the fault. The slip propagation velocity was approximately 0.3 km/s from elevation level (E. L.) −1000 m to E. L. −500 m and approximately 1 km/s from E. L. −500 m to E. L. 0 m in Case (2). The slip propagation velocity is approximately 0.7 times as high as shear wave velocity.[22,23] These values were reasonable when compared to the shear wave velocity (approximately 1.5 km/s). However, we require further investigation on the dependency of slip

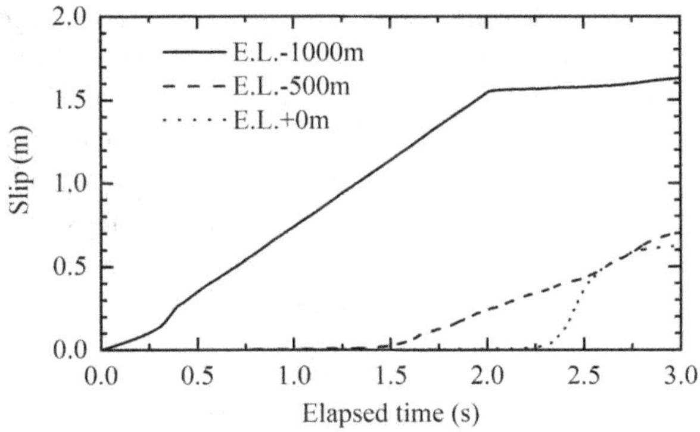

Fig. 19. Slip progression at three depths ($\kappa_d = \kappa_0/1000$).

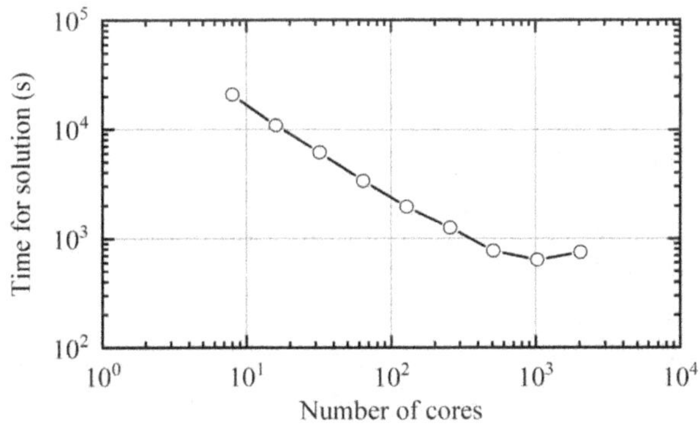

Fig. 20. Scalability of computation.

propagation on the final shear stiffness because shear stiffness is generally dependent on the normal stress.

In the dynamic analysis performed here using symplectic time integration for geotechnical and fault modeling, we confirmed the reduction in computation time as the number of CPU cores increased. The computer used was a cluster system with 2 CPUs/16 cores (CPU clock: 2.6 GHz), 32 GB or 64 GB memory per node, for a total of 2016 nodes. The calculation was conducted using CPU cores from 8 to 2048. The relationship between the number of CPU cores and computation time (strong scaling) is shown in Fig. 20, indicating that up to 512 parallel cores, doubling the number of CPU cores results in good parallel performance, with computation time halved. However, from 512 to 1024 CPU cores, the parallel performance slightly slows down, and the calculation with 2048 CPU cores requires more computation time than that with 1024 CPU cores. Parallel performance should be verified for problems with larger degrees-of-freedom. As described above, we have confirmed that parallel computation can be used for analysis models with 1 million degrees-of-freedom and about 10,000 steps in a sufficiently short computation time.

3.2.2. *Test two: Multi-faults problem*

Model: In test two, we modeled a 5 × 5 × 1 km domain of ground that included two faults. Figure 21 presents the FE mesh of the model. We assumed that the larger fault was a primary fault and the smaller fault was a secondary fault. The two faults had parallel strikes and opposite dip angles of 40 degrees to each other. The model consisted of 7,800 second-order triangle elements for the fault, 157,375 second-order tetrahedral elements, 237,435 nodes, and 699,750 degrees-of-freedom.

In the test, we calculated initial stress distribution by overburdening the pressure of the rock mass prior to the fault-displacement simulation to consider the effect of normal stress on shear stiffness, namely the condition of depth-dependent shear stiffness.

Figure 22 presents the boundary conditions in fault-displacement simulation. We assumed the primary fault was a reverse fault. The footwall was the $-x$ side of the primary fault and was fixed on its bottom (see Fig. 22(a)). The forced displacement shown in Fig. 22(b) was applied on the central 1.25 km of the fault bottom on the side of the hanging wall. The forced displacement was also partially applied

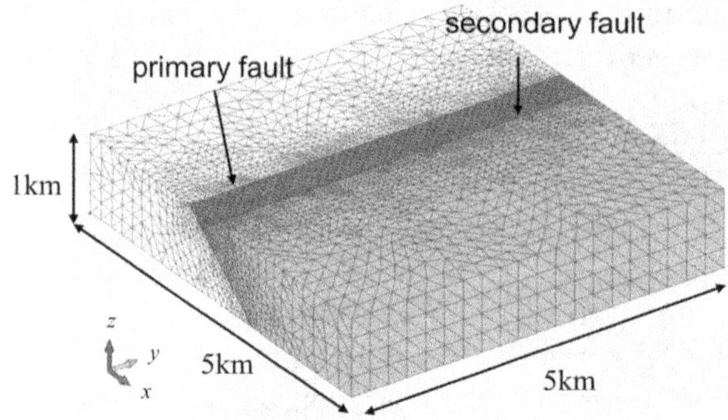

Fig. 21. Model for test simulation 2.

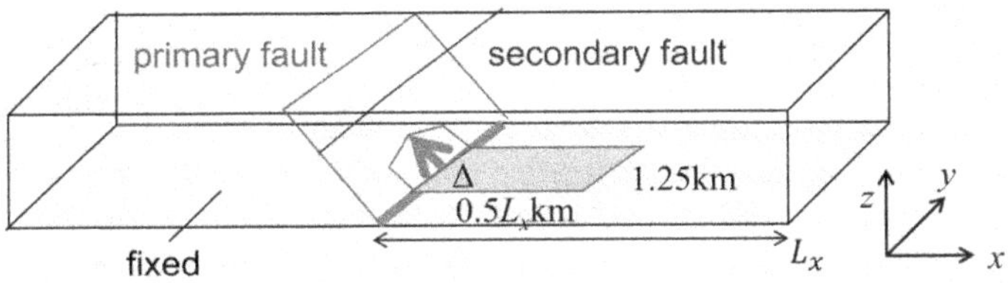

(a) Boundary conditions

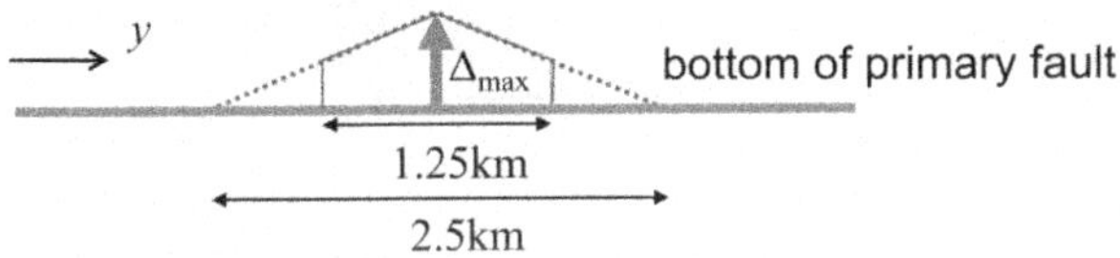

(b) Bottom of the primary fault

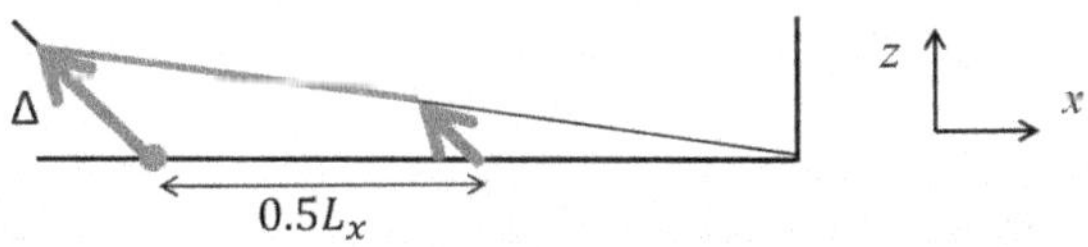

(c) On the plane of y=0

Fig. 22. Boundary conditions. (a) Presents boundary conditions, (b) presents slip distribution forcibly applied on the bottom of the fault, and (c) presents displacement forcibly applied on the plane of $y = 0$.

on the bottom of the hanging wall as shown in Figs. 22(a) and 22(c) in order to avoid the concentration of large shear strain in the rock mass adjacent to the primary fault. When the maximum slip $\Delta_{\max}$ was 3.1 m at the midpoint of the fault bottom, the slip was forced in 2.0 s in dynamic simulations. Thus, the maximum slip rate was 1.55 m/s at the midpoint of the fault bottom. The time increment in dynamic simulations was 10^{-4} s. We also conducted quasi-static simulations in which $\Delta_{\max}$ of 3.1 m was forced in 200 steps at the midpoint of the fault bottom. The other outer boundaries were traction-free.

Table 2 presents the material parameters for test two. Although most parameter values were the same as those in test one, κ_d was 0.01 times as high as κ_0. In the cases of depth-dependent shear stiffness, we defined κ_0 as a linear function of normal stress based on Eq. 23 ($\widetilde{\kappa}_0 = 18.5$ m^{-1}, $b = 0.054$ MPa). These values of $\widetilde{\kappa}_0$ and b were calculated such that the first peak of shear stress in the slip–shear

Table 2. Material parameters for test two.

	Property (unit)	Symbol	Value
Rock	Young's modulus (MPa)	E	1.0×10^4
	Poisson's ratio	ν	0.3
	Density (kg/m^3)	ρ	2.0×10^3
Fault	Initial shear stiffness (MPa/m)	κ_0	(1) 1.5×10^2, (2) $18.5\sigma_n + 0.99$
	Final shear stiffness (MPa/m)	κ_d	$\kappa_0/100$
	Critical slip (m)	u_{cr}	0.1

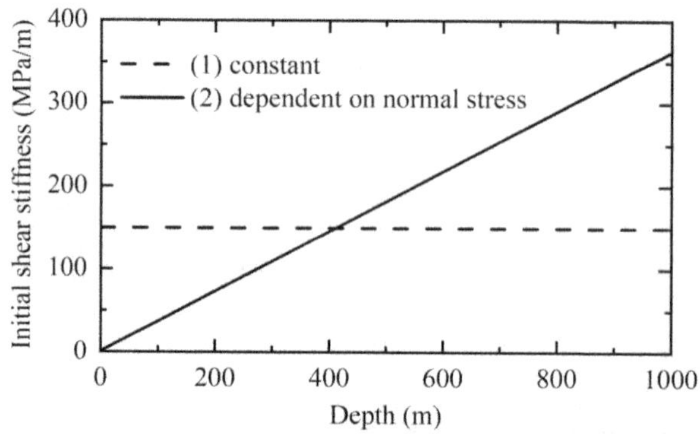

Fig. 23. Relationship between depth and initial shear stiffness κ_0. Normal stress was assumed to be equal to the overburden pressure of ground. The unit of σ_n is MPa.

stress relationship shown in Fig. 10 satisfied the Coulomb friction law $\tau(\text{MPa}) = 0.025 + \sigma_n \tan 25°$, where τ is the shear stress and σ_n is the normal stress. Figure 23 presents the relationship between depth and initial shear stiffness κ_0. In the graph, normal stress was assumed to the equal to the overburden pressure of the ground. Constant shear stiffness of 1.5×10^2 MPa/m almost corresponded to the median value at depths ranging from -1000 to 0 m.

Dynamic Simulation and Quasi-Static Simulation: We commenced test two by comparing the results for the dynamic simulation and quasi-static simulation. In the comparison, we adopted the constant shear stiffness for the joint elements.

We conducted three cases of dynamic simulations with different damping conditions; namely, no damping and damping ratios of 3% and 5% in Rayleigh damping were applied to the cases. It was not possible to set the damping with symplectic time integration in our present FEM program, and thus we used Newmark β method ($\beta = 0.25$ and $\gamma = 0.5$) which was another time integration method implemented in FrontISTR[20] in the cases of damping. A future study should consider implementing damping into symplectic time integration.

Figure 24 presents the slip progression at the midpoint on the surface and E.L. -500 m of the primary fault. The slip of 0.93 m was applied at the slip rate of 1.55 m/s. The results from the cases with no damping and 3% damping were plotted. The input slip profile at the midpoint on the bottom of the primary fault was also plotted. The slip at E. L. -500 m appeared after 1 s, and it progressed with the vibration in the period corresponding to 2 s. The vibration tended to converge in the case of 3% damping although no convergence was observed in the case with no damping. The slip appeared at the surface after 3 s in the case with no damping while it did not appear in the case with 3% damping

Figure 25 presents the relationship between input slip at the depth and surface slip at the surface. The results from quasi-static simulation and dynamic simulation with 3% and 5% damping were plotted. The maximum value in slip history in 3 s was adopted as the slip in dynamic simulations. The results from quasi-static and dynamic

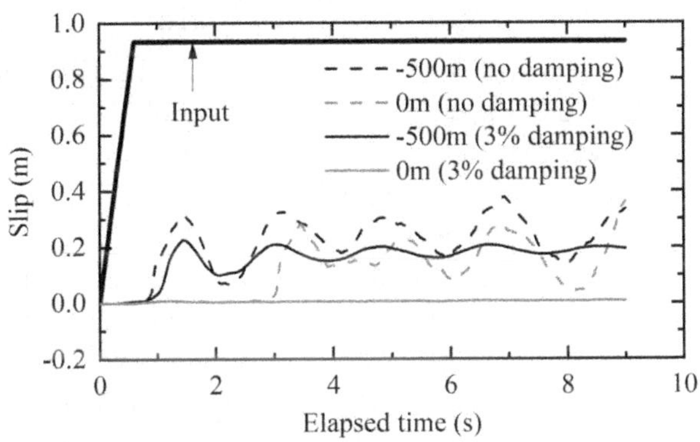

Fig. 24. Effect of damping on slip progression.

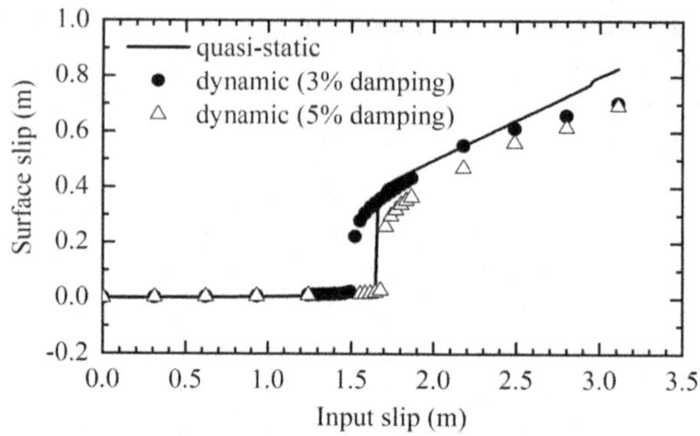

Fig. 25. Relationship between input slip and surface slip.

simulations were almost identical to each other with respect to the critical input slip Δ_c and surface slip δ. It is necessary to focus on the number of calculation cases required to create the relationship shown in Fig. 25. We obtain this only by performing the quasi-static simulation. However, in the dynamic simulation, it is necessary to examine several calculation cases with different maximum values of input slip at depths as implied by the number of dots in the results of the dynamic simulations as shown in Fig. 25.

Effect of Depth-Dependent Shear Stiffness: Subsequently, we investigated the effect of dependency of initial shear stiffness κ_0 in depth. Generally, shear behavior on faults are considered as affected by the normal stress acting on them. We conducted quasi-static simulations in the conditions of constant κ_0 and depth-dependent κ_0.

Figure 26 presents the contour maps of vertical displacement in the condition of constant κ_0 at $\Delta_{\max} = 3.1$ m, the vertical component of which is 2.0 m. The figure on the left-hand side presents perspective view while the figure on the right-hand side shows the map on the plane of A-A'. The surface slip at midpoint of the primary fault was +0.83 m where "plus" implies that the $+x$ side was relatively displaced upwards from the $-x$ side. The surface slip at the midpoint of the secondary fault displayed a slip in the opposite direction although the slip was extremely low. Conversely, surface slips were +1.15 m and −0.09 m at midpoints on the primary and secondary faults, respectively, in the case of depth-dependent κ_0 (see Fig. 27). The slips were higher in the case of depth-dependent κ_0 than in the case of constant κ_0 because the fault slips easily in

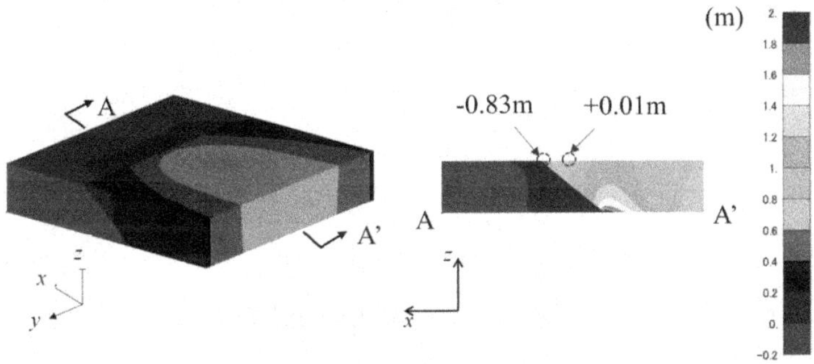

Fig. 26. Vertical displacement contour maps in the condition of constant initial shear stiffness at $\Delta_{\max} = 3.1$ m.

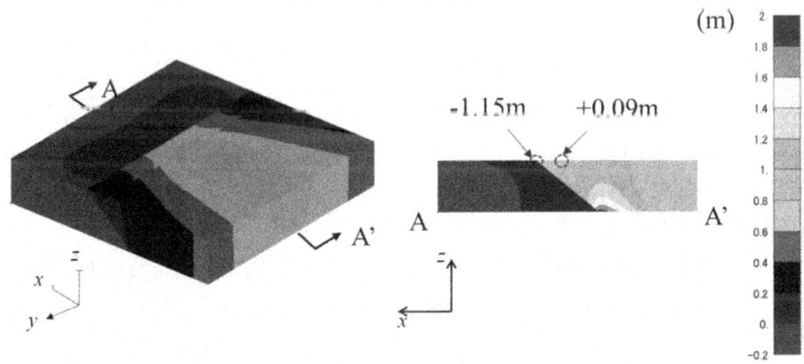

Fig. 27. Vertical displacement contour maps in the condition of depth-dependent initial shear stiffness at $\Delta_{\max} = 3.1$ m.

the shallow zone in the case of depth-dependent κ_0. The tendency that the ground between the primary and secondary fault was lifted (back-thrust like slip) was observed in surface faulting in the 2014 Nagano-ken-hokubu earthquake, Japan.[24]

How to Use HPC for Fault-Displacement Estimation: As shown in Fig. 4, we note the advantage of using high-performance computing for fault displacement problems that inevitably include uncertainties. The uncertainties originated from the loss of stability in a boundary value problem as well as limited quality and quantity of information to construct a numerical model. In order to quantify the uncertainties, it is necessary to perform numerous calculations with different conditions including material properties, input slip at depth, and initial crustal stresses.

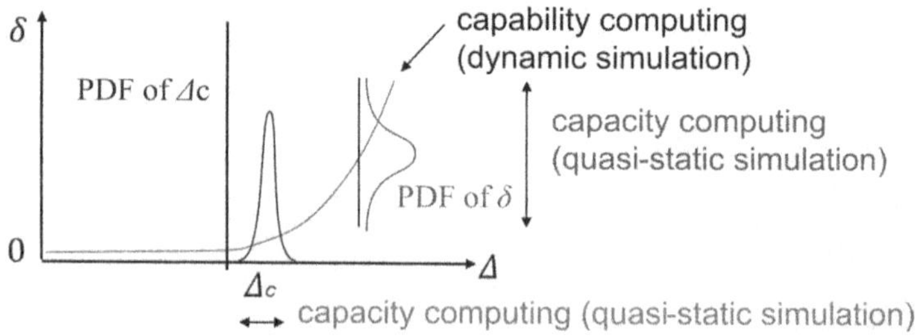

Fig. 28. Application of HPC to estimate the surface fault displacement. PDF means probability density function.

Fault displacement essentially corresponds to dynamic behavior. It is necessary to precisely solve a four-dimensional (spatial and temporal) partial differential equation to obtain the optimal estimate of fault displacement. The parallel FEM program we developed in this study is applicable to the estimation by using a large-scale three-dimensional model (capability computing in Fig. 28).

Conversely, quasi-static simulation exhibits advantages since it obtains a relationship between input slip and surface slip by conducting one case of calculation. This advantage is extremely important, especially in the estimation of Δ_c. The calculated Δ_c and δ in the quasi-static simulation were close to those calculated in the dynamic simulation with damping. Therefore, we propose that we should use quasi-static simulation to implement numerous cases of calculations involving different conditions (capacity computing in Fig. 28).

4. Application to Actual Earthquakes

4.1. *Basic concept of analytical model*

In this section, we describe the basic concept of constructing an analytical model for evaluating fault displacement and apply the developed program to the analyses of real earthquakes with surface earthquake faults (the 2014 Nagano-ken-hokubu earthquake and the 2016 Kumamoto earthquake). The analytical model is validated by comparison with observed surface fault displacements.

As mentioned previously, the purpose of fault displacement evaluation for nuclear facilities is to estimate the surface slip on the sub-faults when slip occurs on the main fault. Slip generated at the epicenter in the deep part of the main fault is considered to propagate on the surface of the main fault and propagate to the sub-fault either directly or through deformation of the surrounding ground at the shallow part. Therefore, a detailed evaluation of the slip propagation from the main fault to the sub-faults at shallow depths is necessary to evaluate the surface slip of the sub-faults.

In this study, we evaluate the spread and dispersion of the slip on the main fault inputted to the bottom of the analysis domain in a shallow area up to about 1 km below the ground surface, including the main fault and the sub-faults. The horizontal extent of the analysis domain is about 5 km on a side, which covers the NPP site where the detailed investigation is conducted and also includes the sub-faults around the main fault. Since the entire main fault is not included in the analytical model, it is necessary to set appropriate boundary conditions based on the analysis of a wider area. One possible method is setting the slip distribution on the main fault and determining the ground displacement over a wide area using elastic theory of dislocation.[2] When considering the source of the earthquake, there are various uncertainties such as the ground structure, the shape of the main fault at depth, asperities, etc. In this study, these uncertainties are considered as uncertainties in the input displacement and boundary conditions.

We determine the crustal structure and mechanical properties of the ground based on the Japan Seismic Hazard Information Station (J-SHIS) database,[25] developed by the National Research Institute for Earth Science and Disaster Resilience (NIED) and covering all of Japan. In the J-SHIS database, elevation data regarding the ground surface and boundaries of elastic wave velocity and density are provided at the center of every grid of 30 arcsec latitude and 45 arcsec longitude (approximately 1×1 km grid). Shape of the land affects the stresses on the fault plane and should be taken into account.

The J-SHIS database also contains data on the geometry of the faults (main faults), and the location, size, strike, and dip of the main faults can be set based on these data. The data may be corrected to match the actual fault traces on the ground surface. In the analyses described in this section, the inversion analysis setup for each earthquake was also referred. Sub-faults are not included in the database, so their location, size, strike, and dip are set based on the results of observations and surveys.

4.2. *The 2014 Nagano-ken-hokubu earthquake*

4.2.1. *Outline of the earthquake and surface faulting*

We applied the numerical method to the November 22, 2014, Nagano-ken-hokubu earthquake with Japan Meteorological Agency magnitude (M_{JMA}) 6.7 (Mw = 6.2).[26,27] The earthquake had a maximum seismic intensity of 6− on the JMA scale, and was of reverse fault type (east side uplift) with a compression axis from west-northwest to east-southeast. A 9-km-long surface rupture was observed on the previously mapped Kamishiro fault. Figure 29 presents the mapped location of the rupture in the northernmost section.[28] Sub-faults were also observed, including Sub-E1, Sub-E2, Sub-N, and Sub-S as shown in Fig. 29. Figure 30 presents the net slip along the surface rupture estimated using airborne Light Detection and Ranging (LiDAR) data obtained at two different times.[28] The maximum net slip values on the main fault and sub-faults were approximately 1.5 and 0.6 m,

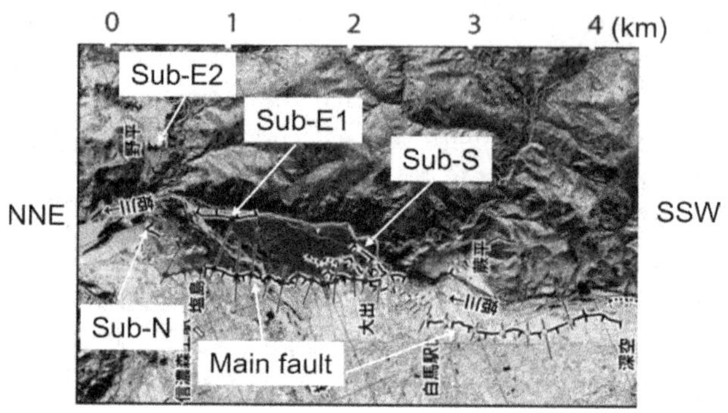

Fig. 29. Distribution of surface ruptures in the northernmost section, modified after Aoyagi.[28]

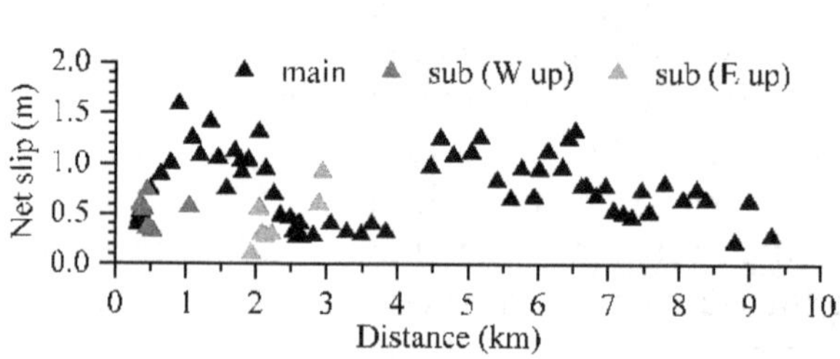

Fig. 30. Surface fault dislocation.[28] The horizontal axis corresponds to the distance from the northern end of the surface rupture, the scale of which is equivalent to that of Fig. 29.

respectively. Sub-E1, Sub-E2, and Sub-N showed west-up slip, which was different from the main fault with east-up slip, and Sub-E1 and Sub-E2 seem to be backthrusts to the main fault.

4.2.2. *Analytical model*

We determined the mechanical properties of the ground and strike angle of the main fault based on the J-SHIS database.[25]

Figure 31 presents a finite element mesh used for the simulation. The target was a 5 × 5 km region around the northern end of the surface rupture. The depth of the domain was approximately 1 km.

We included one main fault and four sub-faults. The dip angle of the main fault was 40°. Sub-E1 and Sub-E2 had the same strike angle as the main fault, and a dip angle of 40° in the opposite direction. Sub-N and Sub-S had strike angles of 45° from the main fault and 45° from the northside and southside of the dip angles, respectively. The ground was discretized by second-order tetrahedral elements and the fault planes were discretized by second-order triangle joint elements. The size of the elements on the fault planes was approximately 50 m and the total number of degrees-of-freedom was approximately 2.58 million.

We considered the shape of the land and the two layers of the ground based on the J-SHIS database. Young's modulus and

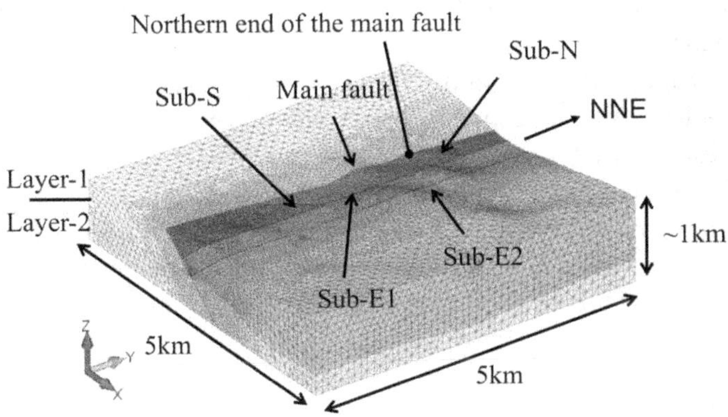

Fig. 31. Finite element mesh. The analytical model is divided into two layers with different material properties. The layers are further divided by five faults. The analytical model is displayed with blocks divided by faults and property boundaries. Its depth depends on the elevation of the ground surface and is approximately 1.0 km.

Table 3. Material parameters.

	Young's modulus (GPa)	Poisson's ratio	Density (g/cm^3)
Layer 1	10.0	0.35	2.2
Layer 2	27.7	0.31	2.4

Poisson's ratio of each layer were determined from the elastic wave velocity and density. Material parameter values for the ground are summarized in Table 3. We assumed that the first peak of the slip–shear stress relation shown in Fig. 10 was identical to the shear strength. The initial shear stiffness κ_0 and coefficients were determined such that the shear strength satisfied Coulomb's law of friction with a cohesion of 0.025 MPa and a 25° friction angle. The ratio between the initial and final shear stiffnesses, κ_d/κ_0, was assumed to be 0.01 and the critical slip, u_{cr}, 0.1 m.

The first step in the calculation procedure is to perform a initial stress analysis and calculate the initial normal stress on the fault plane. The boundary conditions for the analysis are roller boundaries on the sides and bottom and traction-free on the top, and the shear stiffness of the fault is set to a sufficiently large value ($\kappa_0 = 10^5$ MPa/m).

The Geospatial Information Authority of Japan derived a slip distribution on the main fault from the crustal movements estimated by differential interferometric synthetic aperture radar (DInSAR) as shown in Fig. 32.[29] The main fault had dip angles of 40° and 60° in the shallow and deep parts, respectively. We re-evaluated the slip distribution of the shallow part of the main fault based on Fig. 30.[28] We used the elastic theory of dislocation[2] with the slip distribution as the input and determined the forced displacements at the bottom of the FEM model. The maximum slip on the bottom of the fault was 3.0 m, which was twice the estimated slip. The other outer boundaries of the model were traction-free.

4.2.3. *Results*

Figure 33 presents the displacement magnitude contour and deformation of the analytical domain. Figures 33(a) and 33(b) show the results at $\Delta = 1.5$ and 3.0 m, respectively. The deformation is

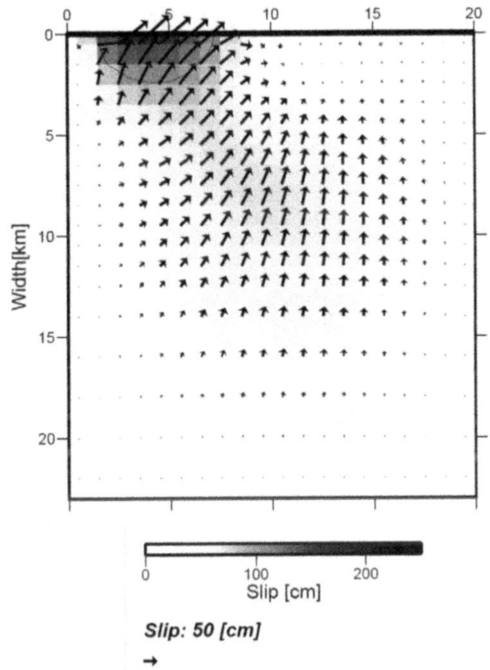

Fig. 32. Slip distribution on the main fault.

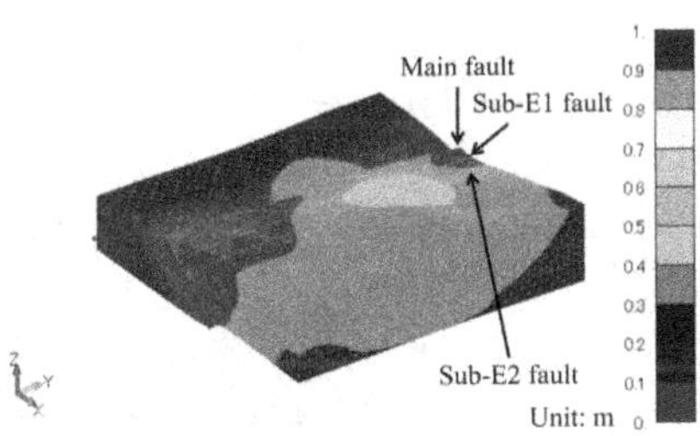

(a) $\Delta = 1.5$ m

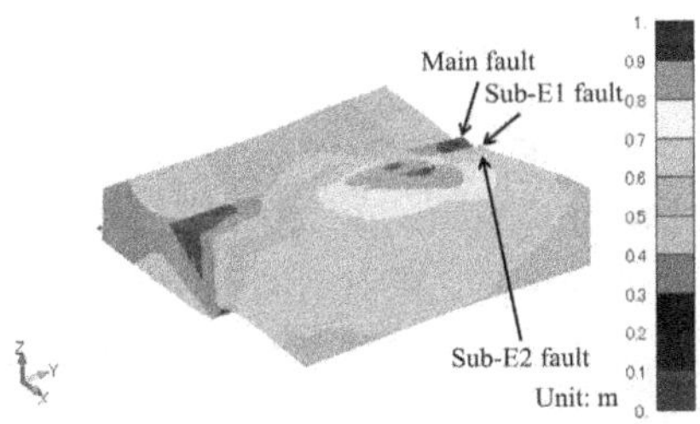

(b) $\Delta = 3.0$ m

Fig. 33. Contours of displacement magnitude and deformation. (a) $\Delta = 1.5$ m, and (b) $\Delta = 3.0$ m.

shown with 300-times amplification. Large deformation occurred at the center of the model where the input slip was large. There was a displacement gap along the main, the Sub-E1, and Sub-E2 faults, showing surface slip.

Figure 34 presents contour plots of the slip at $\Delta = 3.0$ m on the main fault and sub-faults. This figure presents views from the

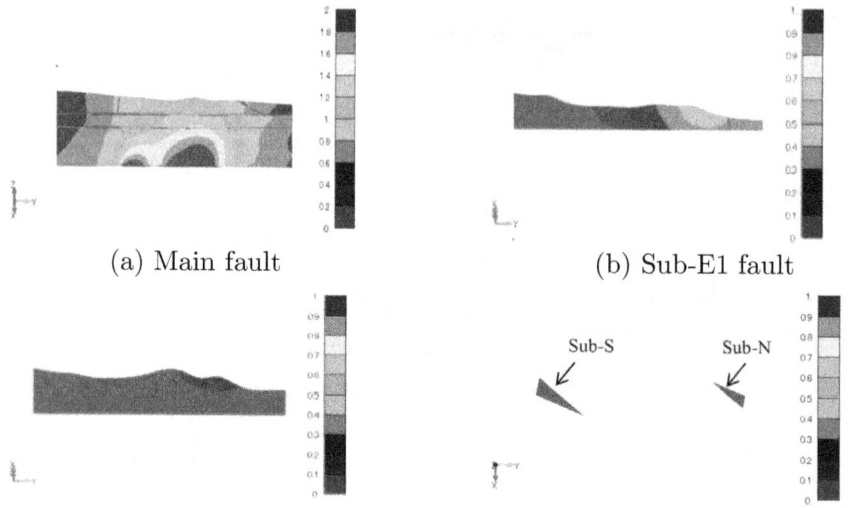

(a) Main fault (b) Sub-E1 fault

(c) Sub-E2 fault (d) Sub-S and Sub-N faults

Fig. 34. Contours of slip on the main fault and sub-faults ($\Delta = 3.0$ m). (a) Main fault, (b) sub-E1 fault, (c) sub-E2 fault, and (d) sub-S and sub-N faults.

$+x$ (east) side. There was a surface slip of greater than 1.0 m on the main fault and a surface slip in the north sections of the Sub-E1 and Sub-E2 faults. In the simulation, the maximum surface slip was approximately 0.5 m on Sub-E1; this value was similar to the measured value shown in Fig. 30. The calculated surface slip was small on Sub-E2, Sub-S, and Sub-N. Overall, the simulated surface slip distribution at $\Delta = 3.0$ m agreed well with the measured distribution.

Figure 35 presents the evaluation points for surface slip. In the legend, "main" and "sub" are abbreviations for the main and sub-faults, respectively, and "Y2000" is an abbreviation for the y-coordinate of 2000 m.

Figure 36 presents the variation in the surface net slip δ as the input slip Δ increases. There was a steep increase in the surface slip at $\Delta = 0.7$ m on the main fault (mainY2500). A steep increase in the surface slip on the Sub-E1 fault (sub-E1Y3500) began at $\Delta = 2.6$ m. Therefore, the critical slip Δ_c was 0.7 m on the main fault and 2.6 m on the Sub-E1 fault. On the Sub-E2 fault (sub-E2Y4000), there was no steep increase, but the slip exceeded 0.1 m at $\Delta = 2.9$ m.

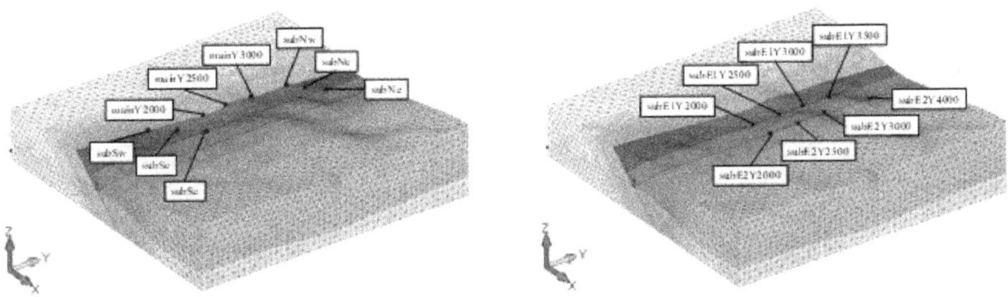

(a) Main, Sub-S and Sub-N faults (b) Sub-E1 and Sub-E2 faults

Fig. 35. Evaluation points. (a) Main, sub-S, and sub-N faults, and (b) sub-E1 and sub-E2 faults.

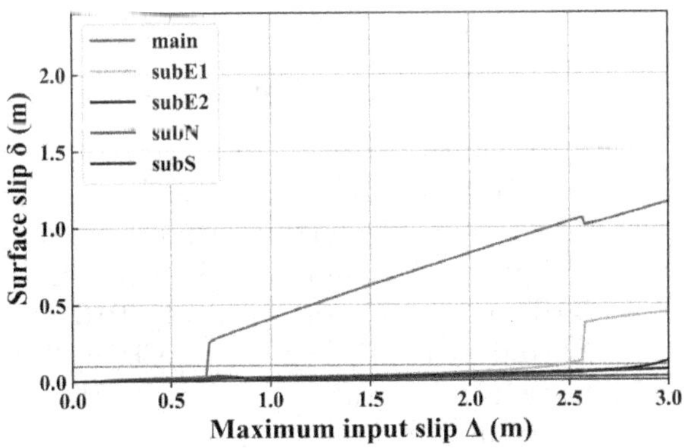

Fig. 36. Surface slip/input slip relationship.

Figure 37 presents the dip and strike-slip components of the surface slip at representative evaluation points of the main fault and Sub-E1 fault. Positive strike slip means left-lateral slip and positive dip slip means normal fault sense. The dip slips on both the main and Sub-E1 faults were negative. The two behaved as reverse faults. The Sub-E1 fault moved as if it was a back-thrust of the main fault. This movement agreed qualitatively with the observations. However, the dip component of the slip on the Sub-E1 fault was less than the measured value.

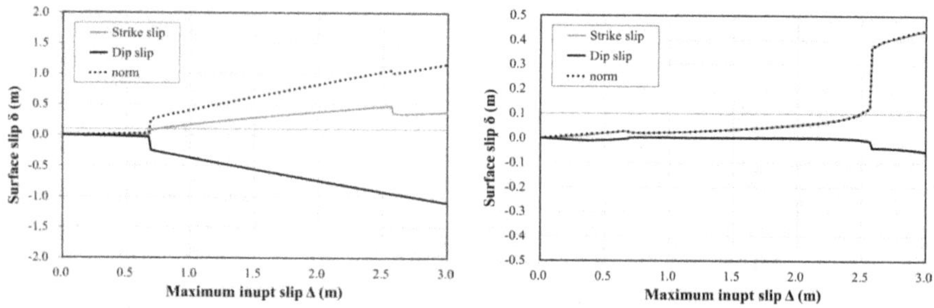

(a) Main fault (mainY2500) (b) Sub-E1 fault (sub-E1Y3500)

Fig. 37. Strike and dip component of the surface slip. (a) Main fault (mainY2500), and (b) sub-E1 fault (subE1Y3500).

4.3. *The 2016 Kumamoto earthquake*

4.3.1. *Outline of the earthquake and surface faulting*

We also applied the developed FEM program to the 2016 Kumamoto earthquake.[30] The earthquake began with a Japan Meteorological Agency magnitude (M_{JMA}) 6.5 earthquake in Kumamoto Prefecture in the central part of Kyushu Island in southwest Japan at 21:26 Japan Standard Time (JST) on April 14, 2016. A larger M_{JMA} 7.3 earthquake occurred at 01:25 JST on April 16, 2016, just 28 h later. A seismic intensity observation station at the Mashiki town hall recorded a seismic intensity of seven on the JMA scale during both events. The M_{JMA} 6.5 event on April 14 was the foreshock and the M_{JMA} 7.3 event on April 16 was the mainshock. The mainshock was the simulation target in this study.

The earthquake occurred along the Futagawa and northern Hinagu fault zones.[31] Surface faults of 28 and 6 km in length were observed along the Futagawa and Hinagu fault zones, respectively (see Fig. 38). Vertically, the north side of the Futagawa fault zone sank ≥ 1.0 m and the south side rose ≥ 0.3 m. Horizontally, the north side was displaced by more than 1.0 m to the east and the south side was displaced by more than 0.5 m to the west. The surface faults were investigated and the fault displacement distribution along the faults was obtained.[32] A right-lateral strike-slip fault of up to 2.2 m was observed at Dozon in the eastern part of Mashiki. The surface fault branched into two faults from Dozon to the west. Although these two branched-out faults exhibited a right-lateral strike-slip fault, a left-lateral strike-slip fault was observed between them. The target area

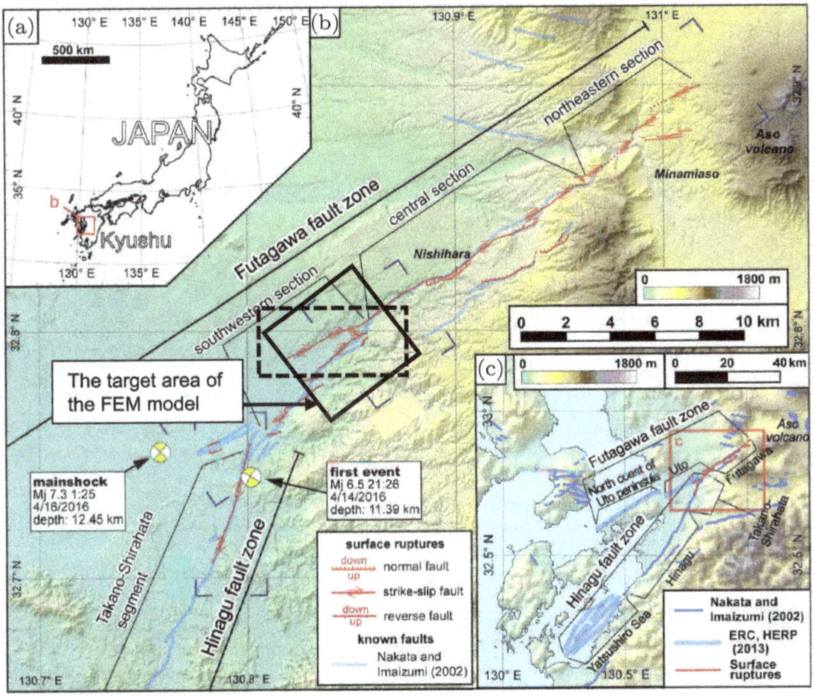

Fig. 38. Map of the surface faults following the 2016 Kumamoto earthquake (modified, Shirahama *et al.*[32]). (a) Location of the study area in Kyushu, southwestern Japan, (b) active faults in the western part of central Kyushu and the 2016 surface ruptures (red line), and (c) distribution of surface ruptures associated with the 2016 Kumamoto earthquake. The dashed-edge box indicates the area of Fig. 40. Blue corners indicate the areas studied in detail by Shirahama *et al.*[32]

of our numerical simulations was a 5 × 5 km square area containing the faults around Dozon as shown in Fig. 38.

Asano and Iwata[33] conducted a seismic source inversion analysis of the 2016 Kumamoto earthquake and obtained slip distributions on the Futagawa and Hinagu faults (see Fig. 39). The Futagawa and Hinagu faults slipped in the form of right-lateral strike-slip faults, including a normal fault component. The slip near the target area of the simulation ranged from 1.7 to 2.6 m.

4.3.2. *Analytical model*

The target of our simulations was a 5 × 5 km square area containing the fault-branching in Mashiki town. The target depth from

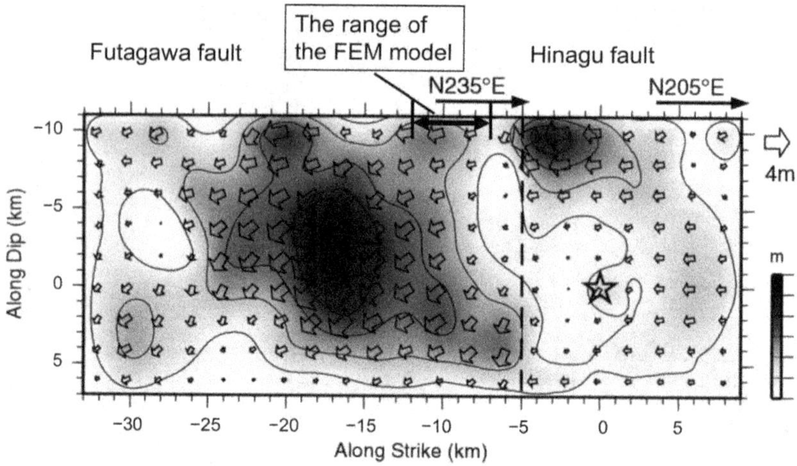

Fig. 39. Slip distribution on main faults (modified, Asano and Iwata[33]).

the ground surface was approximately 1 km. By using the J-SHIS database,[34] we constructed an analytical model that includes the shape of the land and crustal structure. We determined the strike and dip angles of the Futagawa and Hinagu faults by referring to the paper by Asano and Iwata.[33] The dip angles of the Futagawa and Hinagu faults were 65°N and 72°N, respectively. Although the upper edges of the faults did not reach the ground surface, we extended them to the ground surface with the same dip angle.

Figure 40 presents the surface faults along the Kiyama river in Mashiki town, where the Futagawa fault branches into two faults. We treated the line labeled "Main", which is one of the two branched-out faults, as the main fault, which is identical to the Futagawa fault. The other faults, which are labeled as "SubN", were included in the analytical model as sub-faults. Three faults, named Sub2, Sub3, and Sub5, were observed during the 2016 earthquake. Sub2 was the other branched-out fault. Sub3 was a left-lateral strike-slip fault bridging Main and Sub2. Sub1 and Sub4 were faults known to exist before the 2016 earthquake. However, surface faulting was not observed along them during the earthquake.

Figure 41 presents the finite element mesh of the analytical model. Faults are extended to the outer boundary or to other faults. The main fault, Sub4, and Sub5 strikes are parallel to the y axis. The sub-fault dip angles were determined based on the results of a

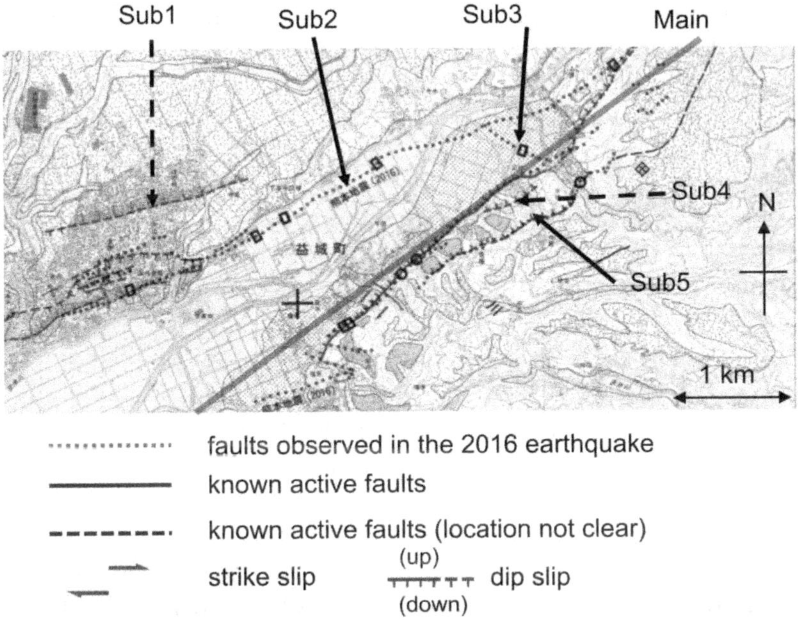

Fig. 40. Surface faults along the Kiyama river in Mashiki town. Sub2, Sub3, and Sub5 with solid arrows were observed during the 2016 earthquake. Surface faulting was not observed along Sub1 and Sub4 with dashed arrows during the earthquake.

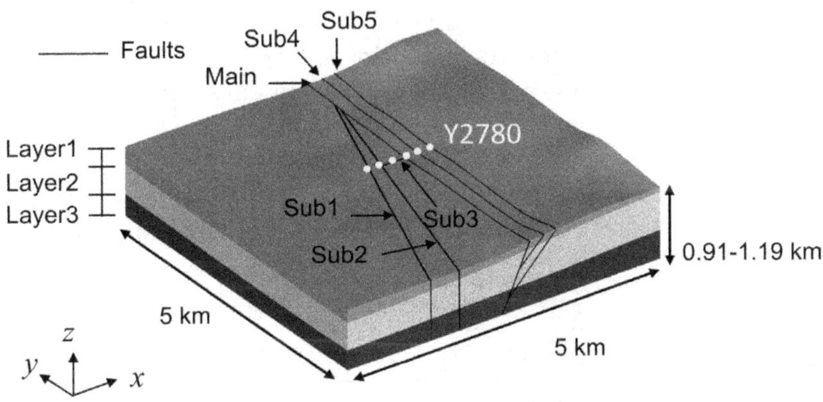

Fig. 41. Finite element mesh. The analytical model is divided into three layers with different material properties. The layers are further divided by six faults. The model depth depends on the elevation of the ground surface and ranges from 0.91 to 1.19 km.

seismic reflection survey by Aoyagi.[35] Sub1, Sub2, and Sub4 are vertical. The ground is divided into three layers based on the J-SHIS database. The ground is discretized using second-order tetrahedral elements and the fault planes are discretized using second-order triangular joint elements. The element size on fault planes is approximately 50 m and the total number of degrees-of-freedom is approximately 3.76 million. The white dots in Fig. 41 indicate the locations of the output points for the fault displacement results. "Y2780" indicates that the y coordinate of these dots is 2780 m.

We assumed that the ground was a linear elastic body and derived Young's modulus and Poisson's ratio of each layer from the density and elastic wave velocity values in the J-SHIS database. Material parameter values for the ground are summarized in Table 4. We conducted an initial stress analysis to determine the values of the parameters $\widetilde{\kappa}_0$ and b on the fault planes using Eqs. (24) and (25) before fault displacement analysis. Normal displacements were fixed at the bottom and side boundaries in the initial stress analysis. We assumed a cohesion of 0.025 MPa and friction angle of 25° on the fault planes by reference to the experimental value of the crushed zone in NPP sites in Japan.[13] The ratio between the initial and final shear stiffness κ_d/κ_0 was assumed to be 0.01 in reference to the analyses conducted using a simple ground/fault model and a real earthquake ground model.[13,19] The critical slip u_{cr} was 0.1 m in reference to the value used in the dynamic rupture simulation of the fault.[6]

The size of the analytical model is presented in Fig. 41 and is smaller than the length of the main faults. The boundary conditions of the analytical model were determined using large-scale analysis including the entire main faults. We used the elastic theory of dislocations[2] with a slip distribution on the main faults, as

Table 4. Material parameters.

	Young's modulus (GPa)	Poisson's ratio	Density (g/cm^3)
Layer 1	7.18	0.38	2.15
Layer 2	27.7	0.31	2.40
Layer 3	63.3	0.27	2.60

shown in Fig. 39, as the input. The upper edge of the main faults is at a depth of 2 km. We extended the main faults to the ground surface and assumed that the slip on the extension was the same as that just below the extension. The ground displacement was calculated and the forced displacement at the bottom of the FEM model was determined. The maximum slip on the bottom of the main fault was 5.2 m, which is twice the estimated 2.6 m slip. The other outer model boundaries were traction-free. We refer to this simulation as Case 1 (Cases 2–4 are defined in Section 5).

4.3.3. *Results*

Figure 42 presents the y-displacement contour. When Δ is 2.6 m, which is the slip estimated by Asano and Iwata,[33] there is a displacement gap along the main fault, Sub1, Sub2, and Sub5. Additionally, there is a displacement gap along Sub4 when Δ is 5.2 m, which is twice the estimated slip value. This gap indicates a surface slip on each fault.

Figure 43 presents contour slip plots at $\Delta = 2.6$ m on the main and sub-faults. There is a surface slip of greater than 1.0 m on the main fault and surface slips of greater than 0.1 m on Sub1, Sub2, and Sub5. The slipped area on Sub1 is limited compared to those on Sub2 and Sub5. The surface slips on the main, Sub2, and Sub5 faults are up to 1.25, 0.3, and 0.8 m, respectively. Figure 44 presents the profiles of the surface slip for the strike-slip and vertical components along Main, Sub2, Sub3, and Sub5. Figure 45 presents the observed slip distributions along the Futagawa fault and Hinagu fault.[32] Comparing

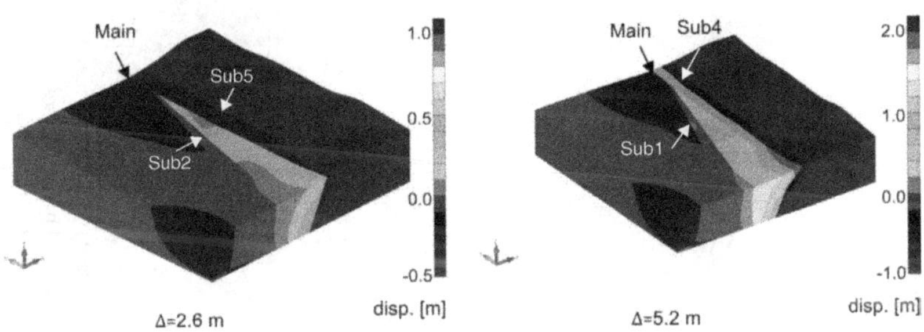

Fig. 42. Contours of the y-displacement (Case 1).

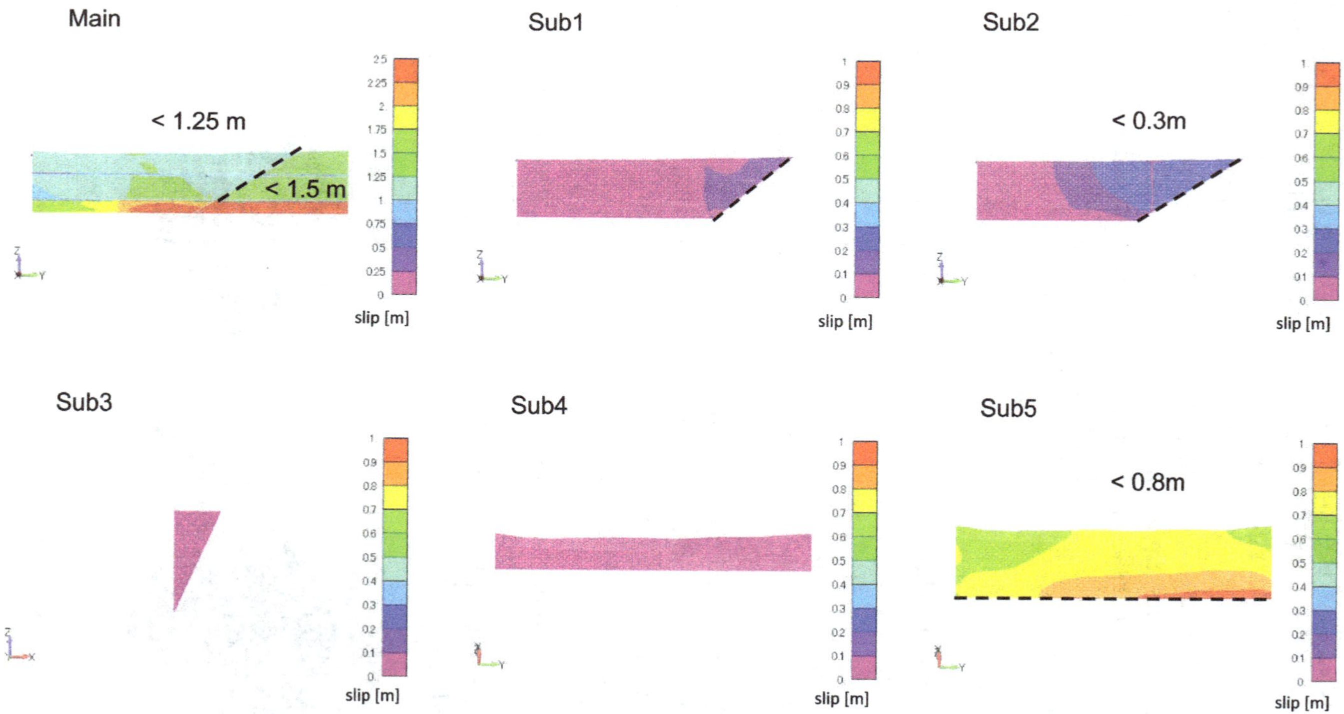

Fig. 43. Contours of slip on the main fault and sub-faults (Case 1, $\Delta = 2.6$ m). The dotted line in the figure of the main fault indicates the intersection with Sub2. The dotted lines in the figures of Sub1, Sub2, and Sub5 indicate the contact edges with the main fault.

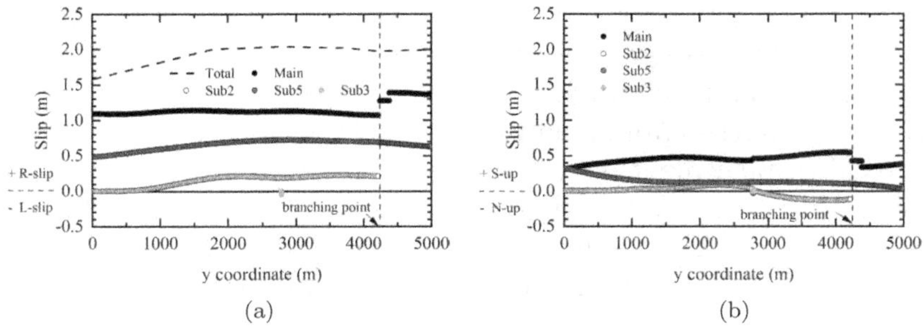

Fig. 44. Profiles of surface slip on main and sub-faults (case1, $\Delta = 2.6$ m). (a) Strike-slip component. The dashed line plot "Total" shows the estimated total slip, and (b) vertical component.

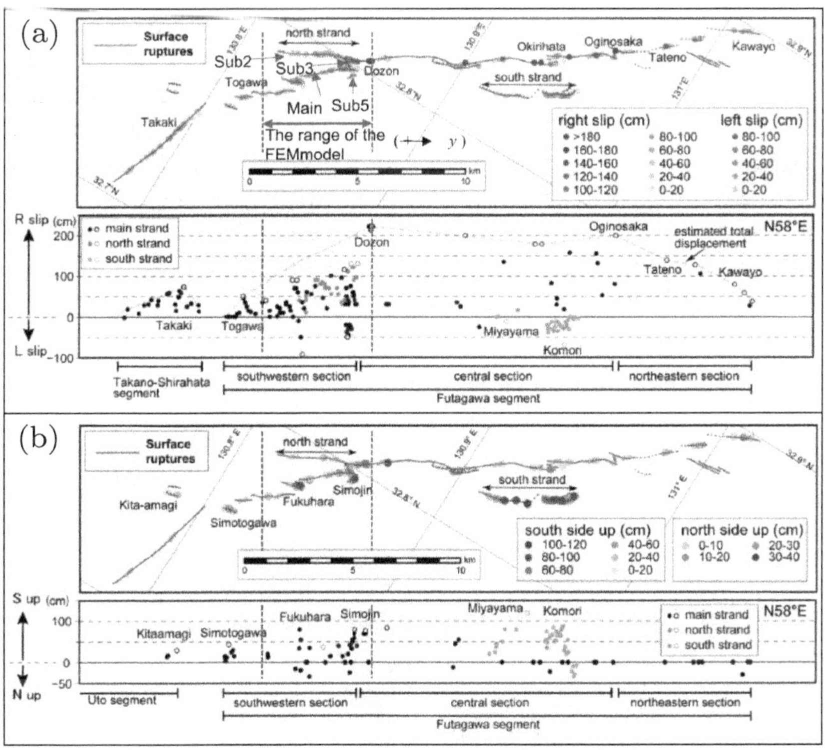

Fig. 45. Distributions of surface ruptures along the Futagawa and Hinagu faults (top) and a diagram of the slip distribution along the fault zone (bottom)[32]: (a) Strike-slip component. The open dots in the bottom diagram show the representative displacement at each sub-strand. The dashed line shows the estimated total displacement. The location of FEM model is shown by vertical dashed lines. Sub2 in FEM model represents the north strand, and (b) vertical component.

the calculated and observed strike-slip, the calculated slip is smaller on Sub2. On the other hand, the surface slip on Sub5 distributes in a longer section in the calculation. The sum of the calculated surface strike-slip values (approximately 2.0 m) is in good agreement with the total displacement of the southwestern section of the Futagawa segment in Fig. 44. The vertical component of the calculated surface slip on the main fault is intermediate to the observed ones, but the sum of the vertical slips on the main fault and Sub5 mostly exceeds the observed ones.

Figure 46 presents the variation in the surface net slip δ as the input slip Δ increases. There is a steep increase in the surface slip at $\Delta = 0.3$ m on the main fault. A steep increase in the surface slips on Sub5 and Sub4 begins at $\Delta = 1.4$ m and $\Delta = 3.0$ m, respectively. Although there is no steep increase on Sub2, Sub1, or Sub3, the slips are greater than 0.1 m at $\Delta = 1.2$ m, $\Delta = 2.7$ m and $\Delta = 3.0$ m, respectively. We define Δ_c as the input slip at which the surface slip exceeds 0.1 m and summarize the values of Δ_c at 3 locations on each fault in Table 5. Δ_c is small for Sub2 and Sub5, where the surface displacement was actually observed in the 2016 earthquake. The sense of the slip at each location is presented in Table 5. The sense of the slip in the simulation was remarkably similar to that of the observed surface faults, as shown in Fig. 45.

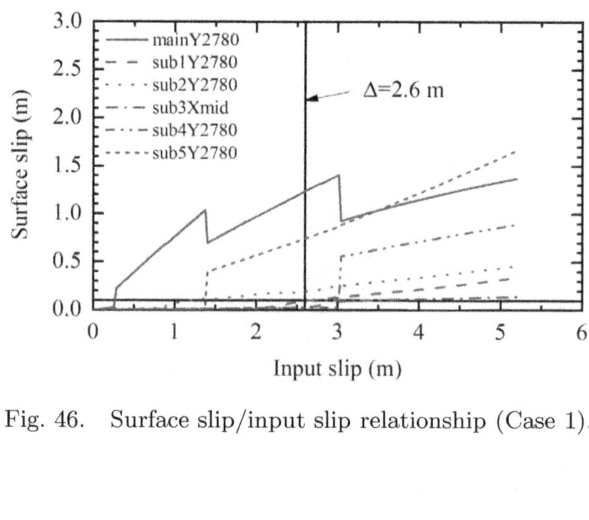

Fig. 46. Surface slip/input slip relationship (Case 1).

Table 5. Slip sense and Δ_c on each fault (Case 1).

Fault	Location[a]	Observation	Δc (m)	Simulation	
Main	1		0.3	Right-lateral	East-up
	2	Right-lateral	0.3	Right-lateral	East-up
	3		0.3	Right-lateral	East-up
Sub1	1		—	—	—
	2	—	2.7	Right-lateral	East-up
	3		2.4	Right-lateral	East-up
Sub2	1		1.7	Right-lateral	East-up
	2	Right-lateral	1.2	Right-lateral	West-up
	3		0.9	Right-lateral	West-up
Sub3	1	Left-lateral south-up	4.3	—	South-up
	2		3.1	Left-lateral	South-up
	3		2.8	Left-lateral	South-up
Sub4	1		3.0	Right-lateral	East-up
	2	—	3.0	Right-lateral	East-up
	3		1.4	Right-lateral	East-up
Sub5	1		1.4	Right-lateral	East-up
	2	East-up	3.0	Right-lateral	East-up
	3		1.4	Right-lateral	East-up

Note: [a]Locations 1, 2, and 3 indicate the positions of the minimum, medium, and maximum x coordinates on Sub3 and the positions of 1780, 2780, and 3780 m in the y coordinates on the other faults, respectively.

4.4. *Discussion*

4.4.1. *2014 Nagano-ken-hokubu earthquake*

We constructed a 5 × 5 × 1 km analytical model and derived the boundary condition from the elastic theory of dislocation using slip distribution on the main fault obtained from inverse analysis of DInSAR and multi LiDAR-DEM data.

When we provided twice the inversely analyzed slip on the main fault to the bottom of the analytical domain, we obtained a steep increase in surface slip on the main fault and a sub-fault. The final surface slip magnitudes agreed well with the measured values on the main fault and Sub-E1 fault. The dip component of the slip on the

Sub-E1 fault was qualitatively identical to the back-thrust-like movement observed at the site. Because the simulation reproduced the main features of the observed surface rupture, the proposed numerical method is applicable to surface fault displacement estimation.

However, some phenomena were not reproduced in the simulation. For example, surface slip did not appear on the Sub-N or Sub-S faults. One possible reason is that slip in the section in which $y > 3855.77$ m on the main fault was not restricted. It may be necessary to consider the spatial distribution of the fault friction properties to reproduce the surface slip on the Sub-N and Sub-S faults.

4.4.2. *2016 Kumamoto earthquake*

We constructed a $5 \times 5 \times 1$ km analytical model and derived the boundary conditions from the elastic theory of dislocation using a slip distribution in the main fault obtained from seismic source inversion analysis. When forced slip occurred at the bottom of the main fault, we observed a steep increase in the surface slip on the main and sub-faults. On the sub-faults that appeared in the 2016 earthquake, the surface slip appeared at a smaller input slip than on the other sub-faults. The sense of slip on each fault in our simulation was remarkably similar to the observations during the real earthquake.

As shown in Fig. 45, the surface fault slip is distributed in an uneven manner. Although there are locations where significant slip occurs, there are many locations where the slip is small or does not occur in the entire section. The calculated surface slip distribution is continuous and agrees well with the measured total slip distribution (enveloped slip distribution curve). The simulation allowed us to make a conservative estimation of surface fault slip. However, some phenomena were not reproduced in our simulations. For example, the surface slip on Sub3 requires a large input slip. In the simulations, the small faults tended to be less likely to slip.

The dashed edges in Fig. 43 represent the Sub1, Sub2, and Sub5 contact edges with the main fault. The slipped sub-fault areas propagate from these contact edges. In the 2014 Nagano-ken-hokubu earthquake simulation, the slip on the sub-fault was larger near the surface than near the contact edge with the main fault (see Fig. 34). The calculated slip distribution on the sub-faults was remarkably different between the two earthquake simulations.

In general, faults are identified from ruptures left on the ground surface. It is difficult to include faults that cannot be identified from the ground surface in the model. Moreover, it is difficult to fully capture the subsurface structure and connectivity of faults. In the analysis of this section, faults were extended to the model boundaries or other faults. The larger the fault plane, the greater the possibility that large displacements will be calculated. On the other hand, if many faults are identified, it is difficult to represent all of them with joint elements. As sub-faults, it is reasonable to include in the model faults that are relatively large or that affect the focused facilities. In general, if the number of faults is limited rather than considering many faults, the fault displacement for each single fault will be calculated as larger. However, this needs to be further investigated.

5. Predictive Simulation

5.1. *Setting of input conditions*

5.1.1. *Basic slip distribution on primary faults*

For predictive simulation, we were not able to use the results from the seismic source inversion analyses shown in Fig. 32 or 39, and we had to assume the slip distribution on the main fault. In this section, we propose a method to set the slip distribution on the main fault and apply it to predictive simulation using an analysis model of the 2016 Kumamoto earthquake. We ignore the Hinagu fault and consider a scenario in which only the Futagawa fault ruptures in the following predictive simulations.

We set the slip distribution using a strong motion prediction method called recipe,[36] which includes the following steps: (1) set the main fault geometry (length, width, depth, strike, and dip angle), (2) calculate the seismic moment, (3) calculate the average slip from the seismic moment and shear modulus of the crust, and (4) set the asperity number and location to calculate the slip on them and on the background. To calculate the seismic moment M_0, the following equation is used for relatively large earthquakes[36]:

$$M_0 = \left(\frac{S}{4.24} \times 10^{11}\right)^2 \times 10^{-7}, \tag{39}$$

where S is the square of the main fault. We adopted the geometry of the Futagawa fault presented by Asano and Iwata.[33] Because $S = 5.04 \times 10^8$ m^2 was derived from a length of 28 km and width of 18 km, M_0 was calculated as 1.41×10^{19} Nm. The average slip D was calculated from S and the shear modulus μ, using the following equation:

$$D = M_0/(\mu \cdot S). \tag{40}$$

When we used $\mu = 30$ GPa, which was based on the assumption of a hard rock mass in the deep underground, $D = 0.934$ m was obtained. We set a uniform slip distribution of 0.934 m on the entire Futagawa fault, as shown in Fig. 47(a), and conducted our first predictive simulation. We refer to this simulation as Case 2 (Case 1 was defined in Section 4.3).

In this recipe, asperities were set on the seismic source fault. Asperity is a region of high strength or stress drop, and if this region is slipped, the slip will be relatively greater than the surrounding area. Subsequently, we set the slip distribution by considering the existence of asperities. Although it is recommended to use an empirical equation for the relationship between the short period level and equivalent radius of the total asperity area, we determined the total asperity area to be 1.28×10^8 m^2 by referring to research on the ratio of the total asperity area to the total fault area introduced in the recipe. We placed two asperities of equal area (64 km^2) on the Futagawa fault, as shown in Fig. 47(b). The average slip of asperity D_a was determined to be ξ times the average slip D as following:

$$D_a = \xi D, \tag{41}$$

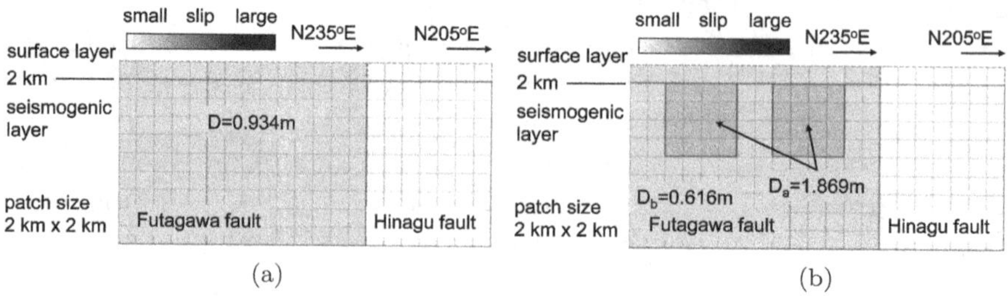

Fig. 47. Setting the slip distribution based on the recipe. (a) No asperity: uniform slip distribution, and (b) two asperities.

where $\xi = 2$ was used based on an analysis of recent crustal earthquakes. The average slip of the background D_b was derived from the following equations:

$$M_{0a} = \mu D_a S_a, \tag{42}$$

$$M_{0b} = M_0 - M_{0a}, \tag{43}$$

$$D_b = M_{0b}/(\mu \cdot S_b), \tag{44}$$

where M_{0a} is the total seismic moment of the asperities, S_a is the total area of the asperities, M_{0b} is the seismic moment of the background, and S_b is the area of the background. We obtained $D_a = 1.869$ m and $D_b = 0.616$ m. We set these slip values on the Futagawa fault, as shown in Fig. 47(b), and conducted a second predictive simulation. We refer to this simulation as Case 3.

5.1.2. *Large surface slip area*

When we set the slip distribution on the main fault based on the recipe, the surface slip was uniform and smaller than that observed during the 2016 Kumamoto earthquake. The slip along the surface fault was not generally uniform. Therefore, we set a large slip area on the surface part of the main fault based on Case 2, as shown in Fig. 48. The following empirically obtained equation proposed by Matsuda *et al.*[37] was used to set the maximum surface slip:

$$D_{\max} = 10^{-4} L, \tag{45}$$

where $D_{\max}$ is the maximum surface slip and L is the surface rupture length. Here we used the 28 km fault length for L and obtained

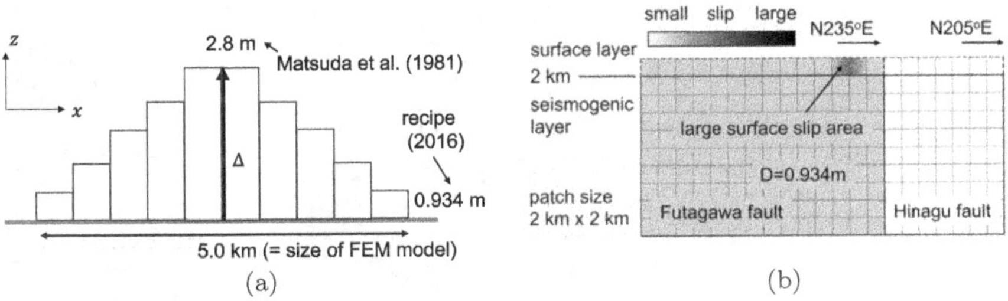

Fig. 48. Setting the slip distribution with a large surface slip area. (a) Slip distribution in the large surface slip area, and (b) slip distribution on the main faults.

$D_{\max} = 2.8$ m. Other equations have been proposed to relate surface fault displacement to the length of faults.[38] We can also use these equations. We assumed that the length of the large surface slip are a was 5 km, which is the same as the size of the FEM model, and that the peak slip position was identical to the center of the FEM model. We refer to this simulation as Case 4.

Additionally, we set up two cases in which the large surface slip area was shifted by +1 km and −1 km in the y direction compared to Case 4, which are referred to as Cases 5 and 6, respectively.

In simulations Cases 2 to 6, the maximum forced slip at the bottom of the FEM model was twice that of the slip distribution.

5.2. *Results*

Figure 49 presents the variation in the surface net slip δ in Case 2 and Case 3 as the input slip Δ increases. In Case 2, there is a steep increase in the surface slip at $\Delta = 0.2$ m on the main fault and only the main fault slips before $\Delta = 0.934$ m. The Δ_c value of Sub5 is 1.25 m, which is similar to the value obtained in Case 1 (1.4 m). In Case 3, only the main fault slips before twice the value of D_b is inputted, despite the fact that twice the value of D_a was inputted in the asperities.

Figure 50 presents the variation in the surface net slip δ in Case 4 as the input slip Δ increases. There is a steep increase in the surface slip at $\Delta = 0.4$ m on the main fault. The Δ_c values on Sub1, Sub2, Sub3, Sub4, and Sub5 are 4.2, 2.7, 2.5, 3.1, and 1.6 m, respectively. In Sub2, Sub3, and Sub5, where surface fault displacement can be

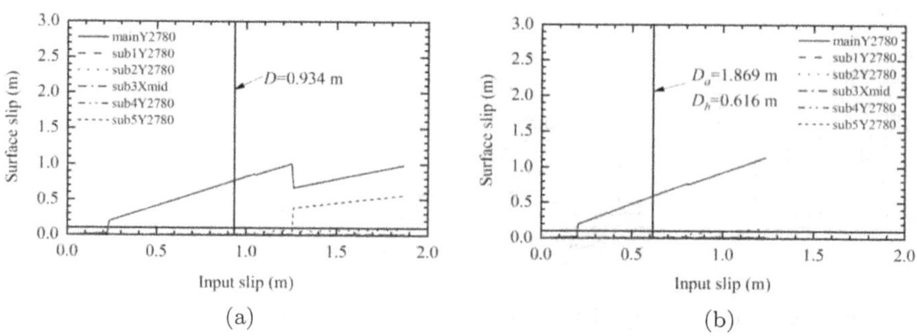

Fig. 49. Surface slip/input slip relationship. (a) Case 2, and (b) Case 3.

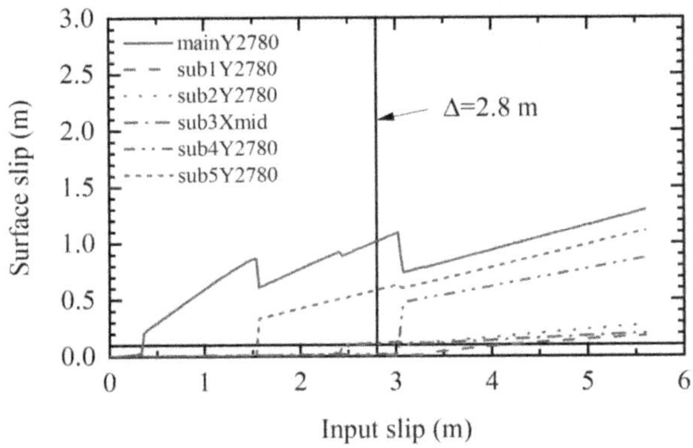

Fig. 50. Surface slip/input slip relationship (Case 4).

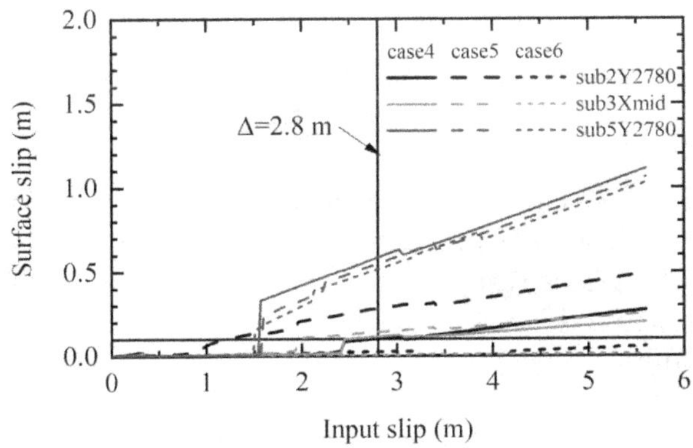

Fig. 51. Surface slip/input slip relationship (Cases 4 to 6).

observed, Δ_c is less than 2.8 m. This tendency is similar to the Case 1 results. The surface slip is slightly smaller in Case 4 than in Case 1. Although the slip distributions at depth are significantly different between Cases 1 and 2, the surface slips on the sub-faults are similar.

Figure 51 presents the variation in the surface net slip δ on Sub2, Sub3, and Sub5 in Cases 4, 5, and 6 as the input slip Δ increases. The results for Sub5 are similar among the three cases. However, the Δ_c values on Sub2 and Sub3 are much smaller in Case 5 than in Cases 4 and 6. As shown in Fig. 52, the peak of the large surface slip area is close to the intersection line between the main fault and Sub2 fault. In contrast, in Case 6, no surface slip occurs on sub-faults Sub2 and

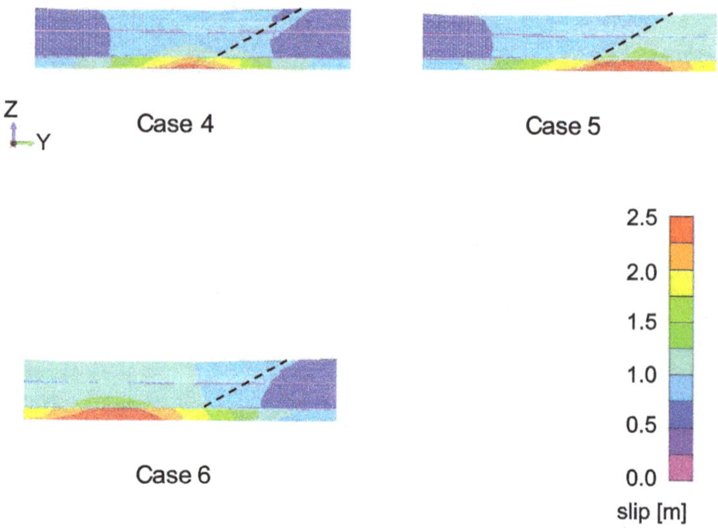

Fig. 52. Slip distribution on the main fault (Cases 4 to 6). The dotted line on the main fault indicates the intersection with Sub2.

Sub3 because the location of the peak of the large surface slip area on the main fault is far from the intersection between the main fault and Sub2 fault. These results indicate that the slips on the sub-faults are strongly affected by the slip distribution on the main fault close to the sub-faults.

5.2.1. *Discussion*

We propose a method for setting the slip distribution on the main fault for predictive simulations based on the method of constructing characterized source models for the prediction of strong motion. In cases without asperities and with two asperities, surface slip on the sub-faults did not appear. These results suggest that surface displacement is less likely to occur on sub-faults. In the case with two asperities, a large slip was inputted into the asperities, but it had little effect on the surface slip on the sub-faults since the two asperities are quite far from the sub-faults.

We also set up cases in which a large surface slip area was located near the sub-faults. We adopted an empirical equation for the relationship between the maximum surface displacement and fault length. We then placed the large surface slip area in a section of the FEM model. From this simulation, we obtained results similar to those of the simulation with the inversely analyzed input slip. The location of this large surface slip area had a strong effect on the surface slip of the sub-faults. In other words, for surface slip to occur on a sub-fault, a large slip must occur on a nearby part of the main fault. The probability of a large slip being very close to a sub-fault is relatively small. Therefore, in setting the large surface slip area, the probability of its occurrence should also be considered.

6. Example of Uncertainty Quantifications

6.1. *Uncertainty in fault displacement simulations*

There are certain limitations to the quality and quantity of available data on underground structures, material properties, stress state, source fault dynamics, and other aspects. Uncertainties can be classified as aleatory and epistemic uncertainty.

Aleatory uncertainty refers to the natural randomness in a process. The aleatory uncertainty corresponds to the uncertainties in the material properties of the rock mass and fault planes in the simulation for surface faulting. Ensemble computing can be utilized to evaluate the variability in fault behavior induced by the aleatory uncertainty which is represented by continuous random variables.

Epistemic uncertainty refers to the scientific uncertainty in the model of a process, which is due to limited data and knowledge. In a surface faulting simulation, the epistemic uncertainty includes the uncertainties in geometry of the underground and fault planes, slip distribution across the earthquake source fault, and its slip value. The slip value of the earthquake source fault is related to the seismic moment or earthquake magnitude. It should be noted that the uncertainty in the slip value can be evaluated by a one-time quasi-static simulation in which the slip is incrementally loaded on the boundary of the analytical model. Other epistemic uncertainties are evaluated using alternative models and parameter values based on the logic tree approach and other approaches.

In this section, we show examples of uncertainty quantification for the simulation of the 2014 Nagano-ken-hokubu earthquake. In the simulation, the crustal deformation caused by the primary fault activity was estimated using the elastic theory of dislocation and it was incrementally imposed on the bottom boundary of the analytical model, as performed in the simulation in Section 4.

6.2. *Uncertainty of material properties*

In this section, we conduct ensemble computing for uncertainties in material properties as examples of the epistemic uncertainty. Herein, the correlation between material properties is considered, but the spatial correlation of material properties is not.

6.2.1. *Sensitivity analysis for material properties*

We confirm the influence of material properties, Young's modulus of the rock mass, friction angle, and cohesion of the fault plane through a sensitivity analysis.[39] We used the analytical model shown in Fig. 53, which corresponds to the model in Fig. 31 ignoring secondary fault E2. We set the maximum value of the primary slip to 4.0 m, which is loaded in 200 steps in this analysis.

First, we conducted a parametric study for Young's modulus of the rock mass and the friction angle of the fault plane of mean $\pm 1\sigma$, in which the mean value is the set value of material properties in Section 4. The COVs of Young's modulus and friction angle are set to 0.3 and 0.15, respectively, and the Young's moduli of the two layers are perfectly correlated. Figures 54 and 55 show the relationship between the input base slip Δ and surface slip at the evaluation points of the primary fault and secondary fault E

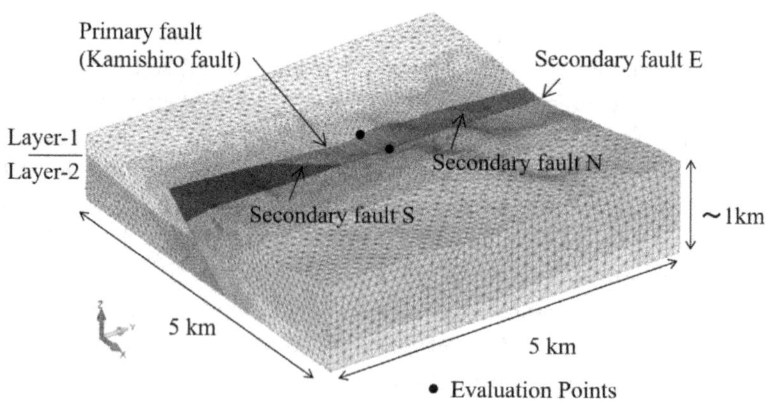

Fig. 53. Finite element mesh with secondary faults E, N, and S.

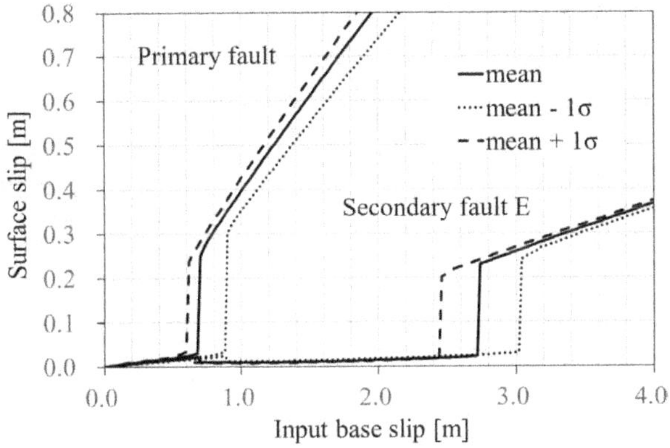

Fig. 54. Relationship between maximum input base slip and surface slip with $E =$ mean $\pm 1\sigma$.

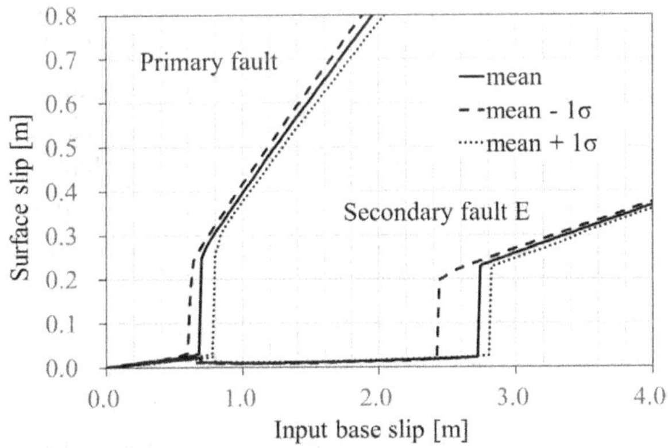

Fig. 55. Relationship between maximum input base slip and surface slip with $\phi =$ mean $\pm 1\sigma$.

shown in Fig. 53. In this figure, the solid line denotes the result of the standard case with mean value, and the dashed and dotted lines are the results of the case with mean $\pm 1\sigma$, respectively. These results reveal that the critical base slip Δ_c decreases when Young's modulus becomes larger or the friction angle becomes smaller. Here, the Δ_c values of secondary fault E are 3.01 and 2.41 m for mean $\pm 1\sigma$ of Young's modulus, respectively, and they are 2.41 and 2.81 m for mean $\pm 1\sigma$ of the friction angle, respectively. The maximum difference from the standard case, $\Delta_c = 2.74$ m, is approximately 0.3 m.

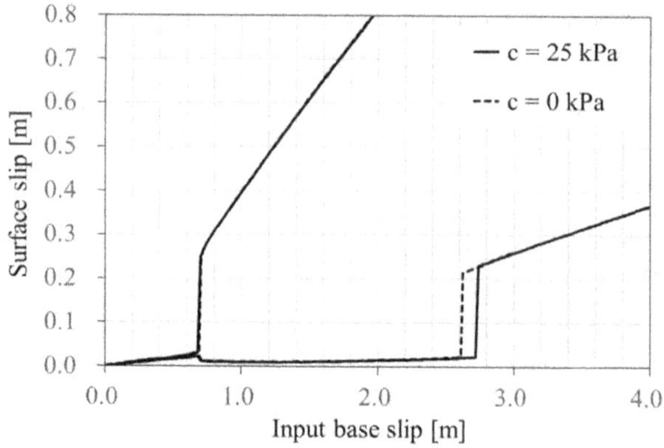

Fig. 56. Relationship between maximum input base slip and surface slip with $c = 0$. The lines for the primary fault are overlapping.

Next, we analyzed the influence of the cohesion of the fault planes conducting an analysis with $c = 0.0$ kPa. Figure 56 shows the relationship between Δ and surface slip at the evaluation points in the case with $c = 0.0$ kPa and the standard case, $c = 25$ kPa. The solid line is the result of the standard case and the dashed line is the result of the case with $c = 0.0$ kPa. Δ_c is 2.64 m in the case with $c = 0.0$ kPa and the difference from the standard case, $\Delta_c = 2.74$, is less than 0.1 m. Figure 56 reveals that the influence of uncertainty in c is considerably smaller than that of the other two material properties, Young's modulus and friction angle.

6.2.2. *Probabilistic response and its convergence with Latin hypercube sampling*

We conducted ensemble computing using numerous models generated based on Latin hypercube sampling (LHS).[39] In LHS, the sample space of each probabilistic variable is divided into n equally probable subspaces, and a sample is randomly chosen from each subspace, as shown in Fig. 57. A sample set having n samples was obtained as a random combination of samples of each probabilistic variable.

We used the analytical model shown in Fig. 53 and the value of material properties in Section 4 as the mean values. Considering the result in Section 6.2.1, we assumed that the three parameters having a potent influence on Δ_c, that is, Young's modulus of the first and second ground layers and friction angle on the fault planes, follow an

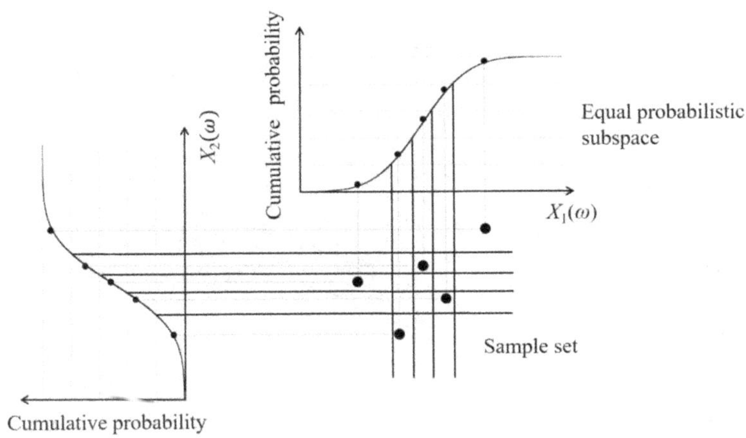

Fig. 57. Overview of LHS with two random variables $X_1(\omega)$ and $X_2(\omega)$. The probabilistic range of each variable is divided into five ranges.

independent log-normal distribution. The COV of Young's moduli and the friction angle are set to 0.3 and 0.15, respectively.

From the point of view of actual practice, it is important to confirm the convergence between the probabilistic response and the number of samples, that is, the number of samples that make it possible to evaluate the probabilistic response with high accuracy. Therefore, we conducted analyses with multiple sample sets having different numbers of samples to confirm the convergence. Table 6 lists the 14 sample sets in which the maximum number of samples was 120. Herein, Set20-3, Set40-2, Set60-2, Set80, and Set120 were generated as a combination of more than a sample set. Figure 58 shows a procedure for generating a sample set combining two sample sets. The component sample sets are also listed in Table 6. We conducted 230 calculations using 230 models which have different values of the material properties, as shown in Fig. 59.

Figure 60 shows the relationship between the input base slip and the surface slip at the evaluation points of the primary fault and secondary fault E shown in Fig. 53 for Set120. The Δ_c of both faults in Fig. 60 has considerable variability. The variability of the secondary fault is greater than that of the primary fault. Figure 61 illustrates a histogram of the Δ_c for Set120. The solid line in Figure 61 is the fitted log-normal distribution, which has an adjusted coefficient of determination R^2 of 0.996 in both faults. These results suggest

Table 6. Sample set.

ID.	Number of samples	Component sample set
Set10-1	10	—
Set10-2	10	—
Set10-3	10	—
Set20-1	20	—
Set20-2	20	—
Set20-3	20	Set10-1 and Set10-2
Set30-1	30	—
Set30-2	30	—
Set40-1	40	—
Set40-2	40	Set20-3 and Set20-2
Set60-1	60	—
Set60-2	60	Set30-1 and Set30-2
Set80	80	Set40-2 and Set40-1
Set120	120	Set60-2 and Set60-1
Total Number of Samples	230	

Note: Set20-3, Set40-2, Set60-2, Set80, and Set120 were generated as a combination of component sample sets.

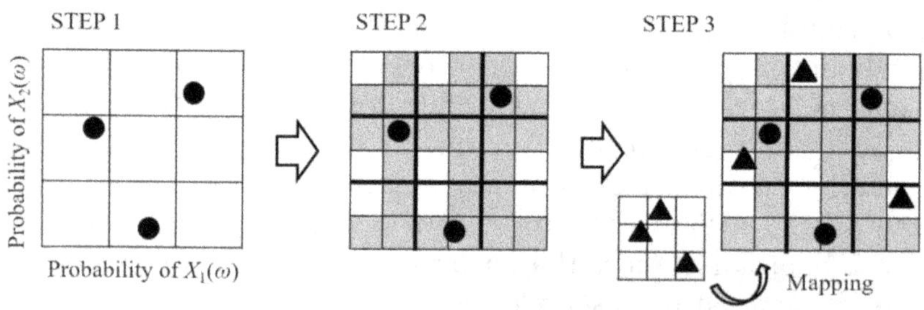

Fig. 58. Overview of how to generate a new sample set from two sample sets. The probabilistic range of each variable, $X_1(\omega)$ and $X_2(\omega)$, is divided into 3 or 6 ranges. In STEP 1, a sample set with three samples is generated. In STEP 2, the sample space is divided into six equally probable subspaces and the samples are allocated in the subspaces. In that case, 3×3 subspaces become blank. In STEP 3, the newly generated three samples are mapped in the 3×3 blank subspaces.

that Δ_c can be assumed by a log-normal distribution, which has a heavy right-tailed unimodal distribution. Therefore, the second-order moment, the mean, and standard deviation of the log-normal distribution provide a good indication of the probabilistic response of Δ_c.

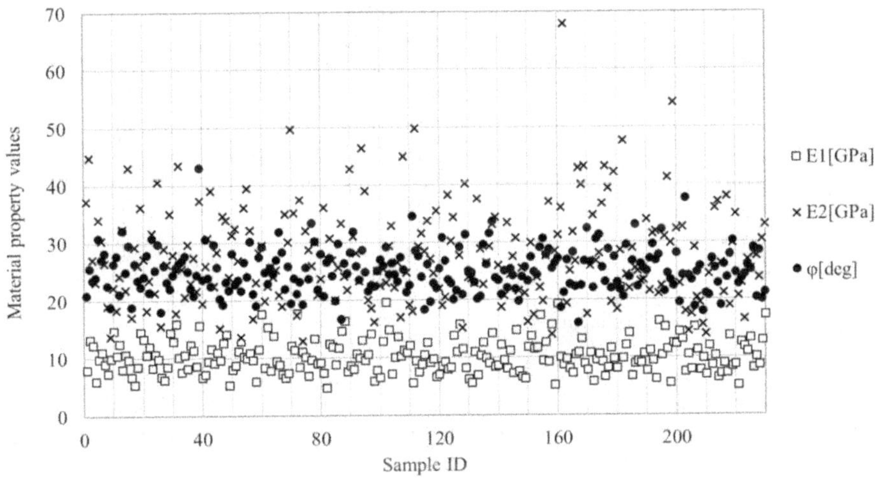

Fig. 59. Material properties of all 230 samples. Herein E_1, E_2, and ϕ denote Young's modulus of the first and second ground layers and friction angle of the fault planes, respectively.

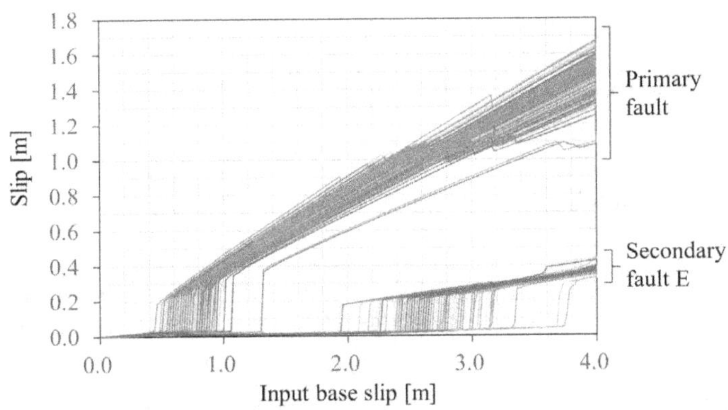

Fig. 60. Relationship between maximum input base slip and surface slip of sample set having 120 samples, set120.

Figure 62 shows the second-order moment of Δ_c of the fitted log-normal distribution in all the cases. The deterministic results are shown as dashed lines in Fig. 62 for reference. The obtained standard deviation suggests that uncertainties in the material properties more strongly affect the Δ_c of secondary faults than that of the primary fault. The variability in the second-order moment of both faults is

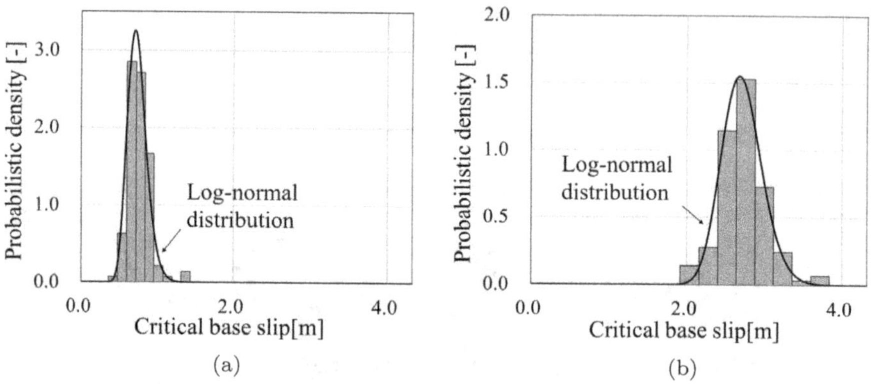

Fig. 61. Probabilistic distribution of the critical base slip Δ_c. The histograms represent the results of 120 samples, and the solid line represents a log-normal distribution fit, for (a) the primary fault, and (b) secondary fault E.

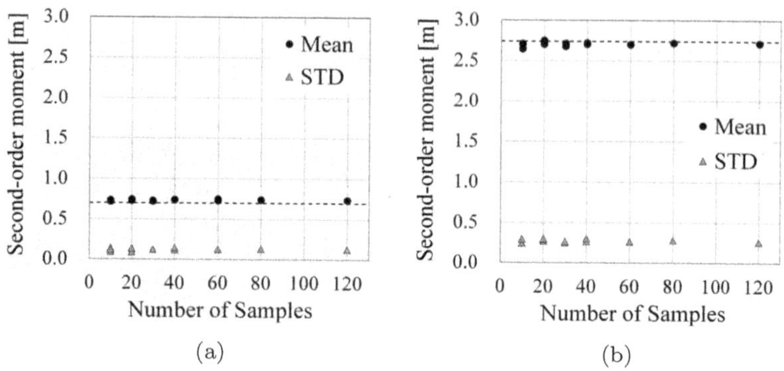

Fig. 62. Relationship between second-order moment of the critical base slip Δ_c and the number of samples for (a) the primary fault, and (b) secondary fault E.

small and the difference between the calculated mean of each sample set and the deterministic result is less than 0.1 m. Figure 63 shows the relationship between the number of samples and error in the second-order moment of Δ_c. Herein, the error of each sample set is defined as the difference between the result of the sample set and that obtained with all 230 samples, which can be viewed as a sample set of the Monte Carlo method. Figure 63 demonstrates that the error is reduced as the number of samples increases. In both faults, the second-order moment converged with at least 60 samples because the error became less than 0.02 m, which is the minimum increment

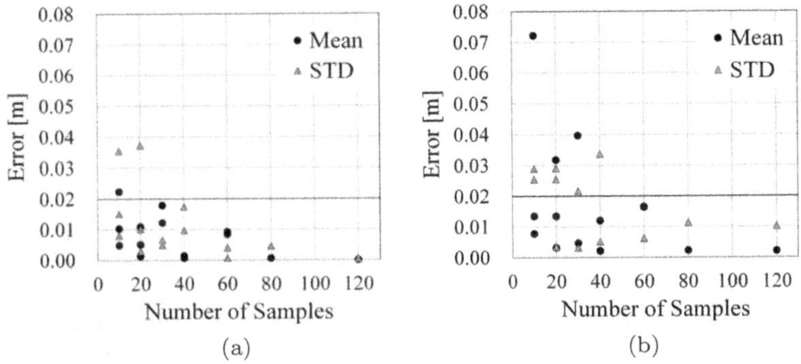

Fig. 63. Relationship between the error of second-order moment of the critical base slip Δ_c and the number of samples, in which the error is the difference to the solution obtained with all 230 samples. The solid line denotes the minimum increment of the input base slip, 0.02 m. (a) The primary fault, and (b) the secondary fault E.

of the input base slip. The errors of mean and standard deviation of secondary fault E are 0.072 and 0.029 m with 10 samples, respectively. That suggested that the effect of material uncertainties on Δ_c is predictable with several tens of samples. We should examine the effect of material uncertainties with a minimum number of samples in each scenario considered epistemic uncertainty.

Next, we analyzed the effect of the material uncertainties on the appearance position of the surface faults,[40] which is defined as the position of maximum surface slip in each step. Figure 64 shows the histogram of the appearance position of the surface faults for set120. Figures 64(a) and 64(b) indicate the appearance position just before and just after a steep increase in the surface slip, respectively. The observed range of surface slip of secondary fault E is also described in Fig. 64 with dotted lines. The appearance position just before the steep increase in the surface slip has very small variation, its standard deviation is 17.9 m, and it agrees with the observed result. On the other hand, the appearance position just after the steep increase in the surface slip varies widely owing to the uncertainties in material properties. The difference between these variations of the appearance position seems to be associated with the stability loss of the mathematical solution, i.e., the appearance position before the steep increase in the surface slip can be evaluated with small variability because the solution is stable. These results suggest that the

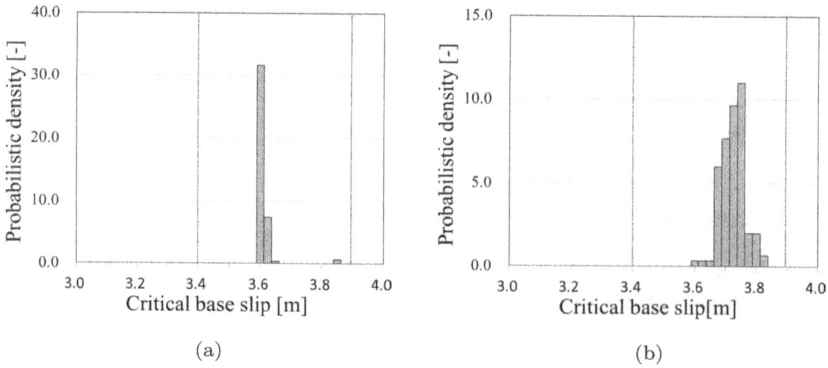

Fig. 64. Histogram of the appearance position of maximum surface slip with 120 simulations for (a) and (b). (a) Just before steep increase in surface slip, and (b) just after steep increase in surface slip.

appearance position of a secondary fault can be stably evaluated by estimating the position before the steep increase in the surface slip, even with uncertainties in material properties.

6.2.3. *Effects of each of the material properties and correlation between material properties with orthogonal sampling*

We revealed the effect of each material property on the probabilistic response of Δ_c by ensemble computing using orthogonal sampling.[41] In orthogonal sampling, the sample points are set systematically at regular intervals. To clarify the influence of the material property values, the sample points were set at equal intervals based on the standard deviation of the material properties, although the sample points are generally set for each sample space with the same probability of occurrence in the ordinary sampling.

For the sake of simplicity, we used the simple rectangular analytical model shown in Fig. 65, which has four fault planes equivalent to those of the model in Fig. 53, but with single geological layer. Each finite element in the analytical model had a length of approximately 50 m, resulting in 1.01 million DOFs. The rock mass exhibited linear elasticity with Young's modulus of 15.0×10^3 MPa, Poisson's ratio of 0.25, and density of 2.0×10^3 kg/m^3. The parameters of the fault planes are the same as those presented in Section 4 except for the cohesion of 0.0 kPa. We assumed that Young's modulus of the

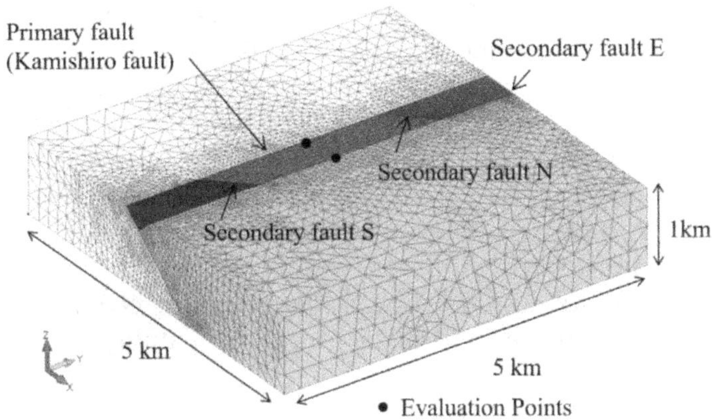

Fig. 65. Finite element mesh with single geological layer.

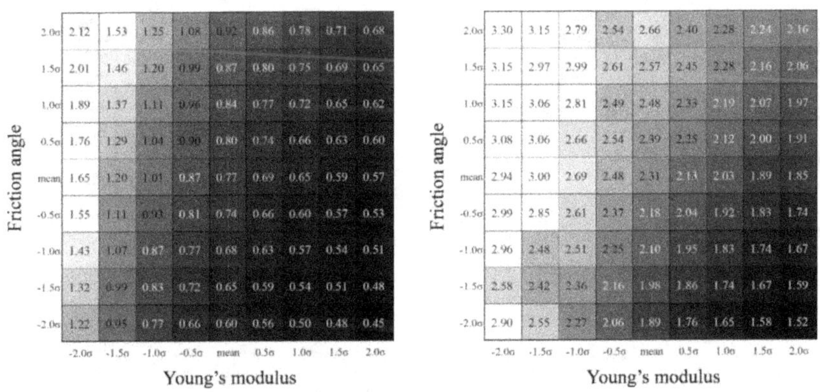

(a) (b)

Fig. 66. Critical base slip Δ_c of all 81 cases for (a) the primary fault, and (b) secondary fault E.

rock mass and friction angle on the fault planes followed a normal distribution and the COV of Young's modulus and friction angle were set to 0.33 and 0.15, respectively. The sample points were set at 0.5 intervals from the mean value up to $\pm 2\sigma$, and 95.45% of the probability of occurrence was included in the target range. As a result, 81 calculations were performed. We set the maximum value of the primary slip to 3.0 m, which was loaded in 200 steps.

Figure 66 shows the results of all 81 cases of Δ_c of the evaluation points of the primary fault and secondary fault E shown in Fig. 65.

Figure 66 indicates that Δ_c becomes small when Young's modulus of the rock mass is large and friction angle of the fault plane is small. That is consistent with the result in Section 6.2.1. These results can be understood as follows. When Young's modulus of the rock mass is small or friction angle of the fault plane is large, the work done by the input forced displacement is more easily converted into strain energy of the rock mass. As a result, the propagation of the slip is suppressed and the critical base slip Δ_c becomes large.

The probability distribution of the response can be estimated summing the response multiplied by the probability of occurrence

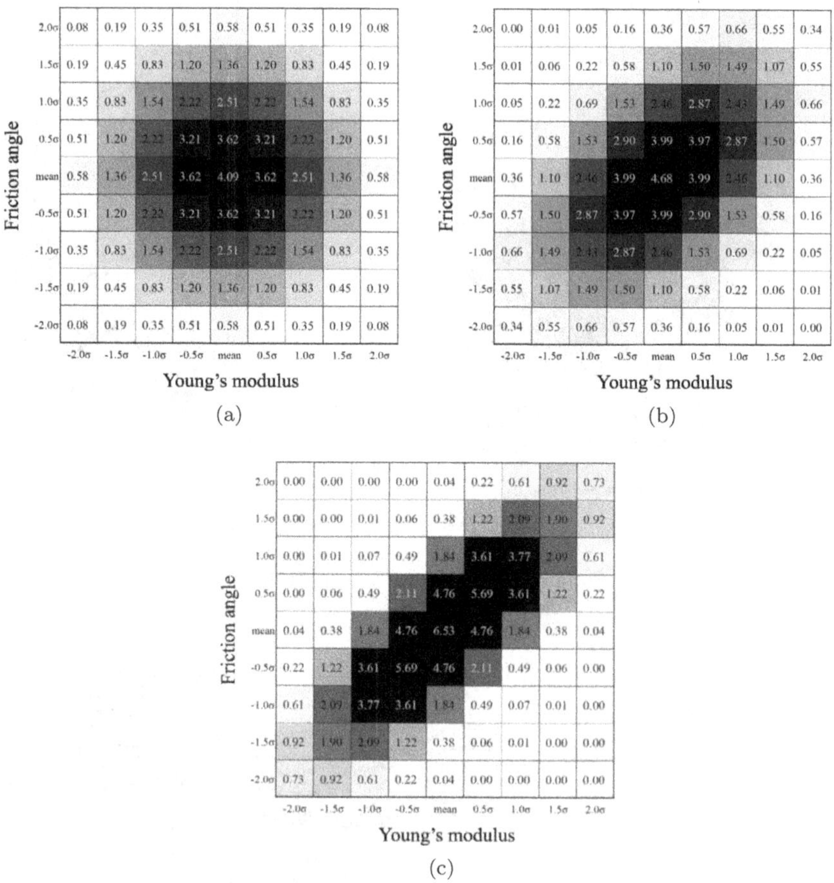

Fig. 67. Probability of occurrence of the sample points for (a) independent random variables and (b) and (c) correlated random variables with correlation coefficient of 0.5 and 0.8, respectively.

of the sample point. In this study, we considered three cases having independent random variables and correlated random variables with correlation coefficients ρ of 0.5 and 0.8. In that case, we considered only the positive correlation between Young's modulus of the rock mass and friction angle of fault planes by reference to the positive correlation between Young's modulus and friction angle of the rock mass. Figure 67 shows the probability of occurrence of the sample points in the three cases. Herein, the total probability of occurrence of all 81 sample points is normalized as 100% because the total probability of sample points is less than 100% by considering only the range within the mean $\pm 2\sigma$. Figure 68 shows the histogram of Δ_c of secondary fault E in the three cases. The dashed lines in the figures show the deterministic results obtained using the mean values of

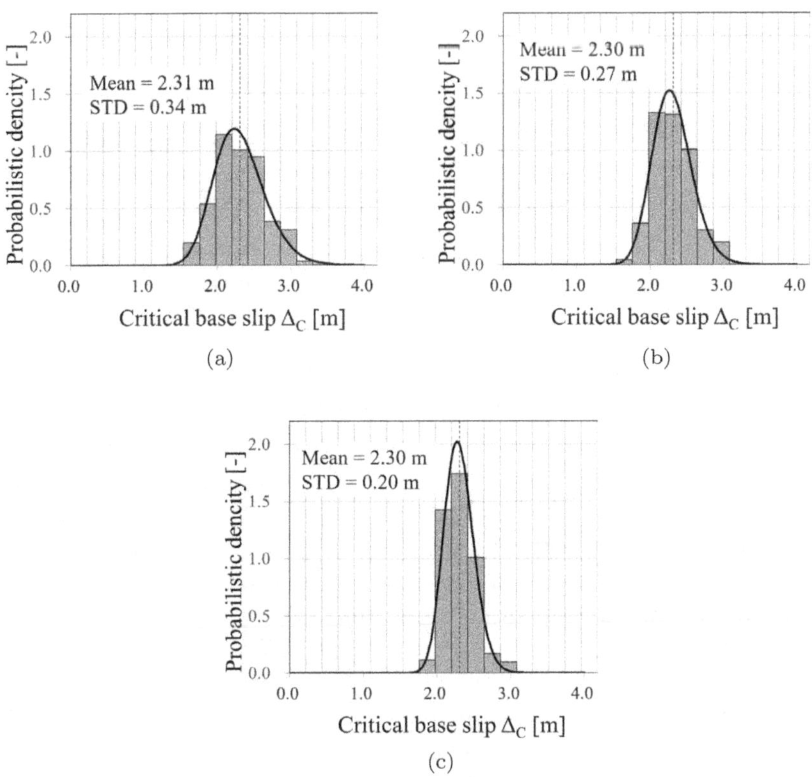

Fig. 68. Histogram of Δ_c of secondary fault E for (a) independent random variables and (b) and (c) correlated random variables with correlation coefficient of 0.5 and 0.8, respectively.

material properties. The solid lines in the figures correspond to the fitted log-normal distributions. Δ_c can be assumed by a log-normal distribution in all cases because the log-normal distribution has an adjusted coefficient of determination R^2 of more than 0.99 in all of them. The mean and standard deviations of Δ_c are, respectively, 2.31 and 0.34 m in the case with $\rho = 0.0$, that is, independent random variables, 2.30 and 0.27 m in the case with $\rho = 0.5$, and 2.30 and 0.20 m in the case with $\rho = 0.8$. These results indicate that the positive correlation between Young's modulus of the rock mass and friction angle of the fault plane does not change the mean value, but decreases the variability of Δ_c.

6.3. *Uncertainty in fault planes*

We conducted sensitivity analyses for uncertainties in slip distribution across the earthquake source fault and configurations of the fault planes in the analytical model as examples of epistemic uncertainty. Here, the influence of the uncertainties was just confirmed by the sensitive analysis. For quantifying the epistemic uncertainty, the models to be considered and their weights have to be set appropriately using, e.g., the logic tree approach based on the expert opinions.

6.4. *Slip distribution across the earthquake source fault*

In the simulation of the 2014 Nagano-ken-hokubu earthquake in Section 4, the slip distribution across the earthquake source fault was set based on the result of source inversion obtained from the Geospatial Information Authority of Japan (GIAJ, 2014), but the slip distribution of the superficial part was corrected based on the observed surface slip distribution. Herein, we carried out an analysis using the result of source inversion not corrected.[42] We used the analytical model shown in Fig. 53 and set the maximum value of the primary slip to 3.0 m, which was loaded in 200 steps. Figures 69 and 70 show the contour of the slip distribution at the input base slip of 2.8 m for the slip distribution corrected and not corrected, respectively. The slip distribution on the primary fault of the not corrected slip distribution is wider than the corrected slip distribution. In the case with the corrected slip distribution, the slips of secondary faults

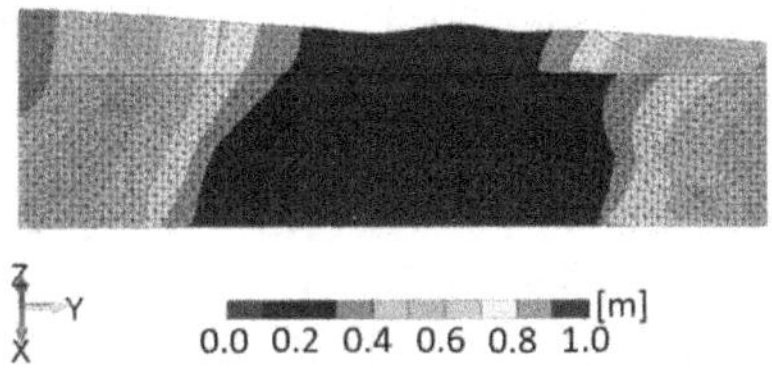

(a) primary fault

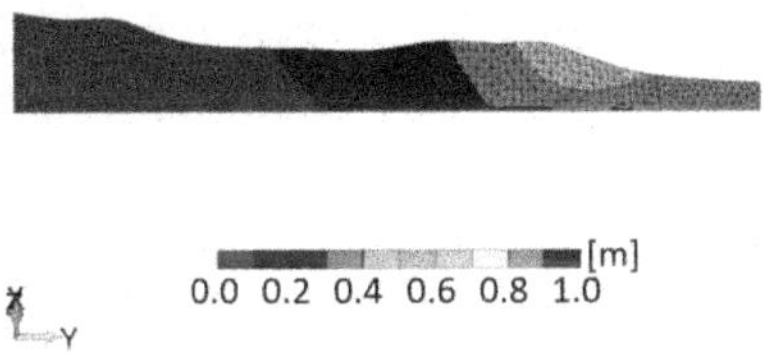

(b) secondary fault E

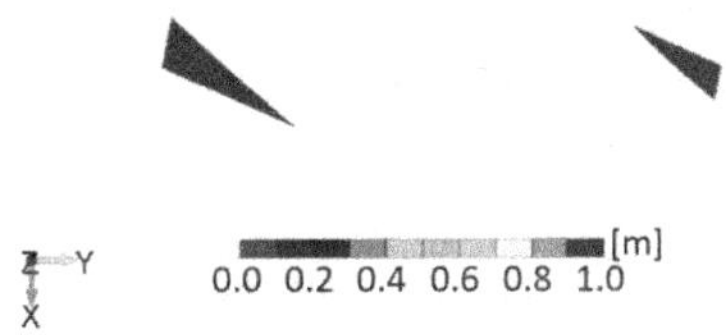

(c) secondary faults N and S

Fig. 69. Contour of slip on fault planes at $\Delta = 2.8$ m in the case with corrected superficial part for (a) the primary fault, (b) secondary fault E, and (c) secondary fault N and S.

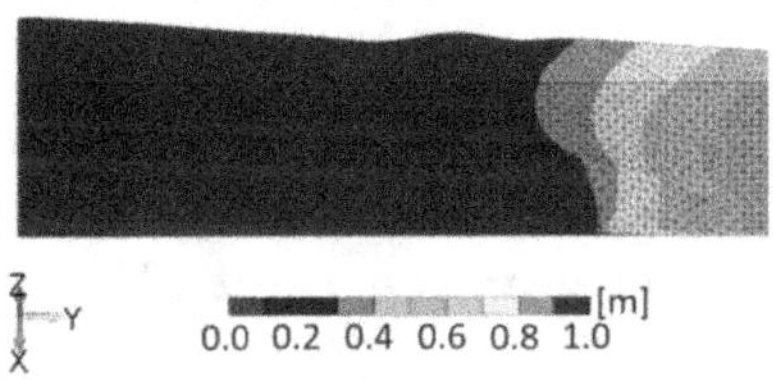

(a) primary fault

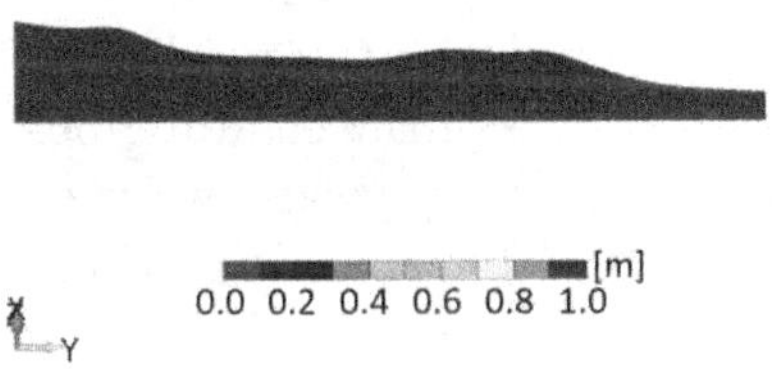

(b) secondary fault E

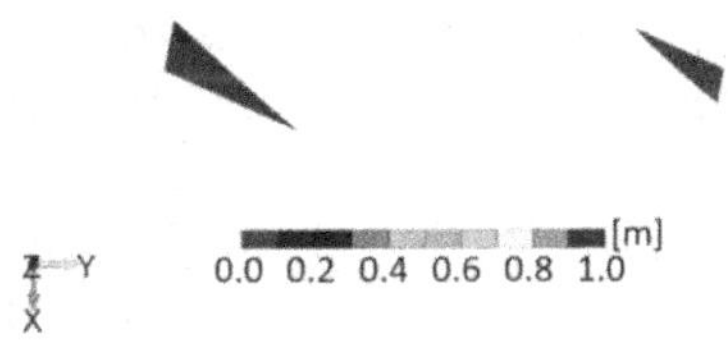

(c) secondary faults N and S

Fig. 70. Contour of slip on fault planes at $\Delta = 2.8$ m in the case of GIAJ (2014) slip distribution for (a) the primary fault, (b) secondary fault E, and (c) secondary fault N and S.

S and N do not appear, but that of the secondary fault E appears. On the other hand, in the case with the not corrected slip distribution, the slips of secondary faults E and N do not appear, but that of the secondary fault S appears near the connection with the primary fault. These results indicate that the resolution of slip distribution of the superficial part strongly affects the occurrence of surface slip of secondary faults.

Furthermore, in the predictive simulation for 2016 Kumamoto earthquake in Section 5, when a large surface slip area across the primary fault was not included, surface slip of any secondary fault did not appear. However, when a large surface slip area was set near the secondary fault, we obtained results similar to observation. This trend is also observed in the simulation of the 2014 Nagano-ken-hokubu earthquake. These results reveal that the location of the large slip area has a strong effect on the surface slip of the secondary fault.

For quantification of uncertainty of the slip distribution across the earthquake source fault, it is important to consider the variation in the position and slip distribution of the large surface slip area, although there is extremely low probability of a large surface slip area.

6.5. *Geometry of secondary fault: Dip angle*

We analyzed the effects of uncertainty in the geometry of secondary faults, especially the dip angle θ_{dip} of secondary fault E.[42] θ_{dip} of the secondary fault E was set to 40° and 50°. The model in the case of $\theta_{\text{dip}} = 40°$ corresponds to Fig. 53. Figure 71 shows the model in the case of $\theta_{\text{dip}} = 50°$. Herein, the position of the surface fault lines in both cases are the same. In the analyses of both cases, we set the maximum value of the primary slip to 3.0 m, which was loaded in 200 steps, and for the material properties, the value given in Section 4 was set except for the cohesion of 0.0 kPa.

Figure 72 shows the relationship between the input base slip and surface slip at the centers of the primary fault and secondary fault E in the cases with $\theta_{\text{dip}} = 40°$ and 50°. The Δ_c values of the primary fault agree with each other. The Δ_c values of secondary fault E in the case with $\theta_{\text{dip}} = 40°$ and 50° are 2.64 and 2.60 m, respectively. This result indicates that the ambiguity of the dip angle of the secondary fault has an insignificant effect on the occurrence of secondary

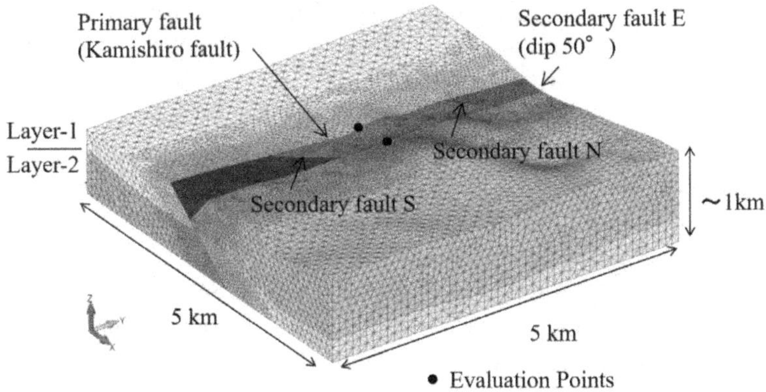

Fig. 71. Finite element mesh with secondary fault E of the dip angle of 50°.

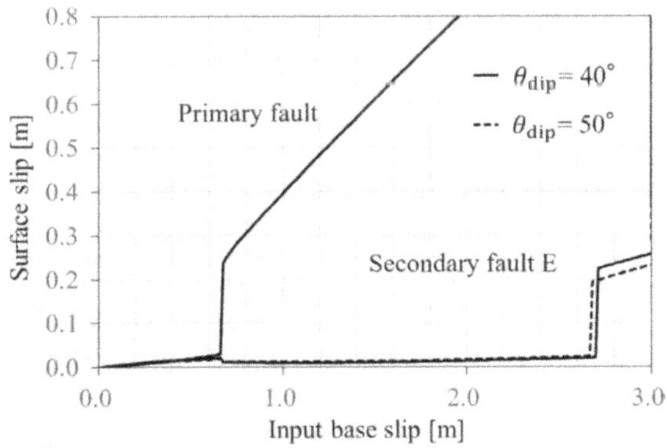

Fig. 72. Relationship between maximum input base slip and surface slip with $\theta_{\text{dip}} = 40°$ and 50°.

fault displacement. If the geometry of a crushed zone that is a candidate for a secondary fault is clear based on a detailed geological survey, the geometry of the secondary fault in the analytical model can be set using this information. However, if such information is not available, the analysis should take into account several candidate geometries.

6.6. *Number of secondary fault planes in analytical model*

We analyzed the effect on a target fault from the number of the other secondary faults considered. In particular, we focused on how Δ_c of the target secondary fault changes considering the other faults in the analytical models. We used the analytical model shown in Fig. 31 having secondary faults E1, E2, N, and S. The four cases listed in Table 7 were calculated, where some secondary faults are fixed to not consider the faults. In the analyses, the maximum slip value of the primary fault of 3.0 m was loaded in 200 steps and the material properties were set as shown in Section 4.

Figure 73 shows the relationship between the input base slip and surface slip at the centers of the primary fault and secondary fault E1. The responses of the secondary fault E1 slightly changed considering the other secondary faults, but the results for the primary fault

Table 7. Analytical cases.

	Secondary fault			
Case ID	E1	E2	N	S
E1	√	×	×	×
E1NS	√	×	√	√
E1E2NS	√	√	√	√
E2NS	×	√	√	√

Note: Here, the check marks mean considering the slip of faults and the cross marks mean that the faults are fixed to ignore the faults.

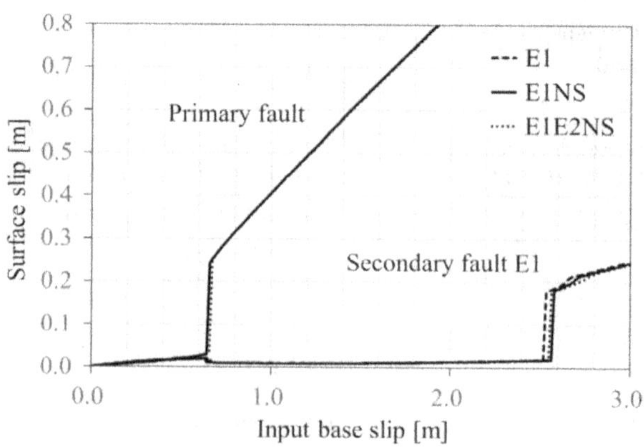

Fig. 73. Relationship between maximum input base slip and surface slip of primary fault and secondary fault E1 in cases E1, E1ND, and E1E2NS.

did not. The Δ_c of secondary fault E1 in the case E1, 2.55 m, is smaller than the ones in the cases E1NS, 2.58 m, and E1E2NS, 2.57 m. Those indicate that the slip of the secondary fault E1 can be inhibited by consuming a part of input energy as the work done by the slip on the secondary faults N and S based on the conservation law of energy. The secondary fault E2 has the smaller influence than the secondary fault N and S because the secondary fault E2 is father away from the secondary fault E1.

Figure 74 describes the change in the surface slip at the center of the primary fault and at the evaluation point subE2Y400, shown in Fig. 35, of the secondary fault E2. The results for the secondary fault E2 vary depending on consideration of the secondary fault E1, in which the Δ_c in cases E2NS and E1E2NS are 2.90 and 2.96 m, respectively. The steep increase in the surface slip of the secondary fault E2 appeared in the case E2NS, but it did not in the case E1E2NS. The Δ_c of the secondary fault E1 is smaller than the one of the secondary fault E2 under these conditions. The large slip on the fault consumes the large energy and the steep increase in the surface slip changes the stress around the fault planes. Therefore, consideration of the secondary fault E1 can have a relatively large influence on the response of the secondary fault E2.

In general, the higher the number of secondary fault planes considered in the analytical model, the less likely it can be that surface slip will occur because of increase of the work done by the slip on the fault

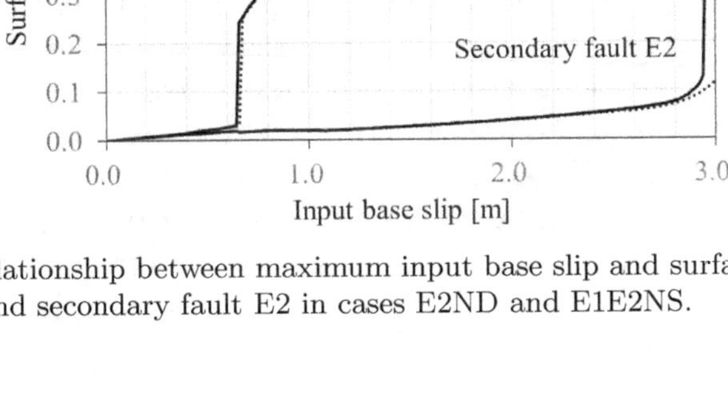

Fig. 74. Relationship between maximum input base slip and surface slip of primary fault and secondary fault E2 in cases E2ND and E1E2NS.

plane. Especially, note that faults near the target fault or/and faults where the surface slips are prone to occur can affect the response of the target faults more strongly.

7. Conclusion

This chapter presents a study on the development of a numerical analysis method using high-performance computing for fault displacement hazard assessment of surface earthquake faults in large inland earthquakes.

At first, physical phenomena related to fault displacement were analyzed to formulate a mathematical problem. The mechanism of fault displacement is the propagation of slip on the fault plane. Therefore, fault displacement is a friction problem that involves rupture of the fault plane. In setting up the mathematical problem, it is necessary to consider the existence of solutions, their uniqueness, and their stability/unstability. From the consideration of the stability of the solution of the friction problem, we decided to use a constitutive equation in which the shear stiffness of the fault decreases with increasing slip from the initial state to the residual state. When this constitutive equation is used, the shear displacement increases significantly with a slight increase in force (jump) in the slip–shear stress relationship. On the other hand, there is a conservation law that holds regardless of whether the solution is stable or unstable. In the case of fault displacement, energy conservation is established where the energy of displacement in the subsurface is converted into frictional work on the fault surface and strain in the surrounding ground.

Quality assurance of the numerical analysis is an important issue in the evaluation of fault displacement hazard by numerical analysis. To evaluate fault displacement on the ground surface, it is necessary to construct a three-dimensional analytical model with appropriate resolution and perform numerical analysis. The number of degrees-of-freedom is on the order of 1,000,000, and the computational resources required are large. When constructing an analytical model of the ground including faults, three-dimensional joint elements are applied to the faults. In order to correctly calculate the slip–shear stress relationship, the above energy conservation must be satisfied throughout the numerical analysis.

For this purpose, a three-dimensional joint element was rigorously derived and symplectic time integration satisfying energy conservation was applied. The three-dimensional joint element was formulated from the Lagrangian of the fault plane. A numerical experiment was conducted to compare the symplectic time integration with the Newmark *beta* method, which is commonly used for time integration, and the superiority of the symplectic time integration was demonstrated.

Next, two functions, the rigorously derived higher-order joint element and the explicit symplectic time integration with excellent energy conservation, were implemented in an open-source parallel finite element method program. This finite element method program enables large-scale fault displacement analysis by parallel computation.

The basic verification of the analysis program was conducted by setting up a one-dimensional fault problem. We also applied the developed program to a relatively simple three-dimensional model of the ground containing faults and verified the program. For fault displacement problems involving solution instability, it was noted that analysis of multiple cases with different conditions is essential to evaluate the uncertainties. More than 100 cases of analysis are required to perform probabilistic studies. We mentioned the superiority of quasi-static analysis, which can evaluate the relationship between input slip and surface slip in a single analysis, and proposed the use of quasi-static analysis for multi-case analysis by confirming the consistency of the solutions from dynamic and quasi-static analysis.

We applied the parallel finite element method to simulate two recent earthquakes which induced surface earthquake faults in Japan, the 2014 Nagano-ken-hokubu earthquake (reverse faulting) and the 2016 Kumamoto earthquake (right lateral strike-slip faulting) for the validation.

After the simulation for the 2014 Nagano-ken-hokubu earthquake, we confirmed that (1) surface slip occurs on the main and sub-faults within the set input slip range, and (2) the analysis results for the slip of the main and sub-faults are consistent with the observation results. When the input slip was set considering up to twice the slip distribution obtained from the inverse analysis of crustal deformation and the observation of displacement of the main fault, steep increase of slip was observed at the surface of the main fault and sub-faults. The slip magnitude was generally consistent with the observation results for both the main and sub-faults. The sub-faults back-thrusted against the main fault and this was consistent with observations.

In the simulation for the 2016 Kumamoto earthquake, as the input slip increased, surface slips appeared on both the main and sub-faults. On the sub-faults that appeared in the 2016 earthquake, the surface slip appeared at a smaller input slip than on the other sub-faults. The sense of slip on each fault in the simulations was in good agreement with that observed during the earthquake. Furthermore, the calculated surface slips agreed well with the envelope curve of the measured surface slip distribution.

Because our simulations reproduced the main features of two different earthquakes with reasonable accuracy, the proposed numerical method should be applicable to surface fault displacement estimation.

We also proposed a method for setting a slip distribution on main faults for predictive simulation based on the strong motion prediction method. However, surface slips did not appear on the sub-faults, even in the case with two asperities. We then set up a large slip area near the sub-faults based on an empirical equation of the relationship between the maximum surface slip and the length of a fault. In this simulation, we obtained results similar to those of the simulation with the inversely analyzed input slip. We consider that the slip distribution on the main fault with a large surface slip area is an extreme case in which surface slip on the sub-faults is likely to occur.

It is essential to consider various uncertainties in the fault rupture process for surface fault analysis. In this study, the effect of material uncertainties on surface fault analysis was examined by performing 230 simulations using Latin hypercube sampling. The effect of the uncertainty of material properties on fault displacement is summarized, and the probability distribution of response values is evaluated. As a result, it was shown that high-performance computing can be used to evaluate the variability of response values in fault displacement evaluation.

It is certain that the points where fault displacement occurs are simultaneously exposed to strong seismic vibration, and the challenge in assessing the fragility of facilities due to fault displacement is to evaluate the effect of the superposition of fault displacement and seismic motion.

References

1. On-Site Fault Assessment Method Review Committee, Assessment methods for nuclear power plant against fault displacement, Computing Science Technical Report JANSI-FDE-03, Japan Nuclear Safety Institute (2013).
2. Y. Okada, Internal deformation due to shear and tensile faults in a half-space, *Bull. Seismol. Soc. Am.* **82**, 1018–1040 (1992).
3. B. T. Aagaard, G. Anderson, and K. W. Hudnut, Dynamic rupture modeling of the transition from thrust to strike slip motion in the 2002 Denali fault earthquake, Alaska, *Bull. Seismol. Soc. Am.* **94** (6B), S190–S201 (2004).
4. R. Harris and S. Day, Dynamics of fault interaction: Parallel strike-slip faults, *J. Geophys. Res.* **98**, 4461–4472 (2008).
5. N. Kame, J. R. Rice, and R. Dmowska, Effects of prestress state and rupture velocity on dynamic fault branching, *J. Geophys. Resear.* **108** (B5), ESE13–1–ESE13–21 (2003).
6. Y. Kase and K. Kuge, Rupture propagation beyond fault discontinuities; significance of fault strike and location, *Geophys. J. Int.* **147**, 330–342 (2001).
7. L. Dalguer, K. Irikura, and H. Wu. Permanent displacement from surface-rupturing earthquakes: Insights from dynamic rupture of mw7.6 1999 chi-chi earthquake. in: *Proceedings of the 16th World Conf. Earthquake Engineering* (pp. paper–1004), Santiago, Chile (2017).
8. L. A. Dalguer, H. Wu, Y. Matsumoto, K. Irikura, T. Takahama, and M. Tonagi, Development of dynamic asperity models to predict surface fault displacement caused by earthquakes, *Pure Appl. Geophys.* **177**, 1983–2006 (2019).
9. L. Bruhat, Y. Klinger, A. Vallage, and E. M. Dunham, Influence of fault roughness on surface displacement: From numerical simulations to coseismic slip distributions, *Geophys. J. Int.* **220**, 1857–1877 (2020).
10. D. D. Oglesby, What can surface-slip distributions tell us about fault connectivity at depth?, *Bull. Seismol. Soc. Am.* **110**, 1025–1036 (2020).
11. Y. Wang and C. Goulet, Validation of fault displacements from dynamic rupture simulations against the observations from the 1992 landers earthquake, *Bull. Seismol. Soc. Am.* **111**, 2574–2594 (2021).
12. R. E. Goodman, R. L. Taylor, and T. L. Brekke, A model for the mechanics of jointed rocks, *J. ASCE.* **94**, 637–659 (1968).

13. M. Sawada, K. Haba, and M. Hori, High performance computing for fault displacement simulation, *J. JSCE.* **73**, I_699–I_710 (2017) (in Japanese).
14. E. Hairer, C. Lubich, and G. Wanner, *Geometric Numerical Integration: Structure-Perserving Algorithms for Ordinary Differential Equations.* Springer, New York (2006).
15. R. Rieben, D. White, and G. Rodrigue, High-order symplectic integration methods for finite element solutions to time dependent maxwell equations, *IEEE Trans. Antennas Propag.* **52**, 2190–2195 (2004).
16. P. W. Vosbeeka and R. M. M. Mattheija, Comtour dynamics with symplectic time integration, *J. Comput. Phys.* **133**, 222–234 (1997).
17. L. Verlet, Computer "experiments" on classical fluids. i. thermodynamical properties of Lennard-Jones molecules, *Phys. Rev.* **159**, 98–103 (1967).
18. R. D. Ruth, A canonical integration technique, *IEEE Trans. Nucl. Sci.* **30**(4), 2669–2671 (1983).
19. M. Sawada, K. Haba, and M. Hori, Estimation of surface fault displacement by high performance computing, *J. Earthq. Tsunami.* **12**, 181003 (2018).
20. FrontISTR. https://www.frontistr.com/ (accessed 7 December 2021).
21. K. Haba, M. Sawada, and M. Hori, On fundamental mechanism and mathematical instability of fault displacement, *J. JSCE.* **75**, I_55–I_67 (2019) (in Japanese).
22. D. J. Andrews, Rupture velocity of plane strain shear cracks, *J. Geophys. Res.* **81**, 5679–5687 (1976).
23. R. J. Geller, Scaling relations for earthquake source parameters and magnitudes, *Bull. Seism. Soc. Am.* **66**, 1501–1523 (1976).
24. D. Ishimura, S. Okada, Y. Niwa, and S. Toda, The surface rupture of the 22 November 2014 Nagano-ken-hokubu earthquake (mw 6.2), along the kamishiro fault, japan, *Active Fault Res.* **43**, 95–108 (2015) (in Japanese).
25. National Research Institute for Earth Science and Disaster Resilience. Japan Seismic Hazard Information Station. http://www.j-shis.bosai.go.jp/en/ (accessed 5 June 2018).
26. M. Sawada, K. Haba, and M. Hori, Evaluation of surface fault displacement in a real earthquake ith surface faulting by high performance computing, *J. JSCE.* **74**, I_627–I_638 (2018) (in Japanese).
27. M. Sawada, K. Haba, and M. Hori, Simulation of the surface fault of the 2014 nagano-ken-hokubu earthquake using high performance computing, in: *Transactions, SMiRT-25.* Charlotte, NC, USA (2019).

28. Y. Aoyagi, Fault displacement distribution of the 2014 nagano-ken-hokubu earthquake based on a differential analysis of multi LiDAR-DEM data, in: *Abstracts, Japan Geoscience Union Meeting*, pp. SSS31–18. (2016) (in Japanese).
29. Geospatial Information Authority of Japan. https://www.jishin.go.jp/main/chousa/14dec_nagano/p.29.htm, (accessed 5 June 2015, in Japanese).
30. M. Sawada, K. Haba, and M. Hori, Predictive simulation for surface fault occurrence using high-performance computing, *GeoHazards.* **3**, 88–105 (2022).
31. S. Fujiwara, H. Yarai, T. Kobayashi, Y. Morishita, T. Nakano, B. MIyahara, H. Nakai, Y. Miura, H. Ueshiba, and Y. Kakiage, Small-displacement linear surface ruptures of the 2016 Kumamoto earthquake sequence detected by ALOS-2 SAR interferometry, *Earth Planets Space.* **68**, 160 (2016)
32. Y. Shirahama, M. Yoshimi, Y. Awata, M. T., T. Azuma, Y. Miyashita, H. Mori, K. Imanishi, and T. Ochi, Characteristics of the surface ruptures associated with the 2016 Kumamoto earthquake sequence, Central Kyushu, Japan, *Earth Planets Space.* **68**, 191 (2016).
33. K. Asano and T. Iwata, Source rupture process of the foreshock and mainshock in the 2016 Kumamoto earthquake sequence estimated from the kinematic waveform inversion of strong motion data, *Earth Planets Space.* **68**, 147 (2016).
34. National Research Institute for Earth Science and Disaster Resilience. Japan Seismic Hazard Information Station. http://www.j-shis.bosai.go.jp/en/ (accessed 7 December 2021).
35. Y. Aoyagi (2018). Personal communication.
36. Strong ground motion prediction method ("recipe"). https://www.jishin.go.jp/main/chousa/17_yosokuchizu/recipe.pdf (accessed 7 December 2021).
37. T. Matsuda, H. Yamazaki, T. Nakata, and T. Imaizumi, The surface faults associated with the Rikuu earthquake of 1896, *Bull. Earthqu. Res. Inst., U. Tokyo.* **55**(3), 795–855 (1981) (in Japanese).
38. W. D. L. and K. J. Coppersmith, New empirical relationships among magnitude, rupture length, rupture width, rupture area, and surface displacement, *Bull. Seism. Soc. Am.* **84**, 974–1002 (1994).
39. K. Haba, A. Hata, M. Sawada, and M. Hori, Estimation of the effect of material uncertainty on surface fault displacement using Latin hypercube sampling, *J. Jpn. Assoc. Earthq. Eng.* **20**, 13–25 (2020).
40. K. Haba, K. Shinohara, M. Sawada, and M. Hori, numerical simulation of surface fault displacement considering material uncertainties,

in: *Proceedings of the 17th World Conf. Earthquake Engineering*, p. C004212. Sendai, Japan (2020).

41. K. Haba, M. Sawada, and M. Hori, Application of high performance computing to probabilistic fault displacement simulation with uncertainties of fault and soil properties, *J. JSCE.* **74**, I_683–I_692 (2018) (in Japanese).
42. K. Haba, M. Sawada, and M. Hori, Surface earthquake fault simulation using high performance computing for consideration on uncertainty, in: *Transactions, SMiRT-25.* Charlotte, NC, USA (2019).

https://doi.org/10.1142/9781800614635_0007

Chapter 7

Traffic Flow Simulator and Travel Demand Simulators for Assessing Congestion on Roads After a Major Earthquake

Takamasa Iryo[*,§], Junji Urata[†,¶], and Riki Kawase[‡,||]

[*]*Graduate School of Information Sciences, Tohoku University, Aoba-ku, Sendai, Japan*

[†]*Department of Civil Engineering, The University of Tokyo, Bunkyo-ku, Tokyo, Japan*

[‡]*Department of Civil and Environmental Engineering, Tokyo Institute of Technology Meguro-ku, Tokyo, Japan*

[§]*iryo@tohoku.ac.jp*

[¶]*urata@bin.t.u-tokyo.ac.jp*

[||]*kawase.r.ac@m.titech.ac.jp*

Assessing the congestion of a road network in a recovery process after a disaster is essential to maintaining the disaster resilience of a city and country. We developed traffic flow and travel demand simulators for this purpose. The traffic flow simulator, called FastDUE, was implemented on high-performance computers (HPCs), which are parallelized for up to about 10,000 processes. The demand simulator called DeSuTA consists of three simulators for different types of travel after an earthquake. Combining FastDUE and DeSuTA, congestions of roads after an earthquake are assessed considering a road network and demand patterns affected by an earthquake. A case study of the Nankai Trough earthquake is presented.

1. Introduction

A road network is likely to be severely congested after a major earthquake. Damages to infrastructure and buildings significantly reduce the capacity of a road network. In addition to the reduction of capacity, changes in demand patterns owing to the damages to the infrastructure can largely affect congestion. For example, damages to a water supply system force residents to seek a site where water supply is provided, causing excess traffic demand that does not exist in a normal situation. Urban road networks are designed to meet traffic demand during normal times. Therefore, both the reduction of road capacity and the change of the demand pattern after a major earthquake can cause non-recurrent heavy congestion that delays a lot of trips, including essential ones such as the rescue of victims, and consequently interferes with disaster recovery.

Assessing the congestion of a road network in a recovery process after a disaster is essential to maintaining the disaster resilience of a city and country. Simulating a traffic flow with a computer is a popular approach to assessing road congestion. However, existing ones basically intend to simulate a normal situation with a single traffic demand pattern that is externally given. As aforementioned, the change in demand pattern also affects road congestion, and consequently, it has to be simulated considering the damages caused by an earthquake. Therefore, we need to simulate both the traffic flow in a road network with a demand pattern that would occur after a major earthquake to properly estimate how the road network is congested and to know to what extent we have to be prepared to alleviate the congestion.

Simulating a traffic flow and demand in a large city incurs an extensive amount of calculation tasks that cannot be accomplished by a normal computer. Using a high-performance computer (HPC) can resolve this issue. We have developed a traffic flow simulator called Fast Dynamic User-Equilibrium (FastDUE) calculator, which is suitable for HPCs.

Parallel implementations of traffic simulators have been investigated in several existing studies, but the degree of parallelism of FastDUE is much greater than most of them. Indeed, using about 10,000 processes to simulate traffic flows in a large-scale network consisting of a few hundred thousand links, in which the traffic flow pattern is extensively skewed spatially, is a challenging task. To improve the degree of parallelism, we have developed a unique calculation algorithm consisting of asynchronized updates of traffic flow patterns.

The demand simulator, called Demand Simulator for Dynamic Traffic Assignment (DeSuTA), consists of three components: (1) Activity Scheduler To Reproduce Observed (ASTRO) behavioral data with the trip-chain condition calculates the normal demand, such as commuting and business/leisure trips affected by the deteriorated level-of-service of a road network; (2) travel demand Simulator for Procurement Activities Caused by Earthquakes (SPACE) calculates the demand caused by trips seeking essential goods such as water and food; (3) rule-based simulator for Humanitarian Logistics Activity after a Natural Disaster (HuLAND) simulates humanitarian logistics activities.

We show the results of a case study in the Kansai area consisting of Osaka city and its surrounding cities. Different earthquake scenarios are considered in the case study. Damages to roads and infrastructure calculated by other simulators under these scenarios are used as input for the simulators developed in the present study.

The structure of the remaining part of this chapter is as follows. The motivations of the study are first explained in Section 2, with studies analyzing the impact of major earthquakes in Japan. Section 3 introduces a framework of the developed simulation system. Section 4 explains FastDUE. Section 5 explains DeSuTA. Section 6 shows a case study of the Nankai Trough earthquake. References introduce papers related to the present study.

2. Motivations for the Study

The motivations for the present study are twofold: one is the practical demand for tools to tackle the disruptions of transport systems that often occur after disasters: the other is the technological challenge of developing the technology for handling the massive amount of computing tasks required to build such tools.

Regarding the practical demand, we introduce details of past major traffic disruptions after major earthquakes in Japan, citing a few studies investigating the impacts of these disasters. Studies related to three disasters are explained: the 1995 Hanshin-Awaji earthquake, the 2011 Tohoku earthquake, and the 2016 Kumamoto earthquake. These earthquakes and subsequent traffic disruptions have different characteristics. Observing these cases is necessary for building a tool that will help practitioners prepare for future disasters.

Regarding the scientific and technological needs, past studies implementing large-scale parallelism for traffic simulators are rare and not very suitable for the needs of simulating region-wide networks damaged by an earthquake. Furthermore, having a traffic simulator is insufficient to assess the impacts of damages to roads — we also have to consider changes in demand patterns, which cannot be accurately estimated before an earthquake happens. The discussion on this point appears in Iryo *et al.*[1] as *Obtaining travel demand data, however, is a more challenging task because travel demand data uncertainty poses a more serious problem. ... In addition, we cannot measure traffic demand changes caused by an earthquake unless it happens. Thus, we need to rely on a model that estimates travel demand after a major earthquake. Such a model, often called an activity-based model, can include many parameters that cannot be directly observed.* This point emphasizes the necessity of large-scale parallelism that accelerates calculations. We briefly review studies of traffic and demand simulators that are relevant to those built by the present study.

2.1. *Past major traffic disruptions after major earthquakes in Japan*

2.1.1. *1995 Hanshin-Awaji earthquake case*

The 1995 Hanshin-Awaji earthquake happened on 17th January 1995 in the Hanshin area, including the city of Kobe and Awaji Island. Damages to transport infrastructure by this M7.3 earthquake were extremely severe as its epicenter was close to a built-up area of the city of Kobe, whose population was 1.5 million. Collapsed buildings blocked many road sections. In addition, bridges and elevated expressways fell or were damaged. These damages to the roads and service suspensions of railways in parallel caused significant changes in demand patterns. Iida *et al.*[2] showed that nighttime traffic volume dramatically increased after the earthquake, implying that many vehicles tried to avoid heavy daytime congestion. Nakagawa *et al.*[3] also reported traffic conditions after the earthquake.

Although quantitative data on traffic situations after the earthquake are extremely rare owing to the limited technologies of this age, the experience suggests two critical points for developing a large-scale traffic simulator: (1) the demand pattern will significantly change after a major earthquake; (2) road closures will be caused not only by damage to the road infrastructure itself but also by the collapse of buildings along arterials. The first finding implies that a demand simulator must be developed to assess road congestion after an earthquake. Since forecasting demand patterns accurately is virtually impossible, we should consider many possible demand patterns that can be realized after an earthquake. The second finding implies that almost all the sections of roads in built-up areas can be damaged, and consequently, the number of possible combinations of road closures can be huge. These two characteristics of road networks challenged us to develop a traffic flow and demand simulator that can accomplish simulation tasks of a large-scale road network as fast as possible.

2.1.2. *Disruption of Tokyo network in the 2011 Tohoku earthquake*

Most of the damages resulting from the 2011 Tohoku earthquake were in rural areas, especially seaside areas, of the Tohoku region, where the population density is basically low. Therefore, road congestion in a large city after an earthquake is not a prominent issue compared to other issues, especially those connected with massive damages and fatalities caused by Tsunamis. Nevertheless, two types of road congestions in a city after the earthquake have been scientifically analyzed, namely congestion by evacuating vehicles due to tsunami immediately after the earthquake (see Ref. 4 for details) and disruption of the Tokyo road network on the day of the earthquake. The latter case is relevant to the present study.

The city of Tokyo was around 400 km away from its epicenter. Still, the performance of transport systems was heavily reduced at night on this day owing to maintenance works for checking possible structural damages to rail tracks and roads. All travel demands tried to use the arterial road network, whose capacity is insufficient to deal with them at the same time, so extraordinary severe congestion occurred. Oyama and Hato[5] reported changes in travel speeds in Tokyo after the earthquake in their case study, indicating that gridlocks happened in the evening. This case study suggests that the insufficient capacity of the road network can cause severe distractions to road transport systems. This implies that assessing them before the earthquake happens is crucial for a better disaster mitigation plan.

2.1.3. *2016 Kumamoto earthquakes case*

The disruption on roads was also caused by the 2016 Kumamoto earthquakes, which hit the city of Kumamoto and its neighboring regions on 14th and 16th April 2016, which were of M6.2 and M7.0 magnitude, respectively. While damages to transport infrastructure were not very severe compared to the 1995 Hanshin-Awaji earthquake case, the capacity of the road network was reduced significantly. This includes the closure of the Kyushu expressway, the major inter-city motorway connecting Kumamoto with Fukuoka, the regional capital city, for about two weeks.

A comprehensive report on disruptions in the road network in Kumamoto was published by Kawasaki *et al.*[6] It indicates that congestions in the city of Kumamoto and on roads entering the city were much more severe than usual. Unlike the Tokyo case, no significant gridlock was observed. In addition, congestion during nighttime was not severe. On the other hand, a large amount of disaster-related demand was found after the earthquakes as residents in the damaged area needed to collect basic supplies such as water and food at a shelter, where such relief goods were distributed, and go to public baths to maintain sanitation during water cut-offs. This implies that the demand simulator should be able to simulate such demands, which are not present in a normal situation. As constructing a road network incurs a huge cost, it is designed to accommodate demand patterns in a normal situation. Therefore, it is natural to consider that the disaster-related demand is essential for a proper estimation of road disruption after an earthquake.

Regarding the demand side, Urata *et al.*[7] have investigated changes in populations staying in a certain size of grids over time before and after the earthquakes. They performed a time-series analysis and found that there was an unusual change in population after the earthquakes, which should be correlated with the change in demand pattern due to the earthquakes. This finding also indicates the existence of disaster-related demands.

2.1.4. *Lessons from three past earthquake cases*

Each of the cases presented in the previous sections has occurred in different eras and under different circumstances and has provided different lessons. However, they all imply common requirements for traffic simulators to predict post-disaster traffic disruption, such as the following:

(1) Besides the collapse of civil engineering structures, such as bridges, many sections of roads within a city can be damaged by the collapsing of buildings.
(2) Not only congestion caused by reduced road capacity due to damage to road infrastructure but also congestion caused by unusual demand has to be taken into account.
(3) Congestion can be extremely severe even if reductions in road capacity are not extreme.

These lessons suggest that not only a traffic flow simulator but also a traffic demand simulator must be developed. In addition, it suggests that the calculation speed should be as fast as possible so as to accommodate the huge amount of calculation tasks for many patterns of road damage and traffic demand. The severe congestion that cannot be observed on normal days also suggests that machine-learning approaches using normal traffic conditions are invalid. These are the reasons that motivate the development of an HPC-based highly parallel traffic flow and traffic demand simulators.

2.2. *Relevant existing studies of traffic and demand simulators*

2.2.1. *Traffic simulators*

Not many parallel implementations for traffic simulators exist in past studies. Scalable implementation of up to 19,200 processes has been shown in a recent study,[8] while it only considers ideal grid networks. Actual road networks are more hierarchized and unevenly distributed than a simple grid, which implies that it is not clear whether this implementation is applicable to a large-scale network in the real world. In addition, this study employs the microscopic traffic flow model, which is not highly relevant to the setting of the present study. Other studies use less than 100 processes (e.g., Refs. 9 and 10).

2.2.2. *Demand simulators*

In transport planning, traffic demand simulators are called activity simulators. Activity simulators predict traffic demand based on the idea that all traffic demand is derived from the activities that individuals perform at different locations within a city during the day. Several models for such simulators have been proposed, such as Bowman and Ben-Akiva's model,[11] PCATS,[12] and Albatross.[13] These describe the mechanism of human behavior in probabilistic models and use them to calculate the daily travel patterns of individuals. However, models such as these have many parameters, and it is not easy to estimate them correctly.

There are also activity simulators that do not create individual daily activity patterns *ab initio*, but instead recombine individual activity patterns observed in normal situations. One of them is called Toronto Area Scheduling model for Household Agents (TASHA).[14,15] TASHA creates a daily schedule for each individual using the activity distribution obtained from a large-scale travel survey in normal situations based on travel diary data. TASHA is an activity simulator based on the activity distribution of the survey data and uses a relatively small number of parameters, making it easy to reproduce normal behavior patterns. In addition, there is a process to check the schedule overlap considering the travel time, and trip cancellations can be evaluated due to longer travel times. In this sense, the simulator is suitable for reproducing activities during the recovery period, when the number of activities changes due to the increased travel times.

3. Outline of the Simulation System

The outline of the simulation system is depicted in Fig. 1. The simulation system receives simulated seismic intensity as input data.

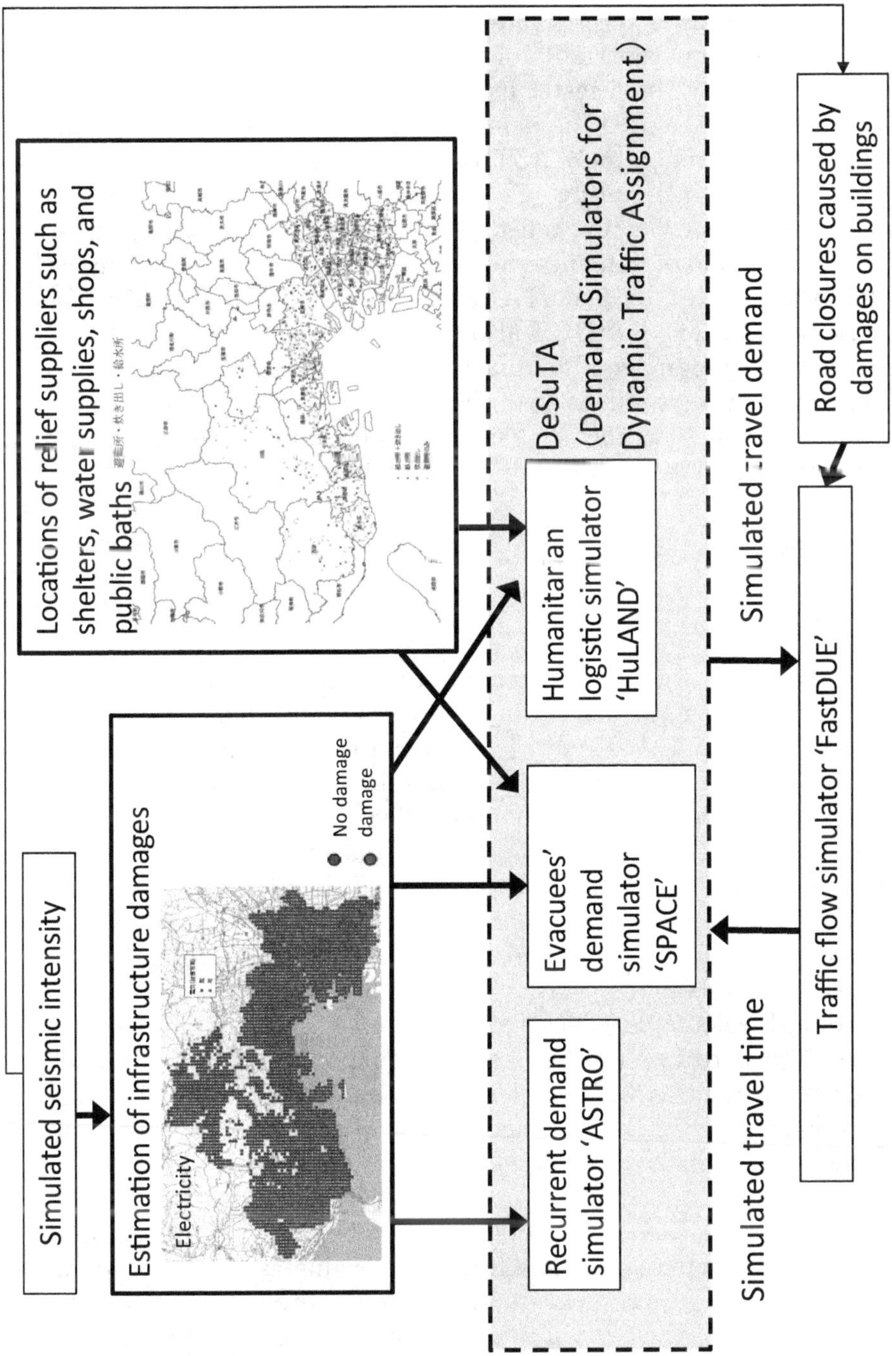

Fig. 1. Outline of the simulation system.

These data are used to estimate damages to the infrastructure for utilities. Based on the result of the estimation, the population of people affected by the disaster using a 500-m grid is estimated. This information is passed to the demand simulator DeSuTA, consisting of three simulators, namely ASTRO for calculating recurrent travel demand (such as commuting), SPACE for calculating evacuees' travel demand (such as collecting relief goods), and HuLAND for calculating traffic demand of humanitarian logistics. Vehicular travel demand is finally calculated by DeSuTA, which is passed to the traffic flow simulator FastDUE. FastDUE also receives information on road closures caused by damaged buildings. Travel times of roads simulated by FastDUE are passed back to DeSuTA because simulations in it are affected by travel time. This iteration should be made a few times to incorporate congestion in the demand simulations.

4. Traffic Flow Simulator — FastDUE

4.1. *Overview*

FastDUE stands for "Fast Dynamic Traffic Assignment", which is a traffic flow simulator based on the macroscopic traffic flow model called the Lighthill-Whitham[16] and Richards[17] (LWR) model. The LWR model regards a traffic flow as a fluid and calculates the mean travel speed of vehicles as a function of vehicles' density on a road section. The LWR model does not have to calculate vehicles' motions within a link at every time step to evaluate congestion in a road network. This is the most significant advantage of the LWR model over microscopic car-following models, which have to calculate individual vehicles' positions and speeds at every time step.

FastDUE calculates vehicles' motions in a network, which consists of links and nodes. A link represents a single road section. A node represents a point at which two or more links are connected. Using the LWR model, FastDUE calculates the travel time of each link and determines the times at which each vehicle passes through nodes of the network.

Typical traffic flow simulators need to have vehicles' origins, destinations, and departure times from the origins as the input data. On the other hand, they calculate vehicles' routes in a network by themselves based on a certain rule. In FastDUE, vehicles are assumed to use routes whose travel costs are smaller than any other routes.

The travel cost is the sum of travel costs incurred by route travel time and tolls imposed on vehicles using the route. The value of time is used to convert travel time to travel cost. The state in which no vehicle can reduce its travel cost by changing its route unilaterally is called equilibrium. Since the route choice of a vehicle can affect the travel costs of other vehicles, finding an equilibrium state is not a trivial task (see, e.g., Iryo[18] for theoretical backgrounds). FastDUE tries to find states that are close to equilibrium, referred to as a near-equilibrium state, by iteratively updating vehicles' route choices. At each iteration, a certain percentage of vehicles running in a network update their routes based on travel times calculated by the last iteration. Then, the travel times of all vehicles, including those which do not change routes, are calculated by the LWR model. This iteration is repeated until the congestion of the network becomes stationary. This process can be considered as a Markov-chain model, and consequently, the existence and uniqueness of the stationary distribution of traffic states are guaranteed.

For the highly parallelized computation of the LWR model, an algorithm called asynchronized update was developed. This algorithm calculates traffic flows in each sub-network independently, then updates them considering their consistency on the borders between sub-networks. Parallelization of route choices can be easily achieved by having each CPU share the computation of different vehicles; FastDUE uses this approach.

4.2. *Models and calculation algorithms*

4.2.1. *Traffic flow model*

The LWR model consists of two equations: flow conservation law and the fundamental diagram (FD). The flow conservation law guarantees that vehicles do not suddenly appear or disappear in the middle of a road section. The FD describes the relationship between the traffic density (the number of vehicles on a road section per unit distance) and traffic volume (the number of vehicles passing through a certain point of a road section per unit time). FastDUE assumes the so-called triangular FD as shown in Fig. 2. The triangular FD implies that vehicles' mean speed is constant in the free-flow region, where the traffic density is not greater than a threshold called the critical density. In the congested region, where the traffic density is

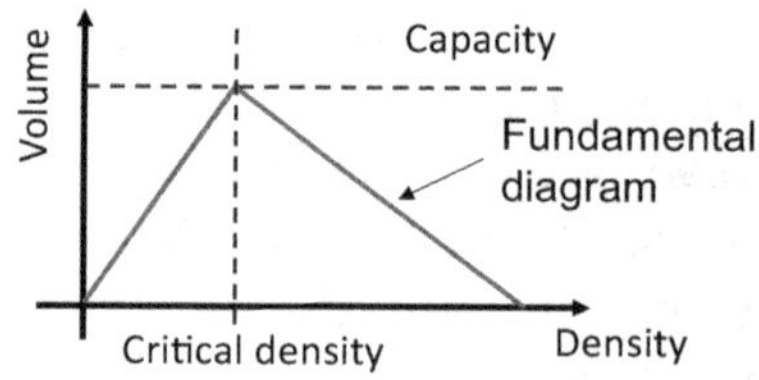

Fig. 2. Fundamental diagram.

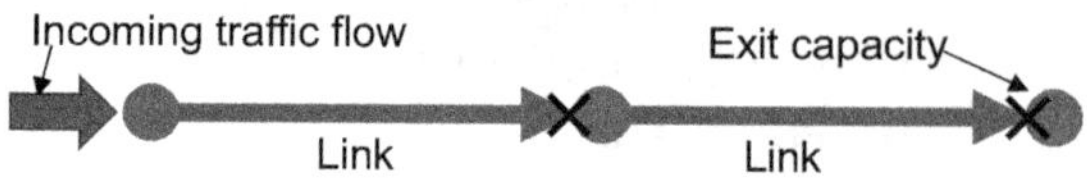

Fig. 3. Links and exit capacity.

greater than the critical density, mean speed decreases with respect to the traffic density. When the traffic density reaches the maximum density (called jam density), vehicles completely stop. The maximum traffic volume, which is the traffic volume at the critical density, is called capacity.

In the LWR model, the typical dynamics of a traffic flow on the road are described as follows. Consider a link representing the road section, as shown in Fig. 3. Vehicles continuously enter this link as a flow. Assume that the capacity of the exit (exit capacity) is not greater than that specified by the FD of this link. When traffic volume is not greater than the exit capacity, vehicles simply run at the free-flow speed without congestion. On the other hand, when traffic volume exceeds the capacity, the traffic volume of an outgoing flow is restricted to the capacity constraint, and vehicles form a queue in front of the exit of the link. The upstream end of the queue is called the shockwave in the LWR model. If the shockwave is inside the link, we can simply calculate vehicles' departure time from the link with respect to their incoming time by calculating the waiting time of vehicles at the end of the road section. In addition, the speed of the shockwave can be calculated by the FD of this road section.

When the shockwave reaches the upstream end of the link, vehicles cannot enter it freely and consequently have to wait in an upstream link. Such a phenomenon is called a queue spillback. When a queue spillback occurs, the exit capacity of the upstream link has to be adjusted so that excess vehicles do not enter the queue.

When two or more upstream links are connected to the upstream end of a link, traffic flows from these links merge when entering the link. In this case, the capacity of the link is split to these upstream links. On the other hand, when two or more downstream links are connected to the link, exit capacity toward each downstream link is independently calculated to reflect its queue spillback. Vehicles' exit times from the link are calculated so as to maintain the first-in-first-out (FIFO) principle, in which no vehicles are allowed to overtake another vehicle.

The LWR model is used to calculate the travel time of each link. FastDUE uses this information to determine the trajectories of individual vehicles running from their origins to destinations. Other indices, such as mean link travel time, are also derived using vehicles' trajectories.

4.2.2. *Route choice model*

Assuming that the travel cost of a route is the sum of the travel costs of links included by the route, finding a route whose travel cost is the smallest is easy by a so-called shortest path search algorithm such as the Dijkstra algorithm. FastDUE also employs an algorithm based on the Dijkstra algorithm with a few extensions.

The shortest path search algorithm in FastDUE finds a route considering changes in travel times of congested links. Since the travel time of a link depends on the time-dependent profile of the traffic flow entering it, the travel time of each link depends on when a vehicle enters it. FastDUE adequately considers this point to find a route on which a vehicle's actual travel cost is minimized.

The travel cost includes not only the link travel time but also tolls. In the real world, tolls are not a simple summation of the toll of each link but depend on combinations of entry and exit points of toll roads, which cannot be calculated as a sum of tolls of links included in a route. As calculating the exact toll of each route requires huge calculation costs owing to the vast number of routes, FastDUE calculates the toll as a sum of fixed tolls, which are applied when a vehicle enters a toll road, and distance-based tolls, which are proportional to the distance of a toll road included by a route.

Assuming that drivers can precisely find a route with the cheapest travel cost is not realistic. In addition, drivers may incorporate

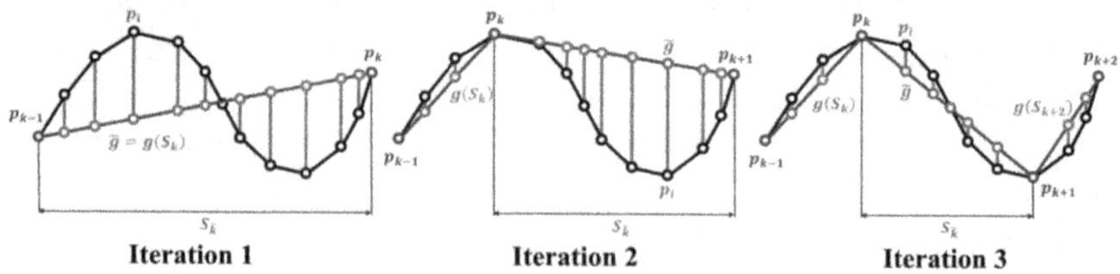

Fig. 4. Compression of travel time profile data by a polyline.

factors other than travel time and tolls to choose a route. To incorporate these effects, a random value is added to a link travel time so that a non-shortest route can be chosen.

Vehicles' route choices are updated based on time-dependent link travel times that are realized on the previous iteration of the simulation. As the time-dependent link travel time is calculated based on the travel times of all vehicles, the data size of the travel time profile can be huge and can cause an out-of-memory error in a large network. To avoid this problem, we have implemented a travel time compression algorithm that extracts an approximated shape of the time-dependent travel time profile.[19] It approximates the travel time profile, described as a function with respect to time, by a polyline with a limited number of vertices. Figure 4 depicts a schematic view of the algorithm, in which vertices are added until sufficient precision is achieved.

We have also developed a more efficient shortest-path search algorithm using the landmark shortest-path search algorithm. The details of the methodology can be found in Peque *et al.*[20]

4.3. *Asynchronized update and its parallelism*

4.3.1. *Asynchronized update*

An algorithm called asynchronized update is developed to calculate the traffic flows of links in FastDUE. The term *asynchronized* implies that the state of a traffic flow in each link is calculated independently over the entire study duration (e.g., between early morning and late night) in the algorithm. Because there is strong interdependence of traffic flows between subsequent links, performing such a calculation just once does not provide a correct solution. After updating the traffic flow state in each link, inconsistency between links is compensated, and link traffic flows are calculated again. This algorithm consists of three subcomponents: internal update, forward relay, and backward relay.

The subcomponent called the internal update calculates the traffic flow of each link with vehicles' entry times and the exit capacity given. The internal update outputs vehicles' outgoing times and the shockwave profiles at the upstream end of the link. The latter describes how many vehicles can enter into the link at every time step when the queue spillback occurs, which should be passed to the upstream links to let vehicles affected by the queue spillback wait on upstream links.

The subcomponent called the forward relay tells vehicles' outgoing times to the downstream links. These outgoing times will become the incoming times of vehicles in the downstream links. On the other hand, the subcomponent called the backward relay provides information on the shockwave at the upstream end to the upstream links. More precisely, the backward relay provides the time-dependent profile of capacity allowed for vehicles entering the link. This information is used to update the exit capacity of the upstream link.

Repeating these processes many times, it is expected that the inconsistency between links will disappear, at which states of traffic flows of all links are correctly calculated. If no queue spillback occurs (or we ignore them), only the iteration of the internal update and forward relay is sufficient to obtain the solution. The convergence of this process with a finite number of iterations is guaranteed at all times.

The convergence of the backward relay is, however, not guaranteed, and small errors in the calculation may not converge but oscillate. To avoid oscillation, the exit capacities of links are not updated at once. Instead, exponential smoothing is adopted, i.e., a new exit capacity is calculated as an average of the current exit capacity and the exit capacity provided by the backward relay.

The iterations of the aforementioned update and relays are combined with updates of route choices. The entire iteration process, including all subcomponents, will be explained in Section 4.3.2.

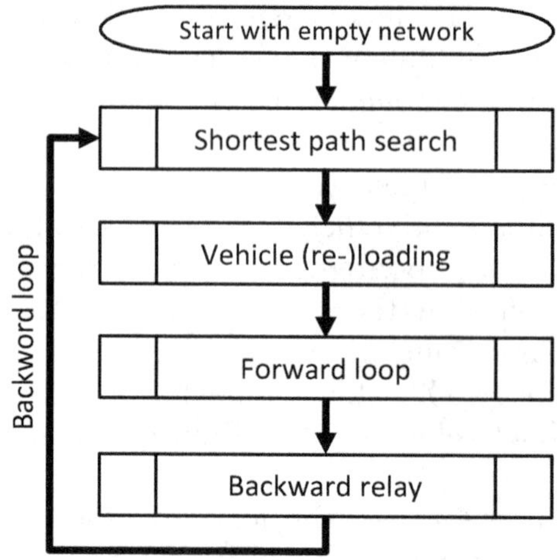

Fig. 5. Outline of forward and outer loops.

4.3.2. *Forward loop and outer loop*

FastDUE iteratively updates vehicles' route choices and the traffic flow of each link simultaneously to find a near-equilibrium state. The iteration process consists of two loops: the forward loop and the backward loop. Figure 5 depicts the outline of these loops.

The outer loop includes four subcomponents: the shortest path search, vehicle reloading, forward loop, and backward relay. In the shortest-path-search component, a certain percentage of vehicles are randomly selected, and then their shortest paths (with random values, as explained in Section 4.2.2) are calculated. In the vehicle-reloading component, vehicles changing their routes are first extracted from the network and then reloaded onto the links in chosen routes. In the forward loop, the internal update and forward relay are repeated until no inconsistency is found, i.e., no vehicle's outgoing time of a link is different from its incoming times of the downstream link. Because the convergence of this iteration process is guaranteed, the forward loop is repeated until it converges. Finally, the backward relay subcomponent updates the exit capacities of links using the result of the internal updates in the forward loop.

The main point of the aforementioned configuration is that the backward relay is not repeated until it converges. Ideally speaking, a combination of the forward loop iteration till its convergence and a single backward relay must be repeated until the backward relay finds no inconsistency (i.e., it updates no exit capacity). However, unlike the forward loop, the convergence of this loop is not guaranteed. Moreover, it increases the calculation time for updating traffic flows. To remedy these problems, FastDUE performs the backward relay just once in the outer loop. Of course, it lets drivers update their routes using a traffic flow state with the inconsistency of exit capacities. Nevertheless, as the limitation of the capacity of the road network is the more dominant reason for congestion compared to the queue spillback, the route choices in such situations would reflect drivers' actual behavior without huge errors.

The vehicles may be loaded to the links gradually during the outer loop iteration. In this case, the network is empty when the outer loop starts. Then, a certain percentage of vehicles (say, 5%) are chosen, and their route choices are calculated. Then, the remaining processes of the outer loop are performed to newly load them into the network. After repeating the iterations of the outer loop several times (e.g., 20 times when 5% vehicles are loaded at each iteration), all vehicles are loaded. Then, routes of already loaded vehicles are updated at every iteration of the outer loop. The gradual loading of vehicles using multiple iterations lets many vehicles avoid congested routes, which is beneficial for faster convergence to a near-equilibrium state.

4.3.3. *Parallelization*

The network is divided into many sub-networks to perform internal updates in parallel with many processes. The calculation of the internal updates of each sub-network is made by a single process. For better load balancing, the network is divided into sub-networks so that the observed calculation times in past iterations are as equal as possible in each sub-network. The partitioning is calculated by METIS.[21] Figure 6 depicts a schematic view of the partitioning.

In the shortest-path searches, vehicles updating their routes are assigned to processes so that the number of assigned vehicles to each process is as equal as possible.

Communications between processes are necessary when the forward and backward updates are made, as well as the shortest-path

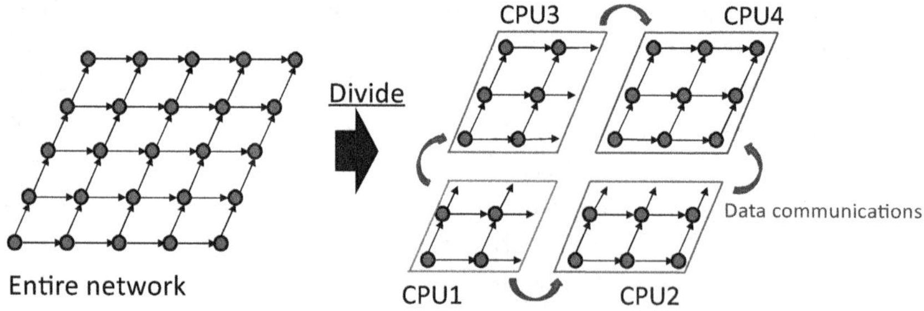

Fig. 6. Outline of forward and outer loops.

searches are performed. For the shortest path searches, information on link travel times in the entire network must be shared by all processes, which is the most important factor for memory consumption. The travel-time compression process mentioned in Section 4.2.2 is vital for this reason.

To accelerate calculations in each sub-network, the calculation order is adjusted so that links on the upstream side are first calculated before calculations of links on the downstream side.[22] While this process does not improve the scalability of the parallel implementation, it is beneficial to reduce the calculation time of the internal updates within each sub-network.

4.4. *Test cases*

4.4.1. *Simple network*

A test case by a simple network is introduced to show how the proposed asynchronized update algorithm works. The configuration of the network is shown in Fig. 7. There exist one origin and two destinations. Because the capacity of Link 3 is substantially smaller than the incoming traffic volume, a queue is built up in front of this link and emerges toward the origin. Note that no route choice is made in this example.

Figures 8–10 show cumulative traffic volumes of Links 1 and 2 during the calculation. At the first iteration, the outgoing flow from Link 2 is restricted owing to the capacity constraint of Link 3. On the other hand, as no queue spillback is incorporated in the first iteration, the incoming flow into Link 2 is not restricted. In the later iterations, the capacity restriction by the queue spillback is passed to Links 1

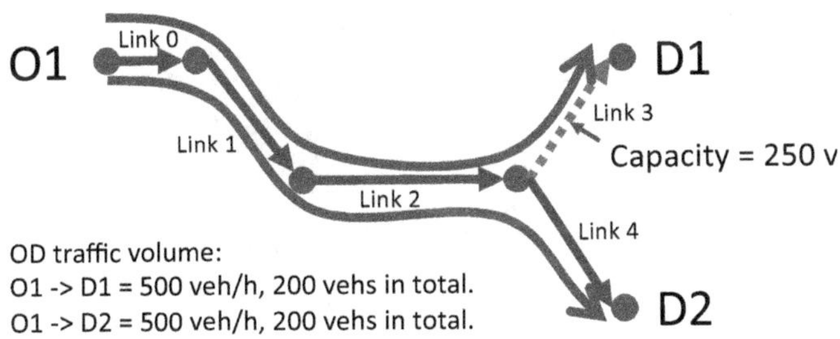

Fig. 7. Simple network.

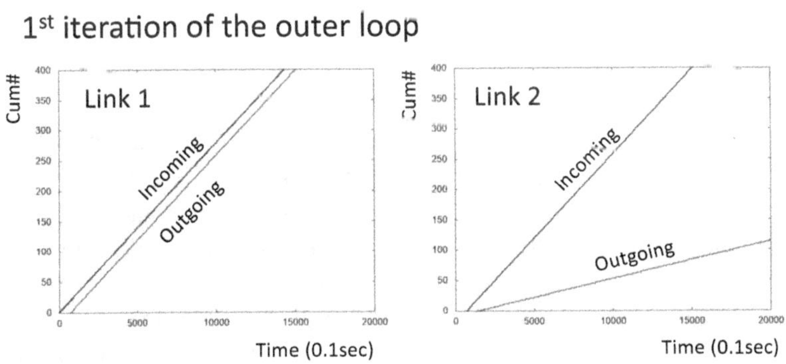

Fig. 8. Cumulative traffic curves at 1st iteration.

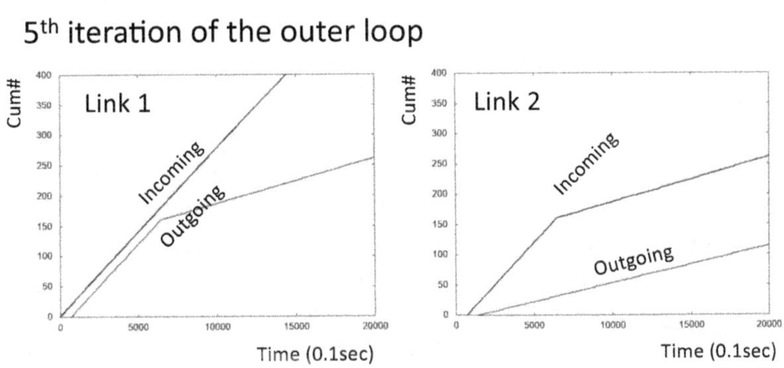

Fig. 9. Cumulative traffic curves at 5th iteration.

and 0. At the 30th iteration, we can observe that the incoming flows into Links 1 and 2 are restricted by the queue spillback, implying that the upstream end of the queue reaches the upstream end of Link 1.

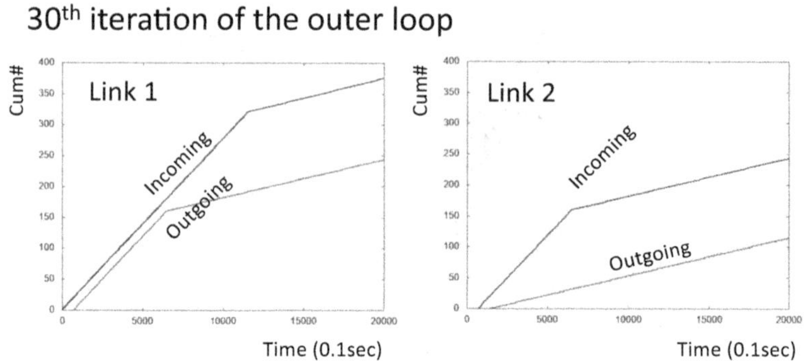

Fig. 10. Cumulative traffic curves at 30th iteration.

4.4.2. *Large-scale network*

The network of the Kansai area in the western part of Japan was used for a large-scale network test case. The number of links was 352,664, of which 49,602 were dummy links, which do not represent actual roads but connect roads to centroids, at which vehicles start/terminate their trips. We also set 23,993 links as intersection links, which were excluded from the forward updates to avoid gridlocks in intersections. As the number of vehicles is extremely large, we combined 10 vehicles (five vehicles for trucks) into one packet. The number of packets was approximately 780,000. The length of the simulation time was 16 hours. A total of 117 iterations were made for 15 hours in the calculation. The number of processes was 576. The network is shown in Fig. 11. Figure 12 depicts traffic volume in the network.

The convergence of the backward relay was also checked in this test case. As explained in Section 4.3.1, the proposed method does not intend to perform the perfect convergence of the backward relay. Therefore, there exist links on which vehicles are overflown owing to incorrect settings of exit capacities on upstream links. Table 1 shows the maximum number of overflown packets in each link. It indicates that the overflow is not significant — it is not greater than 10 in most links. This result implies that the proposed method can handle queue spillbacks with good accuracy.

4.5. *Scalability*

The strong scalability of the proposed method was tested by K computer in Riken, which was in operation till 2019. A similar

Fig. 11. Kansai area network. *Left*: Whole area. *Right*: Enlarged view of Osaka city.

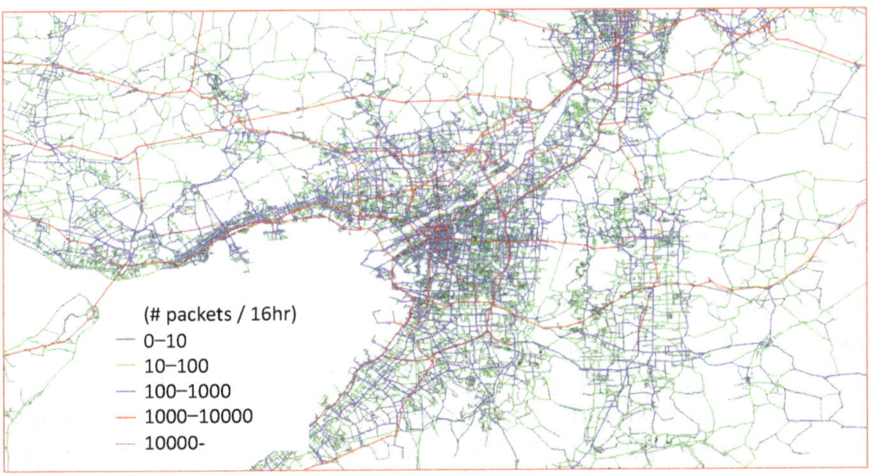

Fig. 12. Traffic flow volume in Osaka area.

configuration of the large-scale network shown in Section 4.4.2. was used. Figure 13 shows the speed of calculation with respect to the number of processes. It indicates that the speed increases up to 18,000 processes, which is a considerably large number compared to existing parallel implementations.

Table 1. Number of overflown packets from each link.

Max. number of overflown packets from a link	Number of links
Less than 1	278,780
1–10	787
10–50	34
50–100	4
Over 100 (max = 127.55)	4

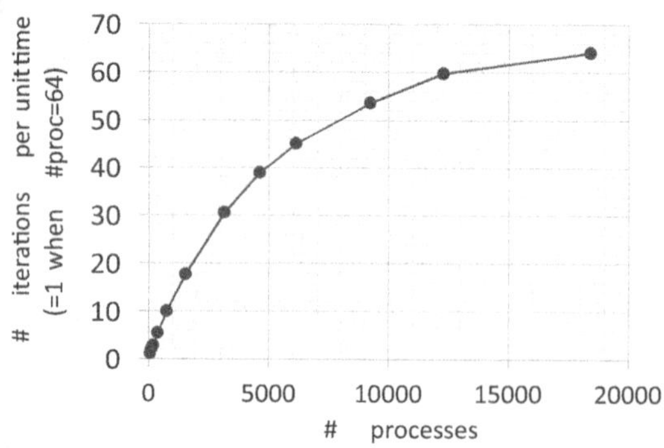

Fig. 13. Calculation speed with respect to number of processes.

5. Travel Demand Simulators — DeSuTA

This section describes DeSuTA, a dynamic travel demand simulator for calculating traffic demand. DeSuTA consists of three simulators, SPACE, ASTRO, and HuLAND. SPACE simulates the disaster-affected travel demand specific to the recovery period, especially travels for procurement of relief goods. ASTRO simulates changes in non-disaster-related activities affected by deteriorations of the road transport system. HuLAND calculates travel demand by trucks transporting relief goods.

5.1. *SPACE*

SPACE is a simulator that describes the dynamics of procurement and consumption of goods during a disaster using a sequential-type decision frame and calculates the transport demand associated with the procurement of goods. It calculates the disaster demand traffic by repeating six processes, namely

(i) supply to the facility,
(ii) departure from the shelter,
(iii) movement,
(iv) receipt of supplies,
(v) return home, and
(vi) consumption of supplies.

For each time step during the study time period, four relief facilities are considered: water stations, soup kitchens, supermarkets, and public bathhouses. Supplies (water and food) are replenished at regular intervals at the three relief facilities, excluding the public baths. Facilities that reach a supply holding capacity of 0 cease to be in service. Three types of supplies are obtained at these relief facilities: water (water stations), food (soup kitchens and supermarkets), and cleanliness (public bathhouses). Evacuees (staying either in a shelter or their own houses) will keep these on hand and consume them over time. They take into account the amount of supplies remaining on hand and the accessibility of the relief facility to make a decision on whether to start traveling toward the facility and, if so, where to go. Mode choices of trips are probabilistically calculated by a multinomial logit model considering the travel time and the proportion of choices by distance zone and time zone obtained from travel survey data. Evacuees arriving at a facility join a queue to receive relief in the order of arrival. The overall process of the simulator is depicted in Fig. 14.

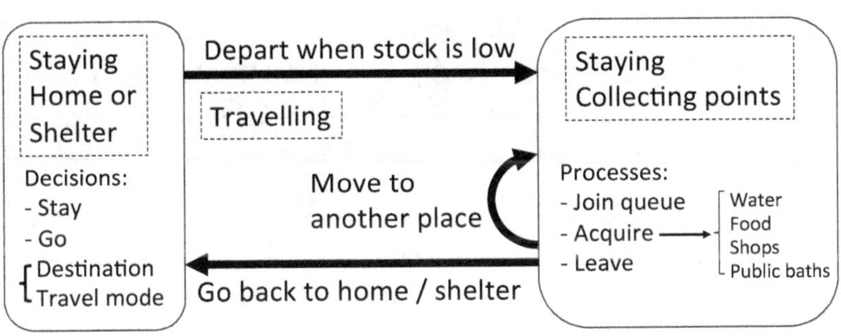

Fig. 14. Overall process of SPACE.

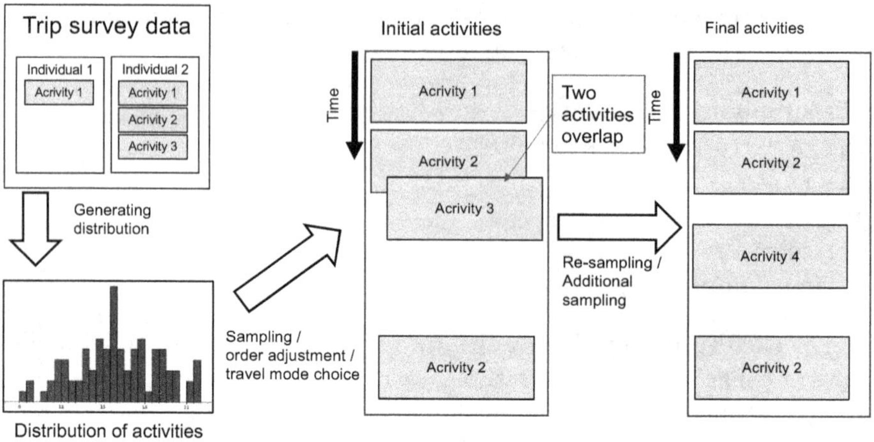

Fig. 15. Overall process of ASTRO.

The three facilities — water stations, soup kitchens, and supermarkets — will experience extended queues and longer stays at the facilities if the service rate is not sufficient. If the supplies that can be provided are depleted, the facility will close, and the evacuees will not be served, forcing them to make a re-destination choice. As public bathhouses do not provide supplies, congestion is described by setting the number of people who can be served. Evacuees who have obtained sufficient supplies return to their shelter or home. Those who have not obtained satisfactory goods or services will select a facility or destination again until they obtain them.

5.2. *ASTRO*

ASTRO consists of six processes: creating activity distributions, generating activities, adjusting activity order, destination and transport mode choices, resampling, and additional sampling. Details of these processes will be explained. A schematic view of the process is shown in Fig. 15.

ASTRO first reads travel survey data such as person-trip survey data (PT data) as input data and creates a distribution of basic personal attributes such as residential and working places and distributions on activities. Based on the created distributions, the attributes and activities of each individual are generated probabilistically. Five activity purpose categories are used: work (activities outside the place of work), work/school (activities at the place of work/school), personal affairs, intermediate return to home (return to home but go outside again), and final return to home. Then, the order of activities is adjusted, as the activities drawn from the distribution may be in an unnatural order (e.g., other activities exist after the final return home).

Destination and transport mode choices are calculated for each activity. Then, travel time is calculated based on the selected origin, destination, and transport mode. Destination choices are calculated only for private and business purposes. Transport mode choices are calculated by the same process in SPACE. After the transport mode choices have been made, they are adjusted to take the trip chains into account. Inside each trip chain, either one of the following transport mode groups is included: public transport and walk, car, or bicycle. One of these groups is determined considering the highest number of selections in the chain and the average trip distance or average trip time in the chain. For a single chain, the above method can be used to simply select a transport mode taking the trip chain into account, but if several chains are involved, the trip chain needs to be taken into account based on further rules. If two chains are inclusive, the transport group of the inner chain should be determined alone and the remaining transport group independently.

Duration of activities or moves between two activities determined by the aforementioned methods may overlap each other. This overlap must be eliminated by re-sampling the activity locations or activity start and end times. Re-sampling of the activity location is carried out for private and business activities, where the activity location is determined by the destination choice model (i.e., not fixed as a working place).

Finally, to avoid underestimation of the number of activities, additional sampling is performed to add activities based on the free time between activities. This is the final process for generating an individual activity in a day by ASTRO.

5.3. *HuLAND*

HuLAND is a simulator for calculating the traffic demand caused by relief goods logistics. It consists of three frameworks: calculation of the amount of relief goods required, calculation of relief goods flow rates, and conversion to heavy vehicle traffic.

For the required volume of relief goods, each relief facility designates the shelters responsible for supporting and calculates the demand for goods based on the total number of people staying in the relevant shelters. Specifically, the relief facility closest to the evacuation center will cover the material needs of the evacuees staying there. Using the criteria for calculating the required volume published by local authorities in their disaster plans (e.g., 3 liter of drinking water per person per day is required), the required volume of relief supplies is calculated from the number of evacuees responsible for material support. In calculating the relief material flow rate, transport is carried out in stages based on Japan's relief material transport system. The primary or secondary collection points closest to the relief facilities will transport the required relief material quantities to the relief facilities, while the nearest primary collection point will transport the materials to the secondary collection points. The locations of the collection points are given by designated collection points in the local authorities' disaster plan or logistics centers of private logistics providers. The transport time between each facility is the shortest route travel time between the zones containing the facilities. In addition, the time taken for unloading and loading at each facility is taken into account to calculate the relief goods flow.

In calculating heavy vehicle traffic volumes, it is assumed that relief materials are loaded up to the maximum truck capacity, and the hourly relief material flows are converted into heavy vehicle traffic volumes. It is assumed that the type of trucks used will be standard ones, according to the disaster plan. Trips back to the originating collection point are also calculated, taking into account unloading times at each facility and travel times between facilities.

6. Case Study of Nankai Trough Earthquake

The results of the case study in the Kansai region are presented. This case study is based on the final report for the Nankai Trough Large Earthquake Countermeasures.[23] Six scenarios are used for the case study: (a) no earthquake, (b) the base case, (c) the east case, (d) the landside case, (e) the landside case with the earthquake ground motion amplified by 1.5 times, and (f) the landside case with the surface ground shear wave velocity increased by 1.3 times.

The traffic demand simulator DeSuTA runs simulations for cities in the Kansai network (Osaka, Kobe, Nishinomiya, Ashiya, and Amagasaki) for the period from 4:00 to 20:00, including the morning and evening peaks.

The damage information, such as evacuation centers used in SPACE and HuLAND, is based on open data published by the public administration. To set the traffic demand data in the traffic flow simulation, centroids of a 500-m grid where traffic demand is generated are added to the Kansai region network as dummy nodes. The demand simulator calculates 31,217 zones in the five cities.

Figures 16 and 17 show the aggregate generated traffic demand in each grid for each case (five disaster cases + one normal case). The basic trend in all the affected cases is that, compared to the normal case, the zones with higher demand have decreased, and new traffic demand has arisen from zones with no trips in the normal case. This implies that the demand for daily activities such as commuting has decreased due to the disaster and that traffic demand specific to the recovery period, such as procurement of goods, has arisen. On the other hand, there are differences in the spatial distribution of the amount of traffic generated in each disaster case, which confirms that the results reflect the spatial distribution of the damage.

The traffic flow simulation is carried out by FastDUE using the traffic demand data calculated by DeSuTA, the road blockage data, and the road network data created based on digital road maps as inputs. FastDUE uses the network shown in Fig. 11. The traffic demand pattern of a normal situation is used for travel demands not covered by DeSuTA. The road closure data are given as the percentage of each link's width that is blocked, and links that are blocked by more than half are considered impassable. In the simulation, we let a minimum number of blocked roads be repaired so that the connectivity between any combination of two arbitrary centroids is always guaranteed.

The results of the congestion calculations are shown in Figs. 18 and 19. Note that no calibration is performed, so the output results have no particular practical significance that is not related to the simulation development itself. The blue color indicates the blockage link, and the red color is the time-averaged velocity relative to the free-flow velocity. No noticeable congestion is observed in the normal case. On the other hand, the five affected cases are generally heavily

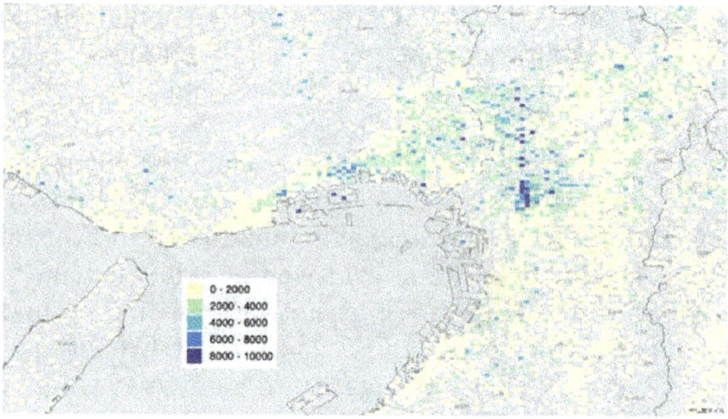

(a)

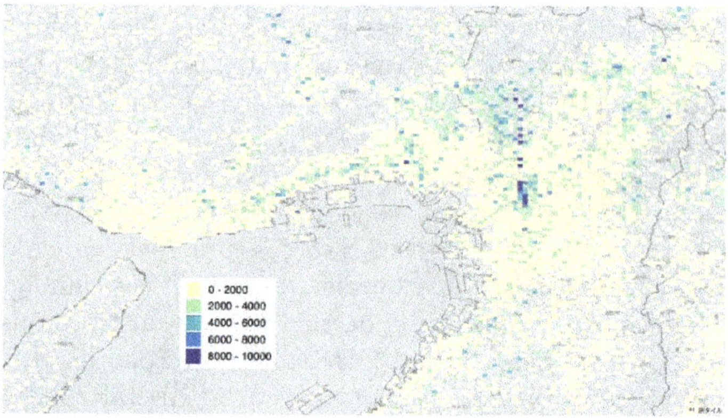

(b)

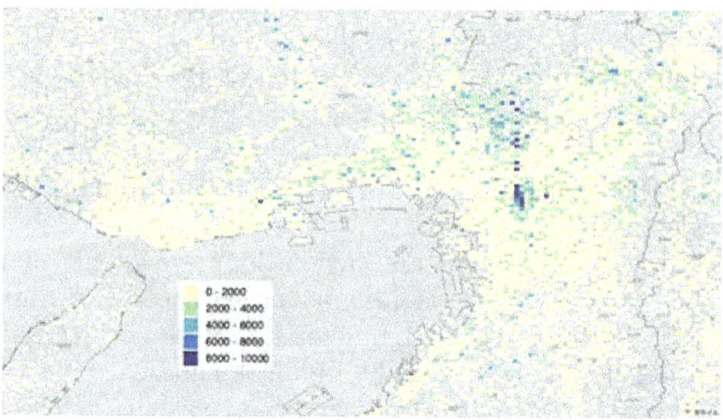

(c)

Fig. 16. Traffic demand in each grid: Cases (a)–(c). (a) No earthquake, (b) base case, and (c) east case.

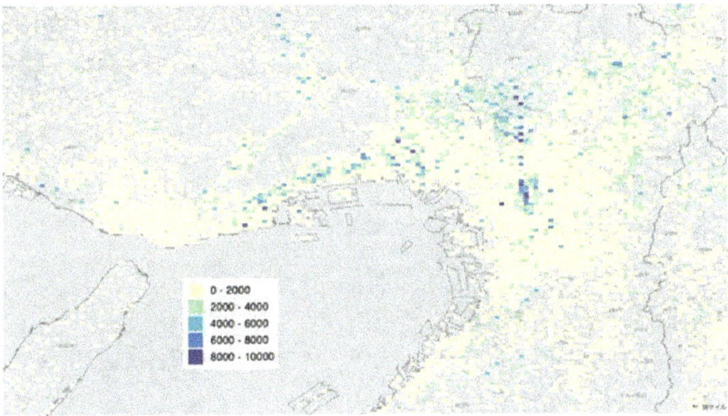

(d)

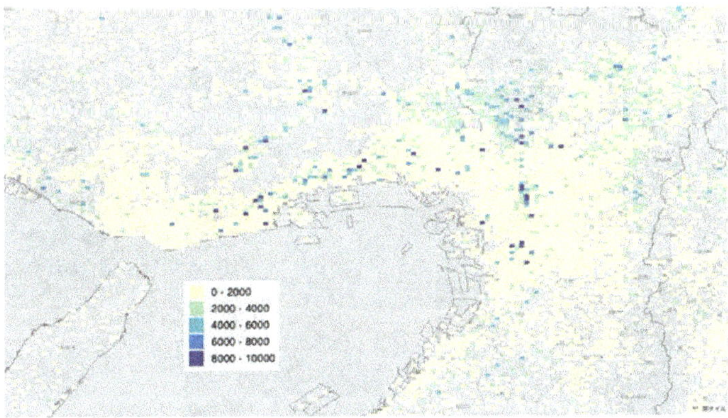

(e)

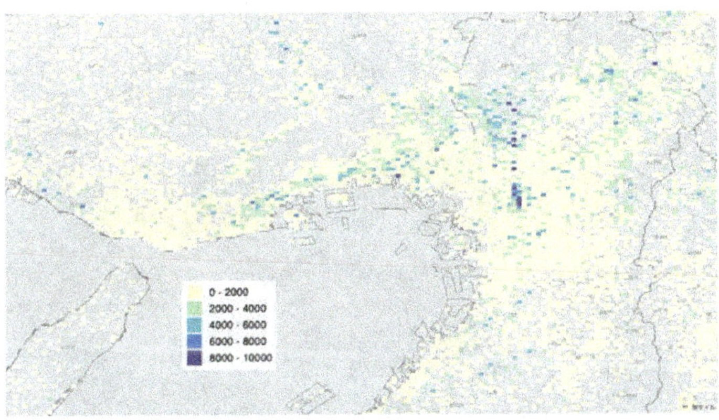

(f)

Fig. 17. Traffic demand in each grid: Cases (d)–(f). (d) Landside case, (e) landside motion ×1.5 case, and (f) landside velocity ×1.3 case.

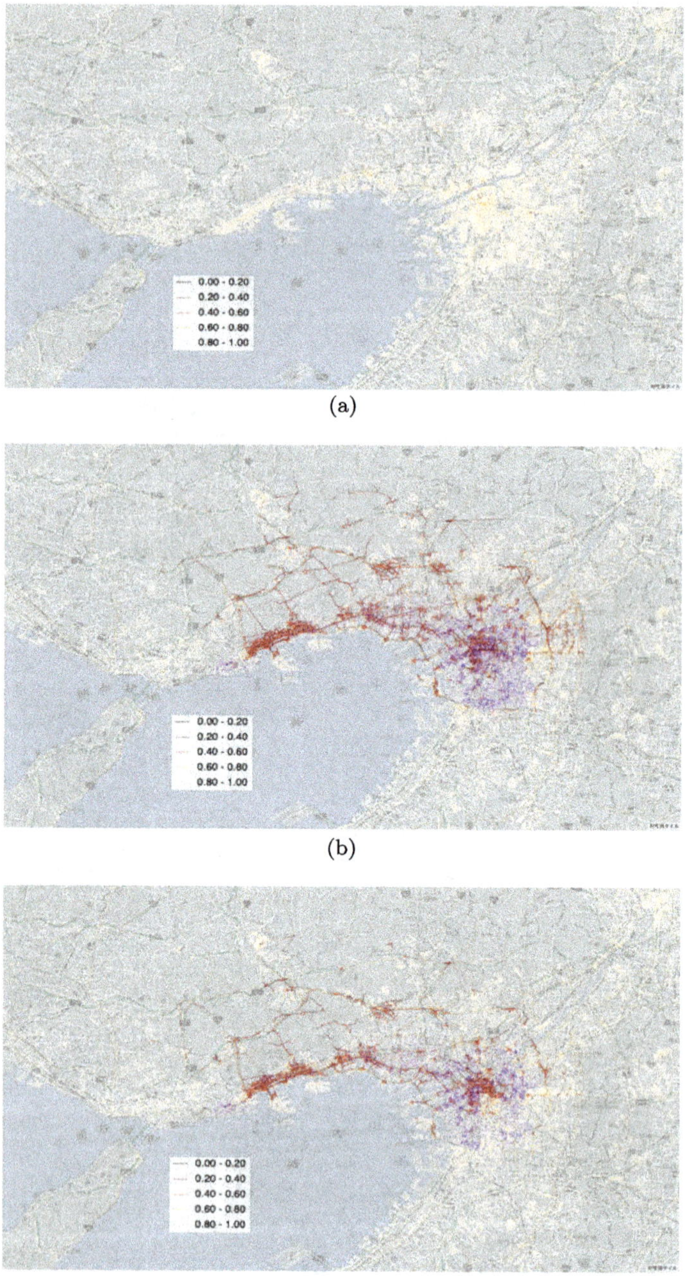

Fig. 18. Road closures and congested roads: Cases (a)–(c). (a) No earthquake, (b) base case, and (c) east case.

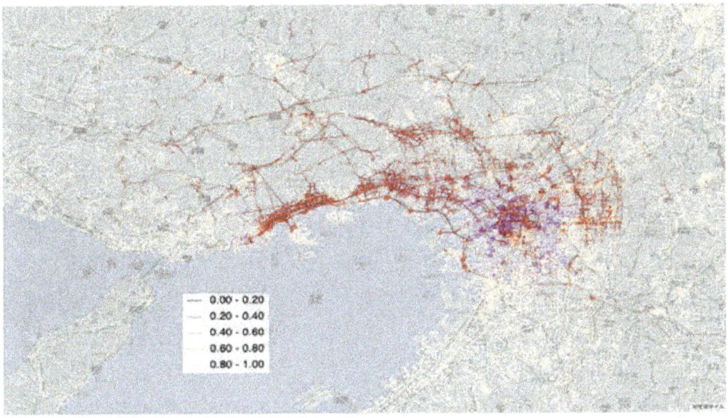

(d)

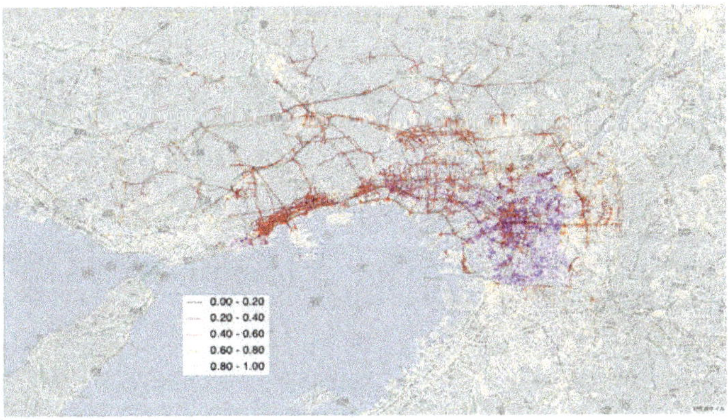

(e)

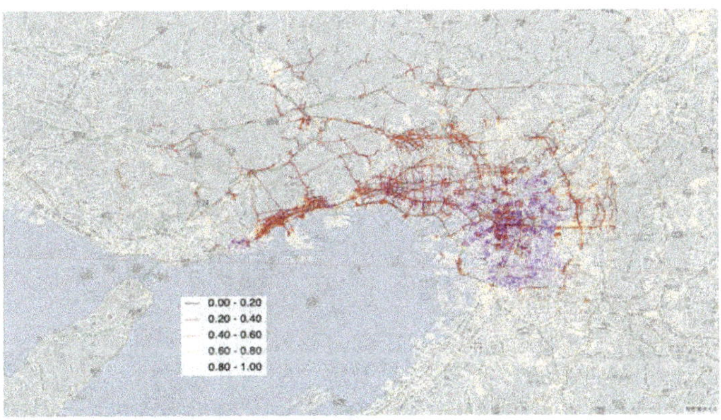

(f)

Fig. 19. Road closures and congested roads: Cases (d)–(f). (d) Landside case, (e) landside motion ×1.5 case, and (f) landside velocity ×1.3 case.

congested in all cases. The heavily congested areas generally correspond to the locations where large seismic motions were assumed in the present case studies. It is considered that the lack of traffic capacity due to large-scale road closures, particularly in the center of Osaka, caused large-scale congestion, which eventually led to gridlock.

Acknowledgments

The development of the simulation system has been accomplished by a number of contributors (see the authors of the publications, including papers and conferences, listed in the References section). The overall developments of FastDUE have been reported in a few conferences.[22,24,25] The near-equilibrium approach used in FastDUE has also been reported separately.[26,27] Techniques related to the shortest path search algorithm have also been reported separately.[19,20,28] The developments of DeSuTA have been reported in two conference proceedings.[29,30]

The studies in this chapter were supported by the Post K computer project (priority issue 3: development of integrated simulation systems for hazard and disaster induced by earthquake and tsunami). This research used computational resources of the K computer provided by the RIKEN Advanced Institute for Computational Science through the HPCI System Research Project (Project ID: hp150283, hp160224, hp170271, hp180197, and hp190191).

References

1. T. Iryo, K. Iiyama, M. Lalith, and M. Hori, On computational costs for monte carlo regional-scale simulation for reliable estimate of city and urban area earthquake disaster, *J. Earthq. Tsunami.* **16**, 2240003 (2022).
2. Y. Iida, F. Kurauchi, and H. Shimada, Traffic management system against major earthquakes, *IATSS Res.* **24**, 6–17 (2000).
3. D. Nakagawa, T. Itoh, K. Yoshikawa, and H. Kobayashi, Characteristics of traffic conditions immediately after the Hanshin-Awaji earthquake, *J. Nat. Disaster Sci.* **20**, 11–20 (1998).

4. Y. Hara and M. Kuwahara, Traffic monitoring immediately after a major natural disaster as revealed by probe data — A case in Ishinomaki after the Great East Japan Earthquake, *Transp. Res. Part A: Policy Pract.* **75**, 1–15 (2015).
5. Y. Oyama and E. Hato, A discounted recursive logit model for dynamic gridlock network analysis, *Transp. Res. Part C: Emerg. Tech.* **85**, 509–527 (2017).
6. Y. Kawasaki, M. Kuwahara, Y. Hara, T. Mitani, A. Takenouchi, T. Iryo, and J. Urata, Investigation of traffic and evacuation aspects at Kumamoto Earthquake and the future issues, *J. Nat. Disaster Res.* **12**, 272–286 (2017).
7. J. Urata, Y. Sasaki, and T. Iryo, Spatio-temporal analysis for understanding the traffic demand after the 2016 Kumamoto Earthquake using mobile usage data, in: *Proceedings of the IEEE Conference on Intelligent Transportation Systems, (ITSC)*, pp. 2496–2503 (2018).
8. W. Turek, Erlang-based desynchronized urban traffic simulation for high-performance computing systems, *Future Gener. Comput. Syst.* **79**, 645–652 (2018).
9. M. Rickert and K. Nagel, Dynamic traffic assignment on parallel computers in TRANSIMS, *Future Gener. Comput. Syst.* **17**, 637–648 (2001).
10. S. Thulasidasan and S. Eidenbenz. Accelerating traffic microsimulations: A parallel discrete-event queue-based approach for speed and scale, in: *Proceedings of the 2009 Winter Simulation Conference (WSC)*, pp. 2457–2466 (2009).
11. J. Bowman and M. Ben-Akiva, Activity-based disaggregate travel demand model system with activity schedules, *Transp. Res. Part A: Policy Pract.* **35**, 1–28 (2001).
12. S. Fujii, Y. Otsuka, R. Kitamura, and T. Monma, A micro-simulation model system of individuals' daily activity behavior that incorporates spatial, temporal and coupling constraints, *Infrast. Plan. Rev.* **14**, 643–652 (1997).
13. T. Arentze, F. Hofman, H. van Mourik, and H. Timmermans, ALBATROSS: Multiagent, rule-based model of activity pattern decisions, *Transport. Res. Rec.* **1706**, 136–144 (2000).
14. E. J. Miller and M. J. Roorda, Prototype model of household activity-travel scheduling, *Transport. Res. Rec.* **1831**, 114–121 (2003).
15. M. J. Roorda, E. J. Miller, and K. M. Habib, Validation of TASHA: A 24-h activity scheduling microsimulation model, *Transp. Res. Part A: Policy Practice.* **42**, 360–375 (2008).
16. M. J. Lighthill and G. B. Whitham, On kinematic waves. II. A theory of traffic flow on long crowded roads, *Proc. R. Soc. A.* **229**, 317–345 (1955).

17. P. Richards, Shock waves on the highway, *Oper. Res.* **4**, 42–51 (1956).
18. T. Iryo, Properties of dynamic user equilibrium solution: Existence, uniqueness, stability, and robust solution methodology, *Transportmetrica B: Transp. Dyn.* **1**, 52–67 (2013).
19. G. Peque Jr., H. Harada, and T. Iryo. Time-dependent link travel time approximation for large-scale dynamic traffic simulations, in: *Lecture Notes in Computer Science*, pp. 562–576 (2019).
20. G. Peque Jr., J. Urata, and T. Iryo, Preprocessing parallelization for the ALT-algorithm, in: *Lecture Notes in Computer Science*, vol. 10862. LNCS (2018).
21. G. Karypis and V. Kumar, A fast and high quality multilevel scheme for partitioning irregular graphs, *SIAM J. Sci. Comput.* **20**, 359–392 (1998).
22. W. Petprakob, L. Wijerathne, T. Iryo, J. Urata, K. Fukuda, and M. Hori, On the implementation of high performance computing extension for day-to-day traffic assignment, *Transp. Res. Proc.* **34**, 267–274 (2018).
23. https://www.bousai.go.jp/jishin/nankai/taisaku_wg/pdf/20130528_houkoku_s1.pdf (accessed 29 October 2022).
24. T. Iryo, J. Urata, K. Fukuda, G. Peque Jr., L. Wijerathne, and W. Petprakob. Fast calculation of dynamic traffic assignment by parallelised network loading algorithm, in: *International Symposium of Transport Simulation & International Workshop on Traffic Data Collection and its Standardization* (2018).
25. T. Iryo. Assessing congestion in a damaged road network after a major earthquake using a high-performance computer, in: *The 17th World Conference on Earthquake Engineering* (2021).
26. K. Fukuda, T. Iryo, J. Urata, and M. Ishihara, Dynamic user equilibrium assignment by Markov chain including queue spillback, *J. Japan Soc. Civil Eng., D3 (Infrastruct. Plann. Manage).* **73**, I_633–I_641 (2017).
27. T. Iryo, J. Urata, M. Ishihara, and K. Fukuda, Fluctuations of queuing congestions by a Markov-chain based day-to-day dynamical model, in: *The 7th International Symposium on Transportation Network Reliability* (2018).
28. G. Peque Jr., J. Urata, and T. Iryo. A link-based probit model for dynamic traffic assignment, in: *The Seventh International Symposium on Dynamic Traffic Assignment* (2018).

29. Y. Sasaki, J. Urata, and T. Iryo, Developing an activity simulator for calculating daily activity demand during disaster recovery, in: *60th Annual Conference on Civil Engineering and Planning* (2019).
30. J. Urata, Y. Sasaki, and T. Iryo, Development and application of activity simulators during disaster recovery, in: *61th Annual Conference on Civil Engineering and Planning* (2020).

https://doi.org/10.1142/9781800614635_0008

Chapter 8

Dynamic Programming of Firms' Activities and Market Interactions After a Disaster

Muneta Yokomatsu[*,‡] and Muneo Hori[†,§]

[*]International Institute for Applied Systems Analysis, Population and Just Societies Program/Advancing Systems Analysis Program, Schlossplatz, Laxenburg, Austria

[†]Japan Agency for Marine-Earth Science and Technology, Research Institute for Value-Added-Information Generation, Kanazawa-ku, Yokohama, Japan

[‡]yoko@iiasa.ac.at

[§]horimune@jamstec.go.jp

This chapter applies parallel computing to numerically solve the Bellman equation of a dynamic optimization problem of firms that engage themselves in reconstruction investment in the aftermath of a natural disaster. The model is characterized by the system of equations where a large number of economic stocks compose the state space, and an optimization problem of each firm is interrelated with programming problems of other firms through price changes in various markets. Non-parametric-value-function iteration is applied to solve the Bellman equation. In numerical analysis, all optimization problems of firms in all industry sectors are solved, in addition to the Monte Carlo simulation of their economic behaviors and resulting market price changes. A challenge in the numerical analysis is to implement communication methods that are associated with efficient use of memories for partitioning the state space into a set of suitable blocks and obtaining values of neighborhood blocks. A case study of numerically solving the Bellman equation is made. The target problem is an economic recovery after the Nankai Trough earthquake. The dynamic optimization problems for 4,320 types of firms are solved, and 5×10,240 disaster scenarios are given by the combination of five disaster patterns and 10,240 restoration processes of lifelines and supply chains. The study shows the potential usefulness of solving the Bellman equation on this scale. This chapter discusses the possibility of various policy analyses using the present method.

1. Introduction

This chapter presents a framework and results of an economic simulation by numerically solving a dynamic optimization problem. The optimization is considered for the process of restoration investment in production facilities and production recovery after a large-scale disaster for mutually dependent firms. A disaster recovery problem and a large-scale simulation are connected by an issue that a large number of stocks such as production facilities, infrastructure, lifelines, and household assets are destroyed by a natural disaster and they ought to be reconstructed in an optimized manner to speed up the recovery of the damaged area. Such dynamic programming faces a "curse of dimensionality".[1] It has prevented us from numerically solving the optimization problem of large dimensions of state variables. The purpose of this study is to apply parallel computing to numerically solve the Bellman equation, which is the most rigorously formulated as a dynamic optimization problem. The Bellman equation actually comprises a system of Bellman equations, each of which is posed for a representative firm of each group (hereafter, "firm", for brevity) in each industry sector (i.e., business sector, but hereafter, just "sector", for brevity) and interrelated with Bellman equations of other firms through the changes in various market prices.

The numerical analysis of the Bellman equation composes the following two steps: (1) the first step is to solve the Bellman equation of each firm; and (2) the second step is the Monte Carlo simulation of the dynamic paths of firms' behaviors and the resulting market price changes. Our simulation strategy is to obtain and store the optimal control array of each firm corresponding to all states, each of which is defined by one combination of all the state variables, in the first step, and apply all the arrays of all firms in the second step. In the second step, a realization of random variables in each time period on each sample path of the Monte Carlo simulation identifies one state (i.e., one combination of all the state variables) in the state space, and the optimal behaviors of all the firms corresponding to that state are extracted from the arrays to be taken as their behaviors. The optimal control arrays are repeatedly used in the Monte Carlo simulation. While many other studies including ones that apply the asynchronous value function iteration do not obtain such exhaustive-optimal-control arrays that cover the entire space, we take the aforementioned strategy considering that random factors could take values of wider ranges in the aftermath of a disaster due to social disruption, and a larger area of the state space could be covered in the Monte Carlo simulation. This two-step calculation is a unique approach that we develop in this chapter.

We apply parallel computing in both steps. It is found that implementation of the above approach faces the problem of the memory limitation in the first step. To cope with this problem, three countermeasures are taken; the implementation of a communication method associated with partitioning the state space into a set of blocks and obtaining values of neighborhood blocks makes the most significant contribution to reducing the size of necessary memory. Other implementations in the parallel computing include (1) introducing a code that balances loads of nodes in both steps; (2) decentralizing the random number generation process in the second step; and (3) parallelizing operations to pick up the optimal control of each firm after realization of each state on each dynamic path in the second step.

Various methods are developed for estimating the optimal value function (simply "value function", hereafter) in dynamic programming which is related to a dynamic optimization problem of this chapter (e.g., Ref. 2). For recent studies of application of dynamic programming, parametric methods, which estimate coefficients of assumed basis functions, are more often used (e.g., Refs. 3 and 4). However, this chapter uses a non-parametric method to directly estimate values of the value function. An advantage of the non-parametric method is that it is possible to incorporate responses that cannot be assumed with simple basis functions of low-dimensional polynomials usually used for high-dimensional-state-space problems. There are various studies in which parallel computing is used to solve dynamic programming (e.g., Refs. 5–7). As mentioned, however, they face the "curse of dimensionality", which often manifests itself as a problem associated with a huge number of calculations caused by a combinatorial explosion (e.g., Refs. 8 and 9). In contrast, in our model, the curse appears as a memory constraint problem. Our numerical analysis is thus unique because it develops a new method of data communication and data storage to mitigate the memory constraint problem.

There is a vast accumulation of models that formulate and analyze economic systems by the recursive framework of dynamic programming (e.g., Ref. 10). A traditional approach to dynamic economic analysis is to create a model with the minimum number of variables in order to derive more general propositions by providing an analytical solution of the model. Numerical analysis was made if the model cannot be analytically solved. Such numerical analysis is employed in the field of applied economics, which is aimed at obtaining a quantitatively accurate solution to each problem. Particularly, applied economics is concerned with global energy and climate change problems, and a series of models that have come in handy in practical policy analysis are developed (e.g., Refs. 11 and 12).

In recent years, there are an increasing number of applied economic applications to natural disaster problems (e.g., Refs. 13 and 14). Large-scale data sets are being used to provide more realistic problem settings and it facilitates the development of an advanced and sophisticated method of numerical analysis. For instance, as we can name at this moment, the representative methods are complex network estimation (e.g., Ref. 15), AI-based (e.g., Ref. 16), and agent-based models (e.g., Refs. 17 and 18).

In this chapter, we consider a large number of disaster scenarios composed of combinations of various input data. For all the scenarios, we seek to solve a multi-agent dynamic decision-making problem using a recursive framework. To the best of the authors' knowledge, there is no study that solves a stock-building problem with such characteristics for disaster recovery.

We carry out a case study targeted at the potential scenarios of the presumed Nankai Trough earthquake in Japan with a special focus on the Kansai area. The model economy is composed of three regions and 36 industry sectors such as "agriculture, forestry, and fisheries", "information and communication", and "commerce". The dynamic optimization problem for 4,320 types of firms is formulated, and the dynamic process of price fluctuations in the 36 sectors in the Kansai market is simulated for a few years. In the case study, the following three elements must be incorporated: (1) the firm capital size; (2) the magnitude of the disaster; and (3) the impact of supply chain and lifeline disruptions within each sector. As for the second and third elements, the disaster scenarios are given by the combination of five destruction cases and 10,240 cases of restoration processes of lifelines and supply chains. We study the effects of a recovery policy on the market, to examine the usefulness of the developed method.

2. Model of Post-Disaster Economy

2.1. *Regions and markets*

The model economy consists of the following three regions: (1) Osaka prefecture; (2) five prefectures in the Kansai area other than Osaka (i.e., Kyoto, Hyogo, Shiga, Nara, and Wakayama prefectures); and (3) other prefectures in Japan, denoted by Region OSK (Osaka), Region ROK (Rest of Kansai) and Region ROJ (Rest of Japan), respectively. The market space is divided into the following two parts: (1) the Kansai market and (2) the ROJ/overseas market. Each of the two markets is further classified into 36 industry sectors.

After a disaster hits, firms in Region OSK and Region ROK start to reconstruct their damaged production facilities by solving dynamic optimization problems. On the other hand, for firms in Region ROJ, the optimization problem is not handled. The process of restoring capital, labor, and infrastructure in Region ROJ is exogenously given by scenarios. In the aforementioned framework, the processes of the sector Gross Regional Product (Sector GRP) are calculated.

2.2. *Lifeline and supply-chain resumption probabilities, bifurcation of firm types*

We consider a discrete-time horizon, starting from $t = 0$ at the time of disaster occurrence. We define unit period as 1 month. Within each sector indexed by h $(= 1, \ldots, H)$, firms are classified into several types according to the capital size, level of damage, and timing of lifeline and supply-chain resumption. A firm is denoted by its index i. In many following contexts, the sector index h is used to represent the sector to which Firm i belongs, and the sector index h' is used to represent the arbitrary sector to which Firm i does not belong. Moreover, although variables of same sector h generally have different values in different regions, for brevity of notation, the index representing the region is omitted in places where variables are not aggregated across regions.

Each lifeline or supply chain is represented by an index ι; for example, $\iota = 1$ represents electricity, $\iota = 2$, gas, $\iota = 3$, water supply, $\iota = 4$, a firm's first supply chain, $\iota = 5$, its second supply chain, and so on. The probability that lifeline or supply chain ι for Sector h in the region to which Firm i belongs has resumed functioning by Period t is denoted by the distribution function $\Phi^1_{\iota h}(t)$. Assuming that the probability of the resumption of each lifeline or supply-chain function is independent, and further assuming that all lifeline and supply-chain functions are completely non-substitutable in production, we can determine the distribution function of production resumption for firms in Sector h in the region as

$$\Phi^1_h(t) = \Pi_\iota \Phi^1_{\iota h}(t). \tag{1}$$

Therefore, the probability of having not resumed production by Period t, denoted by $\Phi^0_h(t)$, and the probability of being out of production by Period $t-1$ and resuming production in Period t, denoted by $P^{\theta h}_{01}$, are given as

$$\Phi^0_h(t) = 1 - \Phi^1_h(t), \tag{2}$$

$$P^{\theta h}_{01}(t) = \frac{\Phi^1_h(t) - \Phi^1_h(t-1)}{\Phi^0_h(t-1)}. \tag{3}$$

The Markov transition probability matrix for the change in production state, which is represented by $\Delta\theta_i = 0, 1$, is expressed as follows:

$$P^{\theta h}(t) = \begin{bmatrix} P^{\theta h}_{00}(t) & P^{\theta h}_{01}(t) \\ 0 & 1 \end{bmatrix}, \tag{4}$$

where the first row and the first column of the transition matrix $P^{\theta h}(t)$ correspond to State $\theta_i = 0$, and the second row and the second column, State $\theta_i = 1$. It holds that $P_{00}^{\theta h}(t) = 1 - P_{01}^{\theta h}(t)$. The transition probability from the production-ready state to the production-stop state is assumed to be zero.

The types of firms within each sector are exogenously and endogenously classified according to the combination: "pre-disaster capital size" × "damage scale" × "production resumption period". We call the type identified by "pre-disaster capital size" × "damage scale" by "pre-type" (i.e., determined before the recovery process), and use j as an index. Moreover, we denote the share of the pre-type j within Sector h as π_{jh}^0, then $\sum_j \pi_{jh}^0 = 1$ holds. We also express the share of firms of the pre-type j as a vector of the share of firms in production stoppage, $\pi_{jh0}(t)$, and the share of firms in production operation, $\pi_{jh1}(t)$, as follows:

$$\pi_{jh}(t) = (\pi_{jh0}(t) \quad \pi_{jh1}(t)). \tag{5}$$

By definition, a recursive formulae of $\pi_{jh}(t)$ is obtained as $\pi_{jh}(t) = \pi_{jh}(t-1) P^{\theta h}(t)$. Thus, the number of firms in each of these categories is as follows:

$$\pi_{jh}(t)\, N_{nh} = (\pi_{jh0}(t)\, N_{nh} \quad \pi_{jh1}(t)\, N_{nh}), \tag{6}$$

where N_{nh} is the number of firms in Sector h. The number of firms that resume production in Period t_s, denoted by $N(j, t_s, h)$, is determined as follows:

$$N(j, t_s, h) = \pi_{jh0}(t_s - 1)\, P_{01}^{\theta h}(t_s)\, N_{nh}. \tag{7}$$

2.3. *Recovery investment problem for firms in Region OSK and Region ROK*

The vector of state variables in Period t in the problem of Firm i is represented by $\boldsymbol{s}_i(t) = (k_i(t), \ell_i(t), \bar{B}_h(t), \boldsymbol{p}(t), \theta_i(t), t)$, where $k_i(t)$ is Firm i's capital stock, which is directly under its control; $\ell_i(t)$ is the labor of Firm i, and $\bar{B}_h(t)$ is the expected value of the infrastructure composite for Sector h. The details of the infrastructure composites are clarified in the following. $\ell_i(t)$ and $\bar{B}_h(t)$ are assumed to be exogenously given. $\boldsymbol{p}(t) = (p_1(t), \ldots, p_H(t))$ is the price vector from Sector 1 to Sector H in the Kansai market, whose dynamics are endogenously determined by the market transactions, while each firm takes it as given. $\theta_i(t)$ denotes a state whether Firm i produces or not, depending on the states of its lifeline and supply chain. If $\theta_i(t) = 1$, Firm i can produce in Period t, and if $\theta_i(t) = 0$, it cannot.

The production technology is formulated as follows. We assume the perfect non-substitutability between the value added $F^i(\cdot)$ and intermediate input goods $m_{h'i}(t)$, and the constant elasticity of substitution (CES) for the value-added function as

$$y_i(t) = \min\left[F^i(\cdot), \frac{m_{1i}(t)}{\kappa_{1h}}, \ldots, \frac{m_{h'i}(t)}{\kappa_{h'h}}, \ldots, \frac{m_{Hi}(t)}{\kappa_{Hh}}\right], \tag{8}$$

$$F^i(\ell_i, k_i^+, \tilde{B}_h, \varepsilon_i, \theta_i, \xi_{di}) = \theta_i(1 + \gamma_i\varepsilon_i)\tilde{B}_h\beta_0(1 + \gamma_{\xi d1}(\xi_{di})^{\alpha_{\xi d1}})^{-1}$$
$$(\beta_h^l \ell_i^{\alpha_h} + \beta_h^k k_i^{+\alpha_h})^{\frac{1}{\alpha_h}}, \tag{9}$$

where ξ_{di} is a state variable representing the debris stock of Firm i, which gives a negative impact on the firm's productivity; ℓ_i, k_i^+, $\tilde{B}_h$, ε_i, θ_i, and ξ_{di} are variables that depend on Period t (the argument (t) is omitted in the equations to avoid notational complexity); γ_i, β_0, $\gamma_{\xi d1}$, $\alpha_{\xi d1}$, β_h^ℓ, β_h^k, and α_h are parameters that are constant throughout time; $\varepsilon_i(t)$ is white noise; and $\tilde{B}_h(t)$ is the level of the infrastructure composite for Sector h that is composed by the following equation:

$$\tilde{B}_h(t) = \sum_b \beta_{bh} B_b(t), \tag{10}$$

where b is an index on the type of infrastructure, and $B_b(t)$ is the level of Infrastructure b in Period t. β_{bh} is the weight on Infrastructure b for each sector h. The formation process of each infrastructure b, denoted by $B_b(t)$, is expressed as follows:

$$B_b(t) = B_b(t-1) + \zeta_b(t-1)\,G_R(t-1)\,(1 + \gamma_b\varepsilon_{Bb}(t)), \tag{11}$$

where $G_R(t)$ and $\zeta_b(t)$ are the total infrastructure investment in Period t and the investment share in Infrastructure b, respectively. They are both policy variables related to infrastructure restoration and are exogenous for firms; $\varepsilon_{Bb}(t)$ is white noise. Note that $\bar{B}_h(t)$ in the state variable vector is the expected value of $\tilde{B}_h(t)$.

The formation process of capital (i.e., production facility) stock $k_i(t)$ is given as follows:

$$k_i(t) = (1 - \delta)k_i^+(t - 1), \tag{12}$$

where $k_i^+(t) = k_i(t) + (1 + \gamma_{\xi d2}(\xi_{di}(t))^{\alpha_{\xi d2}})^{-1}\eta_i(t)$. Here, δ and $\eta_i(t)$ are the depletion rate and investment (i.e., production facility development that includes disaster recovery investment), respectively. $\xi_{di}(t)$ is the debris stock which causes a negative effect on the investment efficiency, which is however decreased according to the following equation:

$$\xi_{di}(t) = \xi_{di}(t - 1) - \eta_{di}(t - 1), \tag{13}$$

where $\eta_{di}(t)$ is the amount of debris removed in Period t. For simplicity, the removal schedule is assumed to be given exogenously.

Firm i's profit in Period t is given as

$$\Pi^i(\ell_i, k_i, \tilde{B}_h, \boldsymbol{p}, \eta_i, \varepsilon_i, \theta_i) = \tilde{p}_{Sh}y_i - \sum_{h'} \tilde{p}_{Dh'}m_{h'i} - w\ell_i - p_\eta \eta_i - p_{\eta d}\eta_{di}, \tag{14}$$

where the argument t is omitted in all the variables here. $\tilde{p}_{Sh}$ is the ex-post effective sale price of goods of Sector h in the Kansai market in Period t. $\tilde{p}_{Dh'}$ is the ex-post effective purchase price. w is the wage rate. p_η and $p_{\eta d}$ are the unit costs of the investment and the debris removal, respectively, which are given to each firm. Note that $\tilde{p}_{Sh}$ and $\tilde{p}_{Dh'}$ are endogenously determined. The first and second terms on the right-hand side are the revenue obtained from the supply of output goods and the cost of purchasing intermediate input goods, respectively. The third, fourth, and fifth terms are wage payments, capital investment costs, and removal costs, respectively.

Each firm aims at maximizing its market value. Denoting the value of Firm i at the beginning of Period t by $\Lambda_i(\boldsymbol{s}_i(t))$, Firm i's problem is expressed in the following form of Bellman equation:

$$\Lambda_i(\boldsymbol{s}_i(t)) = \max_{\boldsymbol{d}_i(t)} \left[E_{\boldsymbol{\varepsilon}}[\Pi^i(\cdot)] + \psi \sum_{\{q, \Delta\theta_i\}} \phi_i(q, \Delta\theta_i)\ \Lambda_i(\boldsymbol{s}_i'(\boldsymbol{s}_i(t), \boldsymbol{d}_i(t))) \right], \tag{15}$$

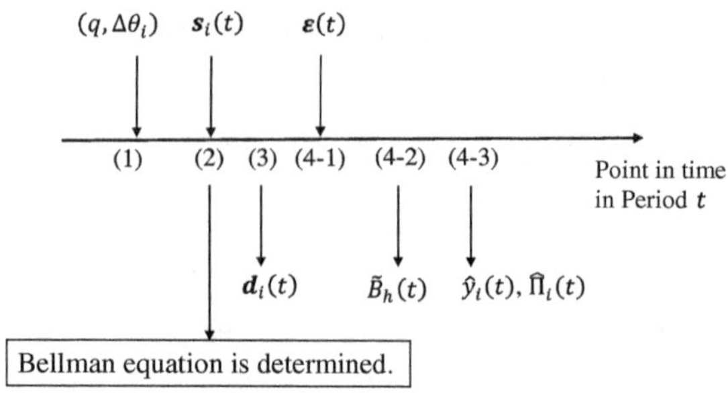

Fig. 1. Event sequence.
Note: The hat of symbols means the realized value of the corresponding random variables.

where $\boldsymbol{d}_i(t) = (\eta_i(t), \{m_{h'i}(t)\})$ is the control variable vector of Firm i; $E_{\boldsymbol{\varepsilon}}[\Pi^i(\cdot)]$ is the expected value of the profit given in Eq. (14) with respect to the white noise vector $\boldsymbol{\varepsilon}(t)$; ψ is the discount factor $(0 < \psi < 1)$; q is the index that indicates the price vector $\boldsymbol{p}$; $\Delta\theta_i = \theta_i(t+1) - \theta_i(t)$ indicates the change in the firm's production state ($\Delta\theta_i = 1$ when Firm i's production state turns from suspended to resumed and $\Delta\theta_i = 0$ when the state remains unchanged, i.e., production remains suspended or remains available for production); $\phi_i(q, \Delta\theta_i)$ is the product of the subjective probability of the realization of the price vector of index q and the probability of the change in production state $\Delta\theta_i$ in the next period ($\sum \phi_i(q, \Delta\theta_i) = 1$ holds); and $\boldsymbol{s}_i'(\boldsymbol{s}_i(t), \boldsymbol{d}_i(t))$ is the state vector in the next period determined by the state $\boldsymbol{s}_i(t)$ and control $\boldsymbol{d}_i(t)$ in the current period. $\Lambda_i(\boldsymbol{s}_i(t))$ is the optimal value function in the dynamic programming problem and determined so that all $\Lambda_i(\boldsymbol{s}_i(t))$'s meet all the sets of Eq. (15) for all $\boldsymbol{s}_i(t)$'s. In other words, Bellman equation of Eq. (15) is a functional equation that determines the optimal value function $\Lambda_i(\boldsymbol{s}_i(t))$.

We assume that the disaster occurs at the end of Period 0. We assume further the sequence of events for each Firm i in each Period t as shown in Fig. 1 and described below:

(1) Supply chain and lifeline resumption $\Delta\theta_i(t)$ and market price $\boldsymbol{p}(t)$ (i.e., index q) are determined.
(2) The state $\boldsymbol{s}_i(t)$ is identified, followed by Eq. (15) for Period t.
(3) Each firm determines the control variable vector $\boldsymbol{d}_i(t)$ so as to maximize the right-hand side of Eq. (15).

(4) The white noise vector $\boldsymbol{\varepsilon}(t)$, which includes $\varepsilon_i(t)$, is determined, being associated with the various infrastructure levels, infrastructure composites, production levels, and profits that depend on $\boldsymbol{\varepsilon}(t)$. The price in the next period is determined according to the excess demand in each market. The rule for determining prices is described in the following.

2.4. *Production in Region ROJ*

Region ROJ is modeled differently from the other two regions. This is because a large proportion of the region is not directly affected by the Nankai Trough earthquake, and the restoration of production facilities is not a central issue of most firms there. We do not consider the heterogeneity of firms within a sector in Region ROJ, but instead deal with aggregate variables of each sector. The process of forming capital, labor, and infrastructure is assumed to be exogenous, and the ROJ market prices are assumed to be constant.

In Region ROJ, the changes of the sector GRPs due to the disaster are caused by (i) damage to and restoration of capital, labor, and infrastructure, (ii) disruption of the supply of intermediate inputs that are to be obtained through the supply chain from Kansai and in the open markets, and (iii) price increases in the Kansai market. The change in the sector GRP is expressed as follows:

$$\Delta \mathrm{GRP}_h^{\mathrm{ROJ}} = \Delta p_h^{\nu\mathrm{ROJ}}\, Y_h^{\mathrm{ROJ}} + p_h^{\nu\mathrm{ROJ}}\, \Delta Y_h^{\mathrm{ROJ}}. \tag{16}$$

We use Δ to express the change from the pre-disaster level. Note, therefore, Δ does not represent the change from the previous period. $\Delta p_h^{\nu\mathrm{ROJ}}$ and $\Delta Y_h^{\mathrm{ROJ}}$ are the changes in the value-added price and the production level, respectively.

2.5. *Market and price change process*

The composition of the markets is illustrated in Fig. 2. As mentioned earlier, prices in ROJ/Overseas markets are assumed to be constant for all sectors because ROJ/Overseas market is much larger than the Kansai market so that the disaster does not impact prices in ROJ/Overseas markets. In addition, it is assumed that firms in Region OSK and Region ROK incur great damage to their

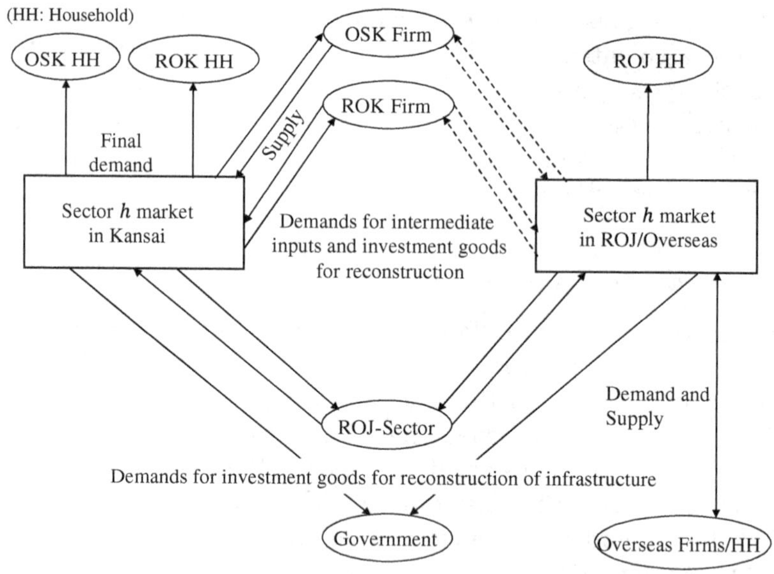

Fig. 2. Markets and flow of goods and services.

transportation means, and thus face larger transportation costs than normal ones. Thereby, they prioritize trading in the Kansai markets. The process of the price change in each sector of the Kansai market is assumed to be as follows:

$$p_h(t) = (1 + \upsilon \,\mathrm{RExD}_h)\, p_h(t-1), \tag{17}$$

where υ is a positive parameter, and RExD_h is the excess demand rate that is defined by

$$\mathrm{RExD}_h = \frac{\mathrm{DM}_{\mathrm{in}\,h} - \mathrm{SP}_{\mathrm{in}\,h}}{\mathrm{DM}_{\mathrm{in}\,h}}. \tag{18}$$

$\mathrm{DM}_{\mathrm{in}\,h}$ is the sum of demand for intermediate goods by firms in Regions OSK, ROK, and ROJ, demand for investment goods by those firms and the government, and final demand by households in Kansai. On the other hand, $\mathrm{SP}_{\mathrm{in}\,h}$ is the sum of supply by firms in Regions OSK, ROK, and ROJ. If the excess demand is positive, the price rises, and if negative, drops. The excess or deficiency of goods in each period is assumed to be adjusted by the supply and demand of foreign firms and households.

2.6. *Policy evaluation function*

Policy evaluation is performed for each disaster scenario by using the following function;

$$
\begin{aligned}
W_\nu = & \sum_i \Lambda_i^{\mathrm{OSK}}(S_i(0)) + \sum_i \Lambda_i^{\mathrm{ROK}}(S_i(0)) \\
& + w_1 \mathrm{Ave}\left[\sum_t \psi^t \sum_h \Delta \mathrm{GRP}_h^{\mathrm{ROJ}}(t)\right] \\
& - w_2 \sum_t \psi^t (G_R^{\mathrm{OSK}}(t) + G_R^{\mathrm{ROK}}(t) + G_R^{\mathrm{ROJ}}(t)), \qquad (19)
\end{aligned}
$$

where ν is the index of the disaster scenario, w_1 and w_2 are the weight parameters, Ave$[(\cdot)]$ is the operation to calculate the average with respect to runs in Monte Carlo simulation, and $G_R^{\mathrm{OSK}}(t)$, $G_R^{\mathrm{ROK}}(t)$, and $G_R^{\mathrm{ROJ}}(t)$ are the total infrastructure investment in Regions OSK, ROK, and ROJ, respectively, in Period t. The first and the second terms on the right-hand side represent the present values of the expected profits of all firms in Regions OSK and ROK, respectively. The third term is the present value of the change in GRP in Region ROJ, and the fourth term represents deduction by present value of the recovery investment costs.

3. Algorithm of Solving Bellman Equation and Simulating Dynamics of Economy

3.1. *Simulation steps*

Figure 3 shows the sequence of calculations. The calculation consists of two steps. In the first step (Step 1), the Bellman equation for each firm type, classified by sector, capital size, and level of disaster damage, is obtained numerically by non-parametric value function iteration (VFI). In VFI, for the purpose of decreasing dimensions of the space of state for Firm i, the price space that is a part of the state space is composed of the prices only of the relevant sectors for Firm i. A list of optimal controls, $\boldsymbol{d}_i = \{\boldsymbol{d}_i^*(\boldsymbol{s}_i^j)|\text{for all } j\}$ and corresponding (optimal-)value-function values, $\boldsymbol{\Lambda}_i = \{\Lambda_i(\boldsymbol{s}_i^j)|\text{for all } j\}$ for all states is produced. As shown in Fig. 3, computations are performed in parallel not only in the problem basis but within each problem where the state space is divided into subspaces to conduct non-parametric VFI in parallel. That is, the list of the value-function values is updated by aggregating the results of the optimization calculation at each node (that is responsible for the divided sub-state space); the information is sent to relevant nodes; and each node carries out the optimization calculation again. Such iteration is repeated until the list of optimal value function values converges.

In the second step (Step 2), a single dynamic path is derived starting from the initial state. The initial state is identified by the

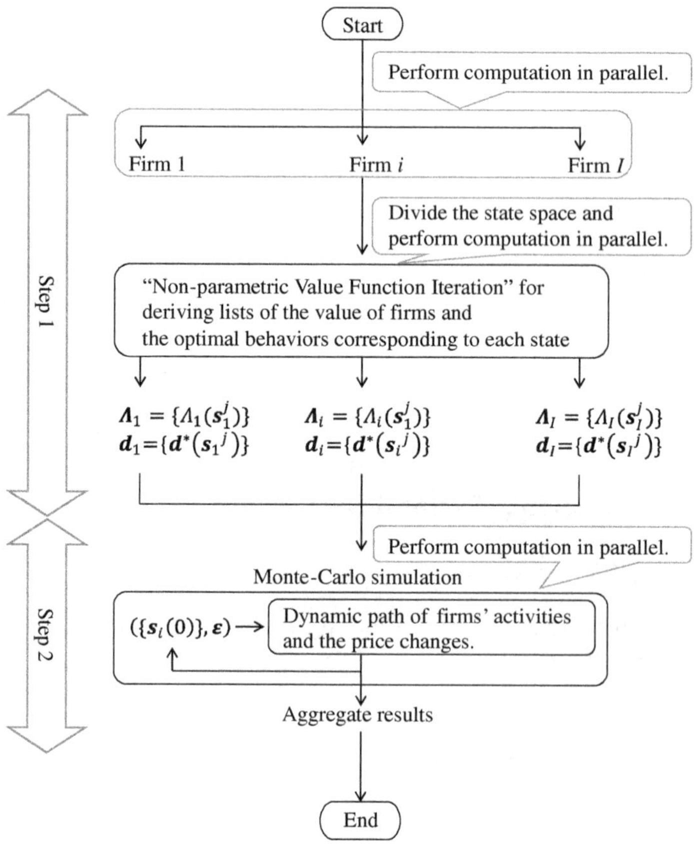

(a)

Fig. 3. Calculation flow. (a) Two-step composition. (b) Non-parametric value function iteration in Step 1. (c) Simulation flow in Monte Carlo simulation in Step 2.

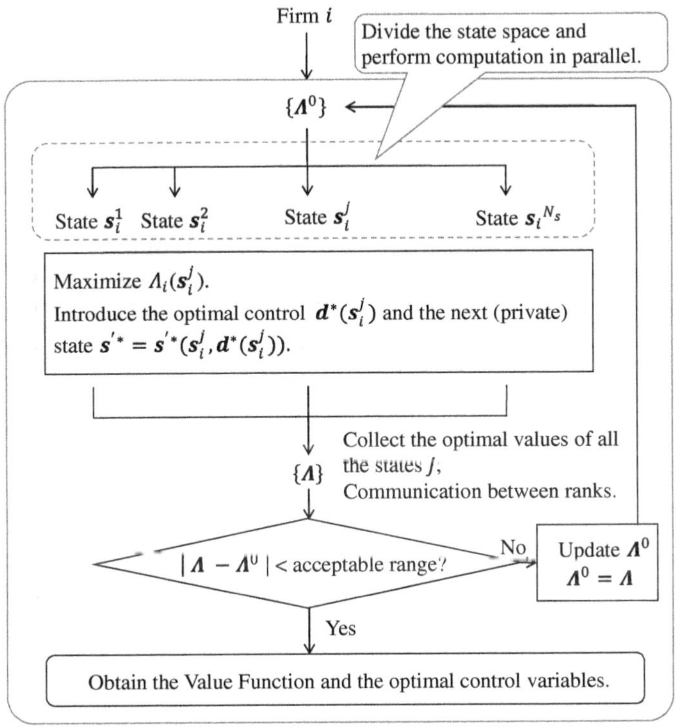

(b)

Fig. 3. (*Continued*)

disaster scenario. The optimal controls of each firm corresponding to the initial state are taken from the list created in Step 1. These are then aggregated to calculate the excess demand in the market. The price vector for the next period is determined according to the level of the excess demand for each good. By combining this with the capital level of each firm and other factors, the overall state of the economy in the next period is determined. The period is updated, and each firm identifies its own optimal behavior in the list corresponding to the next period state. The above process is repeated for the number of periods in the recovery process.

In the recovery process, even if two firms belong to the same type until the time of the disaster, they will follow different processes due to the different timing of resumption of production. In such cases, the advantage of this calculation method is that the same list of

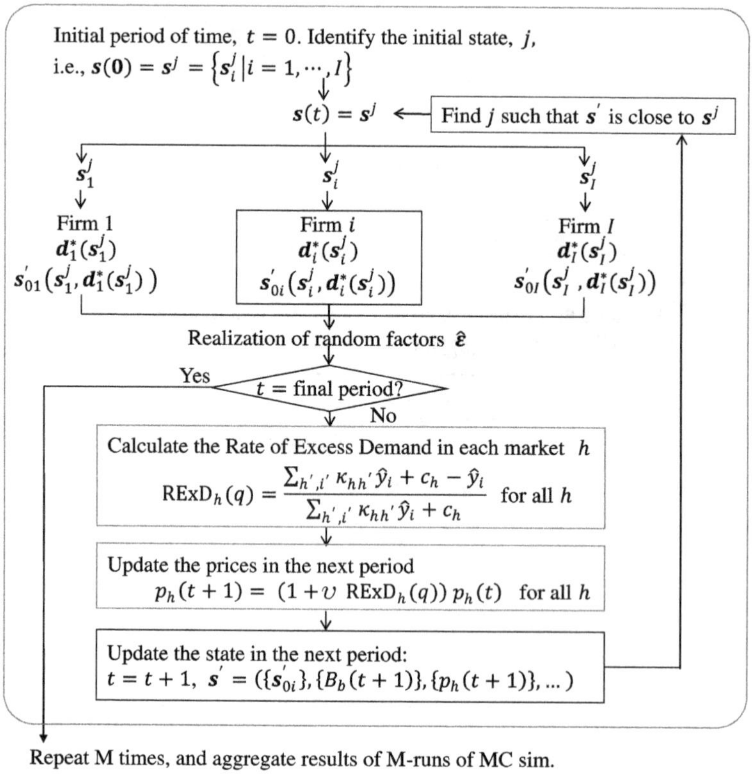

(c)

Fig. 3. (*Continued*)

optimal controls created in the first stage can be referred to. Another advantage of this method is that only Step 2 needs to be repeated for the Monte Carlo simulation.

3.2. *Implementation and scalability*

Parallelization is applied throughout this code. For Step 1 shown in Fig. 3, we implement the code that divides the state space inside each firm's problem and performs parallel computing. In addition to that, we introduce a code that allocates the next job to the node that converges earlier, since there is a difference in the time for the iterations to converge among each firm's problems.

Step 2 consists of the following four parts from a parallelization perspective. Step 2-1 generates all the necessary random numbers, where we parallelize this sub-step and use all nodes and all threads to generate the random numbers. In Step 2-2, the results of Step 1 are used to obtain a series of optimal controls that depend on the path of the realized values of the random numbers for each firm in advance. Here, parallelization with the same load-balancing algorithm as in Step 1 is applied. In Step 2-3, the supply and demand behaviors of all agents are aggregated for the price adjustment. Based on the adjusted prices, control variables for each firm are determined. Here, firms are classified into groups, among which computations are performed in parallel. Final sub-step, Step 2-4, which aggregates and outputs the results, is not parallelized. In summary, the aforementioned three ways of parallel computing are implemented in Step 2.

During the process of implementation of the code, we faced a problem of limitation of memory, and thereby took the following three countermeasures to cope with it. First, we partitioned the state space to blocks and obtained values of neighborhood blocks. To calculate the value of $\Lambda_i(\boldsymbol{s}_i(t))$ by Eq. (15) (i.e., the Bellman equation), data of a set of $\Lambda_i(\boldsymbol{s}'_i)$ with $\boldsymbol{s}'_i$ depending on $\boldsymbol{s}_i(t)$ and $\boldsymbol{d}_i(t)$ given on the right-hand side are procured from other ranks and stored. At the beginning, a large number of domains had been used, resulting in loss of memory efficiency. Then we introduced a new communication method where each datum is included into the sum every time it is collected, and then, only the value of the sum is kept; in other words, when the value regarding the Q'-th state is communicated, only the value of the sum up to Q', i.e., $\sum_{\{q,\Delta\theta_i\}}^{Q'} \phi_i(q, \Delta\theta_i)\Lambda_i(\boldsymbol{s}'_i(\boldsymbol{s}_i(t), \boldsymbol{d}_i(t)))$, is stored.

The change of the communication method is illustrated by Fig. 4 using a case where Rank 0 block collects data from neighborhood blocks of Ranks 1–3. By using the method of Fig. 4(b), the neighborhood cell spaces of Rank 0, which would be needed if the common method were implemented, are no longer necessary; see Fig. 4(a), memory usage of VFI for one firm's problem of a fixed scale was reduced from 3.4GB to 0.21GB. At the same time, the number of communications was increased. In the example shown in Fig. 4, considering that Ranks 1–3 carry out the same process, the total number of communications is 12. Therefore, the choice of the communication

$$\Lambda_i(\boldsymbol{s}_i(t)) = \max_{\boldsymbol{d}_i(t)}[E_\varepsilon[\Pi(\boldsymbol{s}_i(t), \boldsymbol{d}_i(t))] + \psi \textstyle\sum_{\{q',\Delta\theta_i\}}^{Q} \phi_i(q',\Delta\theta_i) \cdot \underline{\Lambda_i(\boldsymbol{s}_i'(\boldsymbol{s}_i(t), \boldsymbol{d}_i(t)))}]$$

Get data from other ranks.

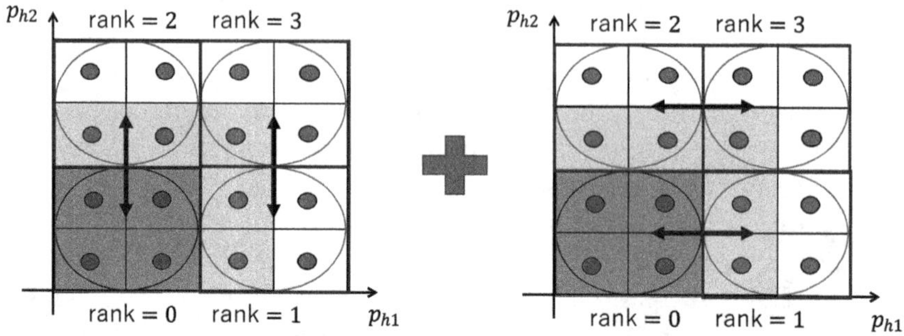

All the data should be retained until the above 8-time communications are completed (in the illustrative case of 2D).

(a)

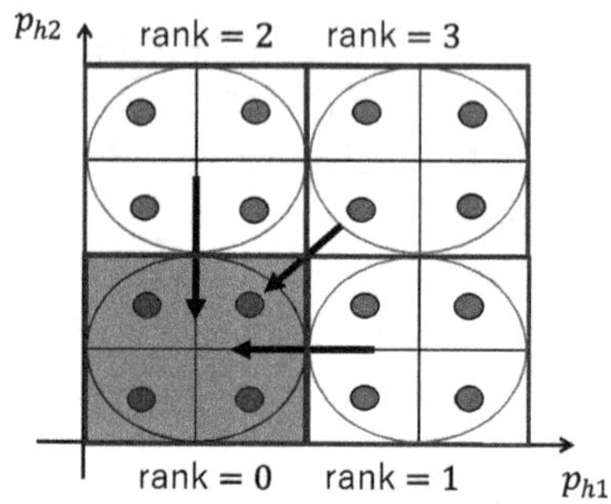

rank = 0 needs 3 communications, hence the total number of communications will be 12 for the four ranks.

Note that each datum is included in the sum every time it is collected, and thus, only the value that is stored is the value of the sum up to Q' whose communication is completed, i.e.

$$\sum_{\{q',\Delta\theta_i\}}^{Q'} \phi_i(q',\Delta\theta_i) \cdot \Lambda_i\left(\boldsymbol{s}_i'(\boldsymbol{s}_i(t), \boldsymbol{d}_i(t))\right).$$

(b)

Fig. 4. Implementation of the memory-efficient communication method. (a) Ordinary communication method, and (b) developed communication method.

methods will be made in consideration of the tradeoff between memory efficiency and computing time.

Second, we reduced the size of data that must be stored in memory by decomposing an array of variables given by polynomials into parts. For example, nine variables in a formula of the form,

$$F(x_1, \ldots, x_9) = F_1(x_1, \ldots, x_7) + F_2(x_6, \ldots, x_9) + F_3(x_1),$$

which originally composed the nine-dimensional array, were decomposed into $F_1(x_1, \ldots, x_7)$, $F_2(x_6, \ldots, x_9)$, and $F_3(x_1)$ to be stored in memory. By such operation, the total number of domains are decreased. For example, memory usage needed for $E_{\varepsilon}(\Pi)$ was reduced from 970MB to 3.22MB by applying this measure.

Third, we set up a special node dedicated to statistical processing and output work. Namely, it generates, at one time, random numbers used in all dynamic paths of the Monte Carlo simulation and holds them, then distributes them to each calculation node each time that each path is calculated. In addition, it collects the simulation results of each path from each node, holds them, and finally aggregates them to create statistical indices. By introducing this special node, other computing nodes no longer need to hold random numbers used in other dynamic paths or calculation results of past paths, and become capable of reusing memory. On this model, the "curse of dimensionality" appeared as a problem of limitation of memory, for which we took the aforementioned three countermeasures.

Figure 5 shows the effect of parallelization measured by taking Region OSK and limiting the numbers of grids of state space to a certain scale. Note that the number of nodes shown on the horizontal axis in Fig. 5 excludes the one node provided for managing load balancing and outputting the calculation results. Note further that, since the structure of parallelization is the same in the three-region model, the effect of parallelization is expected to be the same. On the other hand, because Steps 1 and 2 have separable and completely different calculation processes, the calculation time and the effect of parallelization were measured separately. As a result, the strong scalability of both phases was almost 1. The load-balancing algorithm (code that assigns the next job to the node that converges quickly) introduced in both phases probably had a large impact.

The effect of the decentralized random number generation in Step 2-1 was confirmed by the fact that the introduction of this algorithm

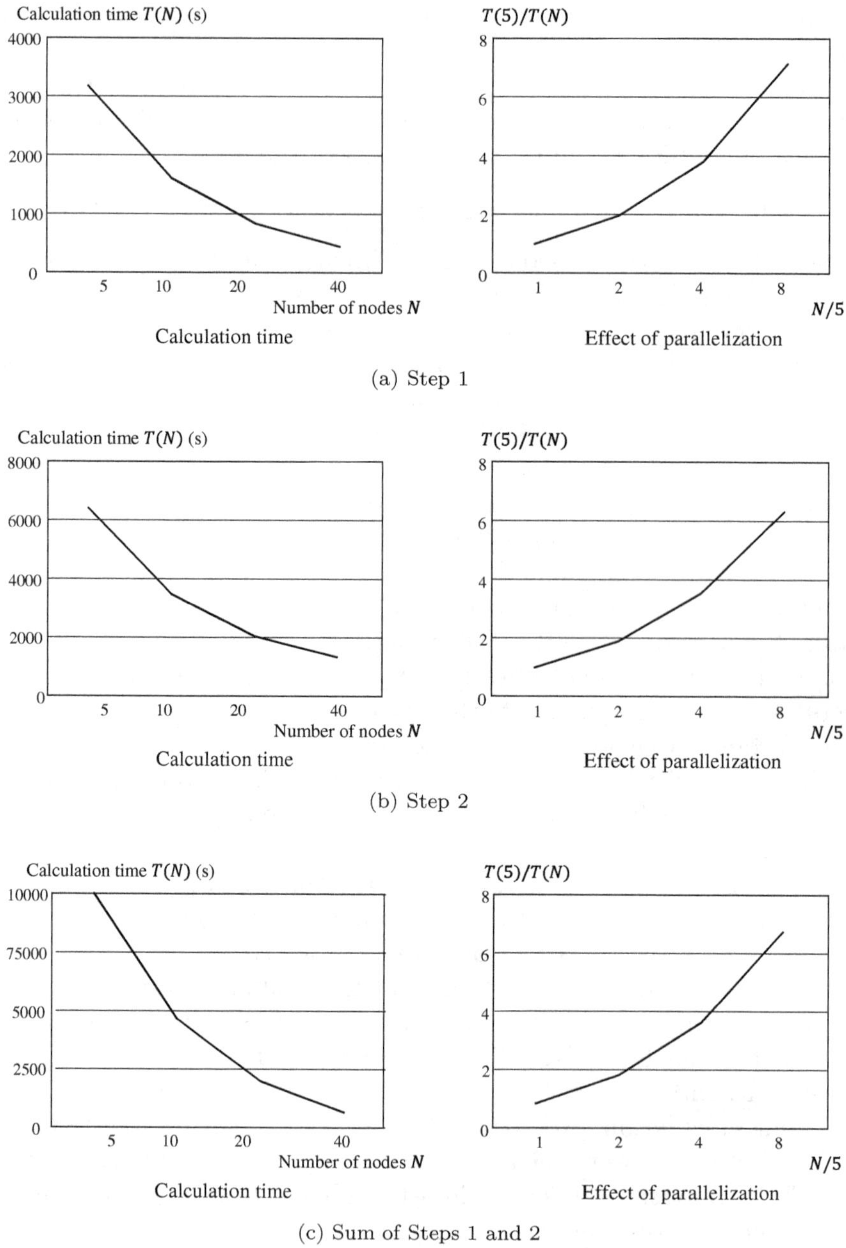

Fig. 5. Calculation time and effect of parallelization for a certain scale of calculation. (Left) Calculation time. (Right) Strong scalability (against the case of the 5-node calculation). (a) Step 1, (b) Step 2 and (c) Sum of Steps 1 and 2.

reduced the computation time for a model of a certain size from 2 h 38 min to 0 h 25 min. On the other hand, the parallelization effect in Step 2–3 may have been mostly offset by the increased time required for communication time in this case. A closer examination of the effects and causes will be the subject of a future study.

4. Case Study

4.1. *Settings*

Each disaster scenario is defined by a combination of the direct damage case of the Nankai Trough earthquake and the lifeline and supply-chain resumption process. The former consists of five cases, and the latter is given by the combination of a period in which three lifelines (electricity, gas, and water supply) resume operation in Regions OSK and ROK, respectively, and a period in which small and large/medium-sized companies in each of the 36 sectors resume production through supply-chain restoration. The total number of combinations is 10,240. The data used to generate those scenarios probabilistically consists of building damage rates estimated by IES, gross capital stock damage due to the Great East Japan Earthquake published by the Cabinet Office, data generated using fragility curves and recovery curves, and corporate data from Teikoku Databank, and so on.

4.2. *Results*

Figures 6–10 show examples of simulation results. Since the scale of transactions and capital originally differs among sectors and firms, the index of "Recovery rate"="Value for each month"/"Value for the month immediately before the disaster" is defined for comparison. Figure 6 shows the variation of the aggregate variables among sectors by using percentile values, and Fig. 7 shows the variation among firms within a specific sector in the same manner. Figure 8 shows the impact on production of Region ROJ. Figure 9 compares a recovery process among the five direct-damage scenarios. In the model, we adopt the assumption that in all regions and sectors, lifeline services and intermediate inputs procured in the supply chain are completely non-substitutable with other inputs. This may have led

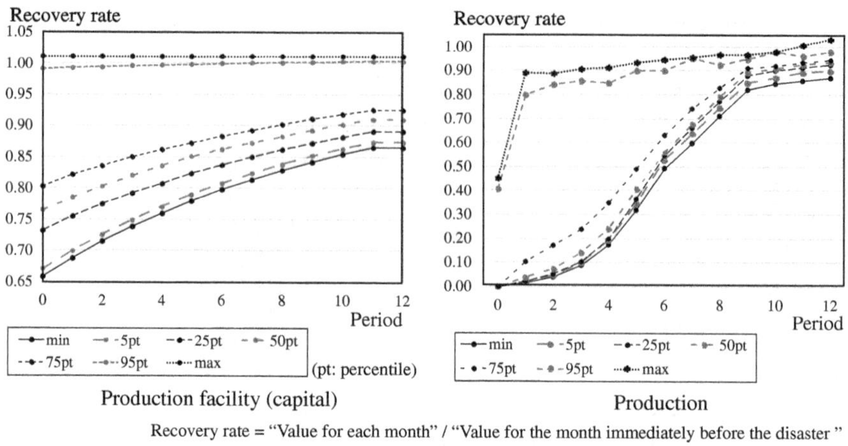

(a)

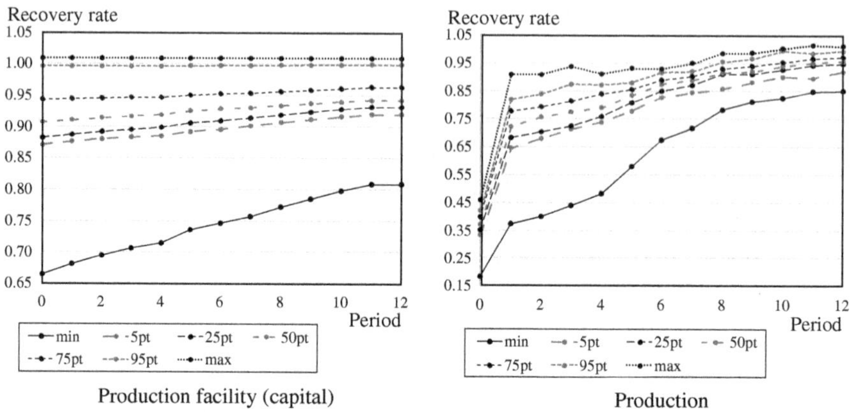

(b)

Fig. 6. Recovery processes and disparity among the 36 sectors in Regions OSK and ROK. (a) Region OSK, and (b) region ROK.

to overestimation of reductions and delays in resumption of production as typically resulted in Region ROJ as shown in Fig. 8. Future work should consider extending the model to account for their partial substitutability with other input goods.

The validity of the policy analysis is directly related to the potential for future practical applications. The model has a structure that allows for the evaluation of a wide variety of policies. Currently, however, due to the limitations of the available data, the values of many

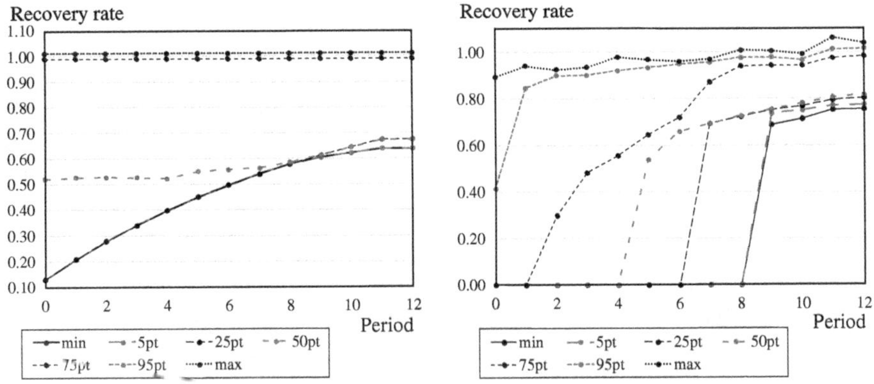

Production facility (capital) of firms of Sector No. 8 in Region OSK

Production of firms of Sector No. 19 in Region ROK

Fig. 7. Disparities of recovery among firms within the same sector.

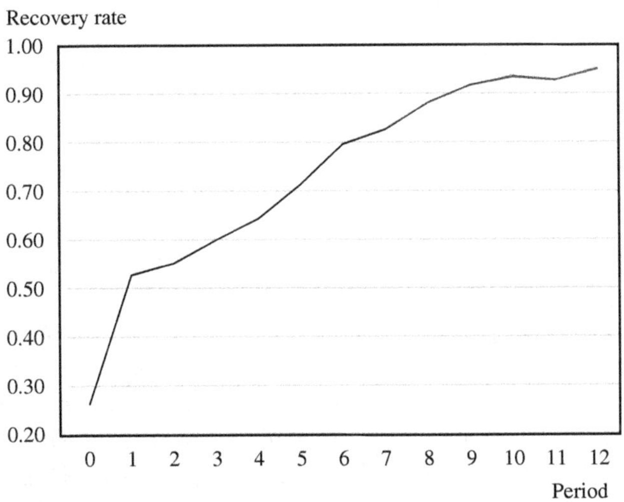

The mean value of productions in the 36 sectors

Fig. 8. Recovery process of production in Region ROJ.

parameters are given by assumption. Therefore, the derived results are not quantitatively validated. Future studies need to develop a variety of policy analyses by acquiring a wider range of data and identifying parameter values.

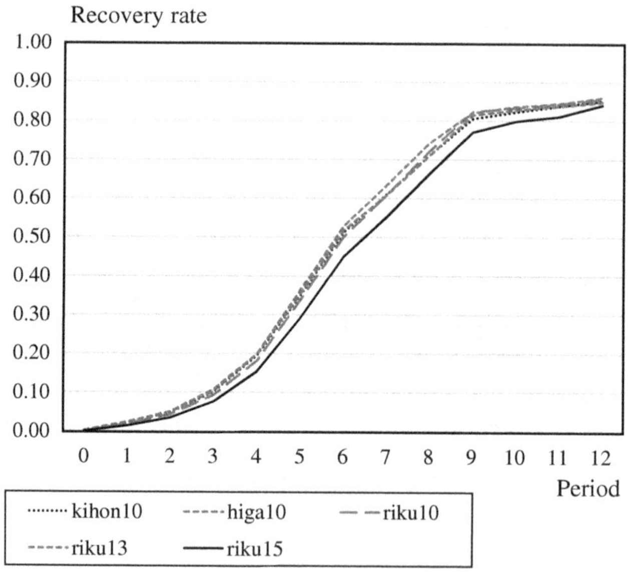

Process of restoring value added in the bottom 25th-percentile-value sectors of the 36 sectors

Fig. 9. An example comparing the recovery process among the five cases of direct earthquake damage.

Figure 10 shows the effect of accelerating infrastructure restoration and development. The basic case is a policy scenario in which damaged infrastructure is restored in 12 months, and the effects of accelerating restoration by 3, 6, and 9 periods are analyzed. Figure 10(a) above shows that the 6-period shortening is the most effective, while the 9-period shortening is inferior to the 6-period shortening because the present value of costs is too large. Figure 10(b) shows that a large proportion of the benefits of 6-period shortening can be attributed to the increase in production in the relatively less damaged sectors, mainly in the first half of the recovery period. Thus, it implies that complementary policies are needed in terms of reducing the damage gap.

Other variables (parameters) for policy analysis include, for example, parameters for allocating investments in 3 regions $\times$ 19 types of infrastructure, parameters for determining how much investment goods to procure from which markets, as well as which infrastructure

Shortening month	Basic case	3	6	9
Values of the policy evaluation function	-1.523×10^8	-1.446×10^8	-1.432×10^8	-1.443×10^8

Recovery of 12 months is used as reference.

(a)

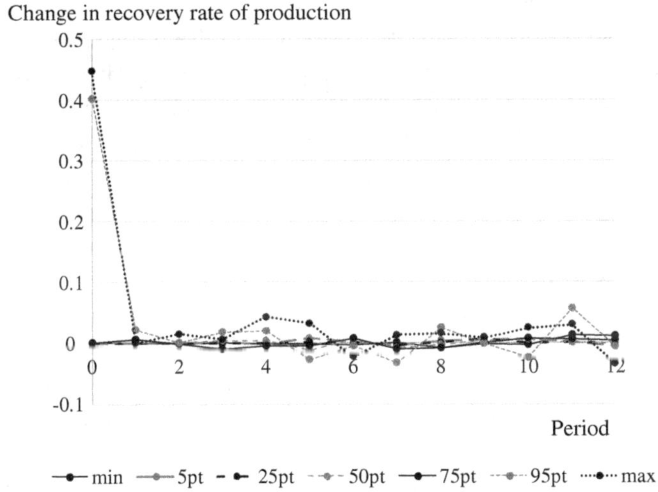

(b)

Fig. 10. Values of the policy evaluation function of the policies for speed-up recovery of infrastructure. Increase in recovery rates of production for seven representative sectors (selected by the above percentile values) in Region OSK, caused by the 6-month shortening policy. (a) Value of policy evaluation function, and (b) change in recovery rate of production.

to focus on in which period within the recovery period and the schedule for debris removal. More diverse data will need to be collected in the future in order to obtain quantitatively meaningful results from these analyses.

5. Conclusion and Future Tasks

This study formulated the process of firms' post-disaster production activities as a dynamic programming problem where a large number of stocks are reconstructed, and firms interact through market prices. The simulation process was composed of two steps; non-parametric value function iteration in the first step, and Monte Carlo simulation of dynamic paths of firms' activities and the price changes in the second step. Parallel computing was applied in several manners that included dividing of the state space inside each firm's problem and introducing of a load-balancing code among nodes that allocate the next job to the node that converges earlier. As a result, the strong scalability was shown to be almost one.

The "curse of dimensionality" problem which appeared as a problem of memory limitation was mitigated by taking three countermeasures for the memory-efficient communication: (1) partitioning the state space to blocks and obtaining values of neighborhood blocks; (2) reducing the size of data that are to be stored in memory by decomposing an array of variables given by polynomials into parts; and (3) setting up a special node dedicated to statistical processing and output work.

The study carried out the case study dealing with 4,320 types of firms, 36 business sectors in Japan that was divided into three regions, and the combinations of the disaster-, the lifeline-, and supply-chain resumption scenarios, which were 10,240 in number, and so on. The model is characterized as an agent-based model in the field of applied economic analysis. In this category, to the best of our knowledge, there has been no study that deals with a dynamic programming problem based on more than 4,000 Bellman equations and computes such a large number of disaster scenarios. On the other hand, this research leaves important topics for the future: the implementation of an efficient algorithm for solving the Bellman equation is the most important issue. The algorithm that this study applied is classified as the synchronous value function iteration, while there is an approach of the asynchronous value function iteration (e.g., Refs. 19 and 20) such as Real-time dynamic programming (RTDP) (e.g., Ref. 2) that includes Bounded RTDP (e.g., Ref. 21), Focused RTDP (e.g., Ref. 22), and value-of-perfect-information RTDP (e.g., Ref. 23). The RTDP approach limits the area of state space to calculate; for example, Bayesian learning is applied to determine the upper and lower bounds of the space, which has efficiently excluded wasteful calculations. In contrast, this study calculated the optimal controls and the associated optimal values over the entire range of the state space of all the agents' decision-making problems, among which only a small part of states were realized as the equilibrium path under one sequence of realized values of random factors. However, in case that the number of runs in Monte Carlo simulation becomes larger, more states are realized as the equilibrium paths, and we could make use of the larger number of information stored in the optimal control array obtained in VFI. Here, additional computing cost may be much smaller compared to other methods that calculate equilibrium solutions on every run of the Monte Carlo simulation. It is also a future topic to compare quantitatively the two approaches.

The second most important task is to enhance the statistical validation of the model. Considering that the point-wise estimation of parameter values is difficult, values of each parameter need to be distributed, and the distribution function needs to be updated through on-line learning. Bayesian data assimilation allows dynamic models not only to be statistically validated over the past years after a specific event, but also to track the realized sequence of stochastic events such as resumptions of lifelines and other socio-economic changes through the filtering process, and furthermore, to update the probability distributions of baseline disaster-damage data through the smoothing process. Improvement of the validation will result in a large increase in the practical applicability of the model whose simulation covers several years or more.

References

1. R. E. Bellman, *Adaptive Control Processes.* New York: Princeton University Press, (1961).
2. A. G. Barto, S. J. Bradtke, and S. P. Singh, Supply chain risk management and artificial intelligence: State of the art and future research directions, *Artif. Intell.* **72**(1–2), 81–138 (1995).
3. K. L. Judd. *Numerical Methods in Economics.* New York: MIT Press, (1998).
4. M. J. Miranda and P. L. Fackler, *Applied Computational Economics and Finance.* New York: MIT Press, (2004).
5. S. L. Chung, F. B. Hanson, and H. H. Xu, Parallel stochastic dynamic programming: Finite element methods (1992).
6. G. Tan, N. Sun, and G. R. Gao, A parallel dynamic programming algorithm on a multi-core architecture, in: *Proceedings of the 19th Annual ACM Symposium on Parallel Algorithms and Architectures,* pp. 135–144 (2007).

7. F. Y. Wang, J. Zhang, Q. Wei, X. Zheng, and L. Li. PDP: parallel dynamic programming. *IEEE/CAA J. Autom. Sinica*, **4**(1), 1–5 (2017).
8. A. P. D. Madrid, S. Dormido, and F. F. Morilla, Reduction of the dimensionality of dynamic programming: A case study, in: *Proceedings of the 1999 American Control Conference, IEEE (Cat. No. 99CH36251)*, number 4, pp. 2852–2856 (1999).
9. J. Rust, Using randomization to break the curse of dimensionality, *Econometrica: J. Econ. Soc.* 487–516 (1997).
10. N. Stokey, R. Lucas, and E. Prescott. *Recursive Methods in Economic Dynamics*. New York: Harvard University Press, (1989).
11. Y. Cai, K. L. Judd, T. M. Lenton, T. S. Lontzek, and D. Narita.
12. W. Nordhaus, Evolution of modeling of the economics of global warming: changes in the dice model, 1992–2017 (2018).
13. B. D. Keen and M. R. Pakko, Monetary policy and natural disasters in a DSGE model, Technical Report 025D, Federal Reserve Bank of St. Louis Working Paper Series (2007).
14. H. Ishiwata and M. Yokomatsu, Dynamic stochastic macroeconomic model of disaster risk reduction investment in developing countries, *Risk Anal.* 38(11), 2424–2440 (2018).
15. Y. Ogawa, Y. Akiyama, M. Yokomatsu, Y. Sekimoto, and R. Shibasaki, Estimation of supply chain network disruption of companies across the country affected by the Nankai trough earthquake tsunami in Kochi City, *J. Disas. Res.* 14(3), 508–520 (2019).
16. G. Baryannis, S. Validi, S. Dani, and G. Antoniou, Supply chain risk management and artificial intelligence: State of the art and future research directions, *Int. J. Product. Res.* **57**(7), 2179–2202 (2019).
17. J. D. Farmer, C. Hepburn, P. Mealy, and A. Teytelboym, A third wave in the economics of climate change, *Environ. Res. Econ.* **62**(2), 329–357 (2015).
18. A. Gill, M. Lalith, S. Poledna, M. Hori, K. Fujita, and T. Ichimura, High-performance computing implementations of agent-based economic models for realizing 1:1 scale simulations of large economies, *IEEE Trans. Parallel Distr. Syst.* **32**(8), 2101–2114 (2021).
19. D. P. Bertsekas, Distributed dynamic programming (1982).
20. D. P. Bertsekas and H. Yu, Distributed asynchronous policy iteration in dynamic programming, in: *48th Annual Allerton Conference on Communication, Control, and Computing*, pp. 1368–1375, IEEE, Allerton (2010).

21. H. B. McMahan, M. Likhachev, and G. J. Gordon. Bounded real-time dynamic programming: Rtdp with monotone upper bounds and performance guarantees, in: *Proceedings of the 22nd International Conference on Machine Learning*, pp. 569–576 (2005).
22. T. Smith, and R. Simmons, Focused real-time dynamic programming for MDPs: Squeezing more out of a heuristic, in: *AAAI*, pp. 1227–1232 (2006).
23. S. Sanner, R. Goetschalckx, K. Driessens, and G. Shani, Bayesian real-time dynamic programming, in: *Proceedings of the 21st International Joint Conference on Artificial Intelligence (IJCAI-09)*, pp. 1784–1789 (2009).

https://doi.org/10.1142/9781800614635_0009

Chapter 9

Toward the 1:1 Scale Agent-Based Simulation of Post-Disaster Economies

Amit Gill*, Maddegedara Lalith†,¶, Sebastian Poledna‡, and Muneo Hori§

Sumitomo Mitsui Construction Company Ltd., Tokyo, Japan
†*Earthquake Research Institute, The University of Tokyo, Bunkyo-ku, Tokyo, Japan*
‡*International Institute for Applied Systems Analysis, Laxenburg, Austria*
§*Research Institute for Value-Added-Information Generation, Japan Agency for Marine-Earth Science and Technology, Yokohama, Japan*
¶*lalith@eri.u-tokyo.ac.jp*

An economy's dependence on infrastructure implies that a destructive earthquake in a major industrial zone, such as the Hanshin Industrial Region of Japan, can lead to disruptions in production and other economic activities cascading through a nation along the supply-chain networks and that last for several years. In order to best utilize the limited resources to recover from such major disasters, it is necessary to comprehensively evaluate all the potential recovery plans using fine-grained integrated models of economy and infrastructure. The lack of such fine-grained and integrated models of economy and infrastructure is a major hurdle in best utilizing limited resources to minimize economic losses caused by major disasters. This chapter presents some details of our attempt to develop a high-performance computing (HPC)-enhanced code for an agent-based economic model and seamlessly integrate it with physics-based earthquake disaster simulators. Our distributed-memory-based HPC-enhanced code is capable of simulating large economies consisting of several hundreds of millions of agents in a few minutes using several tens of CPU cores, making it possible to conduct Monte Carlo simulations to deal with the involved uncertainties utilizing small computer clusters. The details of setting the model parameters using the publicly available Japanese government's statistics, and validation by reproducing observed economy-wide and industry-wide indices, are described. As a demonstrative application of the integrated model, simulations of the Japanese economy in the aftermath of a hypothetical earthquake disaster are conducted, and a comparison of the ordinary state of the economy and the post-disaster economy is presented.

1. Introduction

The economy of a nation is vulnerable to disasters as much as its infrastructure since economic activities heavily depend on healthy supply-chain networks and built environments: mainly buildings and lifeline networks such as gas, water, electricity, sewer, communication, and transportation. This dependence means that severe damages to buildings and lifelines can cause domino-like effects over supply-chain networks causing widespread, long-term economic losses. Especially, major disasters like the impending Nankai Trough earthquake, which is expected to cause significant damages to the infrastructure, can inflict large and long-lasting economic losses. Although the damages to infrastructure are localized, the economic impacts may cascade through the entire nation and even last for a number of years because economic entities depend on each other to carry out various economic activities. Long-term economic impacts can outweigh the direct damages to infrastructure in case of a major earthquake in an industrial region such as the Hanshin Industrial Region of Japan.

Though national and local governments strengthen the existing infrastructure as a preventive measure, only a limited number of infrastructure components can be strengthened in a given period of time due to limited resources. To maximize the benefits, infrastructure components to be strengthened must be selected based on their importance to long-term economic performance, accurate estimation of which requires integrated fine-grained simulations of economic and infrastructural models. Further, when recovering from a destructive earthquake in a major industrial zone, comprehensive evaluation of potential recovery plans considering quantitative estimation of their long-term economic performance is essential to identify the plan which will best utilize the limited human, material, and financial resources to ensure faster recovery with minimum losses. Although a manufacturing firm can recover its own facilities, it will not be economically active unless all the lifeline supplies are restored, it has access to its suppliers, its workers are in a state to come to work, and it has access to the market. A comprehensive evaluation of recovery plans must check all these conditions for each economic entity in the country, according to the weekly or monthly progress of recovery. As discussed, both preventive measures and recovery require comprehensive evaluations using fine-grained integrated simulations. Simulations of ordinary economic conditions, such as inflation or recession, may not require such fine-grained simulations: capturing average behavior is sufficient since the entities of the same economic category experience similar conditions. The lack of fine-grained and integrated models of economy and infrastructures is a major hurdle in best utilizing limited resources to minimize economic losses caused by earthquakes and similar disasters.

Analysis of economic impacts of disasters have been an area of interest for researchers over the past several decades. Rose and Liao[1] use a computable general equilibrium (CGE) model for analyzing economic impacts of a water supply failure due to an earthquake disaster. Tatano and Tsuchiya[2] developed a spatial computable general equilibrium (SCGE) model integrated with a transportation model for estimating indirect economic impacts due to transportation network failure. They showed that indirect economic losses due to transportation network failure after the Niigata–Chūetsu earthquake of 2004 were about 20% to 40% (depending on the region) of direct economic losses. Other examples of the usage of CGE models for estimating economic impacts of infrastructure failures are Brozović *et al.*[3] and Lorenzo *et al.*[4] The CGE models are based on neo-classical economics, and therefore, assume that economic entities are homogeneous and rational, have full information of the market, and take decisions collectively to optimize their utilities. These models find the equilibrium state of the economy for a given economic scenario.

However, the post-disaster economy is in transition state, therefore, neo-classical assumptions cannot describe the behavior of economic entities. Since only a few representative economic entities are considered, the disaster-induced heterogeneous conditions experienced by economic entities are ignored. Further, it is not possible to include the state of each individual economic entity and infrastructure component, and their interdependencies. Therefore, these models fail to estimate the economic impacts at the individual economic entity level, and consequently overestimate the economic performance of a reconstruction plan.

Another commonly used approach for analysis of economic impacts of disasters is the use of input–output models. Hallegatte[5] developed an advanced adaptive regional input–output model for economic impact assessment of natural disasters and demonstrated its application by assessing economic impacts of hurricane Katrina. Markhvida *et al.*[6] use this model to quantify household well-being losses due to a potential large earthquake in the San Francisco Bay area. These input–output models derive their parameters from the input–output table (IO-table) of the corresponding country or region. Therefore, they can easily include the interdependencies of various economic sectors and final-demand sectors. However, these models are highly coarse-grained since the analysis is carried out at the industrial sectors level. As an example, Markhvida *et al.*[6] consider 15 aggregated sectors only. The model requires total infrastructure losses to account for the disaster impacts on infrastructure. The infrastructure losses are then assumed to be capital losses to some industries and households. In addition to the owned capital (i.e., buildings, machines, etc.), the firms are dependent on various infrastructural components like road, electricity, and water supply networks. Since the geographical locations of businesses and infrastructure are not considered, dependency of firms on infrastructure is not completely modeled. Consequently, these models are unable to quantify the impacts due to the failure of each infrastructure component. Quantification of such impacts is of the utmost importance in identifying better performing recovery plans.

An agent-based economic model (ABEM) overcomes the deficiencies of CGE and input–output models by simulating national economies at 1:1 scale without making unrealistic assumptions like fully informed and homogeneous agents, etc. An agent in ABEMs represents an atomic economic entity in the real world and the rules defining the agent's behavior are derived from behavioral economics or real characteristics of the represented economic entity. Recent advancements in big data will enable, within a few decades time, to identify each individual economic entity's characteristics, based on which accurate 1:1 scale ABEMs can be built. Different ABEMs have been developed to study various economic phenomena.[7–12] Poledna *et al.*[11] developed a 1:1 scale ABEM for the Austrian economy to estimate economic impacts of flood disasters in Austria. However, their computer implementation is not fully integrated with infrastructure and relies on pre-estimated infrastructural losses to include their impacts on economic entities. The high flexibility of ABEMs permits one to easily include geographical locations of each economic entity and infrastructure component. A 1:1 scale ABEM seamlessly integrated with a high-fidelity model of infrastructure is an ideal tool to comprehensively quantify direct and indirect, short- and long-term economic impacts of a disaster at a finer scale even including an individual economic entity's reliance on multiple infrastructures. Such numerical models will facilitate identification of critical economic entities and infrastructural components, making it possible to form better-performing disaster-mitigation and recovery plans, and quantitatively compare given set of recovery plans by comprehensively analyzing their short- and long-term economic performances. According to our literature survey, such a seamlessly integrated system capable of simulating 1:1 scale models of large economies does not exist, as of 2022.

Although there are several publications[9,13–15] emphasizing the need for HPC-enhanced implementations of the ABEMs capable of simulating large economies, such as the Eurozone ($\sim$330 $\times$ 10^6 agents) and Japan ($\sim$127 $\times$ 10^6 agents), no such implementations have been reported in the literature to the best of our knowledge. Although there are various HPC-enhanced general-purpose agent-based modeling frameworks like Pandora,[16] Repast HPC,[17] EcoLab,[18] FLAME,[19] etc., they have limited use in agent-based modeling of economies. These general-purpose frameworks are designed for social problems, such as pandemics, pedestrians, traffic, in which the agents' interactions are limited to uni-directional exchanges within small neighborhoods on a few neighbor connectivity graphs. As explained in Section 2, in ABEMs, agents interact on multiple graphs of different characteristics such as scale-free graphs and centralized graphs (e.g., goods market, labor market, bank–customer interactions, government–tax payer interaction). Most interactions, especially on the scale-free graphs, are not limited to small neighborhoods of the agents. As an example, any buyer can buy goods from any preferred seller. Further, a majority of these interactions are random and bi-directional (i.e., the agents originating the interaction wait for the reply from the contacted destination). Therefore, it is necessary to develop a dedicated HPC-framework that facilitates efficient random bi-directional interactions over large scale-free and centralized graphs.

Motivated by the aforementioned discussed needs, we developed a HPC-enhanced framework for ABEMs capable of 1:1 scale simulations of large economies and seamlessly integrated it with a high-fidelity earthquake disaster simulator to perform comprehensive end-to-end simulations of disasters and economies. While the developed HPC-enhanced framework can accommodate any common ABEM, we used the ABEM by Poledna *et al.*[10,12] as the base ABEM. Due to the random nature of the agents' interactions, our HPC-enhanced ABEM is mainly a distributed-memory parallel model based on message passing interface (MPI).[20,21] We further extended it to a distributed + shared-memory hybrid parallel framework targeting the optimal use of many-core computing-nodes that have low per-core memory capacity, such as the Fugaku super-computer.[20] The performance of the implementation has been evaluated by simulating a large economy with 331 million agents, which is similar in size to the Eurozone economy. The current implementation possesses a strong scalability of up to 70%, which is high for this class of problems. Moreover, improved serial algorithms significantly boost the computational performance, enabling large economies to be simulated using a relatively small amount of computational resources, even on workstations, and that too, within minutes.

We calibrated the HPC-enhanced model to simulate the Japanese economy.[22] The model of the Japanese economy consists of a total of 127 million agents comprising 1.8 million domestic firms, 0.9 million foreign firms, 1.8 million investors, 60 millions workers, 62 million inactive households, 1 bank, the central bank, and the central government. The model parameters and initial parameters of the agents are generated from the macroeconomic data of Japan, available from the governmental data sources. Further, the model has been validated by comparing the simulated macroeconomic parameters with their observed values.[22] To account for various uncertainties, Monte Carlo simulations with 1,000 random samples were conducted, and the model reproduced the observed data within $\pm 1\sigma$ limit.

In order to facilitate end-to-end simulations, the developed HPC-enhanced ABEM has been seamlessly integrated with an Integrated Earthquake Simulator (IES)[23,24] and the Performance Assessment Calculation Tool (PACT).[25] IES is utilized to simulate strong ground motion and corresponding seismic response of the buildings located at the sites. PACT is utilized to convert the damage indices from IES into repair costs and repair time. The post-disaster economy in the aftermath of a potential earthquake in the Osaka bay area is simulated as a demonstrative application of the developed integrated system. Specifically, 1.8 million buildings of the Osaka bay area have been analyzed for their seismic performance due to a Nankai Trough mega earthquake as proposed by the Cabinet office of Japan, and the resulting economic impacts have been estimated, under a simple recovery scenario.

The rest of the chapter has been organized as follows. Section 2 provides a brief overview of the ABEM and our HPC-implementation. Step-by-step guide to the model parameter generation process is presented in Section 3, while Section 4 presents the model verification and validation processes. Section 5 discusses the integration with physics-based simulators. A numerical experiment has been presented in Section 6 to demonstrate the application of the integrated system for seamless simulations of disasters and economies. The last section contains a short summary and some future directions.

2. Agent-Based Economic Models and HPC Implementation

Agent-based modeling (ABM) is a widely used technique which takes a bottom-up approach to approximate a system as a collection of interacting autonomous entities called agents.[26–28] Each agent assesses the changes occurring in its surroundings (e.g., neighborhood in the physical space and/or connectivity graph) and autonomously acts according to a set of rules. The set of rules, which defines the agent, can either be fixed or evolving. The evolving rules modeled with machine learning techniques enable one to model learning and adapting behaviors observed in biological or social systems.[26] Agent-based models have been applied to study a wide range of problems spanning from biological systems to complex engineering systems.[26–30]

Various economic systems, in which individual economic entities interact to satisfy their needs and attain their objectives, are natural applications for agent-based systems. During the last two decades, a number of agent-based models for simulating different areas such as banking systems, markets, labor markets, resources, housing market, risk have been proposed. Steinbacher *et al.*[31] and Axtell *et al.*[32] provide detailed reviews of the history, different types, and applications of agent-based models in economics, while there are a number of books providing the details and computer implementations.[33–38]

The main objective of this section is to provide a brief introduction to agent-based macroeconomic models for simulating national or regional economies. Since there exist a large number of studies in the literature (for example, see Refs. 33, 34, and 36 and the citations by Steinbacher *et al.*[31] and Axtell *et al.*[32]) providing the mathematical definition of the agents' behavioral rules, we omit the mathematical details and only provide a very basic introduction to the agents and their activities. Readers are referred to Poledna *et al.*'s work[10,12] for the details of the agent-based model used in this chapter. The latter half of this section provides some details of our HPC-enhanced implementation and its computational performance measurements.

2.1. *Agent-based economic models*

Agent-based macroeconomic models are computer-based empirically founded models for simulating national or regional economies as a time-step-driven dynamic system of heterogeneous interacting agents. For the sake of simplicity, we refer to agent-based macroeconomic models as agent-based economic models (ABEMs) in this chapter. ABEMs consist of a large number of agents representing each individual economic entity of the country or the region being simulated. The behaviors of the agents are modeled based on the observed behaviors of respective economic entities. These models evolve as a result of interactions among the agents mimicking time evolution of an economy.

Though it is not necessary for the objectives of this chapter, let's make a very short comparison of ABEMs with computable general equilibrium models (CGEMs) with a focus on studying post-disaster economies. CGEMs, which are the standard in economics, try to find how to reach an optimal state, defined according to a given objective function, by solving a set of partial differential equations under the following neo-classical assumptions of economics:

(1) agents are homogeneous;
(2) agents are perfectly rational and perfectly informed;
(3) all the agents try to maximize their utility(individuals) or profit(firms).

Unlike the models based on neo-classical assumptions, the agents in the ABEMs are heterogeneous in terms of their decision-making, resources owned, access to information, etc. As an example, each individual firm agent has different amounts of capital and independently makes decisions, such as its next period's production and employment. The agents have partial information of the market, their competitors, and market behavior. Further, the agents are not acting to solve an optimization problem. Instead, they try to take the best decision based on the available information and their experience. Often, the mathematical optimization problems posed by CGEMs are highly nonlinear and the computation cost rapidly increases with the increase in the number of variables. To resolve the excessive computational demand, CGEMs drastically reduce the number of variables by aggregating economic entities (e.g., a large number of independent firms of an industrial sector are included in the model as a few aggregated firms). Such aggregation may not compromise the solution when analyzing ordinary economic conditions such as inflation since such economic conditions affect all the economic entities of any sub-group, such as firms of an industrial sector, almost equally. On the other hand, after a major disaster, different economic entities belonging to the same sub-group can experience highly heterogeneous conditions. As an example, different manufacturers in the disaster-hit area can experience heterogeneous conditions such as long delay in repair of damages, shortage of gas and water due to delay in repairing lifeline networks, shortage of labor, poor access to supply-chain networks, transportation difficulties, etc. Therefore, when simulating post-disaster economies, aggregation can lead to poor and unrealistic results. Further, including the interdependencies of business and infrastructure such as buildings and lifelines, particularly the repair progress of each lifeline, into the CGEM framework can be quite challenging or even not possible.

In comparison to CGEMs, ABEMs do not solve an optimization problem. Instead, ABEMs simulate time evolution of the systems of economic entities according to a set of pre-defined rules and given initial conditions, where the rules are set to make the agents mimic the behavior of their real-world counterparts. Therefore, ABEMs are computationally light enough to include even each individual entity in a large economic zone. Further, the time-step-driven nature of ABEMs makes them highly flexible to include almost any complexity observed in the real world such as the influence of the repair state of lifelines on the production capacity. The agents have their working/living space, and they are linked to other agents by roads and other transportation networks. Though the existing ABEMs in professional economics ignore them, these physical connections are essential for the target post-disaster economic simulations since the actions and interactions of their real-world counterparts heavily depend on these physical connections. An ABEM is an ideal tool for studying a post-disaster economy due to its characteristics such as being designed to simulate the transient state of the economy, computationally light enough to include each individual entity in large economies, ability to model any complex interaction among economic entities over multiple physical or abstract networks, flexibility to be integrated with lifeline networks, and ease of integrating with other physics-based disaster simulators.

Several groups have proposed agent-based models for simulating an economy (Refs. 7 and 8). Most of these models are incomplete proof-of-concept models and vary in terms of level of economic detailing, money flow, etc. The ABEM proposed by Poledna *et al.*[10,12] is the most complete and comprehensive model proposed to date, at least for our target application of simulating a post-disaster economy.

The following are some of the advantages of their model compared to other ABEMs:

- The model is designed in such a way that all the parameters used are derived from the real economic data; mainly IO-tables.
- The model has been verified by comparing with the dynamic stochastic general equilibrium model (DSGM) and other standard economic models.
- The model includes all major economic entities viz. firms, a wide range of industrial sectors, investors, workers, inactive households, banks, the central bank, the central government, foreign buyers, and sellers.

Further, in Poledna *et al.*'s model, the agents do not need to have money at the moment of buying a product or service. Instead, they settle all the transactions at the end of the financial period. For example, firms are dependent on loans, sales, etc., which may not be quantified at the time of buying the raw material. This feature provides advantages in implementation of scalable parallel computing extensions. The use of credit makes it unnecessary to process most of the events like receiving wages, social benefits, buying, in a particular order, making it possible to overlap MPI communications and computations, thereby reducing or even completely eliminating communication overheads in distributed-memory parallel computing.

2.2. *Agent types and their roles*

Figure 1 shows a schematic diagram of a typical ABEM. At the heart of the model are industries, each consisting of a large number of firms. Each of these is connected to banks, governments, households, foreign buyers, and sellers, while the central bank is connected to the government and banks. The basic functions of these agents are briefly explained in what follows. Interested readers are referred to Poledna *et al.*'s work[10, 12] for mathematically precise definitions of each agent's decision-making and actions of our base ABEM. The sequence in which agents' main events are executed is presented in Algorithms 1 and 2, which are included at the end of this section.

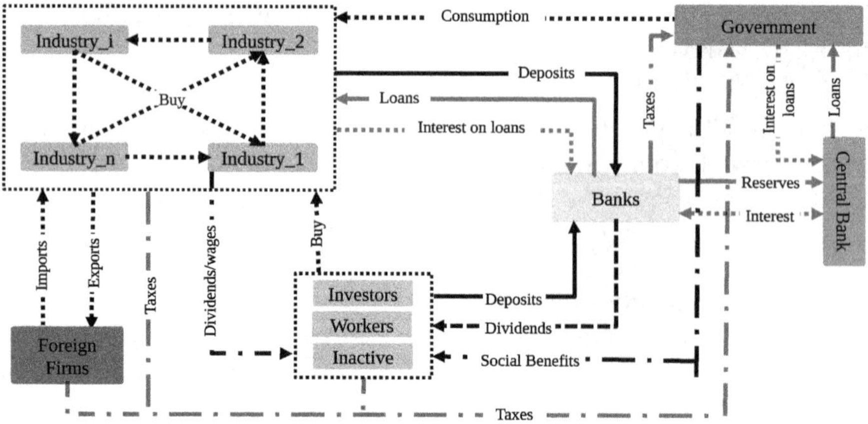

Fig. 1. Schematic diagram of an ABEM.

2.2.1. *Firms*

Firms are categorized into a number of industries depending on the availability of data. Our main data-source for initializing the agents is IO-tables. Hence, we decide the number of industries according to the number of industries in the available IO-tables; refer Section 3 for details. Firms produce goods of a pertaining industry using intermediate goods, capital goods, and labor from the working population. It is assumed that capital goods required by firms are the same irrespective of their industry. Whereas intermediate goods required by firms depend on their industry (i.e., their product). The required amount of intermediate and capitals goods, from each industrial sector, to produce a unit of output, is decided based on the IO-table. As an example, the IO-table for 108 industrial sectors of Japan is provided by e-Stat, a portal site for Japanese Government Statistics.[a] At the beginning of each period t, firms decide their production quantity and price for this period, based on their past experiences: auto regression with lag 1 (i.e., AR(1)) on corresponding historical data is used for making these experience-based estimations. After deciding production quantity, the firms either hire or fire workers to meet the labor requirement to meet their target production. If the firms fail to hire the required number of laborers, their production estimates

[a]https://www.e-stat.go.jp/en, accessed on 13 November 2022.

are revised according to the available labor. Firms use the intermediate and capital goods purchased in the period t for the production in period $t + 1$. Firms estimate the amount of capital to purchase such that they can replenish the depreciated capital and meet their long-term production plan; firms gradually increase their production capacity to meet their long-term plans. Firms requiring additional funds apply for loans from a bank to meet their financial needs. The bank evaluates the loan applications based on the firms' chances of defaulting and its ability to legally extend loans. According to the extended amount of loans, the firms revise their investment plans.

Once production is complete, all the firms enter into the goods market to sell their products of period t and buy the required capital and intermediate goods for the $t+1$ period. In the real world, a majority of customers are not brand loyal but are attracted to large firms with cheaper products. To mimic that, sellers (i.e., firms) are assigned a probability of being visited by a customer (e.g., a household, a firm, a government entity) according to the sellers' production capacities and the selling prices of their goods. The customers randomly choose the seller to visit according to this probability distribution, trying to mimic the real-world's behavior. In the goods market, firms visit firms of other industries to buy the required intermediate and capital goods, until either all the buying firms have satisfied their demands, or all the sellers are sold out. If the total production is lower than the total demand of the firms, some firms will have unsatisfied demand, which they report to some randomly chosen sellers who will try to increase their production in the next period so that there are no unsatisfied firms in the next period.

After the goods market, the firms pay wages to their workers; pay loan installments and the interest on loans to the bank; pay corporate tax, production and products tax, capital formation tax; receive any subsidies from the government; and pay the employers' contribution of social insurance for their workers to the government. At the end of each period, the firms do their accounting to calculate net profit, dividend to be paid, equity, etc. If the bank deposits and equity of any firm fall below zero, it goes bankrupt. Any bankrupt firm is revived by the bank by writing-off a portion of the loans and overdraft. The revived firm functions as a new firm which replaces the bankrupted firm, yet having all the assets (capital) of the old firm and a portion of the old firm's loan.

2.2.2. *Imports and exports*

The region or country being simulated interacts with the rest of the world through foreign sellers and buyers. Foreign sellers provide exogenously defined quantities to the goods market of corresponding industrial sectors at a price of $\bar{P}_g(t-1)$, which is the average price of goods of the respective industry(g) in the period $t-1$. All the foreign firms of an industrial sector are identical in terms of the amount of goods they sell and the unit prices. The random buying and selling process at the goods market determines the actual quantity sold by foreign firms. Consumption budget of foreign buyers is also exogenously given and is assumed to be constant in real terms but adjusted to the last period's average price of the product of the corresponding industry. As mentioned, the amount made available by foreign sellers and the demand of foreign buyers are exogenously set in our current model. When reproducing past economic progress for the validation, these exogenous values can be set according to the real past observations. However, when simulating the future economic progress, these exogenous quantities must be estimated using suitable time-series forecasting techniques.

2.2.3. *Households*

Households are divided into three categories: investors, workers, and inactive households. All households buy consumption and investment goods at the goods market following a behavior similar to the abovementioned purchasing behavior of the firms; they purchase required goods from randomly chosen sellers from relevant industries until either all the households satisfy their demands or sellers are sold-out. All the households pay value-added tax on goods they buy. Households receive/pay interest on their deposit/overdraft from the bank, and any remaining money is saved in banks at the end of each period.

- *Investors*: Investors are the owners of the firms and financial institutions (i.e., banks). The model currently includes only one investor per firm. However, multiple investors to a firm can be assigned to mimic the real world if there is a suitable data set or set of rules to decide the number of investors and amount of their investments in the firm. The main income of investors is dividends paid by the firm they invested in, and they pay tax on their income. Additionally, the investors receive relevant social benefits from the government. The investors decide their consumption budget (i.e., budget to purchase desired consumption goods and investment goods) at the start of each period based on the expected profit of their firms.

- *Workers*: Workers provide labor to the firms and in turn receive wages. Labor is not categorized according to the skills or industry. In each period, unemployed workers look for jobs in the job market by visiting a random set of firms with vacancies. Irrespective of the sizes of firms or wages, the workers accept the first job offer they get. The employed workers receive wages from their employer and social benefits from the government, whereas unemployed workers receive unemployment benefits from the government. The workers decide their consumption budget for consumption goods and investment goods at the start of each period depending on their past income and expected income in this period. The employed workers pay income tax to the government.
- *Inactive households*: Inactive households represent the economically inactive subset such as children, retired people, and disabled. Their main income is the social benefits from the government and they pay only value-added tax on their consumption.

2.2.4. *Bank*

Due to the absence of required data for modeling multiple banks, the ABEM we implemented includes only one bank. The bank accepts deposits from firms and households and pays the customers interest on their deposits. The customers can request overdraft from the bank, and the bank charges interest on the overdraft. The bank extends loans to firms depending on the financial soundness of the firms requesting loans. The bank deposits its reserves at the central bank or receives advances from the central bank. The bank pays dividends to its investors and pays corporate tax to the central government.

2.2.5. *Central bank*

The central bank sets policy rate. It provides liquidity to the banking system by extending advances to the bank and takes deposits from the bank in the form of bank reserves. The central bank is the main creditor to the central government and extends loans to the government unconditionally.

2.2.6. *Central government*

In the model, the government serves two functions: as a consumer in the market and a redistributive entity that levies various taxes and social contributions to provide social benefits to the households or subsidies to the firms. The government is composed of various government entities which represent central governments, state governments, local governments, and social security funds. Total consumption budget of the government for the product of an industry is given exogenously and is assumed to be constant in real terms but adjusted to the last period's average price of the product.

2.3. *Need for an HPC implementation*

A 1:1 scale ABEM of a large economy consists of hundreds of millions of agents. As examples, the number of agents required to simulate the Japanese and the Eurozone economy in 1:1 scale is around 127 million and 331 million, respectively. Compared to physics-based simulations, such as simulation of strong ground motion of a large metropolis or tsunami run-off, these large-scale ABEMs involve significantly small numbers of floating-point operations. However, the complicated interactions among agents and involvement of large data makes large-scale ABEMs computationally demanding, and memory bound. Further, agent-based simulations require stochastic approaches, such as Monte Carlo simulations, to deal with the inherent uncertainties arising from the assumptions and approximations in numerical models, uncertainties in data, etc. This need of Monte Carlo simulations further increases the computational demand by several folds. Hence, our objective of 1:1 scale simulations of post-disaster economies requires an efficient high-performance implementation. To make the implementation useful for a wider range of users, the HPC implementation must be numerically efficient to complete the required Monte Carlo simulations in a reasonably short time (e.g., in a few hours) utilizing widely available amounts of computer resources (e.g., computing clusters consisting of several thousands of CPU cores are increasingly commonplace at most research institutions).

The rest of this section briefly explains how we developed an HPC-enhanced ABEM capable of simulating a single period (i.e., one-time-step iteration) of 1:1 scale model of Japan, consisting of nearly 127 million agents, within 110 s utilizing 64 CPU cores. We refer any interested reader to the paper by Gill *et al.*[20] for the details of the implementation.

2.4. *Challenges in scalable HPC implementation*

Distributed-memory models provide more opportunities to execute agents' actions in parallel since most of the agents' interactions are random. In a distributed-memory parallel model, mutually exclusive subsets of agents (i.e., *partitions* in the terminology of HPC literature) are assigned to CPU cores located on multiple computers connected through a high-speed network. The CPU cores in distributed-memory parallel models, being completely independent (i.e., each has independent execution thread and private memory), need to exchange information via the network to collectively solve a given problem. This exchange of information via the network interfaces is facilitated by MPI.[39] Each computing element (e.g., CPU core) in distributed parallel models is assigned a unique integer ID ranging from 0 to $N-1$, where N is the number of computing units. Following the standard literature, we refer to each of these computing units as an *MPI-rank* or a *rank*. Assigning equal computing workload and minimizing communication overhead are the keys to developing scalable distributed-parallel-computing codes.

Complex interactions of agents are the major hurdles in implementing a scalable parallel-computing extension, hence understanding the nature of these interactions is important in devising strategies to attain high parallel scalability. Specifically, it is important to understand the nature of different types of interactions occurring between any two agents, and the connectivity graphs formed among the agents during these interactions.

The interactions between any two agents can be divided into two categories: uni-directional and bi-directional. Interactions such as paying tax to the government and depositing money in the bank are uni-directional, whereas interactions involving getting a bank loan, subsidy payments by the government, buying goods, job hunting, etc., are bi-directional. Uni-directional interactions involve some data transfer from the source agents to the target agents and can easily be parallelized by clumping together all the related MPI communications. Bi-directional interactions consist of three stages: source agent making a request to target agent, decision-making by the target agent, and informing the decision to the source agent. As an example, when a firm requests a loan, it has to wait until the bank processes its request and informs the decision. Therefore, any bi-directional interaction between agents located in distinct MPI-ranks serializes the corresponding events occurring on those MPI-ranks (i.e., make the events occur in a sequential order), making the run-time of parallel code longer than that of a serial code.

Interactions of agents occur on graphs of different topologies. According to the nature of the connectivity, the involved graphs can be categorized into two: centralized graphs and scale-free graphs. Firms' and households' interactions with the banks, tax payments to the government, and social benefit payments by the government take place over centralized graphs (see Fig. 2(a)). While some of the interactions on centralized graphs are easy to be made parallel, others require strategic planning. Tax paying to the government and depositing money in the bank are uni-directional and can easily be parallelized by making each MPI-rank collect the tax from its agents and then sending those to the rank containing the government agent;

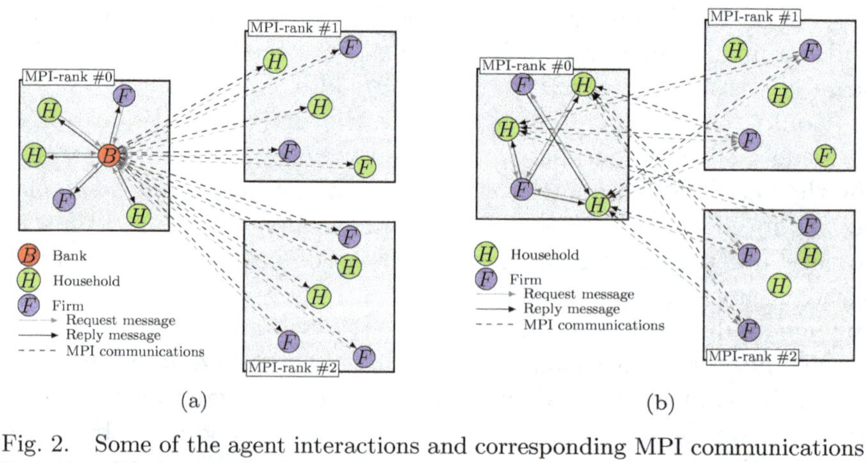

Fig. 2. Some of the agent interactions and corresponding MPI communications. Note that (b) shows only the interactions originating from the households in MPI-rank 0. While which firms to visit is random, the number of maximum visits by a customer is unspecified. (a) Bank-customer interactions, and (b) households on rank 0 purchasing goods from firms.

the same strategy can be applied to depositing money in the bank. Events such as social benefit payments by the government, withdrawing money, and taking a loan are bi-directional, hence cause serialization over all the MPI-ranks.

Bi-directional interactions over scale-free graphs in the goods market and labor market are the major hurdles in parallelizing ABEMs. In each period, all buyers (i.e., firms, foreign firms, households, and government agencies) visit a random set of sellers (i.e., local firms and foreign firms) from each industry to buy capital goods and consumption goods (see Fig. 2(b)). The buyers choose sellers according to the size and the unit price of the sellers; sellers producing large quantities and lower unit prices have a higher probability of being visited by the buyers. The number of sellers visited by a buyer is unpredictable, as it depends on the unit price and the quantity of products available at a seller. Therefore, the goods market can produce a very large number of random and bi-directional communications between ranks. As mentioned previously, these bi-directional interactions make the related events occurring on different ranks happen in a sequential order, making the run-time of parallel implementations longer than a serial implementation. Another random scale-free graph is the job market. According to the labor needed for their planned production, each firm fires or hires workers. The unemployed workers and firms seeking new labor form the job market. Similar to the goods market, the number of firms visited by a worker in search of a job is unpredictable. The random and bi-directional nature of these scale-free graphs pose the main challenge in developing a scalable solution. Goods market being the most computationally involving event, how the activities over the corresponding scale-free graph are treated in the parallel implementation dictates the computational efficiency.

2.5. *Scalable high-performance implementation*

As previously discussed, the complex interactions among agents are the major barriers in developing scalable parallel codes. Further, both the complex interactions and involvement of large volume of data are the main contributors to longer run-times. This section briefly presents the main strategies used to develop a scalable parallel extension and reduce the computation time.

Assigning equal workload to each MPI-rank while minimizing the number and volume of communications among the MPI-ranks are two main strategies used in partitioning. The goods market and the job market being the most communication intensive, a scale-free graph of either of these markets is ideal for partitioning. However, neither of these graphs can be used for partitioning since neither of these exists at the start of the simulation; topological connectivity of these graphs is defined only when the corresponding market interactions are completed, and the connectivity randomly changes from period to period. Therefore, we partitioned the agents based on a representative scale-free graph of employer–employee interactions, and other interaction graphs such as for the goods market are made available at a low communication cost using various strategies.

The representative employer–employee interaction graph is constructed by joining each firm f_i with n_i nearest workers, where n_i is the initial labor demand of f_i. From the remaining agents, inactive households and foreign firms are included in the graph by connecting them to the nearest one or two firms. Neighboring firms are connected to each other by triangulating to ensure that the generated scale-free graph is not segmented. The nodes of the graph are assigned weights according to the computational workload of the associated agents. This hypothetical graph is partitioned using the `METIS_PartGraphKway()` function of the METIS library,[40] such that each partition has a nearly equal sum of nodal weights and the partition boundaries cut the minimum number of edges. Firm investors are assigned to the partitions that include the respective firms. The bank, government, and central bank are assigned to master rank (e.g., rank 0) so that they can interact among themselves without requiring an MPI communication.

2.5.1. *Scalable solutions for the centralized graphs*

Two problems associated with the centralized graphs are the imbalanced workload (i.e., the rank holding the center node of the graph has to process millions of requests received from the rest of the nodes located on other ranks) and the large number of MPI-communications. Both these problems can be effectively addressed by decentralizing the functions of the agents at the center node (e.g., the bank in credit network). As an example, once the credit market is decentralized by establishing local branches of the bank in each rank, it requires only one `MPI_Scatter()` or `MPI_Bcast()` to distribute the

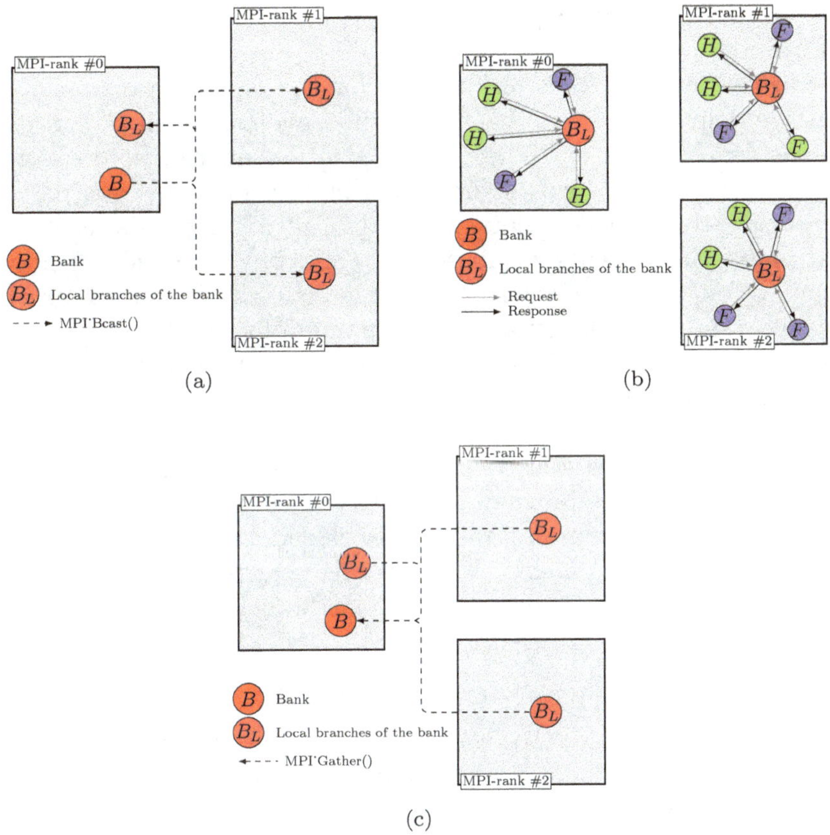

Fig. 3. With local bank branches on each MPI-rank, bank–customer interactions require only two collective communications. (a) Main branch sends financial status to local branches, (b) local branches address the needs of their customers, and (c) local branches send business updates to the main branch.

financial details from the bank's head office to the local branches (see Fig. 3(a)) and one `MPI_Gather()` to collect and deliver information on the business activities of branches to the head office (see Fig. 3(c)). Decentralization enables the customers in each rank to directly interact with the corresponding local branch, thereby distributing the computational workload equally among the ranks (see Fig. 3(b)). Further, the decentralization of the interactions of the centralized graphs significantly reduces the memory requirements of computational nodes since the data (e.g., data of the tens of millions of customers of the bank) can be distributed over all the ranks (i.e., stored at local branches), instead of at the rank holding the central node of the graph.

Similar to the decentralization of bank activities, tax collection and the delivery of social benefits to households, which involve two graphs centralized around the government agent, can be decentralized by establishing local governments in each rank. In each period, the local governments collect taxes from all other agents, calculate and pay social benefits to households, and buy a portion of goods required by the government. At the end of the financial period, local-governments communicate the amount of taxes collected, social benefits paid, and the goods bought to their parent agent, the central government, which is located in the master rank, using a single `MPI_Reduce()`.

As previously explained, decentralization is a perfect solution since it involves a maximum of two collective communications, evenly distributes the computational workload, and reduces the hardware memory requirements. Finding this near-perfect solution was straightforward since it already exists in the real world.

2.5.2. *Scalable solution for scale-free graphs*

As explained in Section 2.4, the random and bi-directional nature of scale-free graphs is the main hurdle in developing a scalable parallel code. Close investigations reveal that it is the agent-based models' highly stylized mathematical models of the goods and labor markets which create these hurdles. As an example, in agent-based models, the customers buy goods directly from manufacturing firms, whereas in the real world, customers buy goods from intermediate entities like supermarkets. Just as in centralized graphs, we can overcome the hurdles in scale-free graphs by decentralizing the involved interactions by mimicking the real world.

The goods market can be decentralized by introducing a sales outlet in each rank for each firm as shown in Fig. 4. A firm $f_i^{\bar{r},s}$ from an industrial sector s located in a given rank $\bar{r}$ is made to sell its products to the customers in any rank r by establishing a sales outlet $o_i^{r,s}$ in rank r. This enables any consumers in rank r to directly buy goods of firm $f_i^{\bar{r},s}$ by visiting the sales outlet $o_i^{r,s}$. While centralized graphs require decentralizing the work of one entity (e.g., bank or government), goods market requires decentralization of millions of

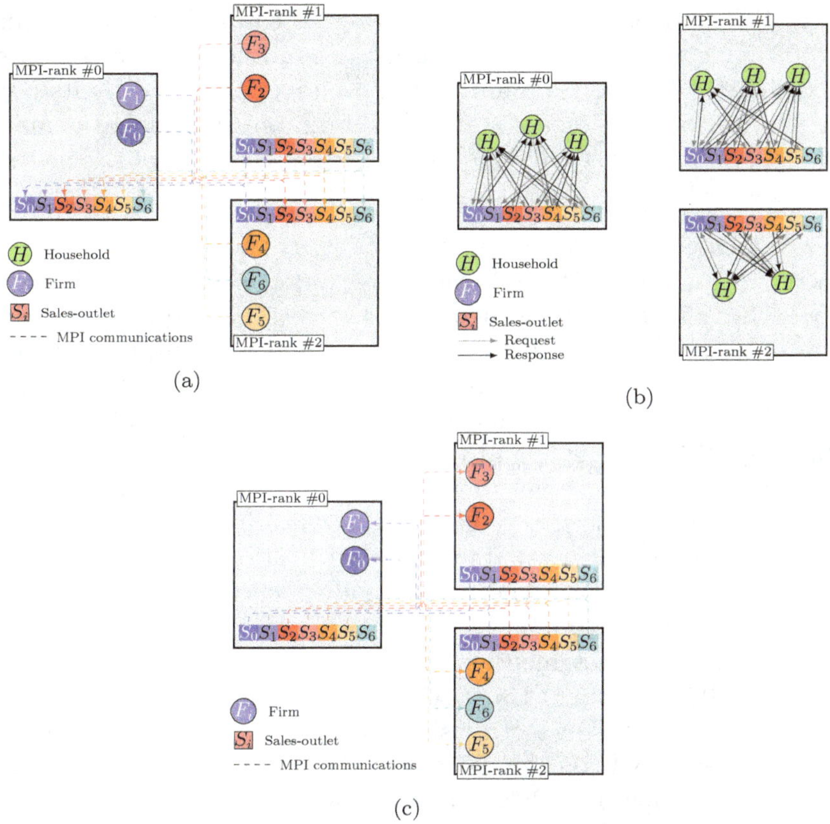

Fig. 4. MPI communication required for the goods market can be significantly reduced by introducing sales outlets. In this figure, only the purchasing of households from the firms of one industrial sector is illustrated. Government entities, foreign sellers, foreign buyers, and local firms themselves buy goods in the goods market. (a) According to the demands in MPIranks, each firm sends their products to corresponding sales outlets, (b) households in each rank purchase from the local sales outlets, and (c) sales outlets send sales data to their parent firms.

firms. Therefore, decentralizing the goods market requires further planning.

In each period t, a firm $f_i^{\bar{r},s}$ and its sales outlets $o_i^{r,s}$ engage in two interactions: $f_i^{\bar{r},s}$ distributes its products to sales outlets $o_i^{r,s}$ prior to the start of the goods market (see Fig. 4(a)), and $f_i^{\bar{r},s}$ collects sales records from $o_i^{r,s}$ at the end of the goods market (see Fig. 4(c)).

Firm $f_i^{\bar{r},s}$ distributes its products to $o_i^{r,s}$ according to the demand $D^{r,s}(t)$ for goods of sector s in each rank r. Each rank r estimates its demand $D^{r,s}(t)$ by summing up the budget allocated by its buyers for the goods of sector s and then estimates the global demand $\hat{D}^s(t)$ $(= \sum_r D^{r,s}(t))$ for sector s using `MPI_Allreduce()`. The firm $f_i^{\bar{r},s}$ assigns $\frac{D^{r,s}(t)}{\hat{D}^s(t)}$ portion of its production $Y_i^{\bar{r},s}(t)$ to the sales outlet $o_i^{r,s}$ located in rank r using `MPI_Scatter()` (see Fig. 4(a)). After receiving the amount to sell, each sales outlet $o_i^{r,s}$ sells to the local buyers in rank r (see Fig. 4(b)), and once sales are completed, it reports sales records, such as amount sold and demand, to its parent firm. The reporting of sales records of each sector s is done in two steps. First, the master rank collects the sales record of all sales outlets using `MPI_Reduce()`, and then it scatters the reduced results to the ranks where the owner firms are located, using `MPI_Scatter()`. What we just discussed is how to decentralize the firms in one industrial sector. Japanese standard classification consists of around 104 industrial sectors, while there are finer classifications with 530 sectors and 1460 sectors.[b]

Obviously, the aforementioned solution for a single industrial sector using `MPI_Scatter()` is not scalable for the goods market involving several hundreds of sectors since the required number of calls to `MPI_Scatter()` will be several hundreds; twice the number of sectors. The organization of the data of the firms and sales outlets of multiple sectors is decided based on the involved algorithms, not the MPI communications. The leftmost part of Fig. 5 shows the desired data layout of firms, while the rest shows that for sales outlets. In the left-hand side, each small box indicates the set of firms from an industry assigned to corresponding MPI-rank. Firms of each industry are split into two groups due to the differences in their constituents: local firms and foreign firms. Out of the variables involved in each local and foreign firm, only a subset is required at their sales outlets. These subsets of data have to be delivered to corresponding sales outlets organized as shown in Fig. 5. Using common MPI collective communication is not the most efficient solution, as it involves at least

[b]https://www.soumu.go.jp/english/dgpp_ss/seido/sangyo/san13-2.htm, accessed on 8 November 2022.

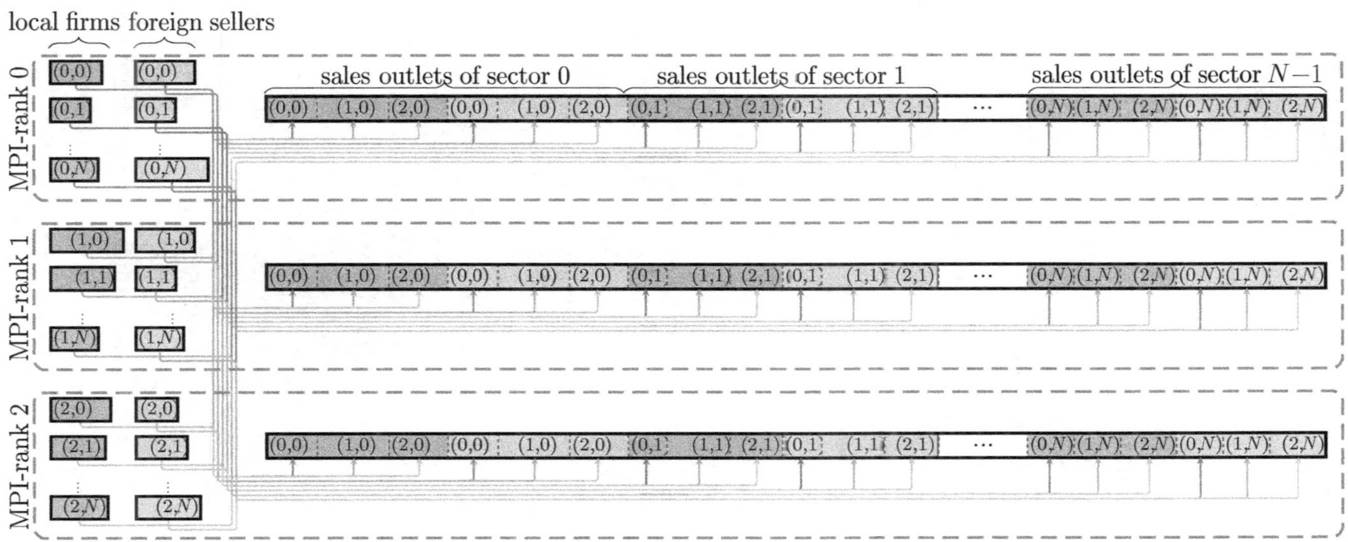

Fig. 5. Layouts of the data of firms and their sales outlets, and the communication pattern required for distributing products of each firm to corresponding sales outlets. Each box bounded by a thick continuous line is an independent array consisting of a large number of firms; all the sales outlets reside in a single array. The pair of numbers inside each box indicates the (MPI-rank, industrial sector). This data communication can be completed using a single call to `Allgatherw()`.

$2S$ number of messages: S calls to MPI_Allgatherv() to distribute products to sell, and another S calls to MPI_Scatterv() to collect the sales record, where S is the number of industrial sectors. This will introduce a significant overhead when it comes to simulating large economies with many industrial sectors. The communication overhead can be reduced by combining these S independent communications into a single communication using MPI_Alltoallw(), which further generalizes the collective communication MPI_Allgatherv() by allowing elements to have different MPI_Datatypes and displacement expressed in bytes. By defining suitable MPI_Datatypes and displacements at the sending and receiving ends, we can use this generalized function to generate Allgatherw() and Scatterw() functions; w indicates the flexibility to communicate buffers with different MPI_Datatypes and specified displacements in bytes. With suitably defined custom MPI_Datatypes, data of local and foreign firms of all the industrial sectors can be delivered in the desired data layout of sales outlets using the newly defined Allgatherw(). At the end of each goods market, rank 0 accumulates the sales data of each sales outlet using an MPI_Reduce(), and sends the calculated total sales data of each sales outlet to the corresponding parent firm using a single call to Scatterw() (see Fig. 6). This use of Allgatherw() and Scatterw() makes it possible to follow through with all the communications involved in the goods market with only three calls to collective MPI functions, instead of $2S + 1$ calls to common MPI functions, improving the parallel scalability.

By introducing recruiting agencies in every rank to match the vacancies and the unemployed agents, we decentralized the job market, restricting the number of communications to a small number and distributing some of the workload involved. The ideal condition that the number of workers in each rank is sufficient to fill the labor demand of firms in the corresponding rank is unlikely to be met. The recruiting agencies are programmed to first fill the vacancies with available labor in the corresponding rank. Lack of labor and high demand for labor can leave some ranks with unfilled vacancies, while the opposite conditions can leave some other ranks with unemployed workers. The recruiting agencies are programmed to collectively decide and assign unemployed workers to firms in the ranks with remaining vacancies. Since the implemented algorithm is somewhat complicated, we omit its details.

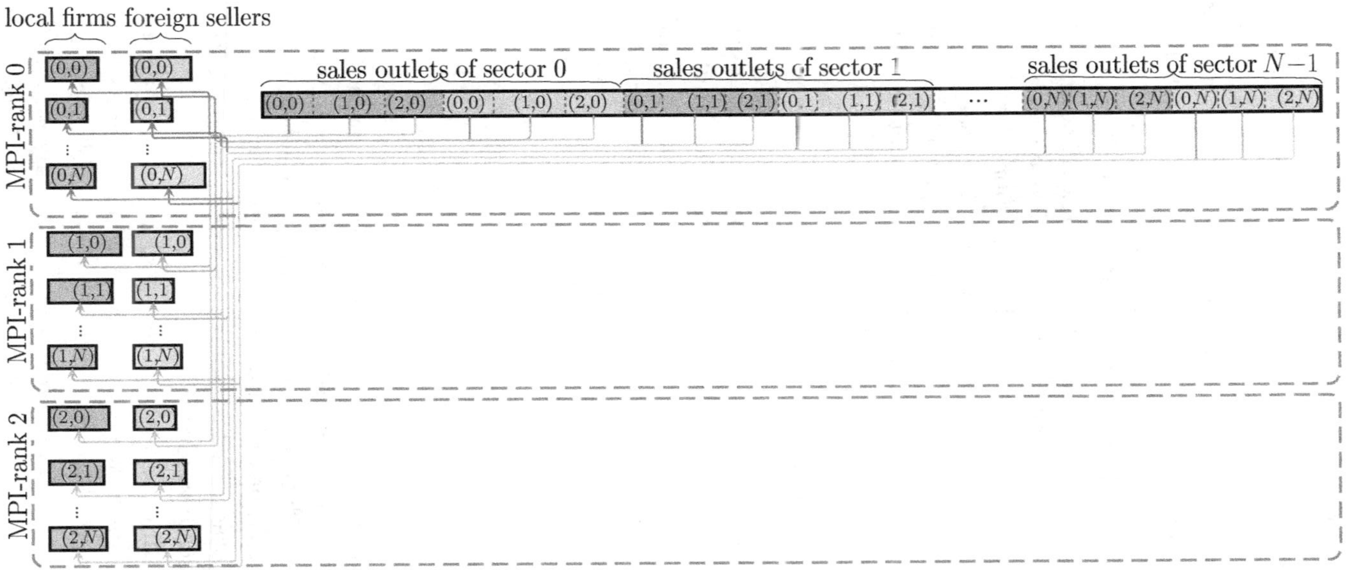

Fig. 6. After the sales outlets in MPI-rank 0 collect the sales data of all the ranks, rank 0 updates the parent firm of each sales outlet with a single call to `Scatterw()`.

2.5.3. *Communication hiding*

Overlapping the communications and computations, or communication hiding, is a standard technique for reducing the overhead of communications and thereby improving the parallel scalability. Even though blocking versions of MPI functions are mentioned in the previous discussions, we used non-blocking MPI functions (e.g., `MPI_Iscatter()` in place of `MPI_Scatter()`). For the goods market, we use the non-blocking version `MPI_Ialltoallw()` to define `IAllgatherw()` and `IScatterw()`. To maximize the overlap time, we post the non-blocking communications as soon as the involved data are available and finalize the reception of the message just before the use of the communicated data. We overlapped almost all the communications with computations. Algorithm 1 provides the details of the communications involved in the main execution loop.

2.5.4. *Algorithmic improvements*

The aforementioned strategies made it possible to develop a distributed parallel code with significant parallel computing efficiency when simulating small economies. As an example, a 9.8 million agent model produced 70% strong scalability on 16 MPI ranks,[20,21] with 18.4s run-time. However, with large economies (i.e., with more than 100 million agents), parallel scalability plummeted to very low values. As an example, a model consisting of 331 million agents on 1024 MPI-ranks required around 5700 s per iteration,[20] which is excessively large compared to the model consisting of 10 million agents; a 34 times larger problem that has consumed 19,826 ($5700/18.4 \times 1024/16$) times more computational resources. Detailed analysis revealed that this poor performance with large-scale models was not arising from any of the previously discussed parallel computing related components. The size of data increases with the model size, and random access to large volumes of data, which occurs when agents in each rank locally interact, is the source of poor computational performance of large-scale models.

In order to reduce the run-time associated with random access to large volumes of data, we made several enhancements to algorithms and data structures. For the details of these enhancements, we refer interested readers to Gill *et al.*'s work,[20] and only a short description of some of the enhancements are presented here. One of the obvious strategies is to minimize the size of data arrays involved in memory-bound operations. As an example, in the goods market, hundreds of millions of customers visit hundreds of thousands of firms of each industry, one industry at a time. This is a highly memory-bound operation, and the size of customer objects and the memory-access patterns will have a notable impact on the run-time. Further, almost 90% of run-time is associated with the goods market. In order to reduce the run-time associated with the goods market, the extensively called memory-bound algorithms are improved to make them cache-friendly. In our implementation, for example, the household objects have at least 15 `doubles`, 2 `floats`, and 4 `ints`. Out of these, only 6 `doubles` are involved in the buying process. To reduce the memory access overhead, new lightweight agents consisting of only these 6 variables are defined as proxies for buyers. Though these lightweight objects have to be stored in a new array C, which increases the memory requirements, it contributes to significantly lowering the run-time since it enables to store three times more customers in cache memory. To provide fair opportunities to all, the buyers are made to go shopping in a random sequence, and to facilitate that, the C array is randomized in each period, prior to the start of the goods market. As element-wise randomizing is highly memory-intensive, we use the Fisher–Yates algorithm, also known as Knuth shuffle[41,42]; as the array C consists of hundreds of millions of elements, block-wise shuffling with a block size of a few thousands is sufficient to provide unbiased opportunities to the buyers.

2.6. *Computational performance*

We conducted parallel scalability tests on two models representing small economies and large economies. The small economy consisted of 9.8 million agents, including 732,289 firms, and the large economy consisted of 331 million agents, including 22.6 million firms. Herein, these two models are referred to as small-economy and large-economy, respectively. Both had 62 industrial sectors. The model of the small-economy and the large-economy are roughly the size of 1:1-scale models of the Austrian economy and the Eurozone economy, respectively. One simulation period corresponds to one financial quarter, and each simulation is conducted for 13 time periods (one initialization period and 12 quarters ahead forecast).

All the simulations are conducted in the Oakbridge-CX supercomputer of The University of Tokyo. Computing nodes of Oakbridge-CX consist of Intel Xeon Platinum 8280 (2.7 GHz with 28 cores)×2 socket, and 192 GB memory with 281 GB/s bandwidth. The interconnection network is Intel Omni-Path Host Fabric Interface (12.5 GBps).

Tables 1(a) and 1(b) show the average run-time of 13 runs and the corresponding strong scalability, which is a standard performance measure in parallel computing and is defined as $(T_n/T_m)/(m/n)$, where T_n is the average run-time taken by n number of ranks, and $m \geq 2n$. Though not close to the ideal, 70% scalability is significantly high for this class of problems. Most importantly, it enables simulating large economies consisting of hundreds of millions of agents using small computational resources, in a short time. Economic simulations require Monte Carlo simulations to address various uncertainties involved which may arise from uncertainties related to agents' actions, model approximations, future trends, data sources, etc. The low computational resource requirements of the developed HPC framework make it possible to conduct Monte Carlo simulations of large economies using a modern computer cluster, in a short time (e.g., less than an hour).

Table 1. Run-time and strong scalability for small-economy and large-economy.

MPI ranks	Run-time (s)	Scalability (%)
(a) Small-economy (9.8 million agents)		
4	22.92	
8	12.03	95.26
16	6.57	91.55
32	3.93	83.58
64	2.53	77.66
(b) Large-economy (331 million agents)		
16	483.75	
32	241.23	100.26
64	152.95	78.85
128	108.95	70.19
256	80.56	67.62

Algorithm 1: Events of the main iteration loop. `Iallgatherw()` and `Iscatterw()` are derived from `MPI_Ialltoallw()`

---

for $t = 0;\ t < T;\ t = t + \Delta t$ **do**

  if $t \neq 0$ **then**

1   Central bank, at the master rank (*rank*-0), estimates growth and inflation rates and posts `MPI_Ibcast()` to broadcast the estimated rates to the other ranks;

2   Central government, at the master rank, decides its consumption budget and posts `MPI_Ibcast()` to broadcast the budget amount to the local governments;

3   Firms decide their production quantity, price, and labor requirements;

4   Firms inform their labor requirements to the recruitment agencies located in their respective ranks. Recruitment agencies fire local workers and post `MPI_Isend()/Irecv()` for informing the non-local fired workers to their respective recruitment agencies;

5   Foreign sellers decide their quantity and price;

6   Recruitment agencies finalize `MPI_Isend()/Irecv()` posted in event **4** and hire workers of their respective ranks;

7   Master recruitment agency, at the master rank, collects extra labor/jobs from all recruitment agencies using `MPI_Gather()`, assigns labor from labor surplus agency to the labor deficient agency, and then informs the cross-ranks labor assignments to each recruitment agency using `MPI_Scatter()`;

8   Recruitment agencies post `MPI_Isend()/Irecv()` to initiate exchanging of labor according to the cross-ranks assignments made by the master recruitment agency;

9   Firms decide the quantities of investment goods and input goods to buy in the current period;

10   Big firms (big buyers) post `Iallgatherw()` to inform their non-local *ProcurementDivisions* about the quantities to buy;

11   Firms apply for loans to cover their financing gap;

---

12 Recruitment agencies finalize the `MPI_Isend()/Irecv()` posted in the event **8**;
13 Recruitment agencies hire external labor received in event **12**;
14 Recruitment agencies post `MPI_Isend/Irecv()` to inform the job details of the remote workers to the respective remote recruitment agencies;
15 Finalize the non-blocking communication posted in event **2**;
16 Recruitment agencies inform the firms about the workers assigned to them at the end of the current period's job market;
17 Firms produce;
18 Share industry-wise total production and price of goods among all ranks using `MPI_Allreduce()`, and each rank calculates industry-specific economic coefficients;
19 Post `Iallgatherw()` to share the products of each firm with their *SalesOutlets* in each rank;
20 Firms pay the wages to the recruitment agencies, and the recruitment agencies pay to the local workers;
21 Finalize `Iallgatherw()` posted in event **10**;
22 Finalize `MPI_Isend()/MPI_Irecv()` posted in event **14**;
23 Recruitment agencies post `MPI_Isend()/Irecv()` to send/receive wage of workers *hired from/working at* other ranks;
24 Finalize `MPI_Isend()/Irecv()` posted in event **23**, and the recruitment agencies pay the received wages to the workers;
25 Households decide their consumption and investment budgets;
26 Big-household buyers post `Iallgatherw()` to distribute their budget to their proxies located in other ranks;
27 Finalize `Iallgatherw()` posted in event **26**;
28 *SalesOutlets* post `MPI_Iallreduce()` to collect total demand of each sector;

29 Central bank calculates the economy-wide indices and posts `MPI_Ibcast()` to share the indices;
30 Finalize `MPI_Iallreduce()` posted in event **28**;
31 Finalize `Iallgatherw()` posted in event **19**, and each sales outlet calculates and keeps its portion of products to sell;
32 Firms buy (i.e., Firms' `buy()`);
33 Post `MPI_Ireduce()` to sum up the quantities bought by *ProcurementDivisions* to the master rank (see event **10**);
34 Consumers, except the firms, buy (i.e., Final consumers' `buy()`);
35 Finalize `MPI_Ibcast()` posted in event **20**;
36 Finalize `MPI_Ireduce()` posted in event **33**, and post `Iscatterw()` to convey the quantities thus gathered by the master rank to the *ProcurementDivisions* of the big firms located in their home ranks;
37 Post `MPI_Ireduce()` to sum up the quantities bought by the proxies of big-household buyers to the master rank;
38 Post `MPI_Ireduce()` to collect the quantities of goods bought by the bank's investor;
39 Post `MPI_Ireduce()` to sum the sales records of all the *SalesOutlets* to the master rank;
40 Local governments distribute social benefits to each household in their respective ranks;
41 Finalize `MPI_Ireduce()` posted in event **39**, and post `Iscatterw()` to convey the sales records thus gathered by the master rank to the *SalesOutlets* located in the firms' home ranks;
42 Firms pay loan installments to the local banks in their respective ranks;
43 Finalize `Iscatterw()` posted in event **41**. *SalesOutlets* deliver the sales record, thus received, to their parent firms;
44 Finalize `MPI_Ireduce()` posted in event **37**, and post `Iscatterw()` to convey the quantities thus gathered by the master rank to the non-local buyers of the big household buyers located in their home ranks;

45 Finalize `Iscatterw()` posted in event **36**. The *ProcurementDivisions* deliver the total quantity, thus received, to their parent firms;
46 Finalize `Iscatterw()` posted in event **44**. The non-local buyers deliver the total quantity, thus received, to their parent households;
47 Do the end of tth period's accounting (see Algorithm 2 for details);
48 Finalize `MPI_Ireduce()`s posted in events **7** and **10** of event **47**;
49 Agents reset their variables to prepare for the next period;
50 Finalize `MPI_Ibcast()` posted in event **20** of event **47**;

Algorithm 2: Events of the end-of-period accounting (i.e., item **47** of Algorithm 1)

1 Firms undertake period-end accounting;
2 Firms pay dividends to their investors;
3 Firms pay taxes to the local governments;
4 Firms settle their accounts with the local banks, and deposit extra money;
5 Foreign buyers pay taxes to the local governments;
6 Post `MPI_Ireduce()` to reduce imports, exports, and GVA to the master rank for the central bank's accounting;
7 Post two `MPI_Ireduce()`s to collect industry summary (i.e., GDP, production, sales, demands, employment, etc.) and foreign buyers' summaries (i.e., consumption budget, quantity of goods bought) at the master rank;
8 Households pay taxes to the local governments;
9 Households settle their accounts with the local banks, and deposit extra money;

10 Post `MPI_Ireduce()` to sum up household summary (i.e., consumption and investment budget, quantities of consumption and investment goods bought, wages received, etc.) to the master rank;
11 Finalize `MPI_Ireduce()` posted in event **6**;
12 Central bank undertakes period-end accounting;
13 Local banks undertake period-end accounting;
14 Post `MPI_Ireduce()` for collecting financial reports from the local banks at the master rank;
15 Variables of the firms and their *SalesOutlets* are reset to prepare for the next period;
16 Finalize `MPI_Ireduce()` posted in event **14**;
17 Main bank undertakes accounting;
18 Bank's investor receives his dividends and updates his/her bank account;
19 Main bank updates its account at the central bank;
20 Post `MPI_Ibcast()` to share the main bank's equity, loan, etc. among all the local banks;
21 Main bank pays tax to the local government;
22 Post `MPI_Ireduce()` to sum up quantity bought, tax collected, and social benefits paid by the local governments at the master rank;
23 Variables of the households are reset to prepare for the next period;
24 Finalize `MPI_Ireduce()` posted in event **22**;
25 Central government undertakes accounting and updates its account at the central bank;

3. Data Generation for Simulating the Japanese Economy

The potential of the developed HPC-enhanced ABEM cannot be realized unless the model parameters and agents' data are derived from the real data so that a given economy can be forecasted to reasonable accuracy. As mentioned in the previous chapters, we choose the ABEM developed by Poledna *et al.*[10,12] which has already been verified and applied it to study the impacts of a flood disaster in Austria.[11] Since our objective is to simulate and compare the pre- and post-disaster Japanese economy, we calibrated it to the Japanese economy. This section presents the procedure to generate the required parameters for the Japanese economy in detail.

3.1. *Required parameters*

The model requires two types of data: parameters which are shared by all the agents of a particular class, e.g., coefficients defining the sectoral decomposition of input goods consumed in an industry, which are the same for all the firms of that industry; and agent specific data which is unique for each agent, e.g., capital, loan, and labor of each individual firm. A detailed list of required input parameters is available in the paper Poledna *et al.*[12]

3.2. *Data sources*

In order to simulate an economy in 1:1 scale, the data of each agent should be set according to the financial data of its real-world counterpart. Though data at such fine scale is commercially available, it is too expensive to buy for academic research. However, such fine-scale data are not available for households, banks, and other types of agents. Due to the unavailability of the data of each individual entity of the Japanese economy, we initialized the agents by disaggregating the macroeconomic data published by the Japanese government and other organizations such as think tanks. When disaggregating the macroeconomic data, we followed observed patterns like the size distributions. As an example, the size of firms in an industry is assumed to follow a power law distribution with coefficient 2.

All data sources used in this study are listed as follows:

- Cabinet Office, https://www.cao.go.jp/index-e.html;
- Ministry of Economy, Trade, and Industry (METI), https://www.meti.go.jp/english/;
- Japanese Government Statistics (e-Stat), https://www.e-stat.go.jp/en;

- Research Institute of Economy, Trade, and Industry (RIETI), https://www.rieti.go.jp/en/;
- Bank of Japan, https://www.boj.or.jp/en/index.htm/.

All these data sources[c] were extensively searched to obtain a consistent set of data from which all the required input parameters for the model can be extracted. The following sections describe the input parameters' identification process or the data disaggregation process for industries, individual firms, individual households, banks, the central government, and the central bank.

3.3. *Industrial parameters*

There are two main sets of parameters required for industries: sector parameters which are common for all the firms of an industry; and the parameters of each individual firm. Sector parameters define the interdependencies between the industries in terms of the input and capital goods. In addition, the consumption behaviors of households, government agencies, exporters, and importers are also captured by sector parameters. The IO-table is the main data source for the model parameters.

3.3.1. *Sector parameters*

For deriving the sector parameters, the 2015 IO-table (IO-table)[43] at producers' price for Japan[d] is used. The IO-table for Japan is complied every 5 years and the 2015 IO-table is the latest available. We opted for the IO-table at producers' price since the input coefficients obtained from it are more stable than those from the IO-table at purchasers' price.[43]

Table 2 shows a representative IO-table at producers' price for Japan. Equations (1)–(9) present the mathematical relationship

---

[c]The web addresses given in the list were accessed on 12 November 2022.

[d]https://https://www.e-stat.go.jp/en/stat-search/filespage=1&layout=datalist&toukei=00200603&tstat=000001130583&cycle=0&year=20150&month=0&stat_infid=000031839447&tclass1val=0, accessed on 31 May 2021.

Table 2. Overview of input–output table at the producers' price.

Supply sector (buyer) / Demand sector (seller)		Intermediate demand						Final demand											
		Sector-1	Sector-2	...	...	Sector-n	Total	Consumption expenditure outside households	Private consumption expenditure	Consumption expenditure of central government	Consumption expenditure of central government (social fixed capital)	Gross domestic fixed capital formation (public sector)	Gross domestic fixed capital formation (private sector)	Increase in stocks	Exports	Total	(Less) Imports	Domestic production
Intermediate input	Sector-1	m_{11}	m_{21}	...	...	m_{n1}	A_1	h_1^1	h_1^2	g_1	c_1^1	c_1^2	c_1^3	s_1	e_1	d_1	i_1	y_1
	Sector-2	m_{12}	m_{22}	...	...	m_{n2}	A_2	h_2^1	h_2^2	g_2	c_2^1	c_2^2	c_2^3	s_2	e_2	d_2	i_2	y_2
	...	...	...	...	...	...	...	...	...	...	...	...	...	...	...	...	...	...
	...	...	...	...	...	...	...	...	...	...	...	...	...	...	...	...	...	...
	Sector-n	m_{1n}	m_{2n}	...	...	m_{nn}	A_n	h_n^1	h_n^2	g_n	c_n^1	c_n^2	c_n^3	s_n	e_n	d_n	i_n	y_n
	Total	M_1	M_2	...	...	M_n	A	H^1	H^2	G	C^1	C^2	C^3	S	E	D	I	Y
Gross value added	Consumption expenditure outside households	w_1^1	w_2^1	...	...	w_n^1	W^1											
	Consumption of employees	w_1^2	w_2^2	...	...	w_n^2	W^2											
	Operating surplus	π_1	π_2	...	...	π_n	Π											
	Consumption of fixed capital	k_1^1	k_2^1	...	...	k_n^1	K^1											
	Consumption of fixed capital (social fixed capital)	k_1^2	k_2^2	...	...	k_n^2	K^2											
	Indirect taxes (except custom duties and commodity taxes on imported goods)	t_1^Y	t_2^Y	...	...	t_n^Y	T^Y											
	(Less) Current subsidies	t_1^P	t_2^P	...	...	t_n^P	T^P											
	Total of gross value added	GVA_1	GVA_2	...	...	GVA_n	GVA											
Total domestic production		y_1	y_2	...	...	y_n	Y											

between the components of the IO-table.

$$M_s = \sum_{g=1}^{n} m_{sg} \quad \forall s \in \{1, 2, \ldots, n\}, \tag{1}$$

where n is the number of industries.

$$\text{Total wage paid in sector } s\text{: } w_s = w_s^1 + w_s^2. \tag{2}$$

$$\text{Capital consumed in sector } s\text{: } k_s = k_s^1 + k_s^2. \tag{3}$$

Gross value added in sector s:

$$\begin{aligned} GVA_s &= (w_s^1 + w_s^2) + \pi_s + (k_s^1 + k_s^2) + t_s^Y + t_s^P \\ &= w_s + \pi_s + k_s + t_s^Y + \iota_s^P. \end{aligned} \tag{4}$$

Total domestic production in sector s:

$$y_s = M_s + GVA_s. \tag{5}$$

Total household consumption of sector s:

$$h_s = h_s^1 + h_s^2. \tag{6}$$

Total fixed capital formation in sector s:

$$c_s = c_s^1 + c_s^2 + c_s^3. \tag{7}$$

Total final demand for the goods of sector s:

$$\begin{aligned} d_s &= (h_s^1 + h_s^2) + g_s + (c_s^1 + c_s^2 + c_s^3) + s_s + e_s \\ &= h_s + g_s + c_s + s_s + e_s. \end{aligned} \tag{8}$$

Also, the total domestic production in sector s satisfies

$$y_s = A_s + d_s - i_s. \tag{9}$$

The number of workers in each industry, N_s, is obtained from Sector-wise Employment Data, which are available at e-Stat.[e] Other

[e]https://www.e-stat.go.jp/en/stat-search/files?page=1&layout=datalist&toukei=00200603&tstat=000001130583&cycle=0&stat_infid=000031839523&tclass1val=0, accessed on 31 May 2021.

important data such as total capital, bank deposits, and bank loans of the firms of each industry are obtained from the System of National Accounts (SNA).[f]

The sector parameters obtained from these data sources are as follows:

(1) Proportion of goods from sector g, in total input goods, consumed by a firm of sector s:

$$a_{sg} = \frac{m_{sg}}{M_s}. \tag{10}$$

(2) Productivity coefficient for input goods for sector s:

$$\beta_s = \frac{y_s}{M_s}. \tag{11}$$

(3) Productivity coefficient for capital goods for sector s:

$$\kappa_s = \frac{y_s}{\omega \times K_s}, \tag{12}$$

where ω is the capacity utilization rate, which is assumed to be 0.85, and K_s, the total capital of sector s, is defined as

$$K_s = \frac{k_s}{\sum_i^n k_i} K^I,$$

where k_s is obtained from Eq. (3) and K^I is total capital of all sectors and obtained from SNA.[g]

(4) Capital depreciation coefficient for sector s:

$$\delta_s = \frac{k_s}{\omega \times K_s}. \tag{13}$$

(5) Production tax rate for sector s:

$$\tau_s^Y = \frac{t_s^Y}{y_s}. \tag{14}$$

[f]https://www.esri.cao.go.jp/en/sna/stock/stock_top.html and https://www.esri.cao.go.jp/en/sna/sddsplus/sddsplus_top.html, accessed on 31 May 2021.

[g]https://www.esri.cao.go.jp/en/sna/sddsplus/sddsplus_top.html, accessed on 2 June 2021.

(6) Products tax/subsidy rate for sector s:

$$\tau_s^P = \frac{t_s^P}{y_s}. \tag{15}$$

(7) Productivity coefficient for labor for sector s:

$$\alpha_s = \frac{y_s}{N_s}, \tag{16}$$

where N_s is the number of workers in the sector s and determined from industry-wise employment data obtained from e-Stat.[h]

(8) Average wage rate paid to workers in sector s:

$$\bar{w_s} = \frac{w_s}{N_s}, \tag{17}$$

where w_s is obtained from Eq. (2).

(9) Proportion of goods of sector g in total household consumption:

$$b_g^{HH} = \frac{h_g^1 + h_g^2}{H^1 + H^2}. \tag{18}$$

(10) Proportion of goods of sector g in total government consumption:

$$c_g^G = \frac{g_g}{G}. \tag{19}$$

(11) Proportion of goods of sector g in total exports:

$$c_g^E = \frac{e_g}{E}. \tag{20}$$

(12) Proportion of goods of sector g in total imports:

$$c_g^I = \frac{i_g}{I}. \tag{21}$$

[h]https://www.e-stat.go.jp/en/stat-search/files?page=1&layout=datalist&toukei=00200603&tstat=000001130583&cycle=0&stat_infid=000031839523&tclass1val=0, accessed on 31 May 2021.

3.3.2. *Parameters of individual firms*

As previously mentioned, since the data are not available for each individual firm, we artificially generated each firm agent's parameters by disaggregating the macroeconomic data. Since the production of a firm is directly proportional to the number of employees, the size distribution of firms can be obtained from the workers' distribution among firms. e-Stat provides the enterprise census data,[i] which contains a classification of the firms according to the number of workers employed. In the data, the firms are classified into 11 groups of increasing number of workers; group 1 consists of firms having 1 to 4 workers and group 11 contains firms with more than 5,000 workers. However, assigning the workers to each firm was not straightforward because of the missing data and inconsistency between different data sets. To resolve these problems, we made the following assumptions.

(1) The latest available IO-table for Japan is for the year 2015. Therefore, the base year of the data set for the current study is set to 2015. However, firm size distribution for 2015 is not available, and the closest available is that for 2016. To resolve this, we assumed the firm size distribution in 2015 to be the same as that in 2016. This assumption may not have a significant impact since the firm sizes do not change much in a short time span of one year.
(2) The industrial sector-classification used in the enterprise census data for 2016 is different from that in the 2015 IO-table. According to the definition of each sector in both the data sets, the firms and workers' distribution for the sectors of 2015 IO-table were derived by merging or splitting some sectors of the 2016 enterprise census data. The sectors were split based on the ratio of workers in the constituent sectors.
(3) Enterprise census data were not available for four sectors: *reuse and recycling*; *public administration*; *nursing care*; and *activities not elsewhere defined.* The following assumptions were made

[i]https://www.e-stat.go.jp/en/stat-search/files?page=1&layout=datalist&toukei=00200553&tstat=000001095895&cycle=0&tclass1=000001106256&tclass2=000001107036&tclass3=000001114545&stat_infid=000031721990&tclass4val=0, accessed on 31 May 2021.

regarding the number of firms and their size distribution in these sectors.

(a) For sectors: *reuse and recycling*, and *activities not elsewhere defined*, the number of firms are assumed to be 1% of the number of workers working in them.
(b) The number of firms in the *public administration* sector is assumed to be 475 since there are 475 districts in Japan.
(c) The firms-to-workers ratio in *nursing care* sector is assumed to be equal to that in the *social insurance and the social welfare* sectors.
(d) The size distribution of firms in all these four sectors is assumed to follow the mean distribution in the economy.

After obtaining the firms size distribution for all sectors as per 2015 IO-table's sector-classification, the number of workers, N_i, is assigned to ith firm of sector s by using Algorithm 3.

Using the N_i and the sector parameters derived in Section 3.3.1, the initial (i.e., at $t = 0$) parameters of individual firms of sector s, such as initial capital, production quantity, input goods, bank deposit, loan, are calculated as explained in the following.[j]

(1) Initial price of products:

$$P_i(t = 0) = 1.$$

(2) Initial production:

$$Y_i(0) = \alpha_s N_i.$$

(3) Initial stock of capital goods:

$$K_i(0) = \frac{Y_i(0)}{\omega \kappa_s}.$$

(4) Initial stock of input goods:

$$M_i(0) = \frac{Y_i(0)}{\omega \beta_s},$$

where ω is the capacity utilization rate.

[j] All these equations except in item 5 are reproduced from Appendix C of Ref. 12 for the sake of completeness.

Algorithm 3: Pseudocode for assigning workers to each firm of a sector s

Input: Number of firms, NF_g, in each of the 11 groups. Minimum and maximum number of workers in each group, $[a_g, a_{g+1})$, where $g \in \{1, 2, \ldots, G\}$. Total number of workers, N_s, in sector s.

Output: Number of workers, N_i, assigned to each firm in the sector s.

```
for g ← 1 to G do                                  // for each group
    NW_g = NF_g × ½(a_g + a_{g+1})   // approximate initial value
    p_g = NW_g / N_s    // probability to distribute imbalance of approximations

// Generate a cumulative probability distribution
P_j = Σ_{i=0}^{i=j} p_i
while N_s ≠ Σ_{i∈{i|f_i in group G}group g} N_i do
    r ← rand(0,1)          // Draw a uniform random number
    g ← g ∈ {1,2,...,G} | (P_g ≤ r < P_{g+1})
    if N_s > Σ_i NW_i then
        NW_g = NW_g + 1
    else
        NW_g = NW_g − 1

// Assign workers to each firm
for g ← 1 to G do                                  // for each group
    foreach f_i in group g do     // for each firm in the group
        N_i = a + rand()%(b − 1)        // rand() generates uniform random numbers, and % is modulo operation

    // Adjust to ensure Σ_i N_i = NW_s
    while Σ_{i∈{i|f_i in group G}} N_i ≠ NW_s do
        r ← rand(1, NF_g)          // Draw a uniform random number
        if NW_s > Σ_i N_i then
            N_r = N_r + 1
        else
            N_r = N_r − 1
```

(5) Initial stock of inventories:

$$S_i(0) = \frac{Y_i(0)}{\sum_i Y_i(0)} s_s,$$

where s_s is total inventories of sector s as reported in "Increase in Stocks" column of the input–output Table 2.

(6) Initial debt:

$$L_i(0) = L^I \frac{K_i(0)}{\sum_i K_i(0)},$$

where L^I is total debt of all the firms in the economy and obtained from SNA.[k]

(7) Unit operating margin:

$$\bar{\pi}_i = 1 - \left(\left(1 + \tau^{SIF}\right) \frac{\bar{w}_s}{\alpha_s} + \frac{\delta_s}{\kappa_s} + \frac{1}{\beta_s} + \tau_s^K + \tau_s^Y \right).$$

(8) Initial bank deposits:

$$D_i(0) = D^I \frac{\max\{\bar{\pi}_i Y_i(0), 0\}}{\sum_i \max\{\bar{\pi}_i Y_i(0), 0\}},$$

where D^I is total deposit of all the firms in the economy and obtained from SNA.[k]

(9) Initial profit:

$$\Pi_i(0) = \bar{\pi}_i Y_i(0) - r(0) L_i(0) + \bar{r}(0) D_i(0),$$

where $\bar{r}(0)$ is the initial policy rate set by the central bank and $r(0)$ $(= \bar{r}(0) + \mu)$ is the initial interest rate set by the bank. μ is risk premium rate of the bank.

[k]https://www.esri.cao.go.jp/en/sna/sddsplus/sddsplus_top.html, accessed on 2 June 2021.

3.3.3. *Foreign buyers' and foreign sellers' data*

In the ABEM, foreign buyers and foreign sellers act as a link between the domestic economy and the rest-of-world's economy. Foreign buyers buy in the domestic market, and their consumption represents exports. On the other hand, foreign sellers sell goods and services produced by foreign economies in the domestic market, and their sales represent imports. Foreign sellers and foreign buyers are represented by foreign firms in the model. Data required for foreign firms are their numbers and their consumption budget or quantity to sell. The following steps describe the process of generating those data.

Foreign Buyers:

(1) The number of foreign buyers is not available, and we assumed their number to be 50% of the number of domestic firms.
(2) The ABEM does not model the consumption behavior of foreign buyers. Therefore, foreign buyers' consumption budget data are given as exogenous values to the model. The SNA data of Japan provides time-series data of GDP calculated using expenditure approach.[1] This data set includes time-series data of exports also. However, the export data for 2015 does not match with the exports as reported in the IO-table. This incompatibility is addressed by normalizing the export data obtained from SNA to the IO-table data.

Foreign Sellers:

(1) No data are available for the number of foreign sellers also. The number of foreign sellers is assumed to be 1 in each sector.
(2) Similar to the case of foreign buyers' data, the ABEM does not model the production and price decision of foreign sellers. Therefore, foreign sellers' quantities to produce are given as exogenous values to the model and generated using a similar approach as for foreign buyers.

[1]https://www.e-stat.go.jp/en/stat-search/files?page=1&layout=datalist&toukei=00100409&tstat=000001014470&cycle=2&tclass1=000001014471&tclass2=000001018314&stat_infid=000031978826&result_page=1&tclass3val=0, accessed on 31 May 2021.

(3) The price set by the foreign seller of a sector is assumed to be equal to the last period's average price of goods of the respective sector.

The steps described above assume that the data for imports and exports are available for all the periods simulated. However, when simulating future economic conditions, for which the observations of imports and exports are yet to be made, we can use techniques like auto-regression and machine-learning to forecast the imports and exports based on historical data.

3.4. *Households' data*

In the model, households are divided into three categories: investors who are firms and bank owners; workers who are active in the labor market and provide labor to the firms; and inactive households who represent non-working population including the retired and children. Parameters required for household agents can be categorized into two main groups: parameters common to all the households; and the parameters private to each household. Identification of households' parameters was more challenging compared to that of the firms mainly because of the limited data. Household parameters are generated from the Census Data,[m] household expenditure survey data,[n] and SNA data[o] by the following steps:

(1) The first step is to get the number of households in each of the three categories. Total population of Japan is obtained from the census data by taking the average of the population reported in the months of October, November, and December in the 2015 census data. The model assumes that there is only one investor

[m]https://www.e-stat.go.jp/en/stat-search/files?page=1&layout=datalist&toukei=00200524&tstat=000000090001&cycle=0&year=20160&month=0&tclass1=000001011678&tclass2=000001091515&tclass3val=0, accessed on 31 May 2021.

[n]https://www.e-stat.go.jp/en/stat-search/files?page=1&layout=datalist&toukei=00200524&tstat=000000090001&cycle=0&year=20160&month=0&tclass1=000001011678&tclass2=000001091515&tclass3val=0, accessed on 31 May 2021.

[o]https://www.esri.cao.go.jp/en/sna/household/household_top.html, accessed on 31 May 2021.

per firm or bank, therefore, number of investors is equal to the number of firms and banks. The number of workers is already known from the industrial data. However, total working population is usually more than the employed because every economy has some degree of unemployment. The unemployment rate is obtained from the employment data for 2015[p] (e-Stat) based on which the total number of workers is calculated. The remaining households are assumed to be inactive households. In this study, we ignore the population growth/decline rate.

(2) Salary or income of each household is required to set parameters both common to all the households and private to each household. The firms pay a portion of their profits to their investors as dividends, which is the main income of the investors. The employed workers receive wages for the labor supplied to firms. Total profit of firms and bank can be obtained from the industrial data ($\sum_i \Pi_i(0)$, where $\Pi_i(0)$ is initial profit of ith firm as mentioned in the previous section) and bank's data (Eq. (29)). However, dividend payout ratio of firms and banks (θ^{DIV}) must be known to obtain the income of investors. The wages received by employed workers are $\bar{w}_s$ as per Eq. (17). The incomes of unemployed workers and inactive households are not readily available. These unavailable values are approximated as later explained in item 6 of this list.

(3) Households' propensities to consume and invest (ψ and ψ^H), dividend payout ratio of firms and banks (θ^{DIV}), income tax rate (τ^{INC}), social contribution rate for workers (τ^{SIW}), unemployment benefits paid by government to unemployed workers (θ^{UB}), social benefits paid by government to inactive households in the form of pension and all other households in the form of subsidies (sb^{inact} and sb^{other}) are not reported in the available data sets. However, household disposable income and

[p]https://www.e-stat.go.jp/en/stat-search/files?page=1&layout=datalist&toukei=00200531&tstat=000000110001&cycle=2&year=20150&month=11010300&tclass1=000001040276&tclass2=000001040283&tclass3=000001040284&stat_infid=000029442046&result_back=1&tclass4val=0, accessed on 2 June 2021.

household saving ratio[q] data are available in the SNA[r] which provides aggregated values of operating surplus and mixed income, compensation of employees, property income, income tax, social contribution payable, social benefits received, disposable income, and consumption of households data. The aforementioned parameters are derived by ensuring that the total income of households, taxes paid, social benefits received, etc., are close to the value reported in the SNA data. θ^{UB} is assumed to be zero since no data were available. As mentioned in the previous item, income of investors and employed workers can be obtained from the industrial data. Total household consumption can be obtained from the IO-table using Eq. (6). Therefore, SNA data are used only for finding the missing data.

(4) Using θ^{DIV}, θ^{UB}, sb^{inact}, and sb^{other}, the income of each individual household can be obtained. In order to obtain the income of investors and workers, a firm is assigned an investor and workers as per its initial labor requirement. Initial income ($Y_h(0)$) for households is then calculated as per Eq. (22)

$$Y_h(0) = \begin{cases} \theta^{\text{DIV}} \max\{0, \Pi_i(0)\} + sb^{\text{other}}, & \text{if investor of } i\text{th firm,} \\ \theta^{\text{DIV}} \max\{0, \Pi_k(0)\} + sb^{\text{other}}, & \text{if investor of } k\text{th bank,} \\ w_s + sb^{\text{other}}, & \text{if employed by a firm of sector } s, \\ sb^{\text{other}}, & \text{if unemployed,} \\ sb^{\text{inact}} + sb^{\text{other}}, & \text{if not economically active.} \end{cases} \tag{22}$$

(5) Households' investment is assumed to be composed of investment in dwellings and durable goods. The IO-table provides total

---

[q]Household Disposable Income is the net income of a household and obtained by subtracting taxes, social security, and contributions from gross income of the household from all sources. Household Saving Ratio is the ratio of total saving by the household to its disposable income.

[r]https://www.esri.cao.go.jp/en/sna/household/household_top.html, accessed on 31 January 2021.

investments by firms and households (c_s) and consumption of capital by firms (k_s). Total household investment is therefore the difference of the two values. Households' investment in dwellings is obtained from GDP time-series data, which are available in SNA.[s] The rest is the investment in durable goods.

(6) Total capital (K^H) and deposits (D^H) of households are obtained from SNA[t] and assigned to the households in proportion to their income. Initial capital ($K_h(0)$) and bank deposit ($D_h(0)$) of hth household are calculated using Eqs. (23) and (24), respectively.[u]

$$K_h(0) = K^H \frac{Y_h(0)}{\sum_h Y_h(0)}. \tag{23}$$

$$D_h(0) = D^H \frac{Y_h(0)}{\sum_h Y_h(0)}. \tag{24}$$

3.5. *Capital consumption coefficients of firms and households*

The previous two sections describe all the steps required to generate almost all parameters required for industries and households. Although there was a discussion on total investments made by firms and households, calculation of capital formation coefficients for firms (b_g^{CFH}) and household investment coefficient (b_g^{HH}) was not discussed. The following steps describe how the values of these coefficients were generated.

(1) From the previous section, we already know households' investments in dwellings and durable goods. Survey data of households' investments[v] in durable goods provide the details of household durable goods. Therefore, based on the source industries of the

[s]https://www.e-stat.go.jp/en/stat-search/files?page=1&query=Nominal%20Gross%20Domestic%20Product&layout=dataset&stat_infid=000032088393&metadata=1&data=1, accessed on 5 January 2021.

[t]https://www.esri.cao.go.jp/en/sna/sddsplus/sddsplus_top.html, accessed on 2 June 2021.

[u]These two equations are reproduced from Ref. 12.

[v]https://www.e-stat.go.jp/en/stat-search/files?page=1&layout=datalist&toukei=00200564&tstat=000001073908&cycle=0&tclass1=000001073965&tclass2=000001074205&tclass3=000001074208&stat_infid=000030780015&tclass4val=0, accessed on 2 June 2021.

durable goods, industry-wise investment in durable goods can be obtained. Sectoral decomposition of investment in dwellings is unknown, therefore, their sectoral decomposition is assumed to be the same as that in Ref. 12. The use of the coefficients adopted from Ref. 12 assigned household dwelling investments to some sectors which did not contribute to capital formation as per the IO-table for Japan. Therefore, household investments in those sectors are assumed to be zero. For some sectors, households' investments as obtained from the survey data and using investment coefficients from Ref. 12 are more than or equal to total fixed capital formation. In such a sector, households' investments are assumed to be 90% of total fixed capital formation in that sector; remaining 10% is assumed to be firms' investments. Ultimately, households' investments in sector g, ι_g^H, are obtained.

(2) Total investment by firms in sector g, ι_g^I, can be calculated as per Eq. (25)

$$\iota_g^I = c_g - \iota_g^H, \tag{25}$$

where c_g is total fixed capital formation in sector g and obtained from Eq. (7).

(3) Capital formation coefficients for firms (b_g^{CF}) and households (b_g^{CFH}) can be obtained from Eqs. (26) and (27), respectively.

$$b_g^{CF} = \frac{\iota_g^I}{\sum_g \iota_g^I}, \tag{26}$$

$$b_g^{CFH} = \frac{\iota_g^H}{\sum_g \iota_g^H}. \tag{27}$$

3.6. *Bank's data*

The banking sector is represented by only one bank since our base ABEM lacks the rules and processes regarding how multiple banks compete and interact in the financial market and how customers choose a bank for their financial needs, which are required to accommodate multiple banks. Further, the required data to initialize multiple banks are not available.

The single bank agent in the model keeps deposits of household and firm agents and provides loans to them. The required parameters for the bank are as follows:

- interest rate (μ),
- dividend payout ratio (θ^{DIV}),
- loan installment rate (θ),
- bank's capital requirement coefficient (ζ),
- loan-to-value ratio for firms (ζ^{LTV}),
- loan-to-capital ratio for new firms after bankruptcy (ζ^{b}).

θ^{DIV} has already been calculated in Section 3.4. μ is assumed to be zero. The values of the rest of the parameters are taken from Ref. 12. Table 3 lists all the parameters and their values.

Initial equity of the bank ($E_k(0)$) is assumed to be zero. Total loans extended to firms (L^I), total deposits of firms (D^I), and deposits of households (D^H) are obtained from SNA. The remaining data of the bank are generated using the following equations.

Reserves held by the bank at the central bank[w]:

$$D_k(0) = D^I + D^H + E_k(0) - L^I. \tag{28}$$

Initial profit of the bank:[w]

$$\Pi_k(0) = r(0)L^I - \bar{r}(0)\left(D^I + D^H - D_k(0)\right), \tag{29}$$

where $\bar{r}(0)$ is the initial policy rate set by the central bank and $r(0)$ ($= \bar{r}(0) + \mu$) is the initial interest rate set by the bank.

Table 3. Bank parameters.

Parameter	Description	Value
μ	Interest rate set by bank	0
θ	Debt installment rate	0.05
ζ	Bank's capital requirement coefficient	0.03
ζ^{LTV}	Loan-to-value ratio for firms	0.6
ζ^{b}	Loan-to-capital ratio for new firms after bankruptcy	0.5

[w]The equation is reproduced from Ref. 12.

3.7. *Government's data*

The functions of the government are to consume goods in the goods market, collect various kinds of taxes from the agents, and provide subsidies to some industries and households. The government takes loans from the central bank to fund its expenses and pays interest on the loans. The following items describe the data generation process in detail.

(1) The government allocates its budget to spend in various sectors through its agencies. The number of agencies is not explicitly available from the data. Therefore, this number is assumed to be 25% of the number of total firms in the economy.
(2) Just like the consumption budget of foreign buyers, governments' consumption budget is also an exogenous parameter. It is obtained from the SNA data in the same way as was adopted for the generation of consumption data of foreign buyers.
(3) The government needs various tax and subsidy rates to collect taxes from the agents. The calculation of all such tax rates is already explained in the industrial data generation and household data generation. The missing tax rates, such as export tax rate and corporate tax rates, are assumed to be zero.
(4) The government debt (L^G) is obtained from the SNA data.[x] The interest rate on loan (r^G) is assumed to be 0 because of the lack of data.

3.8. *Central bank's data*

The functions of the central bank are to set the policy rate ($\bar{r}$), provide liquidity to the commercial bank in the form of advances and keep deposits from the bank in the form of reserves ($D_k(0)$), and provide loans to the central government (L^G). $D_k(0)$ and L^G have already been determined in previous sections. Initial balance of trade with the rest-of-world (D^{RoW}) is assumed to be zero. The policy rates ($\bar{r}(t)$) are also assumed to be zero for all the simulation periods. The central bank's initial equity ($E^{\text{CB}}(0)$) can be determined from the

[x]https://www.esri.cao.go.jp/en/sna/sddsplus/sddsplus_top.html, accessed on 2 June 2021.

following equation.[y]

$$E^{\mathrm{CB}}(0) = L^G - D_k(0) + D^{\mathrm{RoW}}. \tag{30}$$

3.9. *Time histories data for firm agents' production and price decisions*

The production and price decisions of firm agents play a significant role in the accurate modeling of the economy since various parameters, like wages paid, taxes generated, etc., depend on them. In our base ABEM,[10,12] at the start of each period, the firm agents make forecasts of the growth rate of the demand and the inflation rate of the price of their products using the respective time histories data. Based on the forecast rates, the firms increase or decrease their production quantity and price.

In the time-series forecasting-based decision-making, there are two main influencing factors: the forecasting method and the historical time series. There are a large number of time-series forecasting methods ranging from classical auto-regression to advanced machine learning. Just like the original ABEM,[10,12] this study uses an auto-regression model with lag order 1 (AR(1)). AR(1) is a linear model and is not the best one to forecast nonlinear time series influenced by many other factors. Replacing the AR model by an advanced machine-learning model, which is nonlinear, will enable the ABEM to attain higher forecasting accuracy.

We have three main options regarding historical data for firms.

(1) use each individual firm's historical data,
(2) use sector-wise historical data,
(3) use economy-wide historical data.

Obviously, the first option is the ideal, and can open up the potential of making economic forecasts with significant degree of confidence up to an individual firm level. However, as mentioned in the beginning of this section, we don't have access to firm-level data to prepare this highly desired ideal input. In the second option, sectoral time histories data are used by all the firms of a sector, which implies

[y]The equation is reproduced from Ref. 12.

that the firms of a sector make homogeneous decisions. Nonetheless, this option has the potential of forecasting individual sector-level economic indices to a higher degree of confidence. Moreover, sector-wise time histories are widely available free of charge. In the third option, all the firms in the economy use the economy-wide time-histories data. A major motivation for the ABM of the economy is to take the real-world economic entities' highly heterogeneous decision-making into account. Obviously, the third option, which makes all the firm agents' decision-making homogeneous, contradicts this major motivation. In addition to the firm agents, households, foreign buyers, and the government also use forecast of inflation rate to decide their consumption and investment budgets. In the current model, these agents always use economy-wide forecasts. For making forecasts of growth and inflation rates, time-series data of total real production and gross domestic product (GDP) deflator are required, respectively. The data preparation processes under each of the aforementioned three options are explained as follows.

- **Economy-wide forecast model:** In this case, all the firms use economy-wide forecasts of growth and inflation rates. Therefore, economy-wide total production and GDP deflator data are required. As mentioned in previous sections, the SNA data of Japan provides time-series data of GDP calculated using expenditure approach. GDP deflator data can also be obtained from the SNA.[z] However, time-series data of economy-wide total production ($Y^I(t)$) are not explicitly available. Since GDP is directly proportional to total production, GDP time-series data are used to derive production time series ($Y(t)$) as per Eq. (31).

$$Y(t) = \frac{Y(t = Q1\ 2015)}{\text{GDP}(t = Q1\ 2015)} \text{GDP}(t), \tag{31}$$

where $\frac{Y(t=Q1\ 2015)}{GDP(t=Q1\ 2015)}$ is the normalizing factor.

The GDP deflator data are also normalized so that the price is 1 for Q1 2015, which is the base quarter for the IO-table.

[z]https://www.e-stat.go.jp/en/stat-search/files?page=1&layout=datalist&toukei=00100409&tstat=000001014470&cycle=2&tclass1=000001014471&tclass2=000001018314&stat_infid=000031978853&result_page=1&tclass3val=0, accessed on 31 May 2021.

- **Sector-wise forecast model:** Sector-wise data of nominal production ($y_s^N(t)$) and real production ($y_s^R(t)$) can be obtained from RIETI (Research Institute for Economy Trade and Industry).[aa] However, the data cannot be used in the raw form and has to be preprocessed as follows.
  (1) Sector-wise classification used in the RIETI data is different from that in the IO-table. According to the definitions of each sector, some industries were split or aggregated to obtain the industrial data as per the IO-table. Further, data were not available for some industries. Such industries are assumed to follow the macroeconomic trends and, therefore, economy-wide data are used for such industries.
  (2) The frequency of RIETI data is yearly whereas the ABEM requires quarterly data. Therefore, yearly data are converted to quarterly data using linear interpolation.
  (3) Sector-wise GDP deflator data are obtained using Equation (32):

$$\text{GDP Deflator(t)} = \frac{y_s^N(t)}{y_s^R(t)}. \tag{32}$$

  (4) Finally, the real production data are normalized to real production as recorded in the IO-table and GDP deflator data are normalized so that the price for Q1 2015 is 1.
- **Firm-level forecast model:** Firm-level data are not available as a part of data published by the Japanese government agencies and other freely available data sources.

4. Verification and Validation

Verification and validation are essential steps in scientific methods to ensure that a numerical implementation is devoid of errors and the underlying mathematical model is capable of reproducing the physical phenomena being modeled, respectively. Numerical implementation can be verified by ensuring that the code can reproduce either

[aa] https://www.rieti.go.jp/en/database/JIP2021/index.html, accessed on 31 May 2021.

an analytical solution or a solution from another verified numerical implementation to a sufficient accuracy. We verified our HPC-enhanced object-oriented implementation by comparing its outputs with the results from Dr. Poledna's[12] matrix-based MATLAB implementation of the same ABEM. Model validation is the process by which model outputs are compared to the independently observed data generated by the underlying real system being modeled to determine if the model is capable of replicating the real system to a sufficient accuracy. After verification and validation, the model can be safely used to study the underlying system in detail and even predict its future behavior. Though the ABEM that we are using has been already validated for the Austrian economy,[12] it is essential to re-validate it for the Japanese economy to ensure that the ABEM with any modification, that we introduced, and with the input parameters for the Japanese economy, that we prepared in Section 3, can reproduce the Japanese economy to a sufficient accuracy.

4.1. *Verification*

For the verification, we compared the results from our HPC-enhanced implementation with the results provided by Dr. Poledna from his MATLAB implementation of the same ABEM. The following problem settings were used for the verification test.

(1) The model parameters and initial data of the agents are taken from Poledna *et al.*[12] There are 9.92 million agents comprising 0.61 million firms divided into 62 industries, 4.11 million workers, and 4.13 million inactive households.
(2) Simulation is conducted for 13 periods and each simulation period represents one financial quarter.
(3) Only one simulation is conducted without randomness (e.g., excluding the error part in the AR(1) process).

Figure 7 presents a comparison of four macroeconomic variables. As the figure shows, our HPC-enhanced implementation reproduces the results of Dr. Poledna's implementation, concluding that our HPC-enhanced implementation of his ABEM is free of any errors.

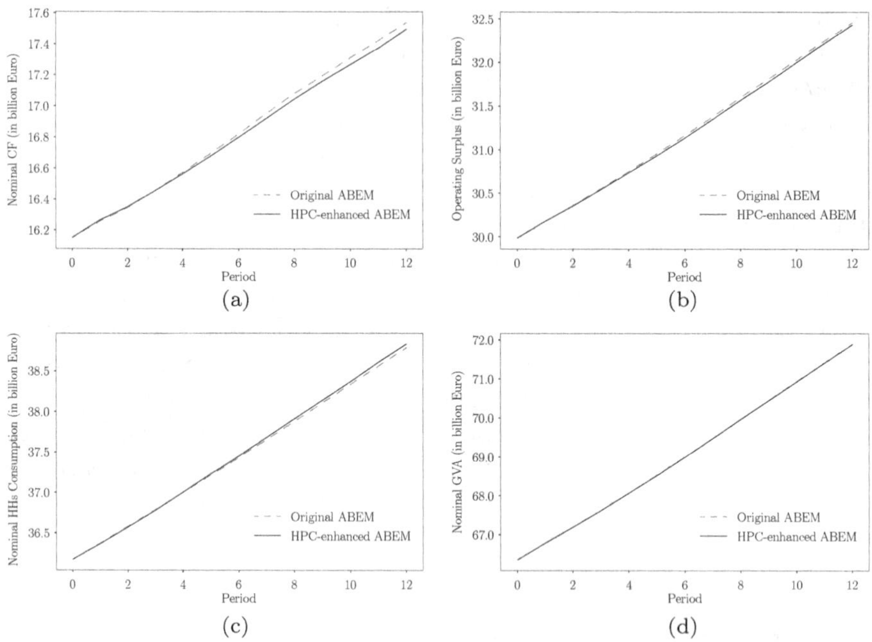

Fig. 7. Comparison of outputs of the HPC-enhanced ABEM with that of original ABEM.[12] (a) Nominal capital formation, (b) operating surplus, (c) nominal consumption by households, and (d) nominal gross value added.

4.2. *Validation*

We validate our HPC-enhanced implementation for the Japanese economy using the model parameters and agents' data generated in Section 3. As explained in Section 3.9, the forecasts of growth and inflation rates used in firm agents' decision-making can be made with data at three levels of resolutions: economy-wide; sector-wise; and firm level. Since ABEMs are micro-founded, the model performance is supposed to improve with the use of finer resolution data. In order to check the corresponding degree of improvement in forecasting accuracy, we conducted simulations using both economy-wide and sector-wise time-series data and compared results with real-world observations. The initial conditions (i.e., all the input parameters) are set according to the state of the 1st quarter of 2015 of the Japanese economy, and simulations are conducted for consecutive 22 periods (i.e., 22 quarters until the Q2 of 2020) without any additional inputs.

Monte Carlo simulations are conducted to take the involved uncertainties into account, and the statistical results of the simulations are compared with the real-world observations.

4.2.1. *Problem settings*

The problem for the validation is set as follows:

(1) The model parameters and initial parameters of each individual agent are generated according to the process mentioned in the last section and using Japanese economy data of Q1 2015. Table 4 shows the number of all the types of agents in the model.
(2) One simulation period represents one financial quarter (i.e., 3 months).
(3) Simulation is conducted for 22 periods: from Q1 2015 to Q2 2020, and the results are compared with the observed data. The observed economic data for the simulation period, i.e., Q1 2015 to Q2 2020, is taken from the SNA data of Japan. The observation data were normalized to the IO-table data, just like the GDP deflator and other data during the input data preparation.
(4) No external inputs or instructions are given to the model during the simulations.

Table 4. Number of different types of agents in the model of the Japanese economy used for the validation test.

Agent type	Number of agents
Industrial sectors	104
Domestic firms	1.84 million
Foreign buyers firms	920,000
Foreign sellers firms	104
Investors	1.84 million
Workers	60 million
Inactive households	62 million
Bank	1
Government agencies	460,000
Central bank	1
Total	127 million

(5) As mentioned in Section 3.3.2, the sizes of all the firms in a sector are not available in any of our data sources, and we randomly generated the firms' sizes according to Algorithm 3. Further, the error term in the AR(1) process used for forecasting growth and inflation rates (see Section 3.9) also introduces randomness in the simulations making the firms of the same size in different simulations make different decisions. In order to take these uncertainties into account, Monte Carlo simulations are conducted.
(6) To determine the required number of samples for Monte Carlo simulations, the distribution of nominal gross value-added in the 8th and 16th periods were plotted for different number of samples. Figure 8 presents the distribution plots in the case of economy-wide forecasts when time-series data from Q1 2011 are used to estimate the AR(1) model. The plots show that the distribution converges at 800 samples, and thus all the Monte Carlo simulations were conducted using 1,000 random samples.
(7) All the simulations were conducted in the Oakbridge-CX supercomputer of The University of Tokyo, using 64 MPI processes.

Out of several major economic indicators, we could find compatible observation data for only real production, nominal production, real households' consumption, nominal GDP, exports, and imports. We compared these observations with the statistical results of our Monte Carlo simulations. Real production, nominal production, and real household production are endogenous variables, whereas nominal GDP is partially endogenous since it includes government consumption, exports, and imports, which are exogenous variables. Although imports and exports are exogenous inputs to the model, these only define the maximum inflow or outflow capacities, and the amount bought or sold by the agents can be lower than these upper bounds.

4.2.2. *Validation: The model with economy-wide forecasting for agents' decision-making*

This section presents the results of the validation of the model in which forecasts of growth and inflation rates are made using economy-wide data. As mentioned in Section 3.9, firm agents and others are set to use AR(1) for evaluating their expectations of future

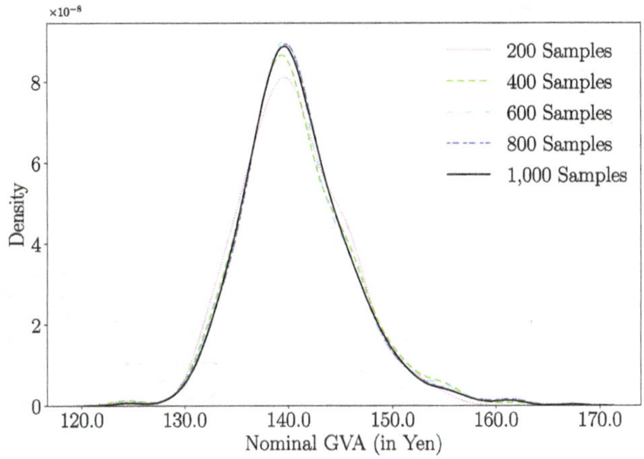

(a)

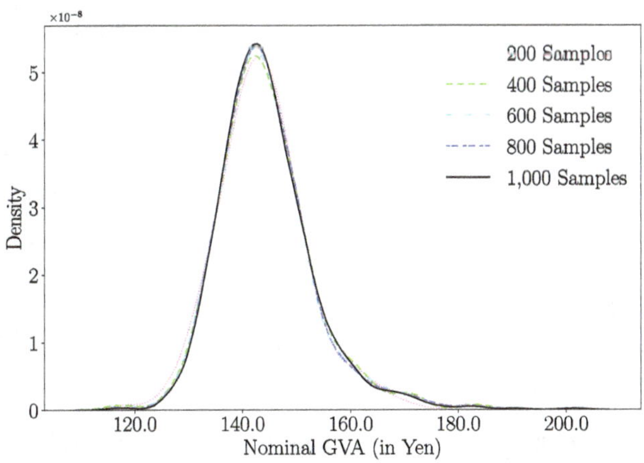

(b)

Fig. 8. Distribution of nominal gross value-added in 8th and 16th periods for Monte Carlo simulations with different number of samples. (a) Distribution in the 8th period, and (b) distribution in the 16th period.

inflation rate, etc., to make decisions. AR is a linear model which represents the system data by a line determined using Ordinary Least Square Estimator. Therefore, its performance is highly dependent on the type and length of time-series data used to estimate its parameters. Figure 9(a) presents a plot of annual real production in the Japanese economy for 22 years starting from 1994. The plot shows

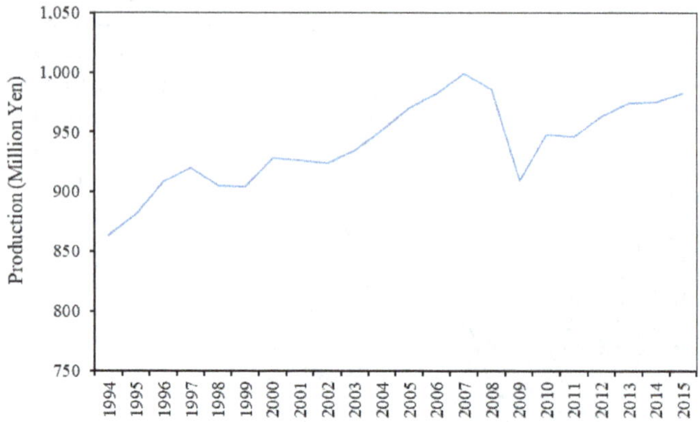

(a)

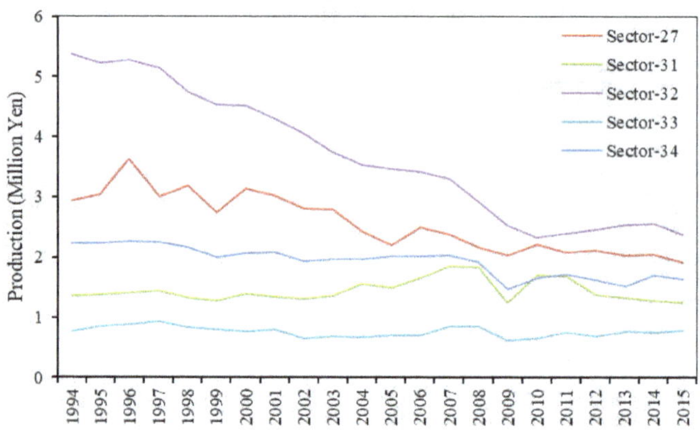

(b)

Fig. 9. While different sectors of the Japanese economy have different production trends, none have a trend similar to the economy-wide trend of production. (a) Annual economy-wise real, and (b) annual sector-wise real production.

an overall upward trend with some localized ups and downs. The big drop in 2009 is a result of the financial crisis of 2008. The growth rate of the period 2011–2015 is lower than the period 2003–2007. Sudden fluctuations and different growth rates in different periods can make the AR forecasts highly dependent on the range of data used for evaluating the parameters of AR.

In order to check the influence of the length of the time-series data used for identifying AR parameters, three simulations with time-series data from Q1 2000, time-series data from Q1 2005, and

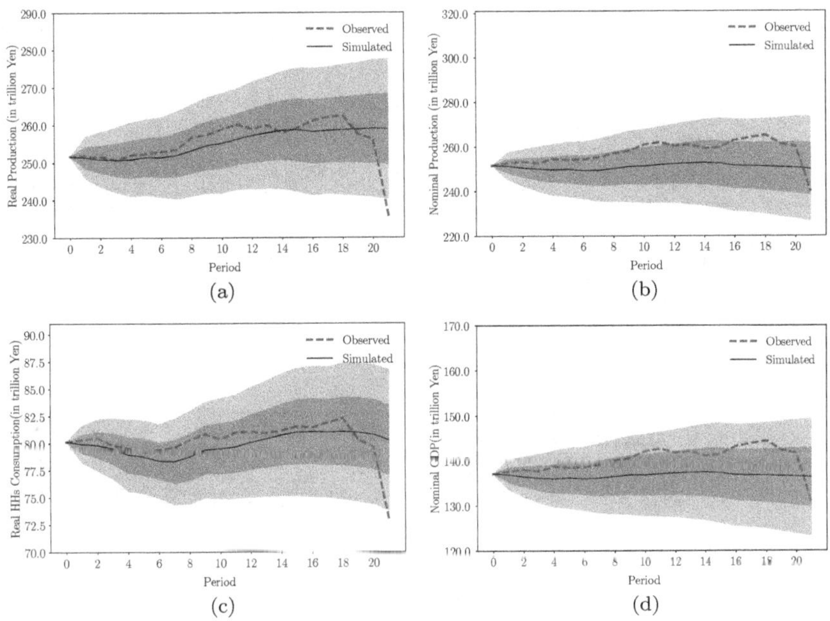

Fig. 10. Validation using economy-wide forecasts when time-series data from Q1 2000 are used to estimate the AR(1) model. (a) Real production, (b) nominal production, (c) real consumption by households, and (d) nominal gross domestic product.

time-series data from Q1 2011 were conducted. The results of the simulations are shown in Figs. 10–12, respectively. The dark and light gray regions represent the 68% and 95% confidence intervals, respectively.

Comparison with Observations (Macroeconomic Parameters): As seen in Figs. 10 and 11, the model underestimates all macroeconomic parameters under first two scenarios (i.e., time series starting with Q1 2000 and Q1 2005); some of the observations fall outside the 65% confidence interval. On the other hand, as shown in Fig. 12, the simulation results are in good agreement with the observed data, except for the last two periods, in case of the third scenario with time-series data from Q1 2011. The last two periods correspond to the corona pandemic and the model agents are not aware of it. There are many reasons for poor performance of the ABEM for the first two scenarios. The main reason is the limitations of the AR(1) model. Another reason is that the economic structure

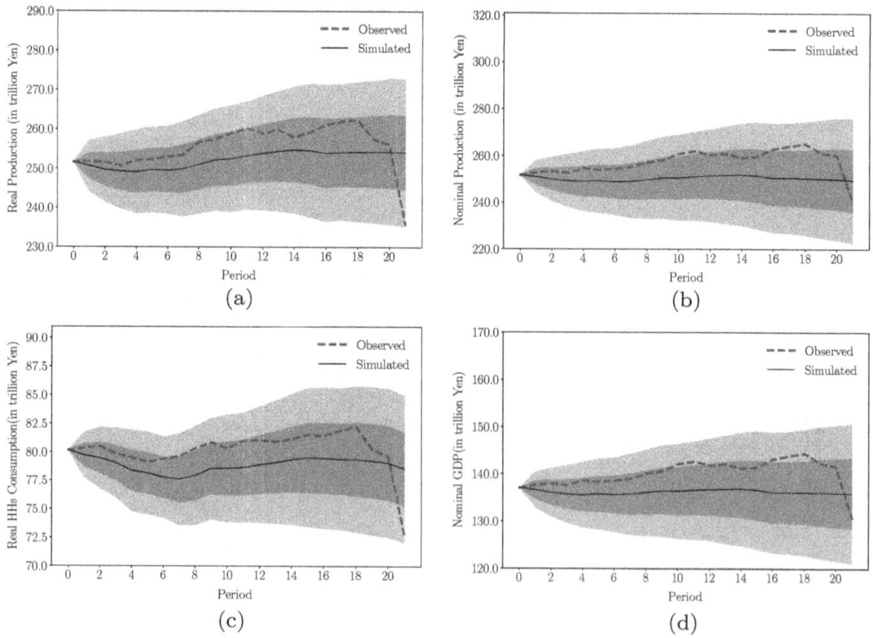

Fig. 11. Validation using economy-wide forecasts when time-series data from Q1 2005 are used to estimate the AR(1) model. (a) Real production, (b) nominal production, (c) real consumption by households, and (d) nominal gross domestic product.

of a country keeps on changing with time. Because of the technical innovations, new industries enter, and some industries become obsolete. The change in consumer behavior and global events, like the financial crisis of 2008, have profound effect on economies. A linear model like AR cannot capture such nonlinearity of the underlying economic processes.

Comparison with Observations (Sector-Wise Parameters): The main advantage of the ABEM is that it provides economic performance of each individual agent. Since we have the observed data for sector-wise real and nominal production up to Q1 2018, i.e., period 12, we compare observed values of these parameters with the simulation results obtained by using time-series data from 2011 for the AR(1). The model performance is not equally good for all sectors. In this section, plots for one of the best and one of the worst performing sectors are presented in Fig. 13. The model is capable

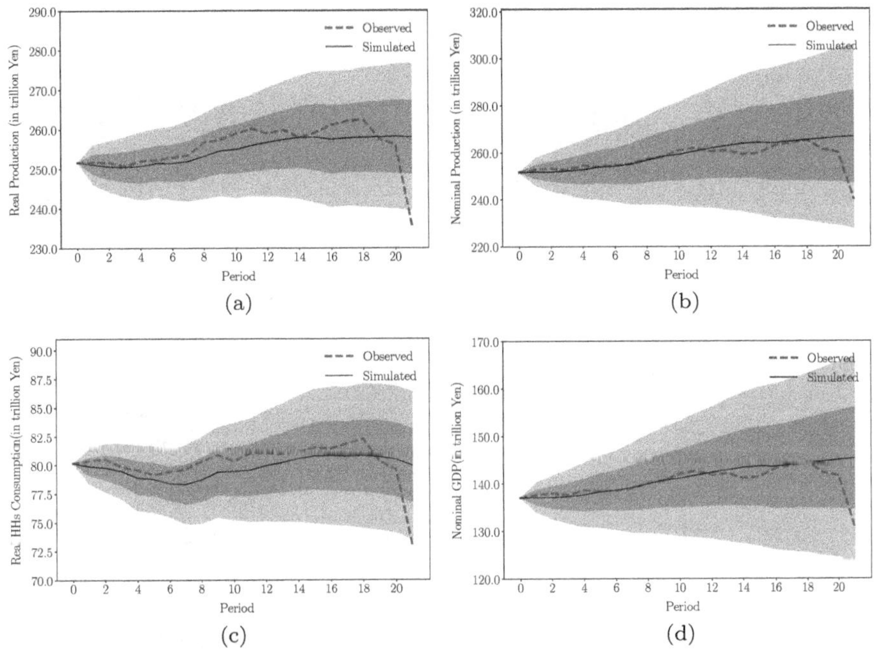

Fig. 12. Validation using economy-wide forecasts when time-series data from Q1 2011 are used to estimate the AR(1) model. (a) Real production, (b) nominal production, (c) real consumption by households, and (d) nominal gross domestic product.

of replicating the observed data for the "Commerce" sector as presented in Figs. 13(c) and 13(d), whereas the model outputs deviate from the observed data for "General-purpose Machinery" sector as shown in Figs. 13(a) and 13(b). One reason for the poor performance of the ABEM for some sectors is that the data source for the observed sector-wise data (RIETI) is different from the data source for IO-table (e-Stat). The consumption behaviors of all the agents are derived from the IO-table. Economic performance of the sectors is derived by the demand for the goods and services produced by them. Therefore, sector-wise production is influenced by the consumption behaviors of the agents, which are derived from the IO-table. A better way to compare the model performance is to obtain the observed sector-wise data from e-Stat. However, e-Stat does not provide sector-wise statistics.

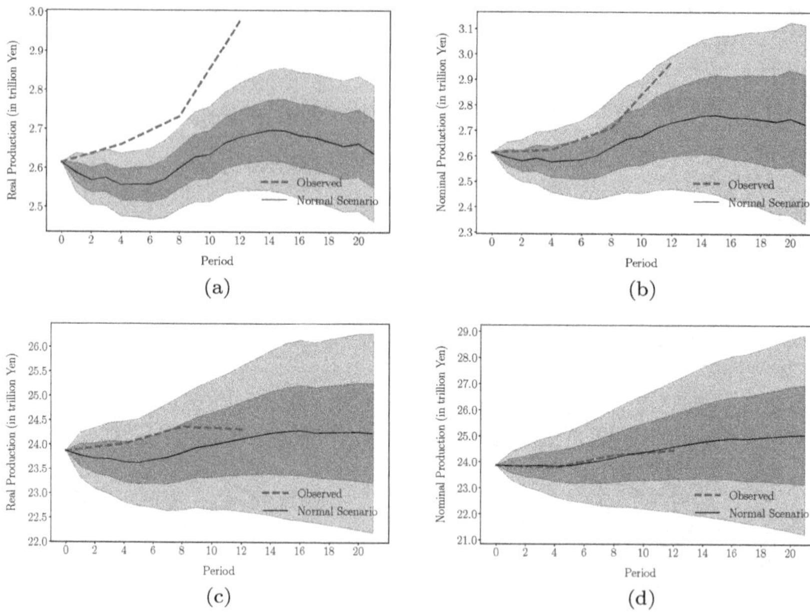

Fig. 13. Sector-wise validation using economy-wide forecasts when time-series data from Q1 2011 are used to estimate the AR(1) model. (a) Real production of "General-purpose Machinery" sector, (b) nominal production of "General-purpose Machinery" sector, (c) real production of "Commerce" sector, and (d) nominal production of "Commerce" sector.

4.2.3. *Validation: The model with sector-wise forecasting in firms' decision-making*

Figure 9(b) shows that many of the sectors have different production trends, and a comparison with Fig. 9(a) shows that the sector-wise trends are significantly different from the economy-wide trend. The use of sector-wise data ensures that firms of an industry follow the trends of the corresponding industry, thereby enabling the ABEM to forecast the sector-wise economic indices accurately. Compared to the use of economy-wide forecasts in firms' decision-making as discussed in the previous section, this can produce a significant improvement in ABEM's forecasting accuracy not only for economy-wide indices like GDP but also down to the sector-level economic indices. This section presents the results of the validation of the model whose firm agents use sector-wise forecasts of growth and inflation rates in their decision-making.

Model changes: The model has been originally developed to use economy-wide forecasting in firms' decision-making. Therefore, some changes are necessary to incorporate sector-wise forecasts of growth and inflation rates. One major change required is in the production and price decisions by the firms. Eqs. (33) and (34) are original equations for production and price decision of firms (reproduced from Ref. 12). These equations use $\gamma^e(t)$ and $\pi^e(t)$, which are time-series forecasts of economy-wide growth and inflation rates.

$$Q_i^s(t) = Q_i^e(t) = Q_i^d(t-1)(1+\gamma^e(t)). \tag{33}$$

$$\begin{aligned} P_i(t) = {} & \frac{\overbrace{\bar{w}_i(1+\tau^{SIF})\bar{P}^{HH}(t-1)(1+\pi^e(t))}^{\text{Unit Labor Costs}}}{\bar{\alpha}_i} \\ & + \overbrace{\frac{1}{\beta}\sum_g a_{sg}\bar{P}_g(t-1)(1+\pi^e(t))}^{\text{Unit Material Costs}} \\ & + \overbrace{\frac{\delta_i}{\kappa_i}\bar{P}^{CF}(t-1)(1+\pi^e(t))}^{\text{Unit Capital Costs}} + \overbrace{\tau_i^Y P_i(t-1)(1+\pi^e(t))}^{\text{Unit Net Taxes/Subsidies Products}} \\ & + \overbrace{\tau_i^K P_i(t-1)(1+\pi^e(t))}^{\text{Unit Net Taxes/Subsidies Production}} + \overbrace{\bar{\pi}_i P_i(t-1)(1+\pi^e(t))}^{\text{Target Unit Operating Surplus}}. \end{aligned} \tag{34}$$

To use sector-wise forecasts of growth and inflation rates, $\gamma_s^e(t)$ and $\pi_s^e(t)$, these equations are modified to Eqs. (35) and (36). Now, ith firm of sector s estimates expectations of demand for its products in period t using estimated growth rate of its sector, $\gamma_s^e(t)$. For determining price, the firm needs to estimate expectations of unit labor costs, material costs, capital costs, unit taxes, and subsidies and operating surplus. Since the increase in unit labor costs, unit taxes, and subsidies do not depend on inflation in a particular sector, the firm uses forecast of economy-wide inflation rate for making a realistic estimation of these costs. However, unit material cost and capital cost depend on the inflation in their constituent sectors, therefore, sector-wise inflation rates $\pi_g^e(t)$ are used for them.

$$Q_i^s(t) = Q_i^e(t) = Q_i^d(t-1)(1+\gamma_s^e(t)), \tag{35}$$

$$\begin{aligned}
P_i(t) = {} & \overbrace{\frac{\bar{w}_i(1+\tau^{SIF})\bar{P}^{HH}(t-1)(1+\pi^e(t))}{\bar{\alpha}_i}}^{\text{Unit Labor Costs}} + \overbrace{\frac{1}{\beta}\sum_g a_{sg}\bar{P}_g(t-1)(1+\pi_g^e(t))}^{\text{Unit Material Costs}} \\
& + \overbrace{\frac{\delta_i}{\kappa_i}\sum_g b_g^{CF}\bar{P}_g(t-1)(1+\pi_g^e(t))}^{\text{Unit Capital Costs}} + \overbrace{\tau_i^Y P_i(t-1)(1+\pi^e(t))}^{\text{Unit Net Taxes/Subsidies Products}} \\
& + \overbrace{\tau_i^K P_i(t-1)(1+\pi^e(t))}^{\text{Unit Net Taxes/Subsidies Production}} + \overbrace{\bar{\pi}_i P_i(t-1)(1+\pi^e(t))}^{\text{Target Unit Operating Surplus}} .
\end{aligned} \tag{36}$$

Comparison with Observations (Macroeconomic Parameters): Just as in the previous section, three simulations with time-series data from Q1 2000 (scenario-1), time-series data from Q1 2005 (scenario-2), and time-series data from Q1 2011 (scenario-3) were conducted. Figures 14–16 present a comparison of the ABEM's output with the observed data under the three scenarios. In contrast to economy-wide forecasting, the model outputs are in good agreement, except for last two periods which correspond to COVID-19 pandemic, with the observed data under all the three scenarios. Under scenario-1 and scenario-2, model outputs for both real and nominal variables follow their observed data closely. Whereas in case of scenario-3, model performance is better for real variables compared to nominal variables as nominal variables are a little underestimated (Figs. 16(b) and 16(d)). Further, scenario-1 and scenario-2 have lower standard deviation for both real and nominal variables than scenario-3. This shows improvement in model performance with the use of high-resolution data.

Comparison with Observations (Sector-Wise Parameters): Just as in the former case, not all the sectors performed equally well.

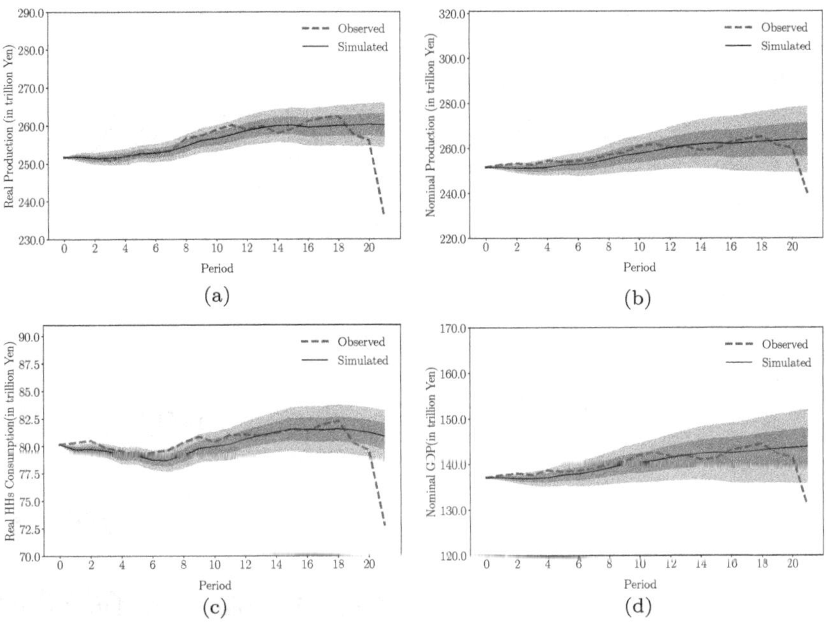

Fig. 14. Validation using sector-wise forecasts when time-series data from Q1 2000 are used to estimate the AR(1) model. (a) Real production, (b) nominal production, (c) real consumption by households, and (d) nominal gross domestic product.

Figure 17 compares simulated and observed values of real and nominal production for "General-purpose Machinery" and "Commerce" sectors. From Figures 16 and 17, we can see that the performance for macro-economic parameters is much better than for sector-wise parameters. One reason for this is the incompatibility of data as explained in Section 4.2.2. Another reason is that the model outputs not only depend on the AR(1) process but also the interactions between agents. The agents use forecasts of growth and inflation rates to make preliminary estimates. The values of various parameters realized by the agents are a result of agents' interactions and model dynamics. The agents' interactions serve as a feedback mechanism to the AR process and corrects its path. Moreover, the model performance for large sectors like the "Commerce" sector is not bad. The smaller sectors have little impact on the macro values.

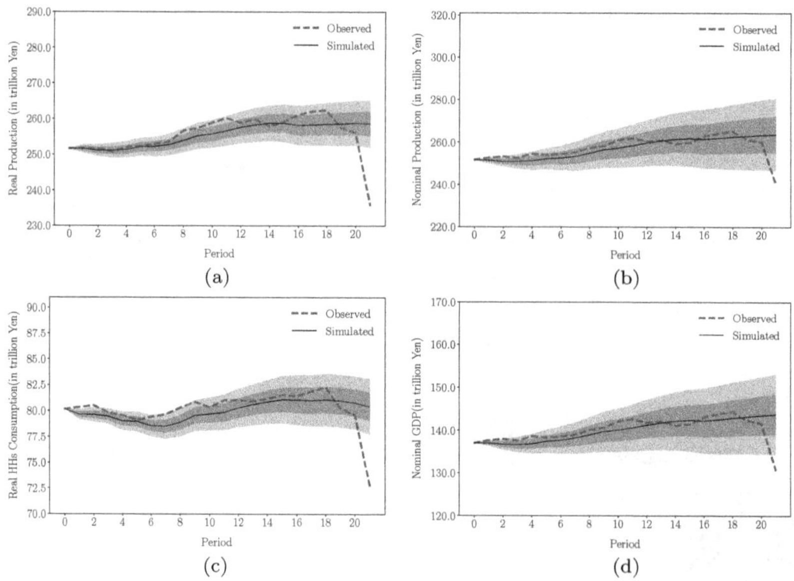

Fig. 15. Validation using sector-wise forecasts when time-series data from Q1 2005 are used to estimate the AR(1) model. (a) Real production, (b) nominal production, (c) real consumption by households, and (d) nominal gross domestic product.

4.2.4. *Comparison of model performance under economy-wide and sector-wise forecasts*

Figures 18 and 19 compare model outputs under economy-wide forecasts and sector-wise forecasts made using time-series data from Q1 2011. As it can be seen, in the case of sector-wise forecasts, the standard deviation in the model outputs is almost half of that in case of economy-wide forecasts. This clearly shows an improvement in the confidence level of the model outputs, as expected. The ABEM is a micro-founded model which models each individual agent's behavior. In order to mimic the aggregated economic behavior, the model agents need to mimic the behavior of their real-world counterparts. In the case of sector-wise forecasts, the firms' behavior is more accurate compared to the case with economy-wide forecasts. These results indicate that sector-wise forecasts induce better performance than economy-wide forecasts. Likewise, we expect that the model performance will further improve if firm-level data are used to model each firm's behavior.

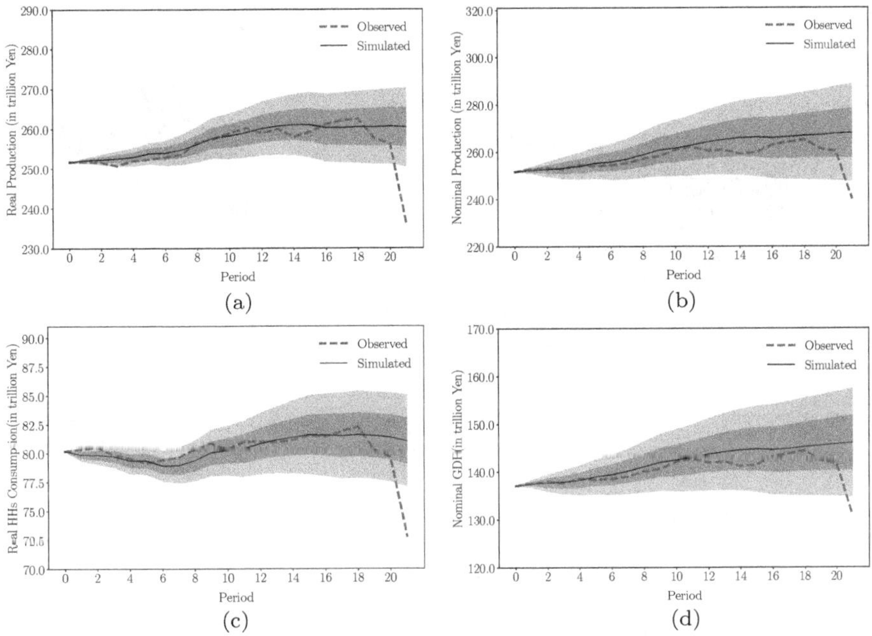

Fig. 16. Validation using sector-wise forecasts when time-series data from Q1 2011 are used to estimate the AR(1) model. (a) Real production, (b) nominal production, (c) real consumption by households, and (d) nominal gross domestic product.

5. Integration of the ABEM with Physics-Based Simulators

The end goal of developing the HPC-enhanced ABEM is to simulate a post-disaster economy under different recovery scenarios and quantitatively evaluate those scenarios considering their long-term economic impacts. To that end, the ABEM has to be tightly integrated with a detailed model of the built environment, supply chains, and physics-based numerical tools to simulate the damages to the built environment due to the target natural disasters. Ideally, the built environment should include high-resolution models of each building, lifeline networks like transportation, communication, power, water, sewer, and gas, while the physics-based simulators must be equipped with numerical tools to simulate the target disasters in sufficiently high spatial and temporal resolutions so that the damages to each

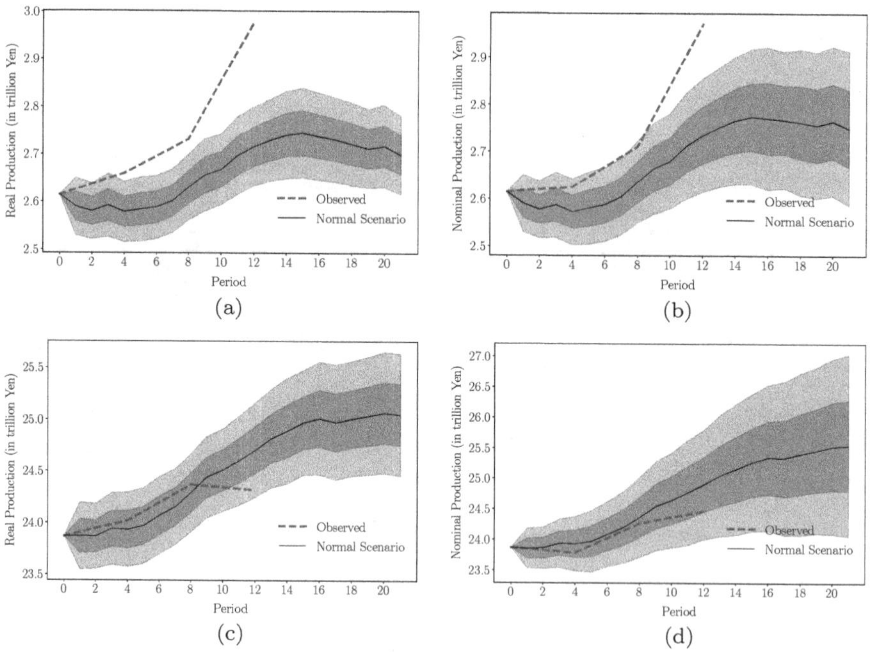

Fig. 17. Sector-wise validation using sector-wise forecasts when time-series data from Q1 2011 are used to estimate the AR(1) model. (a) Real production of "General-purpose Machinery" sector, (b) nominal production of "General-purpose Machinery" sector, (c) real production of "Commerce" sector, and (d) nominal production of "Commerce" sector.

element of the built environment can be estimated to a desired level of confidence. Such integration of ABEM with the built environment will have the potential to reasonably capture the post-disaster economy of a country if the agents are programmed to engage in economic activities only when they have access to the necessary infrastructures for their activities (e.g., a firm can start production only if it has access to all the required lifelines such as gas, water, transportation, in addition to the existing constraints such as the availability of labor, capital goods, and intermediate goods). The integrated system, thus developed, can perform end-to-end simulations of a disaster and the impacted economy. Such a system can be used to study vulnerability to disasters, identify critical parts of infrastructure and the economy, take preventive actions, and devise recovery strategies. Moreover, the integrated system can be used to evaluate

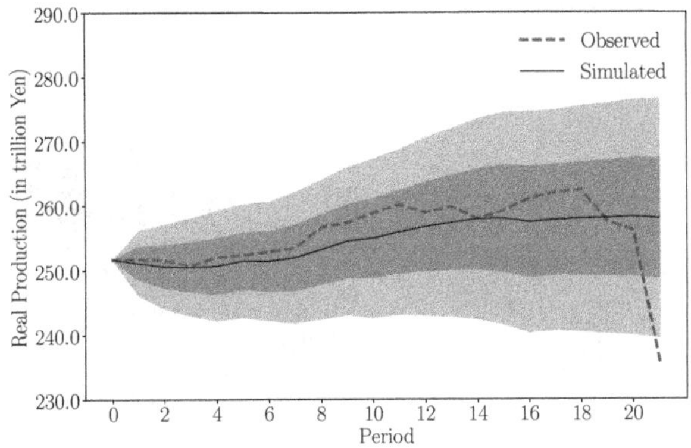

(a)

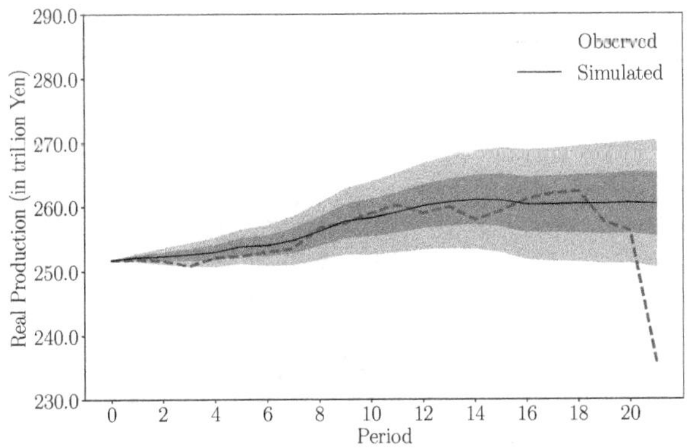

(b)

Fig. 18. Comparison of real production obtained using time-series data from Q1 2011 for estimating AR(1). (a) Real production when firm agents use economy-wide forecasts, and (b) real production when firm agents use sector-wise forecasts.

different recovery plans based on their short- and long-term economic performance.

In order to achieve such end-to-end simulations of earthquake disasters and their impacts on the economy, the ABEM is integrated with the Integrated Earthquake Simulator (IES),[23,24] which can simulate the strong ground motion and the corresponding seismic response of each building in large cities, and a repair cost estimation tool, which can convert the seismic performance of buildings into corresponding estimates of repair costs and repair time (see Fig. 20). The current integrated system does not consider the constraints such

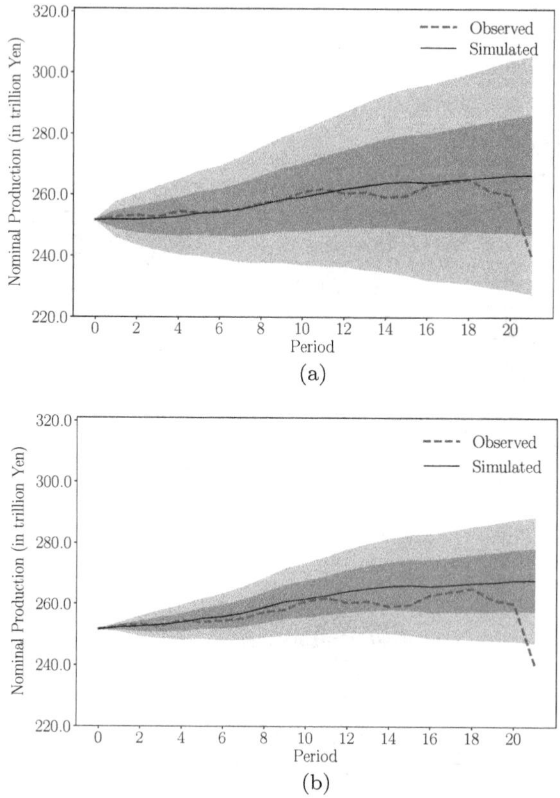

Fig. 19. Comparison of nominal production obtained using time-series data from Q1 2011 for estimating AR(1). (a) Nominal production when firm agents use economy-wide forecasts, and (b) nominal production when firm agents use sector-wise forecasts.

as limited manpower and material for repairing and the accessibility to lifelines for starting economic activities of the buildings' occupants; we plan to include these constraints in future developments. The rest of this section presents some details of the current state of the integrated system for post-disaster economic simulations.

5.1. *Integrated earthquake simulator*

We used the integrated earthquake simulator (IES)[23,24] to conduct soil amplification analysis (SAA), and seismic response analysis

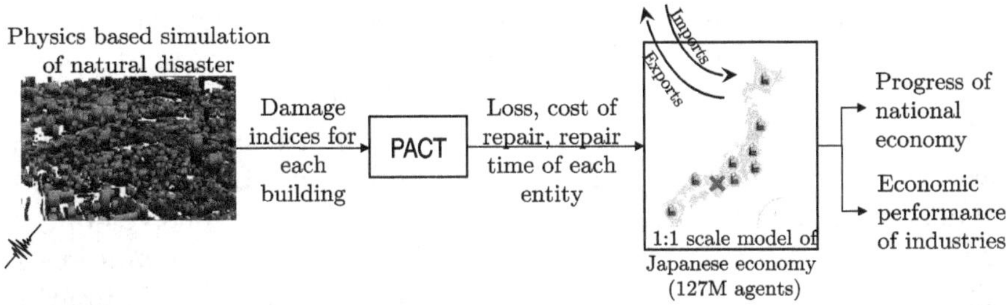

Fig. 20. Schematic diagram of the integrated system for seamlessly simulating earthquake disasters and post-disaster economies.

(SRA) of the target domain. The SAA simulates the amplification of the seismic waves in soft soil layers beneath each building and the SRA simulates the seismic response of each building due to the amplified seismic wave. The inputs to the SAA are the profile of the soil layers beneath each building and the desired seismic wave at the bedrock. The soil profile data are generated based on the data available at G-Space Information Center.[bb] The inputs required for the SRA are the amplified seismic waves from the SAA and the structural details of buildings such as the configuration of load-carrying skeleton and material properties. Such details of buildings are not openly available due to privacy reasons. In the absence of buildings' data, the IES is capable of making a reasonable estimate of the structural details and material properties of each building based on the outer shape and the dimensions available in the GIS data and based on the fact that structures are designed to comply with the design standards. The level of damages in each building can be estimated based on the high-resolution outputs from IES, such as the time histories of each floor's displacements, drift angles, velocities, accelerations, and forces.

5.2. *Estimation of repair costs*

To analyze the economy's response to an earthquake disaster, the results of the IES need to be quantified in monetary terms. Repair cost and repair time are the most suitable parameters which indicate total costs and total time required to restore the infrastructure

[bb]https://www.geospatial.jp/gp_front/.

to its pre-damaged state. Performance Assessment Calculation Tool (PACT)[25] is one such tool which can convert the seismic response of a building into repair costs required to restore the building. It has been developed by the Federal Emergency Management Agency, USA, and uses a database of building components' fragility and consequence functions. The database is developed by extensive surveys of various types of buildings in the USA and by experiments in the laboratory. PACT comes with the Normative Quantity Estimation Tool (NQET),[44] which provides estimates of the types and quantities of non-structural components typically present in a building based on its size and occupancy type. To the best of our knowledge, no such tool exists for Japan. Hence, we used PACT to estimate the damages to buildings in Japanese cities. Such direct usage of PACT introduces a certain amount of error, and our application of PACT is only intended for demonstrative purposes; detailed studies on utilizing PACT for Japanese buildings are necessary prior to any practical application. We expect that the PACT's estimation of cost and time of repair are not far from the reality since both the USA and Japan have similar design standards, construction materials, and construction practices.

PACT and NQET are designed to receive inputs via a GUI requiring manual operation. According to our experience, it takes about 4 min to analyze a 3-story building. Obviously, it is necessary to automate and speed up these tools since our target is to analyze millions of buildings in large cities. We eliminated manual inputs by modifying their .NET source code to read the required inputs from files and speeded up the analysis process by utilizing shared memory parallelization to accelerate most time-consuming functions. In addition, several performance improvements were implemented: Excel files used for passing information among different functions were replaced with text files since accessing Excel files is extremely slow, turns off display updates during the analysis, prevents the launch of new user objects, etc. Table 5 presents the mean run-time (mean of time taken for 20 3-story buildings) taken to simulate a 3-story building. As the table shows, the performance improvements speed-up the applications by 120 times. This enables us to analyze millions of buildings using PACT.

Table 5. Mean run-time taken to simulate a 3-story building by different versions of PACT.

PACT version	Mean run-time for a 3-story building (s)
Unmodified code: Manual analysis	240
Automated code	60
With performance improvements	20
Parallelized code	2

Table 6. Building occupancy types and occupying agents' types.

Building occupancy type	Occupying agents' type
Commercial office	Firms of all sectors except the sectors represented by other occupancy types as presented in the following rows
Educational institutes	Firms of education sector
Healthcare	Firms of medical service and health and hygiene sectors
Hospitality	Firms of hotels sector
Residential buildings	Households
Research laboratories	Firms of research sector
Retail	Firms of commerce sector
Warehouse	Firms of storage facility services sector
Public offices	Firms of public administration sector
Production facilities	Firms of miscellaneous manufacturing products sector

5.3. *Integration with the ABEM*

PACT estimates the repair costs (i.e., infrastructure losses) based on buildings' occupancy type (e.g., residential, healthcare, educational). Since we don't have information of the type of occupancy of each building, we made a reasonable guess of the occupancy type and estimated the repair costs based on the following assumptions.

(1) Depending on the occupancy type, the buildings are assumed to belong to firms of an industry or households as shown in Table 6.

(2) The repair costs for buildings are assigned as losses to individual firms and households in proportion to the floor area occupied by them. Because of the absence of the relevant data, the following assumptions are made regarding floor area required for an employee and household.

   (a) Required floor area per employee of "Education", "Medical Service", and "Health and Hygiene" sectors is $83.6\,\mathrm{m}^2$ ($900\,\mathrm{ft}^2$).[cc]
   (b) Office floor area required per employee for all other sectors is assumed to be $13.9\,\mathrm{m}^2$.[dd]
   (c) Living space required for a worker or inactive household is assumed to be $22.3\,\mathrm{m}^2$ ($240\,\mathrm{ft}^2$), and $33.45\,\mathrm{m}^2$ ($360\,\mathrm{ft}^2$) for an investor.[ee]

(3) Based on the values of unit floor area required for a worker, the number of affected workers is calculated for each occupancy type. Accordingly, loss per worker is also calculated.

(4) Some occupancy types represent more than one sector, as an example, Commercial Office. For such occupancy types, Equation (37) is used to get the number of workers affected in each constituent sector, k.

$$N_{\mathrm{aff}}(k) = \frac{N_{s=k}}{\sum_s N_s} N_{\mathrm{aff}}^T, \tag{37}$$

where N_s is the total number of workers in sector s and N_{aff}^T is total number of workers affected for that occupancy type.

Total floor area damaged is divided among investors, workers, and inactive households in proportion to their salaries. Based on damaged floor area and floor area per person for each type of household, numbers of affected investors, workers, and inactive households are determined.

[cc] The assumption is based on the size of a basic classroom or hospital room.

[dd] The approximate floor area per employee is in the range 9.3–$16.7\,\mathrm{m}^2$. https://www.reitasiapac.com/tokyos-office-space-shrunk-to-its-smallest-on-record-in-2019/.

[ee] $22.3\,\mathrm{m}^2$ is average living space per person in Japan. https://resources.realestate.co.jp/living/how-much-living-space-does-the-average-household-have-in-japan/.

(5) From the previous step, we can determine the total number of workers affected in each sector and the number of affected investors, workers, and inactive households. To assign the losses to firms of sector k, n_f firms of that sector are randomly selected such that $\sum_{j=0}^{n_f} N_j \geq N_{\text{aff}}(k)$, where N_j is the number of workers of firm j. $N_j \times U_w$ is the loss assigned to a randomly selected firm j, where U_w is the unit loss per worker.

To assign the losses to individual investors, n_i number of investors are randomly chosen and assumed to be affected by the disaster, where n_i is the number of affected investors as determined in the previous step. Average value of loss is assigned to each affected investor. Similarly, losses are assigned to affected workers and inactive households.

6. Demonstrative Simulation of a Post-Disaster Economy

The preceding sections described the development of the ABEM, and the process of integrating it with physics-based numerical tools such as IES for performing end-to-end simulations of earthquake disasters and the corresponding economy. In order to demonstrate the application of the integrated system in simulating post-disaster economies, we simulated the Japanese economy in the aftermath of a potential earthquake disaster in the Osaka bay area caused by the Nankai trough mega earthquake scenario, proposed by the Cabinet office of Japan. Ground-shaking and earthquake-induced damages to 1.8 million buildings of the Osaka bay area are estimated using IES and the seismic response-based damage estimations are used in the ABEM to analyze their impact on the economy. This section presents the simulation process in detail and the analysis of the simulation results.

6.1. *Modifications to the ABEM*

In the original ABEM,[10,12] agents' decision rules are defined to simulate their economic behavior under the normal economic scenario. After a disaster, many agents suffer losses in the form of damages, which have a significant effect on their decision-making. The rules defining agents' behavior have to be modified to model their post-disaster economic behavior. The following sections present the modified rules and the rationale for various agents.

6.1.1. *Firms*

The capital goods owned by firms consist of buildings and machinery. Firms require capital goods to produce goods and services. After an earthquake, the damages to the buildings reduce the capital stock of firms limiting their production capacity. Therefore, the firms would like to repair or rebuild damaged buildings to bring back their operations to pre-disaster levels. Assuming that the earthquake disaster occurs at the start of period t and ith firm loses X_i capital because of damages to its buildings, the firm's rules are modified as follows.

(1) Available real capital after disaster: $K_i(t)$

$$K_i(t) = \bar{K}_i(t) - \frac{X_i}{\bar{P}^{CF}(t)},$$

where $\bar{K}_i(t)$ is the firm's real capital before disaster and $\bar{P}^{CF}(t)$ is capital formation price index in period t.

(2) Firm profit: $\Pi_i(t)$

$$\Pi_i(t) = \bar{\Pi}_i(t) - X_i,$$

where $\bar{\Pi}_i(t)$ is the firm's profit without considering the disaster.[ff]

(3) Investment required for reconstruction: $I_i^{\text{reconstruct}}(t)$

$$I_i^{\text{reconstruct}}(t) = \frac{X_i}{\bar{P}^{CF}(t)},$$

where $\bar{P}^{CF}(t)$ is capital formation price index in period t.

[ff] This is a standard procedure, as per accounting rules defined in System of National Accounts (SNA), to account for losses caused by disasters.

(4) Desired investment in capital stock: $I_i^d(t)$

$$I_i^d(t) = \frac{\delta_i}{\kappa_i} \min\left\{Q_i^s(t), \kappa_i K_i(t)\right\} + \phi_i(t) I_i^{\text{reconstruct}}(t) - \frac{\Delta L_i^d(t) - \Delta L_i(t)}{\bar{P}^{CF}(t-1)(1+\pi^e(t))},$$

where $Q_i^s(t)$ is total supply decided by the firm, $K_i(t)$ is capital stock, $\phi_i(t) I_i^{\text{reconstruct}}(t)$ is the planned investment toward recovery with $\phi_i(t) \in [0,1]$, $\Delta L_i^d(t)$ is the amount of loan demanded from a bank, and $\Delta L_i(t)$ is the amount of loan granted by the bank.

Desired investment in capital stock for non-disaster periods, t': $I_i^d(t')$

$$I_i^d(t') = \frac{\delta_i}{\kappa_i} Q_i^s(t') + \phi_i(t') I_i^{\text{reconstruct}}(t') - \frac{\Delta L_i^d(t') - \Delta L_i(t')}{\bar{P}^{CF}(t'-1)(1+\pi^e(t'))}.$$

(5) Desired amount of intermediate goods: $\Delta M_{ig}^d(t)$

$$\Delta M_{ig}^d(t) = a_{sg} \frac{\min\left\{Q_i^s(t), \kappa_i K_i(t)\right\}}{\beta_i}.$$

Desired amount of intermediate goods for non-disaster periods, t': $\Delta M_{ig}^d(t')$

$$\Delta M_{ig}^d(t') = a_{sg} \frac{Q_i^s(t')}{\beta_i}.$$

(6) Desired number of workers: $N_i^d(t)$

$$N_i^d(t) = \max\left\{1, \text{round}\left(\frac{\min\left\{Q_i^s(t), \kappa_i K_i(t)\right\}}{\bar{\alpha}_i}\right)\right\}.$$

Desired number of workers for non-disaster periods, t': $N_i^d(t')$

$$N_i^d(t) = \max\left\{1, \text{round}\left(\frac{Q_i^s(t')}{\bar{\alpha}_i}\right)\right\}.$$

(7) After buying capital in the goods market, the updated value of investment required for reconstruction is as follows:

$$I_i^{\text{reconstruct}}(t) = I_i^{\text{reconstruct}}(t-1) - I_i(t) + \frac{\delta_i}{\kappa_i} Y_i(t),$$

where $I_i(t)$ is the capital bought, and $Y_i(t)$ is quantity of goods produced by the firm.

6.1.2. *Households*

The households are assumed to suffer capital loss in the form of building damages. Households may have earthquake insurance for their homes. For simplicity, we assume that the government pays total amount of damages to the households as disaster aid. The households use the disaster aid to repair/rebuild their damaged buildings. If X_h and X_h^{aid} are the amounts of damages and disaster aid for a household h, the household's decision rules are modified as follows:

(1) Disaster aid received: X_h^{aid}

$$X_h^{\text{aid}} = X_h.$$

(2) Deposits:

$$D_h(t) = \bar{D}_h(t) + X_h^{\text{aid}},$$

where $\bar{D}_h(t)$ is the household's deposit before disaster.

(3) Capital:

$$K_h(t) = \bar{K}_h(t) - X_h,$$

where $\bar{K}_h(t)$ is the household's capital before disaster.

(4) Investment required for reconstruction:

$$I_h^{\text{reconstruct}}(t) = X_h^{\text{aid}}.$$

(5) Desired investment:

$$I_h^d(t) = \frac{\psi^H Y_h^e(t)}{1 + \tau^{CF}} + \phi_h(t) I_h^{\text{reconstruct}}(t),$$

where ψ^H is the household's propensity to invest, $Y_h^e(t)$ is expected income of the household, τ^{CF} is the tax rate on investment goods, and $\phi_h(t) I_h^{\text{reconstruct}}(t)$ is the planned investment toward recovery with $\phi_h(t) \in [0, 1]$.

(6) After buying investment goods, the updated value of investment required for reconstruction is

$$I_h^{\text{reconstruct}}(t) = I_h^{\text{reconstruct}}(t-1) + \frac{\psi^H Y_h^e(t) - I_h(t)}{1+\tau^{CF}},$$

where $I_h(t)$ is the amount of investment goods purchased by the household.

6.1.3. *Government*

The government takes loans from the central bank to provide disaster aid to households. Therefore, its total loan changes as follows:

$$L^G(t) = \bar{L}^G(t) + \sum_h X_h^{\text{aid}},$$

where $\bar{L}^G(t)$ is the loan amount prior to the disaster.

6.2. ***Simulations of post-disaster economy***

As a demonstrative application of the integrated system of IES and ABEM, we simulated the economy following a hypothetical earthquake disaster. The earthquake disaster and the affected area considered for this hypothetical scenario are as follows.

6.2.1. *Disaster and impacted region*

Earthquake scenario: The Ministry of Land, Infrastructure, Transport, and Tourism (MLIT) of Japan provides a list of impending large-scale earthquakes[gg] in Japan. According to MLIT, there is 70%–80% probability of occurrence of a Nankai Trough earthquake of magnitude M8-M9 within 30 years. The Nankai Trough earthquake is anticipated to impact economically important regions of western Japan. Therefore, we decided to estimate the impacts of the Nankai Trough earthquake. In order to evaluate seismic performance of the infrastructure in the region, we used the seismic waves proposed by the Cabinet Office of Japan.

[gg] https://www.mlit.go.jp/river/earthquake/en/future/index.html, accessed on 10 June 2021.

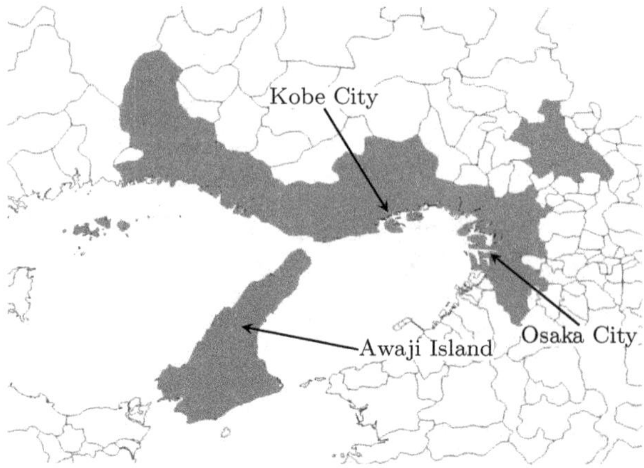

Fig. 21. Seismic damages to the buildings located in the highlighted areas in Kansai region of Japan are considered in this study.

Impacted region: Hanshin industrial region is expected to be one of the economically important regions impacted by the expected Nankai Trough earthquake. Being one of the largest industrial regions of Japan and an economic powerhouse of western Japan, seismic damages in the region will have significant impacts on the national economy. In order to estimate the corresponding economic impacts, we evaluated seismic performance of approximately 1.8 million buildings of the region as shown in Fig. 21 and calculated their seismic damages.

The main steps involved in end-to-end simulations of earthquake disasters and their impact on the economy, which are described in Section 5, are explained as follows.

6.2.2. *Infrastructure data from GIS*

The first step is to obtain the buildings' properties listed as follows:

(1) Number of floors
(2) Floor area and floor height
(3) Occupancy type
(4) Construction material
(5) Load-resisting system

The first three items are obtained from the GIS data. Information about construction material and load-resisting system is not available, so, these values were assumed depending on the number of floors and floor area.

A total of 1,786,845 buildings of Osaka city, Kobe city, Awaji island, and surrounding areas, as shown in Fig. 21, are considered in this study. The buildings are categorized into 17 occupancy types. The occupancy types are commercial office, kindergarten, elementary, middle, and high school, clinic, hospital, research laboratory, retail, warehouse, production facility, university, miscellaneous educational institutes, houses, hotels, and public offices. The number of stories in the shortest and tallest buildings are 1 and 55, respectively.

6.2.3. *Seismic performance evaluation using IES*

In the second step, seismic performance evaluation of the buildings is conducted using an IES as explained in Section 5.1. The IES provides floor-wise seismic displacement, velocity, acceleration, and story drift for all buildings.

6.2.4. *Loss estimation using PACT*

After obtaining the seismic demand parameters of the buildings, their associated repair costs are estimated using PACT and NQET (Section 5.2). First, we use NQET to estimate normative quantities of various non-structural components typically present in a building. In order to make building models using PACT, types and quantities of structural components of the buildings are required. However, the data regarding structural components are unavailable. PACT provides a list of standard load-resisting systems and their components. According to the definition of various load-resisting systems provided by PACT, we assigned load-resisting systems to buildings depending on their construction material and height. All these assumptions are presented in Table 7.

Using the building dimensions, occupancy type, quantities of structural and non-structural components, and seismic demand parameters obtained from the IES, building models are generated for each building type using PACT Building Modeler. The building models are then analyzed using PACT Engine to obtain repair costs, repair time, and the probability of being assigned an unsafe placard.

Table 7. Assumptions regarding types and quantities of structural components for different types of buildings.

Number of stories	Construction material	Structural components[a]	Quantity
≤ 4	Wood	B1071.031: Wood walls with diagonal let-in bracing	All exterior walls
	Steel	B1033.101a: Steel Buckling Restrained Brace (BRB), Chevron Brace, Weight of brace < 40 plf	One BRB bay per 30 feet of exterior walls
	Concrete	B1041.001a and B1041.001b: Simple Moment Frame (SMF) with columns and beams of size $24'' \times 24''$	One SMF per 30 feet of exterior walls
$4<$ and ≤ 20	Steel	B1033.101b: Steel Buckling Restrained Brace (BRB), Chevron Brace, Weight of brace > 41 plf and < 99 plf	One BRB bay per 30 feet of exterior walls
	Concrete	B1041.002a and B1041.002b: Simple Moment Frame (SMF) with columns and beams of size $24'' \times 36''$	One SMF per 30 feet of exterior walls
>20	Steel	B1033.101c: Steel Buckling Restrained Brace (BRB), Chevron Brace, Weight of brace > 100 plf	One BRB bay per 30 feet of exterior walls
	Concrete	B1041.003a and B1041.003b: Simple Moment Frame (SMF) with columns and beams of size $36'' \times 36''$	One SMF per 30 feet of exterior walls

[a]Code and definition of the Structural Components are as defined in PACT.

The PACT provides repair costs in USD, which need to be converted to JPY. We assume that the disaster occurs in Q1 2015, the base quarter of the IO-table and the first simulation period of the ABEM. Therefore, an USD/JPY exchange rate of 121, which is the average exchange rate for 2015, is used to convert the repair costs into JPY. The cost of constructing a new building is assumed to be 323,000 JPY per unit floor area. A summary of repair costs is provided in Table 8. Further, Fig. 22 presents the distribution of estimated values of repair cost per unit floor area and repair time. We can see that the repair cost is in the range of 10,000 to 100,000 JPY/m^2 with an overall average of 25,510 JPY/m^2. The average repair cost is less than

Table 8. Summary of repair costs estimated by PACT.

Occupancy type	Number of buildings	Total floor area (m^2)	Total repair cost (million JPY)	Repair cost per unit floor area (JPY/m^2)
Commercial office	137,630	7.2×10^7	1.79×10^6	24,826
Elementary school	491	1.53×10^6	18,903	12,382
Middle school	277	1.16×10^6	14,192	12,255
High school	143	954,281	9,396	9,846
Clinic	11,582	1.25×10^7	260,760	20,903
Hospital	76	793,606	14,436.40	18,190
Research laboratory	44	157,538	2,005	12,729
Retail	140,343	1.40×10^8	1.92×10^6	13,661
Warehouse	2,887	6.25×10^6	189,079	30,228
Production facility	52,293	5.20×10^7	1.93×10^6	37,149
Educational institutes	2,583	3.77×10^6	45,106	11,968
University	75	493,106	5,294	10,737
House	987,406	1.54×10^8	5.38×10^6	34,856
Others	447,013	1.34×10^8	3.32×10^6	24,810
Hotels	918	2.58×10^6	21,000	8,145
Kindergarten	1,345	1.70×10^6	26,902	15,783
Public offices	1,739	4.51×10^6	79,389	17,620
Total	1,786,845	5.89×10^8	1.50×10^7	25,510

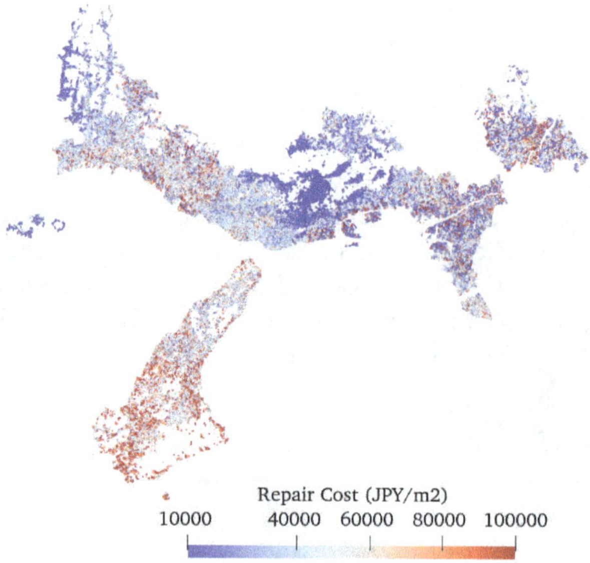

(a)

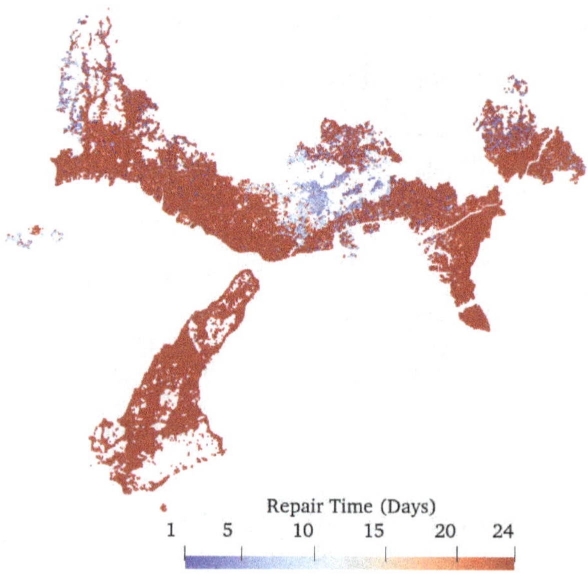

(b)

Fig. 22. Distribution of repair cost and repair time estimated by PACT. (a) Repair cost per unit floor area, and (b) repair time in days.

10% of the cost of a new building. The repair time for most of the buildings is 24 days. However, it should be noted that PACT does not consider resource constraints and resource mobilization time in estimating repair time. The repair time estimated by PACT reflects only the time taken to replace/repair the damaged components of the building with sufficient availability of labor and raw materials.

6.2.5. *Post-disaster economic simulations using ABEM*

In the last step, repair cost and repair time estimated by the PACT are assigned to the ABEM agents as explained in Section 5.3. The repair costs are capital loss suffered by various agents. The following settings are used for the ABEM to simulate a post-disaster economy.

(1) The disaster is assumed to occur at the end of the first simulation period i.e., Q1 2015 or in other words, in the beginning of the second simulation period i.e., Q2 2015.
(2) The agents' behavioral rules are modified as explained in Section 6.1. The values of $\phi_i(t)$ and $\phi_h(t)$ are assumed 1, which means that all the agents allocate recovery budget equal to their unrecovered capital.
(3) A worker affected by the disaster is able to work half-time only. Unaffected workers can work overtime also but with a maximum limit of 1.5 times their normal time.
(4) The model parameters and initial parameters of ABEM agents are the same as those used in the validation of the ABEM (see Section 4.2.1).
(5) AR(1) process is used to make forecasts of growth and inflation rates. Economy-wide forecast rates are used.
(6) The simulation is conducted for 22 periods and Monte Carlo simulations are conducted for 1,000 random samples.

Figure 23 compares the mean values of various macroeconomic indicators under a disaster scenario and a normal scenario (i.e., no disaster); the dashed and continuous lines represent the disaster scenario and the normal scenario, respectively. As we can see in Fig. 23(b), the firms' capital decreases in the disaster period because of the capital loss. In order to recover their lost capital, all the affected firms allocate investment budget toward recovery, which is evident from Fig. 23(a). These two figures also show that the firms are able to recover their lost capital by the end of the 3rd period. Similarly, the households also allocate higher investment budget to recover their lost capital as shown in Fig. 23(c). Figure 23(d) shows that the households cannot spend their planned investment budget

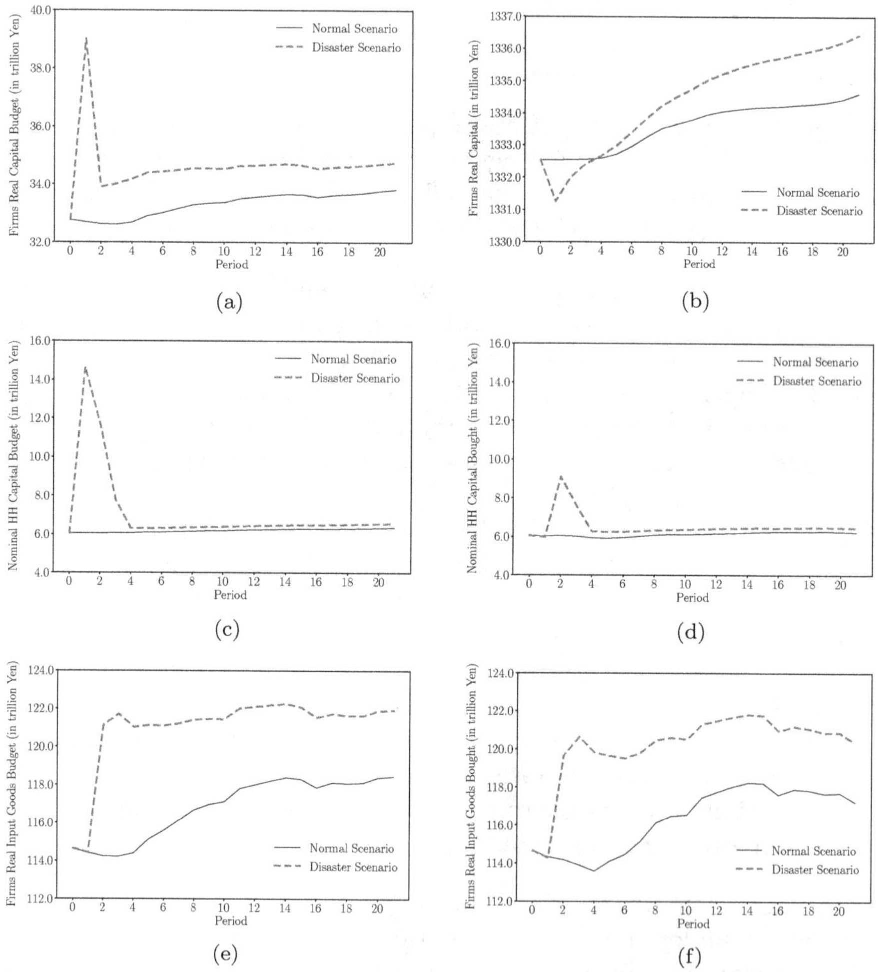

Fig. 23. Comparison of mean values of various economic indicators under the disaster scenario and the normal scenario. (a) Firms' real capital budget, (b) firms' real capital, (c) households' capital budget, (d) capital bought by households, (e) firms' real input goods budget, (f) real input goods bought by firms, (g) real production, (h) nominal production, (i) employment, and (j) nominal gross value-added.

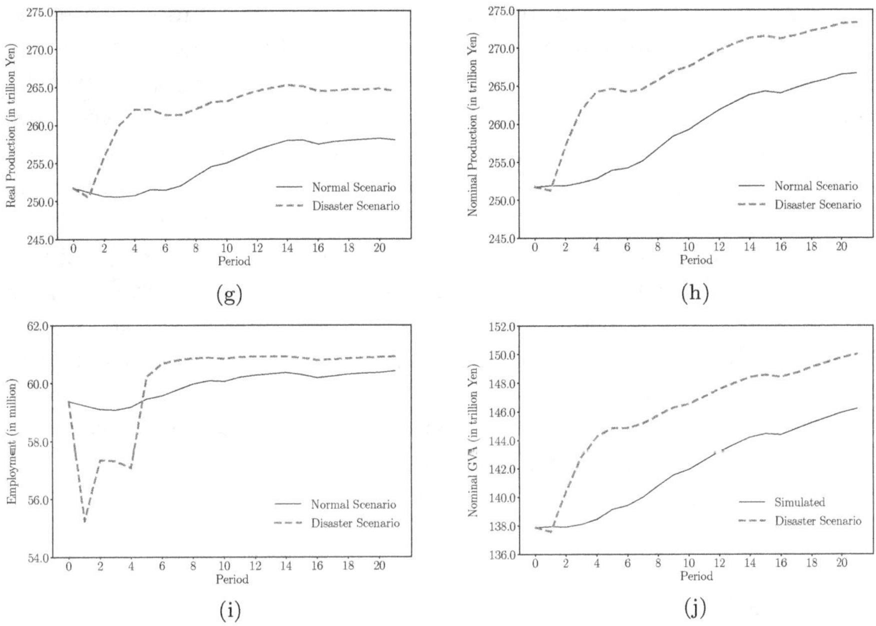

Fig. 23. (*Continued*)

completely in the disaster period and a few periods after the disaster. This is because of the limited supply of investment goods in the economy. However, the households recover their damaged capital in 4 periods (i.e., 4 quarters or 1 year).

Because of higher demand of investment goods in the economy, the firms producing investment goods plan to produce more. Since the firms of different industries are interdependent on each other for their input goods and capital goods, the demand for all goods increases. In order to scale up their production to satisfy higher demands, the firms buy more input goods as shown in Fig. 23(e). Figure 23(f) shows that the firms are able to buy almost 99% of their planned budget. Looking at the real and nominal production plots, Figs. 23(g) and 23(h), total production is lower than the normal scenario level in the disaster period because of the lost capital. However, from the second period onward, the firms start producing at full capacity to satisfy higher demands in the economy.

Figure 23(i) presents a comparison of employment under normal and disaster scenarios. As mentioned in the model settings previously, an affected worker is able to work only for half the time. Therefore, affected workers are counted as half workers. From the plot, we can see that the employment is lower than the normal scenario level till the 4th period. This implies that all workers are able to work full time only after the 4th period. In other words, the workers recover their damaged capital in the 4th period, which is the same conclusion as drawn from Figs. 23(c) and 23(d) previously.

The plot of nominal gross value-added (NGVA), Fig. 23(j), shows that NGVA decreases a little in the disaster period and maintains a higher level for all periods following a disaster. The reason for higher values of NGVA is the higher demand in the economy because of disaster and increased employment.

It must be noted that a very simple recovery scenario was considered in this numerical experiment. More complex settings, considering the availability of repair crews, availability of transportation networks, repair time, etc., are necessary for realistic estimations of the economic impacts and to evaluate the effectiveness of different recovery plans.

7. Concluding Remarks

This chapter presented an HPC framework for agent-based simulations of large economies in 1:1 scale, how to identify model parameters based on publicly available Japanese government statistics, and validation of the model by reproducing the observations for the period of 2015–2019. To seamlessly evaluate the economic progress in the aftermath of any potential earthquake disaster, we integrated the HPC-enhanced ABEM with physics-based earthquake disaster simulators utilizing PACT[25] to convert the degree of seismic-induced damages to economic losses. As a demonstrative application of the integrated system, we simulated the seismic response of 1.8 million buildings in Hanshin Industrial region for an anticipated scenario of the Nankai-Trough earthquake as proposed by the Cabinet Office of Japan, and the following economic progress under a simple recovery plan.

The current model is far from complete and requires a number of advancements to the economic simulator and integration with lifeline networks for simulating a post-disaster economy with realistic details. Some of the advancements required for the economic model are: capturing the agents' consumption behaviors at prefecture level; setting the firm agents' parameters based on fine-grained data and validating at individual firm-level at least for major firms; implementing an active insurance industry; and including the influence of delays in the transportation of inputs goods, etc., on production. As mentioned in the introduction, economic entities of the same category can have different consumption requirements and expenses depending on their geographical location. As an example, the household expenses of water, gas, and food may depend on the prefecture or city. These heterogeneities are important in accurately capturing the economic impacts of a localized disaster and its domino-like effect over the supply-chain, and hence the need of simulating the national economy with prefecture-level consumption behavior of agents, which can be set according to state-level or prefecture-level IO-tables. Though there is an insurance sector in the current model, it is inactive in the sense that it does not pay any compensation. This inactive insurance sector must be improved to pay at least earthquake insurance claims, and make the government pay the compensations if the total claims exceed the threshold. According to historical observations, a major earthquake disaster in a local region can influence the transportation capacity over almost the entire country, causing widespread delays in production and their domino effects. Therefore, it is essential to develop a nation-wide transportation simulator at a suitable resolution to estimate the transportation capacity and time required for transporting goods between firms according to the recovery progress of the transportation network. Seamless integration of such a transportation simulator and the economic simulator will be useful not only in increasing the accuracy of post-disaster economic simulations, but also for non-disaster-related applications such as evaluating long-term economic benefits of expanding the transportation network with new highways.

Acknowledgments

This work was supported by the JSPS Kakenhi grant 18H01675. Parts of the results were obtained using the Oakbridge-CX supercomputer at the Univ. of Tokyo., and the supercomputer Fugaku at the RIKEN Center for Computational Science through the HPCI System Research Projects (Project IDs: hp200256 and JPMXP1020200203).

References

1. R. Adam and L. Shu-Yi, Modeling regional economic resilience to disasters: A computable general equilibrium analysis of water service disruptions, *J. Reg. Sci.* **45**(1), 75–112 (2005).
2. H. Tatano and S. Tsuchiya, A framework for economic loss estimation due to seismic transportation network disruption: A spatial computable general equilibrium approach, *Nat. Hazards.* **44**(2), 253–265 (2008).
3. N. Brozović, D. L. Sunding, and D. Zilberman, Estimating business and residential water supply interruption losses from catastrophic events, *Water Res. Res.* **43**(8) (2007).
4. C. Lorenzo, S. Gabriele, B. Francesco, and M. Jaroslav, Assessing direct and indirect economic impacts of a flood event through the integration of spatial and computable general equilibrium modelling, *Environ. Modell. Software.* **63**, 109–122 (2015).
5. S. Hallegatte, An adaptive regional input-output model and its application to the assessment of the economic cost of katrina, *Risk Analy.: An Int. J.* **28**(3), 779–799 (2008).
6. M. Markhvida, B. Walsh, S. Hallegatte, and J. Baker, Quantification of disaster impacts through household well-being losses, *Nat. Sustain.* 1–10 (2020).
7. T. Assenza, D. D. Gatti, and J. Grazzini, Emergent dynamics of a macroeconomic agent based model with capital and credit, *J. Econ. Dyn. Control.* **50**, 5–28 (2015).
8. A. Caiani, A. Godin, E. Caverzasi, M. Gallegati, S. Kinsella, and J. E. Stiglitz, Agent based-stock flow consistent macroeconomics: Towards a benchmark model, *J. Econ. Dyn. Control.* **69**, 375–408 (2016).
9. EURACE. Eurace: An agent-based software platform for European economic policy design with heterogeneous interacting agents: New insights from a bottom up approach to economic modelling and simulation, https://cordis.europa.eu/project/id/035086.
10. S. Poledna, M. G. Miess, C. Hommes, and K. Rabitsch, Economic forecasting with an agent-based model, *Eur. Econom. Rev.* **151**, 104306 (2023). doi:10.1016/j.euroecorev.2022.104306.
11. S. Poledna, S. Hochrainer-Stigler, M. G. Miess, P. Klimek, S. Schmelzer, J. Sorger, E. Shchekinova, E. Rovenskaya, J. Linnerooth-Bayer, U. Dieckmann, *et al.*, When does a disaster become a systemic event? estimating indirect economic losses from natural disasters, (2018), arXiv preprint arXiv:1801.09740.
12. S. Poledna, M. G. Miess, and C. H. Hommes, Economic forecasting with an agent-based model (2020); available at SSRN 3484768.

13. J. Farmer and D. Foley, The economy needs agent-based modelling, *Nature.* **460**, 685–686 (2020). doi:10.1038/460685a.
14. S. Chen, *Agent-Based Computational Economics: How the idea originated and where it is going.* Routledge Advances in Experimental and Computable Economics, Taylor & Francis (2017). ISBN 9781317560920. doi:10.4324/9781315734422. URL: https://books.google.co.jp/books?id=uno2DwAAQBAJ.
15. ICT article. The futurict knowledge accelerator:creating socially interactive information technologies for a sustainable future. https://cordis.europa.eu/project/id/284709.
16. X. Rubio-Campillo, Pandora: A versatile agent-based modelling platform for social simulation, *Proceedings of SIMUL.* pp. 29–34 (2014).
17. N. Collier and M. North, Parallel agent-based simulation with repast for high performance computing, *Simulation.* **89**(10), 1215–1235 (2013).
18. R. K. Standish, Going stupid with ecolab, *Simulation.* **84** (12), 611–618 (2008).
19. S. Coakley, L. Chin, M. Holcomb, C. Greenough, and D. Worth. Flexible large-scale agent modelling environment (flame). University of Sheffield and Rutherford Appleton laboratories, stfc, license: Lesser gpl v3 (2012).
20. A. Gill, M. Lalith, S. Poledna, M. Hori, K. Fujita, and T. Ichimura, High-performance computing implementations of agent-based economic models for realizing 1:1 scale simulations of large economies, *IEEE Trans. on Parallel and Distrib. Syst.* **32**(8), 2101–2114 (2021). doi:10.1109/TPDS.2021.3060462.
21. M. Lalith, A. Gill, S. Poledna, M. Hori, I. Hikaru, N. Tomoyuki, T. Koyo, and T. Ichimura. Distributed memory parallel implementation of agent-based economic models. In *International Conference on Computational Science* (pp. 419–433). Springer (2019).
22. A. Gill, M. Lalith, Y. Ogawa, H. O-tani, T. Ichimura, K. Fujita, F. Gatti, and M. Hori, Seamless simulations of earthquake disasters and economy at 1:1 scale utilizing big-data, *2020 IEEE/WIC/ACM International Joint Conference on Web Intelligence and Intelligent Agent Technology (WI-IAT)* (pp. 101–108). Melbourne, Australia (2021). doi:10.1145/3498851.3498937.
23. M. Hori, T. Ichimura, and L. Maddegedara, *Integrated Earthquake Simulation.* CRC Press (2022). ISBN 9780367711856.
24. M. Hori and T. Ichimura, Current state of integrated earthquake simulation for earthquake hazard and disaster, *J. Seismol.* **12**(2), 307–321 (2008).
25. A. T. Council, Performance assessment calculation tool (pact), https://femap58.atcouncil.org/pact.

26. E. Bonabeau, Agent-based modeling: Methods and techniques for simulating human systems, *Proc. Natl. Acad. Sciences.* **99**(suppl_3), 7280–7287 (2002), doi:10.1073/pnas.082080899, https://www.pnas.org/doi/abs/10.1073/pnas.082080899.
27. K. H. Dam, I. Nikolic, and Z. Lukszo, *Agent-Based Modelling of Socio-Technical Systems.* Springer Dordrecht (2013), doi:978-94-007-4933-7.
28. S. J. E. Taylor, *Agent-based Modeling and Simulation.* Palgrave Macmillan London (2014), doi:9781137453648.
29. S. F. Railsback and V. Grimm, *Agent-Based and Individual-Based Modeling: A Practical Introduction*, 2nd edition. Princeton University Press, (2019).
30. U. Wilensky and W. Rand, *An Introduction to Agent-Based Modeling; Modeling Natural, Social, and Engineered Complex Systems with NetLogo.* Cambridge, Massachusetts London, England: The MIT Press, (2015), ISBN 9780262328111.
31. M. Steinbacher, M. Raddant, F. Karimi, E. C. Cuena, S. Alfarano, G. Iori, and T. Lux, Advances in the agent-based modeling of economic and social behavior, *SN Business & Economics.* **1** (2021), doi:10.1007/s43546-021-00103-3.
32. R. L. Axtell and J. D. Farmer, Agent-based modeling in economics and finance: Past, present, and future, *INET Oxford Working Papers 2022-10, Institute for New Economic Thinking at the Oxford Martin School, University of Oxford* (2022), https://www.inet.ox.ac.uk/files/JEL-v2.0.pdf.
33. D. D. Gatti, S. Desiderio, E. Gaffeo, P. Cirillo, and M. Gallegati, *Macroeconomics from the Bottom-up.* Springer (2011), ISBN 978-88-470-1971-3.
34. L. Hamill and N. Gilbert, *Agent-Based Modelling in Economics.* Wiley, (2016), ISBN 978-1-118-45607-1.
35. D. Gatti, G. Fagiolo, M. Gallegati, M. Richiardi, and A. Russo, *Agent-based Models in Economics: A Toolkit*, Cambridge, MA: Cambridge University Press, (2018), ISBN 978-1108400046.
36. M. Gallegati, A. Palestrini, and A. Russo, *Introduction to Agent-Based Economics.* Academic Press (2017). ISBN 978-0-12-803834-5.
37. S.-H. Chen, *Agent-Based Computational Economics.* Routledge (2016), ISBN 9781138499713.
38. R. Boero, M. Morini, M. Sonnessa, and P. Terna, *Agent-based Models of the Economy.* London: Palgrave Macmillan, (2015), ISBN 978-1-137-33980-5.

39. Message Passing Interface Forum. Mpi: A message-passing interface standard version 3.1. (2015). https://www.mpi-forum.org/docs/mpi-3.1/mpi31-report.pdf
40. G. Karypis and V. Kumar, A fast and high quality multilevel scheme for partitioning irregular graphs, *SIAM J. Sci. Comput.* **20**(1), 359–392 (1998).
41. D. E. Knuth, *Seminumerical Algorithms. The Art of Computer Programming.* (Vol. 2). Addison-Wesley Longman Publishing Co., Inc. (1997), ISBN 978-0-201-89684-8.
42. R. Durstenfeld, Algorithm 235: random permutation, *Commun ACM.* **7**(7), 420 (1964).
43. IO Table: E-stat. 2015 input-output tables for Japan, https://www.soumu.go.jp/main_content/000732915.pdf (2021).
44. Applied Technology Council. Normative Quantity Estimation Tool (NQET), https://femap58.atcouncil.org/index.php?option=com_content&view=article&id=5:supporting-materials-and-background-documentation&catid=2:uncategorised.

https://doi.org/10.1142/9781800614635_0010

Chapter 10

Study on Improved and Advanced Urban Information Data for Implementation into Social Science Simulations of Earthquakes

Yoshiki Ogawa* and Yoshihide Sekimoto†

The University of Tokyo, Center for Spatial Information Science, Meguroku, Tokyo, Japan

**ogawa@csis.u-tokyo.ac.jp*

†sekimoto@csis.u-tokyo.ac.jp

This chapter introduces the development of advanced urban data to support, develop, and analyze social science simulations of earthquake and tsunami disasters using supercomputers. Toward this, basic real-world data from cities, such as human flows, buildings, and business activities, are essential. However, there are problems with observed data (such as global positioning system (GPS) data, images, etc.) and big data (such as building and corporate activity data, etc.) related to the real world, including missing data and noise, which render the utilization of raw data impractical for social science simulations, including evacuation and economic damage simulations. Therefore, a method to overcome these problems was developed, which includes the advancement of sophisticated urban data related to human flow, buildings, and companies. Specifically, data were developed on transactions between company offices, human flow data (using big data on cell phone movement histories), and building data with the associated age of the structures. This improved urban information was combined to conduct a numerical analysis on infrastructure damage, economic consequences, and evacuation behavior in the event of an earthquake or tsunami, using a supercomputer. Subsequently, a time-based visualization of the results was created using GIS-based software.

1. Introduction

The accumulation and application of urban data can capture the distribution and activities of the people, buildings, and the economy of a city. This can be accomplished by combining micro urban spatial and location information to support, develop, and analyze social science simulations. It can also be applied to earthquake and tsunami simulations using supercomputers. In recent years, digital technology and advanced data have become important tools to support the creation of solutions for increasingly complex and diverse urban problems. Therefore, it has become a crucial element of service and accountability for citizens to understand and attain their rights, and to have access to basic service deliveries and resources. As the informatization and digitization of cities through smart cities continues, there is a growing need to use basic urban mapping information and measured sensing data.[1] Natural disasters, such as earthquakes, can cause extensive damage to urban operations; therefore, social science simulation for mitigating damage is considered one of the most important application areas of urban data. Thus, the sophisticated urban data that are necessary to improve urban resilience are introduced in this chapter.

To develop a city which is resilient to earthquakes and tsunamis, it is necessary to understand the mechanics of seismic motion, the prediction of occurrence, and social impacts of these events. A digitized reproduction of a city's reaction to an earthquake is an effective tool for this.[2] Therefore, it is necessary to construct urban data that can be included in a digital simulation of a city to simulate urban behavior.[3]

A simulation of a city will aid in the prediction of the impacts of earthquakes and tsunamis on urban buildings and economic activities there. Therefore, urban data, including data on buildings, people, and the economy, can play an important role in simulating disasters such as earthquakes. However, individual companies which obtain global positioning system (GPS) data and supply-chain network data (which can capture daily people flow) have rarely been used in research on evacuation measures or in economic simulations. Therefore, in social science simulations of disasters such as the Nankai Trough earthquake, which impacts a large area, some models are predominantly based on statistical data. This is because urban development and its associated data are lagging when compared to the speed of the development of sophisticated simulation models.

Raw data, such as observed data and big data, cannot be used for simulations because it often includes missing data and noise. High-resolution and accurate simulation methods for earthquake and tsunami run-up data have been studied with the development of high-performance computing (HPC), providing an accumulation of high-tech results. However, although the models are becoming more sophisticated, social science simulations are still insufficient regarding the utilization of urban data. Additionally, these simulations are often conducted using virtual data. However, it is possible to evaluate the robustness of the computational algorithms used, by conducting experiments on complex real-world urban data. The aim of the study was to contribute to the utilization of urban data in computer science, through research and development. Additionally, a "digital twin" of earthquake and tsunami disasters was developed. This was achieved by preparing sophisticated urban data that can reproduce real-world situations over a wide area (Fig. 1).

Real-world data are necessary to construct a simulation that reproduces the actual situation on the ground. Therefore, improvements in various observation technologies and the accumulation of different types of data have led to big data. Big data on spatiotemporal information have been generated and accumulated and have been used in many studies. Such data include: transaction networks among 1.6 million companies nation-wide in Japan, data on people flow nation-wide, and logistics data (from ETC 2.0).

Therefore, a real-world database related to "people," "buildings," and "economy" was developed, by combining observation data, existing statistics, and GIS data. However, this poses a challenge regarding the integration of data, and subsequent use in analyses, owing to the lack of algorithms for combining heterogeneous data.[4] For social science simulations of disaster scenarios specifically, it is

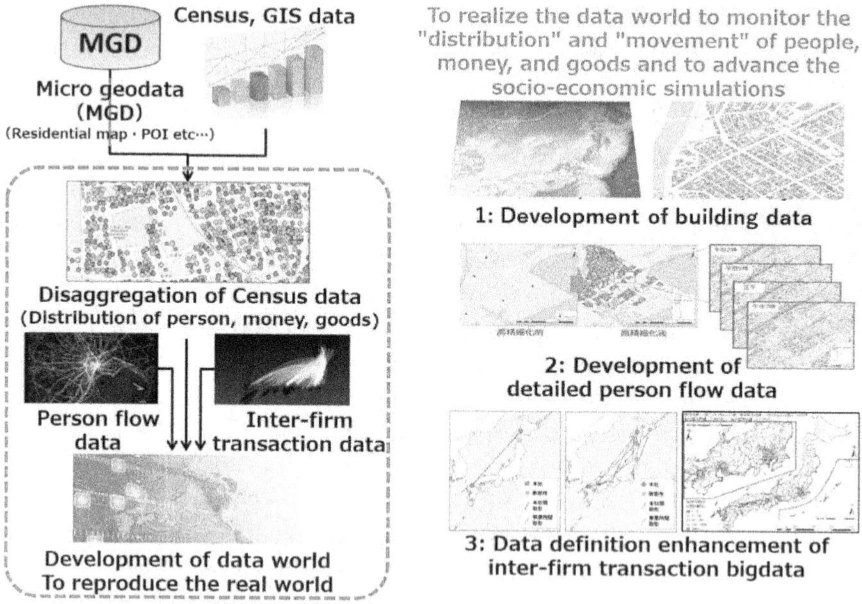

Fig. 1. Development of advanced urban data to support social science simulations of earthquakes and tsunamis.

necessary to integrate data ranging from urban building and infrastructure damage, to corporate and human activities, all of which vary in scale and details. To overcome this challenge, spatial information technology was applied to understand the connection between earthquake and tsunami simulations and social science simulations. The urban data were combined with the integrated earthquake simulation (IES) for prediction simulations[5,6] and social science simulations. Consequently, it was possible to carry out a numerical analysis of economic activities and evacuation simulations for a hypothetical Nankai Trough earthquake and tsunami, by using people flow data.

To estimate the damage caused by an earthquake for each building on a city scale, building data are required. This includes the building age, structure, and footprint. However, even local government buildings generally do not provide open data. In recent years, the 3D building models of the "PLATEAU" software have made open data available in Japan. However, they frequently do not include attributes such as building age or structure; therefore, they are often used for visualization rather than analysis.[7] Building information modeling (BIM) data, which are detailed structural data, are available for the newer apartment buildings but are rarely available for older, detached buildings. Building footprint data are available nation-wide as residential maps. Therefore, the building age and structure were added to the existing building footprint data nation-wide, into the IES. Thus, this study can contribute to building modeling and building response analysis.

Numerical analyses of economic activity have been conducted in many studies using the dynamic stochastic general equilibrium model, combined with the inter-regional input–output table (IO) table, which is Japanese government statistical data.[8–10] However, individual buildings were not considered in many natural disaster analyses, despite the fact that damage in the same region varies considerably, depending on building location and structure. Recently, credit reporting companies in Japan have been collecting and processing information on companies. This process has provided data on business-to-business (B2B) transactions, with the associated latitude and longitude.[11] Therefore, in this study, using these data, a B2B transaction database was developed that is decomposed and highly detailed, down to the branch or business office level. Therefore, it can be used at multiple scales, including the company and office level, mesh level, and prefecture level.[12] Using these data, multi-scale economic simulations can be performed for companies throughout Japan. Furthermore, it is possible to consider not only the spatial distribution of earthquake damage, but also the company attributes (sector, sales, etc.) and supply-chain networks.

The movement history data was obtained from cell phone GPS data, and was subsequently examined. This data requires spatiotemporal interpolation and map matching because of the non-uniformity in observation errors and observation time.[13,14] Furthermore, because demographics such as age and gender have a significant impact on evacuation speed and behavior, a method for estimating the demographics was investigated. A refined method for evacuation simulations was developed, aiming not only to guarantee accuracy in aggregate units for age and gender, but also to provide additional details at an individual level.[12]

This study demonstrates how the advanced urban data, developed in the study, can be combined and input into social science simulations of earthquakes and tsunamis. These can further be applied in economic and evacuation simulations.

The remainder of this chapter consists of five parts. Section 2 presents a method for estimating the age of individual buildings for input into building damage simulations. Section 3 explains the development of a method for estimating data on people flow (including age and gender), using big data from cell phones. Section 4 presents a method for estimating the number of people in an earthquake. Section 5 presents an example case study of collaboration between an earthquake/tsunami simulation and a social science simulation, using the data which have been developed. Finally, Section 6 describes the conclusions and limitations.

2. Development of Building Data

Building data are required for city models to perform numerical response analyses of buildings in an IES.[15,16] In an IES, the building data used to simulate entire cities are highly versatile. In this study, digital residential map data (Zmap TOWN-II, Zenrin, Inc.) and building data from local government urban planning surveys were used. Based on this information, a 3D model was automatically generated, considering the number of floors and structure of the buildings. Subsequently, an integrated simulation was performed from initial ground motion to collapse.

The seismic strength of a building varies greatly, depending on its age and structure; therefore, this information is required. However, although building data are owned by each local government, generally, they are not open data.[17] To solve this problem, the age and structure of each building was estimated, using machine learning data which included real estate data (At Home, Co. Ltd.) as the training data, and attributes, such as features, were obtained from residential maps.

The residential maps used in this study contain polygon data on the building footprints of approximately 60 million buildings nationwide in Japan. When estimating the structure and construction year of a building, only the necessary attributes are extracted from each polygon, such as location, number of floors, and area. These are then converted into point data with latitude and longitude.

According to the survey, from a total of 49.60 million housing units, 29.23 million (58.9%) were wooden, with the remaining 20.37 million buildings (41.1%) were made from other materials. The percentage of wooden houses was 81.7% in 1978, showing a considerable increase in non-wooden houses over the years. Consequently, the risk of earthquake damage has decreased. However, some areas have a high density of wooden buildings, such as the Sumida and Arakawa wards in Tokyo. Therefore, to assess earthquake risk, it is important to understand the detailed distribution of wooden and non-wooden buildings. In addition, the construction year of the building also has a significant impact on the risk of building collapse. It is particularly important to evaluate the years before and after the 1981 and 2000 revisions to the Building Standard Law Enforcement Ordinance. The first revision (June 1, 1981) was concerning new earthquake resistance. Data from the Great Hanshin–Awaji Earthquake confirmed that earthquake resistance against building collapse was significantly different before and after the 1981 revision.[18] The next revision, performed on June 1, 2000, introduced performance design including not only conventional allowable stress calculations but also strength limit calculations. Using the Niigata Chuetsu earthquake data, a comparison of buildings was performed on those completed in the 1990s and those of the 2000s. Results showed that the total destruction of buildings completed in the 2000s was less than that of those completed in the 1990s. Therefore, it is important to consider both the building structure and the year of construction when estimating earthquake damage risk.

2.1. *Data*

2.1.1. *GIS building data*

For this study, the construction year and structure data for modeling were required. In Japan, there are several GIS building footprint datasets covering areas country-wide. These include building maps from municipalities, open street maps, PLATEAU software (Ministry

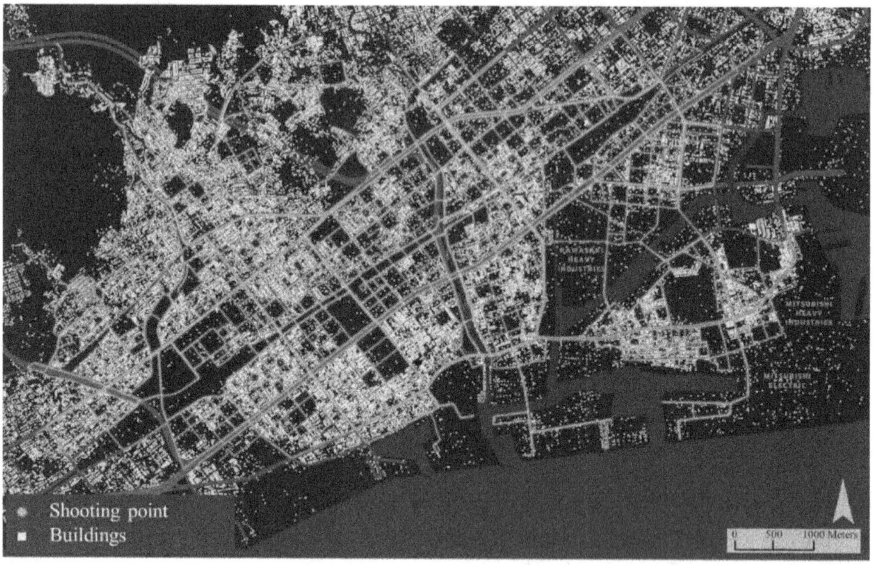

Fig. 2. Example of GIS building data distribution at Kobe City, Hyogo prefecture, Japan (white represents the GIS building data distribution).

of Land, Infrastructure, Transport and Tourism), and residential maps from private companies (such as Zenrin Co. Ltd.). The real estate data were provided by At home Co. Ltd. Additionally, the results of the 2015 City Basic Urban Planning Survey were provided by the G-Space Information Center and were used as the GIS building data. Attributes such as building structure, year of construction, function, and number of floors were included. Additionally, residential maps were used as GIS building footprint data, provided by Zenrin Co. Ltd. The building distribution for Kobe City is shown in Fig. 2 (in white).

2.1.2. *Street view image data*

In this study, 195,105 street view images (SVIs) were obtained from Zenrin Corporation for 2013 in Kobe City, Japan, as it has a comprehensive collection of SVIs for Japan. The images were captured at 2.5-m intervals using a 360° camera mounted on a vehicle roof while driving through the city. Each image was annotated and geotagged with textual information such as latitude and longitude (measured using a GPS), vehicle azimuth, and shooting time. The original images were panoramic in a jpeg or tar format, at 2,700 (height) × 5,400 (width) pixels (which is a higher resolution than that of Google Street View).

2.2. *Built year and structure prediction*

2.2.1. *Construction year and structure estimation using building attributes*

The method of Ref. 19 was used to estimate the year of construction and structure of a building, by using attributes from a residential map, such as ground surface area and number of floors. This method uses real estate data with latitude and longitude as training data for the year of construction and the building structure. Additionally, machine learning uses attributes obtained from residential maps as features. The flow of this method is shown in Fig. 3. The Random

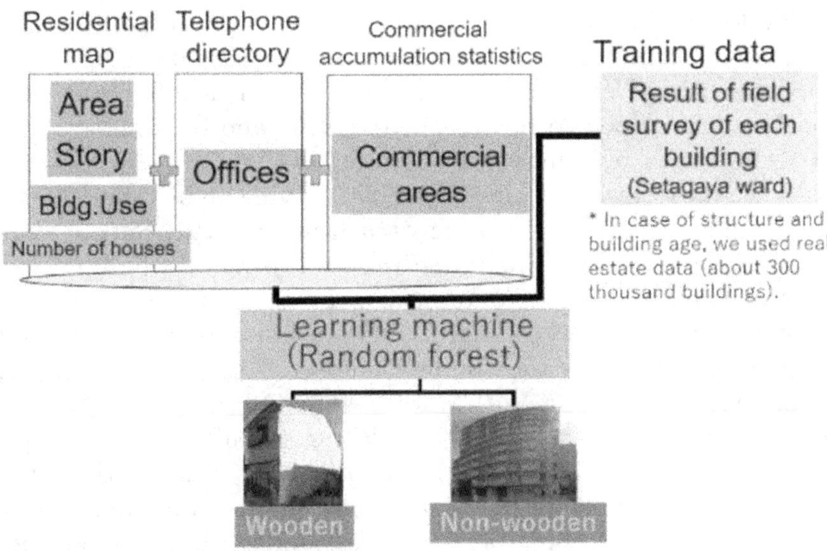

Fig. 3. Machine learning method to estimate building age and structure by using building attributes.

Forest[20] machine learning model was used, as it has the advantages of being computationally inexpensive, able to handle large datasets, and us robust to noise.

To create the training data, each building point data from the residential map and the real estate data were spatially joined. The real estate data are assigned latitude and longitude in the same way as the residential maps; therefore, buildings with matching addresses were first combined using the text addresses. Next, based on latitude and longitude, the At Home data were combined with the building point data of the nearest neighbors. As described above, the residential map and real estate data were spatially combined, and training data were prepared.

The number of floors, floor space, total floor space, building use, commercial concentration area, and densely inhabited district [DID (1970, 1980, 1990, 2000, 2005)] for each year were selected as features for use in the model. The models were constructed by learning these features.

2.2.2. *Built year and structure prediction using street view*

To develop highly accurate data with ease, an automatic method for annotating the construction year and structure of buildings was proposed, using SVIs and GIS building data. A deep learning (DL) model was developed to classify the structure and building age using a well-developed building image dataset.

A training dataset was developed by applying an instance segmentation model to extract building data from the SVI. Subsequently, a method was developed to match the two, using GIS building data and street image location information. Furthermore, the latest DL classification models were used to classify individual buildings into different classes, using the training dataset. There were six classes of built years: 1962, 1963–1971, 1972–1980, 1981–1989, 1990–2001, and post-2002). There were three classes of different seismic performances: pre-1980; 1981–2001; and post-2002. Finally, there were three classes of structures: RC; steel, and wooden building and two classes of different seismic performances: wooden building and non-wooden building.

(1) *Development of training data*

An automated method for the development of training data was produced by the automated combination of SVIs with building footprint GIS data[21]:

(a) Individual buildings were extracted from panoramic SVIs through instance segmentation, which is a DL approach. The positions and angles of view of both edges of the individual buildings in the SVIs were calculated.
(b) The theoretical angles of view of individual buildings were estimated from the image capture point using GIS building data and image point information (latitude, longitude, and azimuth) of each SVI observed by the onboard GPS.
(c) Finally, the two angles were compared, and the most adjacent GIS building data were merged with the bounding box (BBOX) of the target building.

The GIS data of the target building were assigned simultaneously to ensure that each building image was assigned an individual building identification (ID) from the GIS data when combined. An image dataset was then prepared, annotated with building structure and year.

(2) *Image classification with deep learning algorithms*

DL has significantly outperformed the traditional algorithms in the field of computer vision. The ImageNet Large Scale Visual Recognition Challenge (ILSVRC) is a benchmark competition in the field of image classification and has been held since 2010. It contains a dataset of 1.4 M natural images labeled across 1,000 classes. In this competition, the application of deep convolutional neural networks (DCNN) significantly improved the score of the ILSVRC since AlexNet.[22] In addition to the excellent performance of the DCNN, ViT, which is another type of DL algorithm,[23,24] achieved better results in several computer vision tasks.

The resulting building-level façade panorama image dataset was split into six classes based on year: pre-1962, 1963–1971, 1972–1980, 1981–1989, 1990–2001, and post-2002. It was further classified into three classes based on structure: RC; steel, and wood (see Fig. 4). The building structures and built-year classification models were then created.

Fig. 4. Example of benchmark dataset for building structure and year of construction. (a) Wooden (built in 1962 or earlier), (b) wooden (built from 1963 to 1971), (c) wooden (built from 1972 to 1980), (d) wooden (built from 1981 to 1989), (e) wooden (built from 1990 to 2001), (f) wooden (built in 2002 or later), (g) steel (built 2002 or later), (h) RC (built from 1990 to 2001), and (i) RC (built in 2002 or later).

2.3. *Results*

2.3.1. *Estimation of building structure material and year of construction using building attributes*

(1) *Building structure classification*
The estimation accuracy of building structures using the error matrix for buildings in Tokyo, Saitama, Chiba, and Kanagawa prefectures

Table 1. Validation results of the structure material estimation by random forests algorithm in Tokyo, Saitama, Chiba, and Kanagawa prefectures.

		Ground truth			
		Non-wooden	Wooden	Total	Acc.
Estimation	Non-wooden	179,003	27,543	206,546	0.867
	Wooden	23,850	82,560	106,410	0.776
				Total acc	0.836

was over 80%, and the estimation accuracy by building structure material was 78% for wooden structures, and 87% for non-wooden structures (Table 1).

To determine the hyperparameters of random forest, the parameters were optimized by obtaining the (out-of-bag OOB) for each parameter. Consequently, it was found that the number of decision trees and the minimum number of data points for each branch were 200 and 2, respectively, which were appropriate for this method. Using the Gini coefficients for each feature, the feature with the highest Gini coefficient was the number of building stories, followed by total floor area, building use, and floor area. This is due to the fact that the non-wooden construction ratio increases as the number of stories and total floor area increases.

(2) *Construction year classification*

The accuracy of the construction year estimation using the error matrix for buildings in the Tokyo, Saitama, Chiba, and Kanagawa prefectures was 62% in the pre-1980s, 64% in the 1990s, and 72% in the 2000s, resulting in an overall estimation accuracy of 64% (Table 2). Identification of the year of construction becomes increasingly difficult because of factors such as rebuilding and reform. It is necessary to find data and factors that will allow for the consideration of rebuilding, which will lead to an improvement in accuracy.

In estimating the construction year, the number of decision trees of the random forest was set to 500, and the number of branches for each was set to three. The feature with the highest Gini coefficient was the total floor area, followed by building area, number of stories, and building use. In particular, the Gini coefficients for the building area were the most important. This indicates that buildings with a larger total floor area were built at a newer age, but the factors

Table 2. Validation results of built-age estimation by random forests in Tokyo, Saitama, Chiba, and Kanagawa prefectures.

		Ground truth				
		Pre-1980	1981–2000	Post-2000	Total	Acc.
Estimation	Pre-1980	1,828	389	338	2,555	0.715
	1981–2000	30,903	188,581	76,256	295,740	0.638
	Post-2000	1,417	4,107	9,137	14,661	0.623
					Total acc.	0.638

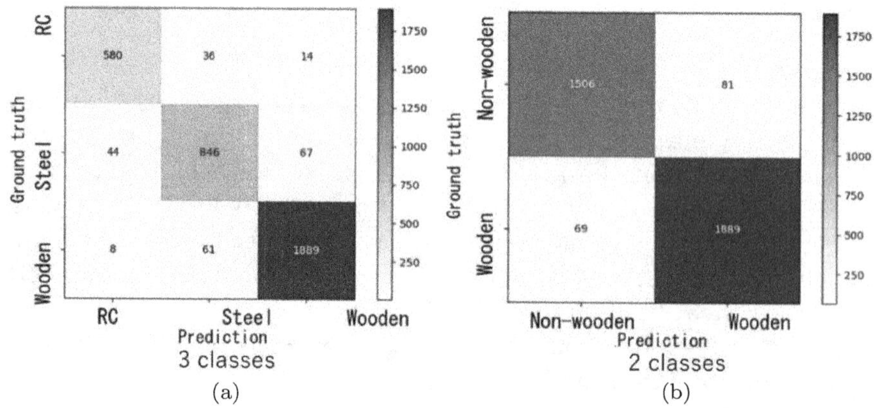

Fig. 5. Confusion matrices of the building structure classification evaluated on the test data. (a) Three classes, and (b) two classes.

behind this need to be examined. In addition, the estimation method was constructed based on the assumption that the DID is effective, but its contribution is not very high.

2.3.2. *Building structure and construction year estimation using street view images*

(1) *Building structure classification*

Figure 5 shows the confusion matrices of the structures using the Swin Transformer based on the test data. Table 3 details the recall, precision, and F1 scores for each structure of the swine transform

Table 3. Overall accuracy, recall, precision, and F1 scores based on the three and two structure classifications of the test data.

	Recall	Precision	F1
RC	0.920	0.918	0.919
Steel	0.884	0.897	0.891
Wooden	0.965	0.959	0.962
Overall accuracy (3 classes)		0.935	
Non-wooden	0.949	0.956	0.952
Wooden	0.965	0.959	0.962
Overall accuracy (2 classes)		0.958	

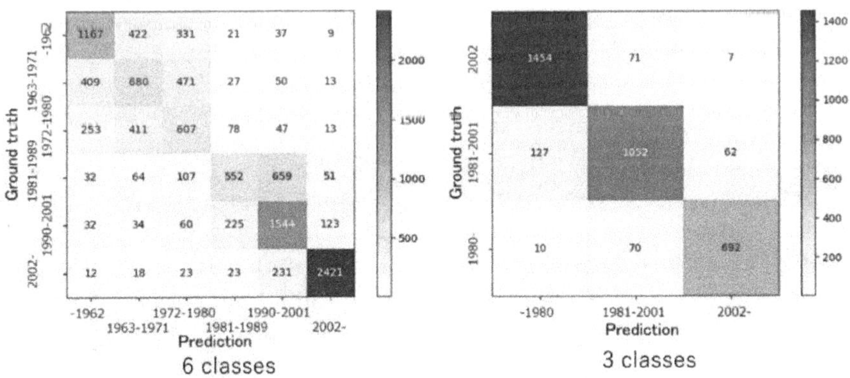

(a) (b)

Fig. 6. Confusion matrices of built year classification evaluated on the test data. (a) Six classes, and (b) three classes.

base. All the classes showed a high accuracy of approximately 89% or higher. The specific classification accuracies for wood, RC, and steel were 0.962, 0.919, and 0.891, respectively.

(2) *Construction year classification*

Figure 6 shows the confusion matrix of the image and building bases of the construction year evaluated using the test data. Table 4 provides the recall, precision, and F1 scores for each year of construction. Low scores can be attributed to the fact that the same building in the SVI is used for both classes, which suggests that the ability to acquire multiple SVIs of the same building could

Table 4. Recall, precision, and F1 scores on the six construction year classes and three built year classes of the test data.

	Recall	Precision	F1
Pre-1962	0.634	0.675	0.654
1963–1971	0.421	0.417	0.419
1972–1980	0.466	0.364	0.409
1981–1989	0.332	0.453	0.383
1990–2001	0.781	0.757	0.769
post-2002	0.896	0.909	0.903
Overall accuracy (6 classes)		0.662	
Pre-1980	0.949	0.914	0.931
1981–2001	0.848	0.882	0.864
Post 2002	0.896	0.909	0.903
Overall accuracy (3 classes)		0.902	

improve classification accuracy. The construction year with the highest accuracy rate varied significantly, depending on the period. The classification accuracy was highest post 2002, followed by 1990–2001, and prior to 1962. The classification accuracy for 1963–1989 was low; however, it was high (0.902) for 1981 and 2001, which was when the seismic structural standards changed significantly.

Kobe City was selected for this study to evaluate the feasibility of mapping building structures and years of construction over a wide area, using SVIs to demonstrate the efficiency of the proposed method in urban environments, with high building densities. Figure 7 shows the predicted city-scale building structure and construction year classification map, as well as the ground truth.

A total of 195,105 panoramic images were used to classify 22,392 buildings. A comparison of the spatial distribution of the ground truth against the prediction revealed that the distribution of the building structure and building year were similar. The newer RC and steel buildings formed a coherent cluster along the railroad tracks in the center of Kobe City. The older wooden buildings were distributed in suburban residential areas, located far from the city center. The trend in the change in the construction year and structural class varied according to the street location, which in turn depended on the area.

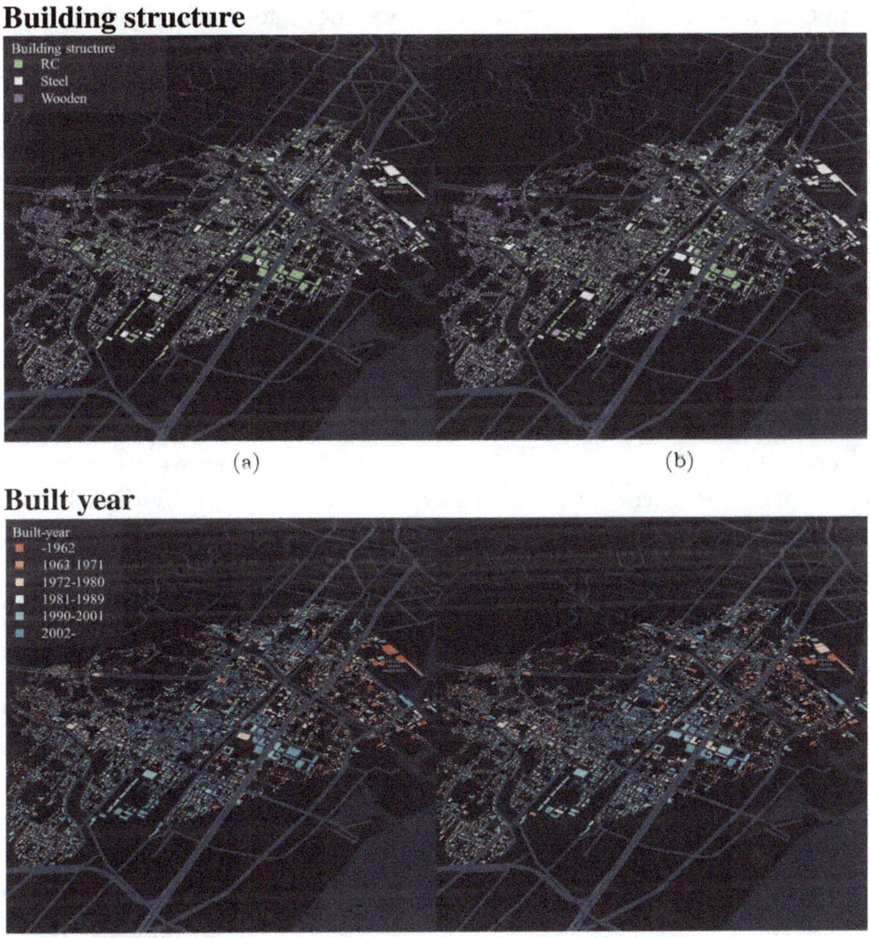

Fig. 7. City-scale (a) ground truth building structure, (b) predicted building structure, (c) ground truth construction year, and (d) predicted construction year. Different colors represent different building-structure classes.

2.4. *Evaluating earthquake damage*

As discussed, the structure and age of building units are important for assessing the risk of damage caused by collapse during an earthquake. However, it is difficult to collect such data at a national scale. The method proposed in this study allows the structure and age of each building to be accurately classified, thereby resulting in a higher resolution and more accurate earthquake damage prediction. Furthermore, street images required can be collected with relatively little effort by vehicles equipped with omnidirectional cameras.

Once the year of construction and structure type are determined, researchers and national and local governments can assess earthquake and tsunami risks, at an individual building scale. This can contribute to the establishment of more detailed and thorough disaster prevention plans and resilience assessments for the anticipated future Nankai Trough earthquakes. For example, using the proposed method, the fragility model can be applied to building damage simulation. This study suggests that the geographic distribution of the construction year and the structure of individual buildings should be considered in the seismic tsunami risk assessment, instead of considering the structure construction year as aggregate data. A well-developed structural and age database on the individual building level scale would be particularly useful for countries that sustain damage from earthquakes and other natural disasters.

3. Development of People Flow Data

In recent years, people flow studies have been conducted in disaster management, marketing, transportation, and urban planning, using data that can capture the dynamic and ever-changing movements of people.[25–27] During the COVID-19 pandemic, studies were conducted using people flow data to geographically analyze the impact of infection spread and lockdown on people flow, which received much attention.[28,29] Technologies to capture such data include GPS, cellular communication history, or call detail records (CDR), cameras, and integrated circuit tags, in addition to static data from statistical surveys. Recent technological innovations have made detailed measurements possible. In Japan, the GeoSpatial Information Center provides a platform where anyone can easily search for and purchase spatial data on people flow data; Table 5 lists the commercial datasets on people flow available. These datasets provide information on the population distribution and people flow over time. Recently, the author's research group has also developed an open dataset on pseudo-people flow for Japan, by combining official statistical and travel survey open data.[30]

Table 5. Variety of commercial data on mass people movement in urban areas.[30]

Data type	Contents	Data method	Coverage area	Data source organization
Traffic volume data	Hourly volume of vehicle traffic on each link	Pioneer car navigation user's GPS log	All parts of Japan	Pioneer Corporation
Vehicle travel data	Average travel speed, hourly traffic volume, and transit time on each link	Navitime application user's GPS log	All parts of Japan	Navitime Japan Co. Ltd.
Population flow data	Hourly floating population data in 250-m mesh or 500-m mesh or Point format	Agoop smartphone application user's GPS log	All parts of Japan	Agoop Corporation
Congestion degree data	Hourly 250-m mesh population data	Mobile phone GPS log of NTT DOCOMO users	All parts of Japan	ZENRIN DataCom Co. Ltd.
Mobile spatial statistics	Hourly 500-m mesh population distribution data	Call Detail Records of NTT DOCOMO users	All parts of Japan	NTT DOCOMO, Inc.

However, to use these data for disaster simulation, it is necessary to consider a wide variety of data processing techniques. This is particularly relevant for tsunami evacuation behavior, evacuation speed, and people behavior, as these all differ depending on the date and time of the disaster, evacuees' age and gender, and other factors.[31] Therefore, in tsunami evacuation simulations, it is necessary to consider the initial location conditions of the evacuee agents model, evacuation speed, and other variables. However, GPS movement history data are sparse because the observation time is not uniform, there are errors in the observed positions; consequently, the data cannot be used as is. Therefore, to introduce real-world people flow into a disaster simulation model, this study examined a method for spatiotemporal interpolation of movement history data, as well as a method for estimating age and gender.

3.1. *Data*

3.1.1. *GPS data*

Data from the 2011 mobile phone GPS "Congestion Analysis®" logs were provided by ZENRIN DataCom Co., LTD. Konzatsu–Tokei® refers to people flow data collected by individual locations sent from mobile phones with enabled auto-GPS function, by consent, through the "DOCOMO map navi" service provided by NTT DOCOMO, Inc. These data are processed collectively and statistically to conceal private information. The original location data are GPS data (latitude and longitude) sent in a minimum period of 5 min, exclusive of information on gender or age. This is a large database constructed of text data from approximately 9 B records (2012) belonging to approximately 1.5 M users throughout Japan. In this study, data from 2012 were used. The data processing method devised for this research was applied to the GPS data provided by NTT DOCOMO, Inc. The data are available for ages 15–89 years.

3.1.2. *Parson Trip data*

The Parson Trip (PT) survey data were used as training data for the estimation of demographics. Large-scale people movement datasets of PT in Japanese cities have been developed as open data.[13] The method of creating person flow data using PT data comprises geocoding the spatiotemporal position of the starting point of the sub-trip base location and performing a route search on the shortest path. The PT data can be used to understand people's flow, including age and gender. Personal trip data were used for the year 2000, from Kochi City (approximately 47,000 samples).

3.2. *Development of people flow data*

The flow of creating people flow data with demographics is shown in Fig. 8. First, map matching was performed using the latitude and

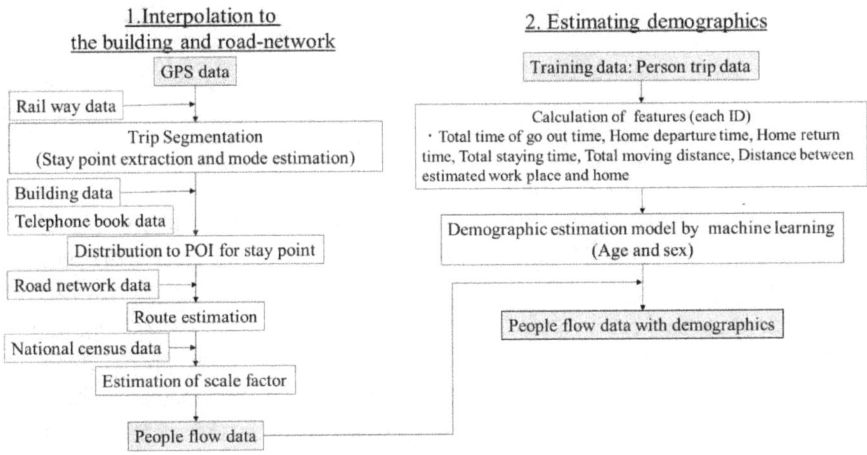

Fig. 8. Workflow of developing demographics using people flow data.

longitude and observation time of the GPS movement history data, to correct for observation errors and estimate each person's trip. This was sample data; therefore, a scale factor was assigned to each person to match the actual population. Next, demographics were estimated. In this study, estimating age and gender was considered using machine learning, with the PT data as the training data. Details of each process are explained in the following Sections 3.2.1–3.2.4.

3.2.1. *Geocording to point-of-interest data*

First, each point of the GPS data was assigned a flag representing whether it was stationary or moving, at 5 min intervals. Data were extracted on the stationary points of each user, from the location coordinates and observation times. The stationary points represent points from which a user remains still for a certain time and within a certain range. Stagnation was estimated from the observation point by dividing the trip and judging the stationary point against the threshold value of the group, within a 300 m radius for 15 min or longer. In this study, GPS data from multiple days were used; therefore, there were several points with limited or no movement (for example, individuals near their respective residences). Therefore, clustering was used to combine houses with similar locations. Density-based spatial clustering of applications (DBSCAN) was used as the clustering algorithm. It uses two parameters: distance threshold and target number threshold minimum points. It then defines the connection relationship of objects within the radiusPoints were classified into the same cluster if there were more points than the minimum number of points of connected targets. The Eps was set to 150 m, the minimum number of points was set to 2. The calculations were then performed, and the central point of the maximum cluster from the column of points of the stationary period were set as stationary points.[32,33] Work and home locations were estimated at each stationary point, based on the observation times. Thus, using building data [Z-map Town II (Zenrin Co. Ltd.)] for the points-of-interest (POI), each location point was established by taking the center of gravity of the stationary point (from midnight to 16:00, during 2010), and estimating the latitude and longitude of each building in that neighborhood. Subsequently, by focusing on observation points before and after each observation, the velocity and traffic modes of each point were estimated and added. Traffic modes considered in this study were cars, walk, bikes, buses, and trains.

3.2.2. *Road network interpolation*

Considering the estimated traffic mode of the GPS data, road and railroad network data were used to interpolate the spatiotemporal data. The routes taken at the starting points were also processed. If the connection between the observation point and the road network was large, the error increased. Therefore, the link connection was measured by searching for the nearest neighbor on the link, It was then possible to reduce the error via the spatiotemporal interpolation method of Sekimoto *et al.*[13] (Fig. 9).

3.2.3. *Scaling factor*

To match the number of points of GPS data to the population, the expansion of the sample data from the estimated work and home locations was determined using the data on the number of employees and statistical data, respectively. This included 250-m mesh data from the 2010 residential population census for the population, and the 250-m mesh data from the 2010 economic census for the number of employees. Figure 10 shows the method of estimating the expansion, which was determined by dividing the residential population by

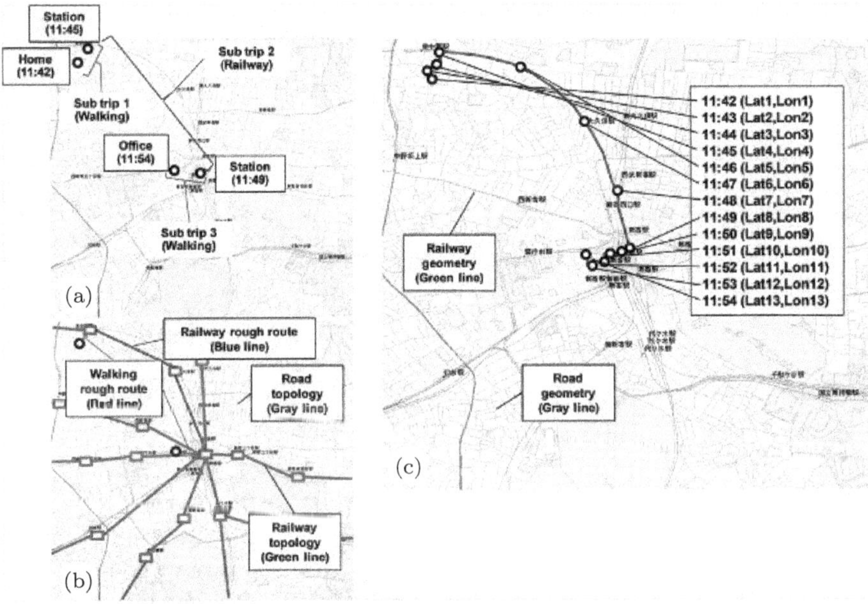

Fig. 9. Spatiotemporal interpolation method of trajectory data.[13]

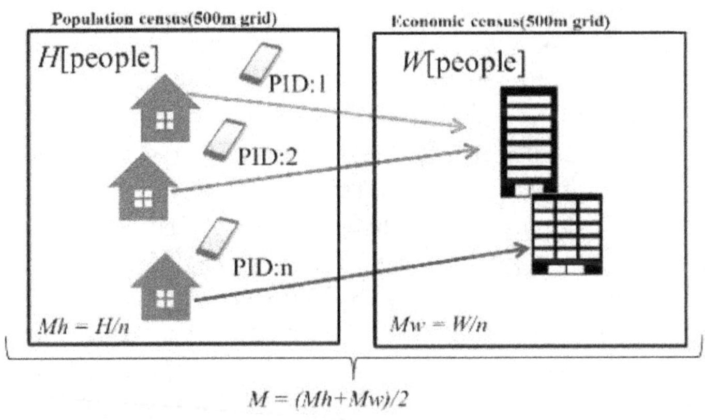

Fig. 10. Method of estimating scaling factor (using people and cell phones).

the number of home terminals, and the number of employees by the number of terminals at a company. From the average of the expansion magnification coefficients (Mh and Mw), the final magnification coefficient was obtained. A magnification factor was assigned to each

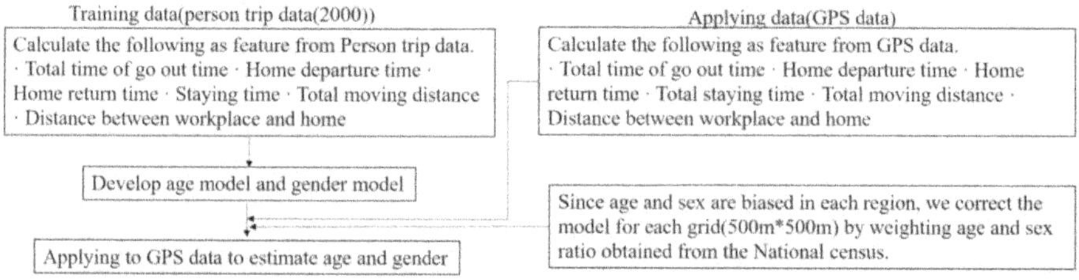

Fig. 11. Workflow of estimating individual trajectories of age and gender demographics.

data point, making it possible to estimate the distribution of the population rather than the distribution of a sample.

3.2.4. *Estimation of attribute*

The flowchart of the estimation method is shown in Fig. 11. First, a model was constructed using the Parson Tip survey data (47,000 samples) as the training data. Next, the model was applied to the GPS data. The data from the PT were questionnaire data and for the GPS were observational data; thus, the properties of each are different. Therefore, when applying a model to GPS data, some areas did not match. As a countermeasure, the attributes among groups were weighted using census data when applied to the construction model. However, variations occurred; therefore, attributes with a low probability of matching the census were revised. The learner model adopted the machine learning method of least absolute shrinkage and selection operator (LASSO) support vector machine (SVM), which displayed excellent classification performance by internally performing nonlinear mapping (i.e., kernel trick).[30] Variable selection and model construction were performed simultaneously by selecting LASSO (i.e., L1 regularization). The features used for the model construction were: the total time to leave; home departure time; return home time; total time in one place; distance from home to the workplace, and total distance traveled. The same features were used for gender and age, and a learning model using LASSO SVM was created. The SVM parameters were fine-tuned by ten cross-validations, a set radius for the SVM center, and the SVM type to EPS regression (where gamma was set to 0.125, and epsilon was set to 0.1).

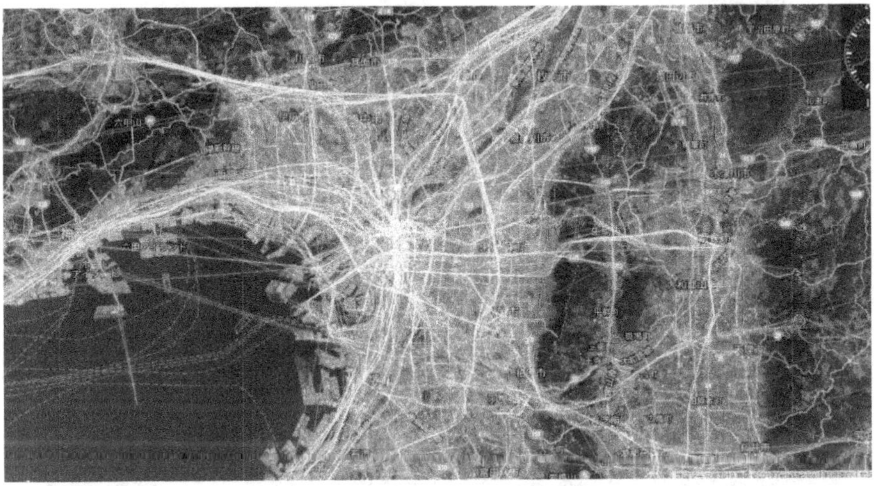

Fig. 12. Example of people flow data visualization in Osaka City (August 27, 2012, at approximately 14:00; white line is the movement trajectory, and gray points are the stationary points).

3.2.5. *Results*

Figures 12 and 13 show examples of people flow data for Kochi and Osaka. Figure 14 presents the estimated population with demographics (for the population age ranges of 5–20, 21–64, and 65) in Kochi at the 500 m mesh by using the GPS data. Over time, it became possible to determine the time at which a person with an attribute moved to a location (Fig. 13). For example, various scenarios such as weekdays, holidays, and events can be generated. By using mobile phone GPS with age and sex, as described previously, it is now possible to follow the flow of people every day, which cannot be grasped by conventional census and PT data. By conducting evacuation simulations using people flow data based on actual observation data, it is possible to grasp the vulnerability of each area more accurately.

3.3. *Validation*

The reliability of the model with PT data was checked using ten cross-tests and error matrices; the results are shown in Tables 6 and 7. The results of the error matrix for the age estimation model showed that the accuracy varied with age, with 40% accuracy for

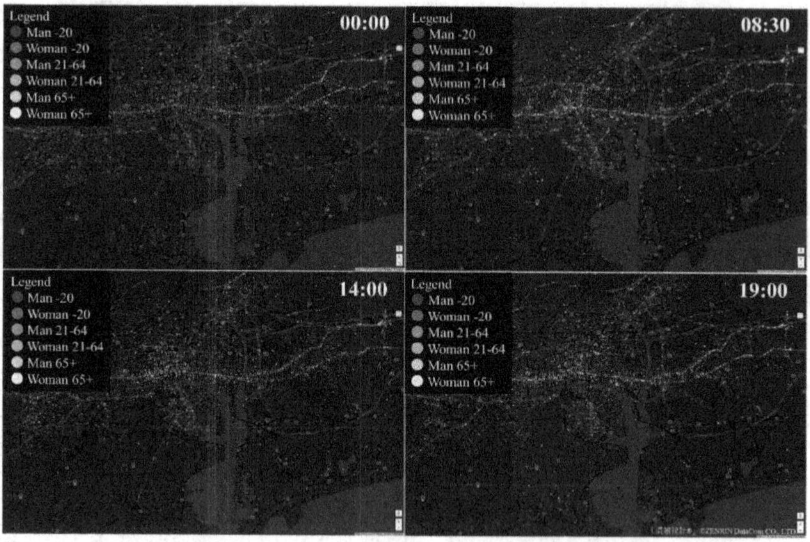

Fig. 13. Example of people flow data with individual gender demographics on October 17, 2012, in Kochi City.

those aged 20 and under, and 54% for those aged 65 and over. However, the estimation accuracy for those aged 21–65 was generally higher, at 80% (Table 6). This may be due to the fact that many college and high-school students, even those under 20 years of age, behave in the same way as working adults (21–65 years). The results of the error matrix for the gender estimation model showed that the overall accuracy was 63%. However, the estimation accuracy for gender was lower for males than for females (Table 7). This is due to the fact that men and women have similar lifestyles in present times, indicating that it is difficult to discriminate between the two using only the features in this study. After applying the data to GPS coordinates, the data were corrected to fit the census (500-m mesh). The results are shown in Fig. 15, which confirmed that each age and sex category was correctly fitted. However, there are two issues which need to be considered. First, it is estimated that a considerable amount of GPS data only used population data from night-time, which was used in this study to estimate the scaling factor of population distribution. Second, there is existing data for the train stations for the numbers of passengers per train in Japan; it is possible to estimate population data more reliably by using these data. In future, to

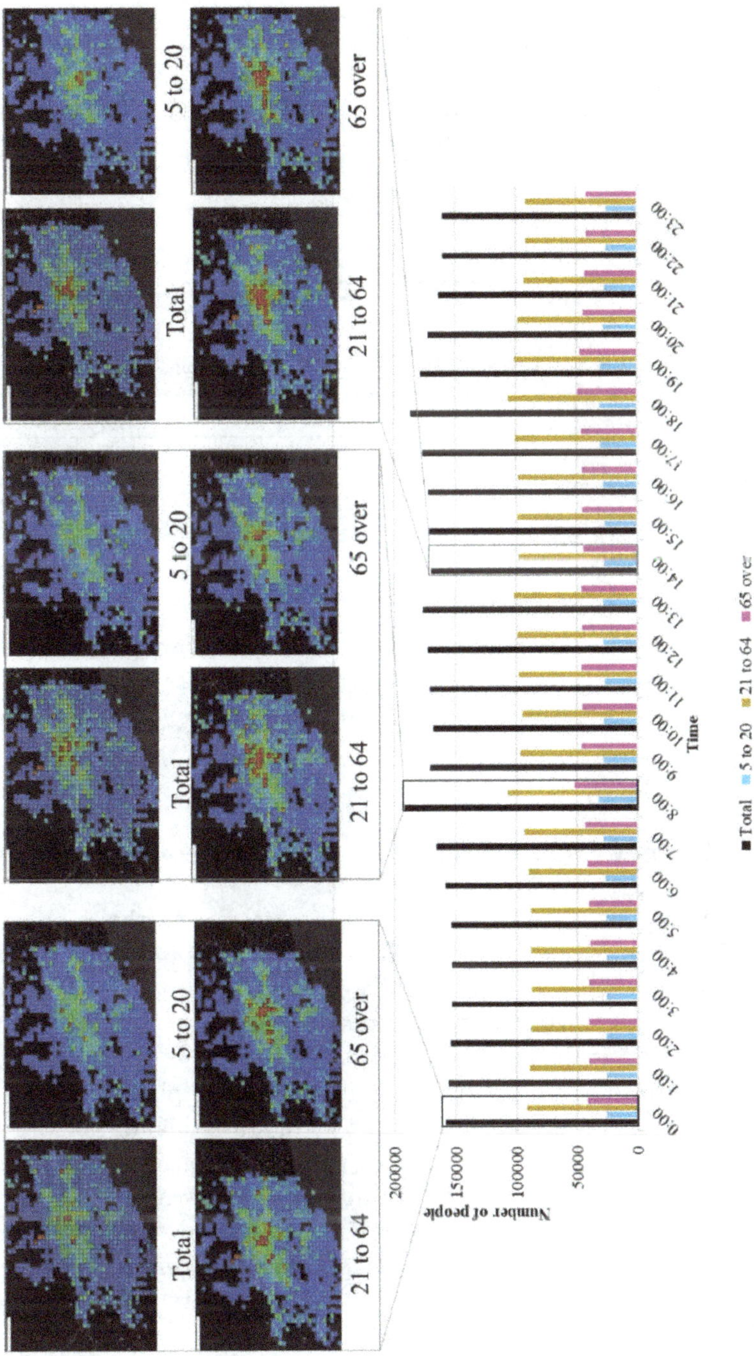

Fig. 14. People flow data with age demographics on October 17, 2012, in Kochi City.

Table 6. Error matrix of age estimation.

	Prediction				
	Age	Under 20	2165	Over 65	User's accuracy
Ground truth	Under 20	589	835	66	40.0%
	21–65	183	3798	767	80.0%
	65 over	77	929	1173	53.8%
	Producer accuracy	70.0%	68.3%	58.5%	Overall: 66.0%

Table 7. Error matrix of gender estimation.

	Prediction			
	Gender	Man	Woman	User's accuracy
Ground truth	Man	1803	2056	45.7%
	Woman	1072	3486	76.5%
	Producer accuracy	62.7%	62.9%	Overall: 62.8%

improve the estimation accuracy, traffic modes should be considered, such as whether or not light motor vehicles are used, as well as the point of interest.

4. Development of Business Transaction Economic Data

IO tables have been used in many studies to assess the economic impact of natural disasters.[34–36] However, these tables divide the nation into nine regions, therefore representing statistics of these regions, thereby ignoring the business relationships among firms. Additionally, the analysis assumes that firms in the same industry within a region are homogeneous. In Japan, the compilation of inter-region IO tables is based on a combination of statistical data from 12 ministries and agencies. This includes the census of logistics, which is published five years after the actual survey, owing to the data update costs. These tables have not been updated since the 2005 inter-regional IO table was released in 2010. Therefore, it is necessary to develop data that will enable the understanding of not only inter-industry relations but also supply-chain relations among companies.

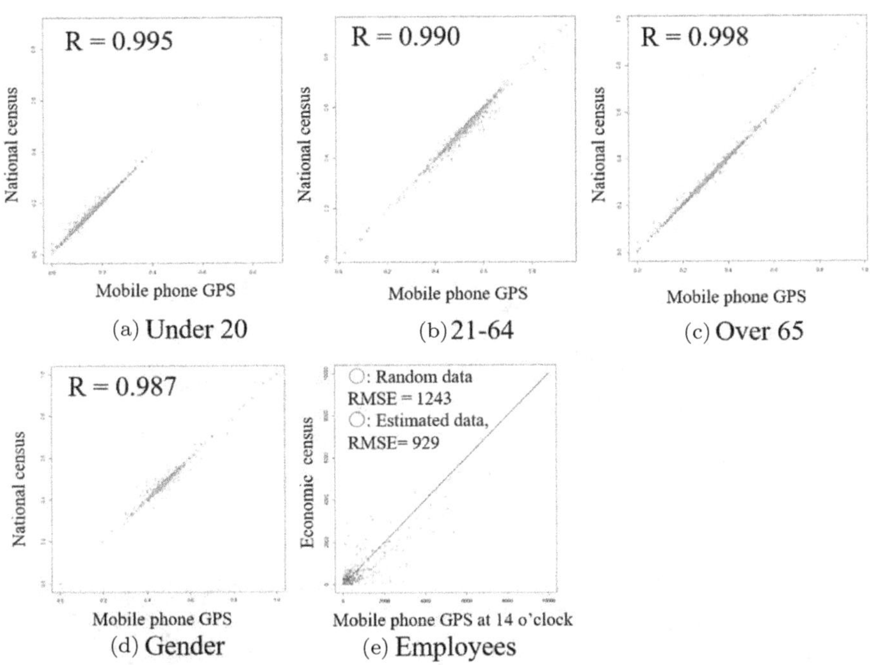

Fig. 15. Comparison of estimation results with national census (500-m grids). (a) Under 20, (b) 21–64, (c) over 65, (d) gender, and (e) employees.

In recent years, economic analyses have been performed using various micro data, such as nation-wide residential maps and mobile phone movement history data. Therefore, there is a need for inter-regional IO tables that can be used on a spatiotemporal, more detailed micro unit basis. It has also become possible to obtain information on transactions between companies using data on B2B transactions, which are maintained for the purpose of credit research on companies. The data provides a high-resolution picture of capital movements among companies on a company-by-company basis. However, the transactions recorded in the data are only for transactions between the head office of a company, which receives and places orders. The actual flow of money and goods transactions occurs not only between head offices but also from branch offices, such as production facilities, sales offices, and factories. Therefore, in this study, the transactions between business establishments were estimated using business inter-company transaction data, and data that will enable the understanding of micro-transactions were developed using GIS.

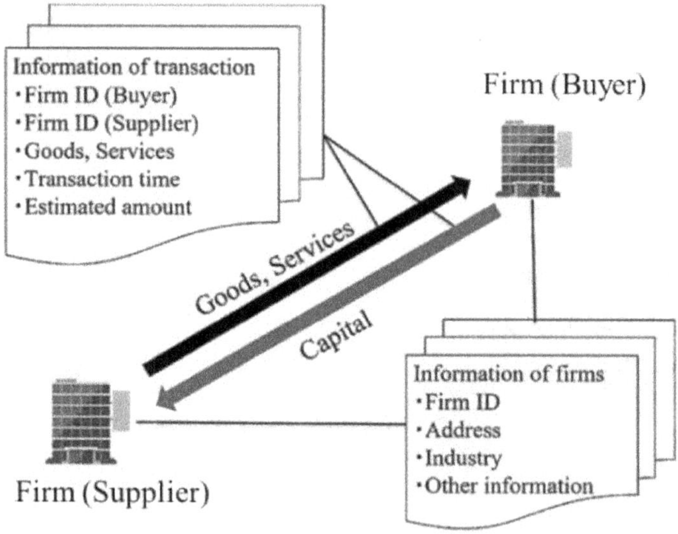

Fig. 16. Inter-company transaction data.

4.1. *Inter-company transaction data*

Teikoku Databank, Ltd., uses data obtained from corporate credit research. These data can be divided into two groups: data containing establishment information, and transactional network data containing supply-chain network data, as shown in Fig. 16. It is possible to create links between the two using specific IDs. This study used approximately 2.23 million items of establishment data (1.65 million headquarters and $\sim$ 580,000 establishments that belong to those headquarters) on Japanese enterprises nation-wide, as of 2015. This data set covers $\sim$ 90% of Japan's 1.87 million incorporated enterprises (1.62 million registered companies and 250,000 other types of incorporated enterprises [2016 Economic Census]). It is very comprehensive because it was sourced in a survey of establishments in all 47 prefectural and city governments.

In addition, the value of the transactions included is $\sim$ 500 trillion yen, which is close to the real value of Japan's GDP ($\sim$ 520 trillion yen in 2016). Therefore, it can be considered to broadly include all transactions related to production activity. Table 8 shows the sector names and sector codes used in this study.

Table 8. Sector classification.

Sector	Sector code
Agriculture, forestry, and fishery	1
Mining	2
Construction	3
Manufacturing	4
Wholesale, retail, and restaurants	5
Finance and insurance	6
Realty	7
Transport and telecommunications	8
Utilities	9

4.2. *Mythology of development of business transaction economic data*

This section provides an outline of the method for estimating inter-establishment transactions using inter-headquarters transaction data. The estimation was carried out in three steps:

(1) The total number of employees was estimated for all establishments and each enterprise.
(2) The number of employees at each establishment was estimated from the total number of employees at all establishments.
(3) Allocation of inter-headquarter transactions carried out using a gravity model that uses the number of employees at each establishment/headquarters, distance between establishment/headquarters, and inter-regional IO tables.

4.2.1. *Estimation of number of employees of each establishment*

To estimate the number of employees at each establishment, the total number of employees for each enterprise was estimated first. Approximately 60% of the data included employee numbers, which were used as training data in the construction of a linear regression model. The total number of employees at all establishments correlates with sales value, capital, and headquarters employees; therefore, these were used as input variables when constructing the multiple regression model for each sector using these as variables. In addition, models were constructed to consider the interaction between variables, thereby creating more accurate models.

Next, the probability distribution of the number of employees in each class of establishment was determined for each sector. Given the probability distribution of the total number of employees, the establishment classes were estimated. When there were multiple establishments in the same class, the allocation was weighted according to the surface area of the establishment. In addition, an estimation was made for each class of establishment from data containing the respective employee numbers. The maximum number of employees at an establishment was added as a limiting condition, from the existing data on number of employees.

4.2.2. *Estimation of transaction value*

There were no available training data for the direct estimation of transaction value in this study; therefore, a gravity model was used to estimate parameters and obtain transaction value weightings. In the gravity model, transaction value is considered to be larger for firms with larger head offices, whereas it is smaller when the distance between offices is greater. This was modeled as equivalent to the law that the gravitational force between celestial bodies is proportional to the weight of the celestial bodies and inversely proportional to the distance between them. Therefore, the following relationship is considered true, and the weight (Wij) of the transaction value between a firm i and a firm j is given by Eq. (1).

$$\ln W_{ij} \propto \alpha \ln Q_{ij}, \beta \ln D_{ij}, \gamma \ln P_{ij}, \tag{1}$$

where Q is the number of branch offices, D is the distance between the branch offices, and P is the number of employees. Industrial sectors are denoted by h and $h'(h' = 1, \ldots, 9)$, which represents the sector to which the establishment j belongs. However, h' represents arbitrary sectors (to which establishment j does not belong).

The inter-company transaction data were then processed into transactions between the branch offices (Fig. 17). Using the four methods in Table 9, the transaction value was disaggregated into branch office units, and the differences in transaction trends between head offices were analyzed.

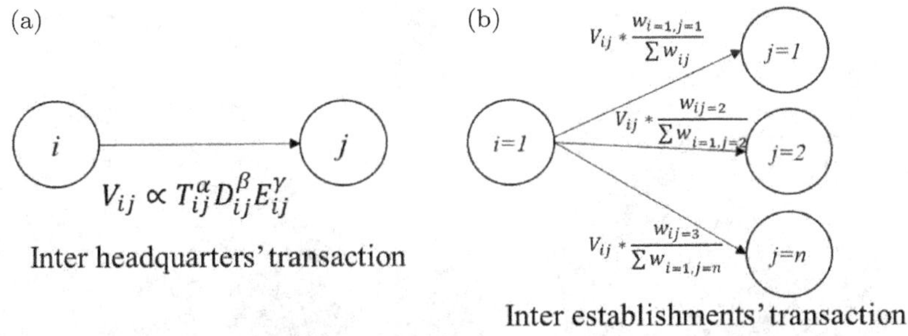

Fig. 17. Diagrams of transaction value. (a) Transaction strength between firms, and (b) the transaction value weighting based on establishments' share ratio.

Table 9. List of methods for redistribution of transaction amounts.

Method	Overview
1	Equal distribution of the transaction amount between the head offices, using the number of branch offices on both sides of the transaction order.
2	Allocation of the transaction amount between the head offices by the number of employees on both sides of the branch offices.
3	The ordering side is allocated by Method (1), and the receiving side is allocated by the distance from the ordering company's branch offices as a weighting.
4	The ordering side is allocated using Method (2), and the side receiving the order is allocated using the distance from the ordering branch offices as a weighting.

Proportional Allocation using Method 1 is the simplest method in which the entire transaction amount is first equalized by the number of branch offices on the ordering side, and the transaction amount allocated to each branch office is then equally allocated to each branch office on the receiving side. Method 2 uses the number of employees at each branch office as a weighting when allocating the amount to the ordering and receiving offices, rather than allocating it equally. In Method 3, the transaction amount is equally allocated to the ordering branch office, and then allocated to the receiving branch office using the distance between the ordering and receiving branch offices as a weighting. Method 4 changes the method of allocating the transaction amount to ordering branch offices from Method 3 and uses the number of employees weighting. In both Methods 3 and 4,

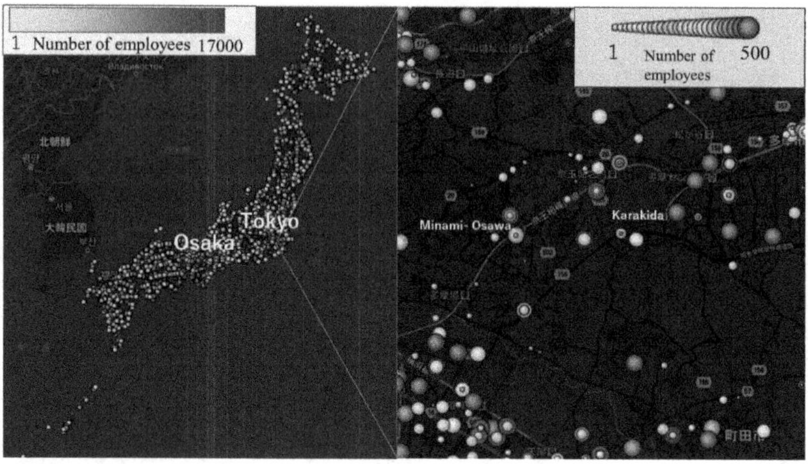

Fig. 18. Results of estimating the number of employees at each establishment throughout Japan.

the transaction amounts between the head offices are allocated using the weights of the transaction amounts in Eq. (2),

$$S_{ij} = H * (W_{ij}/\Sigma W_{ij}). \tag{2}$$

4.3. *Results*

4.3.1. *Number of employees*

Figure 18 shows the results of the estimation of the number of employees in each establishment. The number of employees in all establishments or offices in Japan can be estimated using a model constructed from existing data. Many companies with large numbers of employees have been estimated in the Tokyo metropolitan area.

4.3.2. *Transaction value*

The estimated number of branch office transaction networks is increasing compared with inter-firm transactions, and money flow can be assessed in detail (Fig. 19). For instance, the mining sector has 700 links for inter-headquarter transactions, but 2,500 links for inter-establishment transactions.

Figure 20 shows an example of estimated transactions, as each establishment in the agriculture, forestry, and fishery sectors

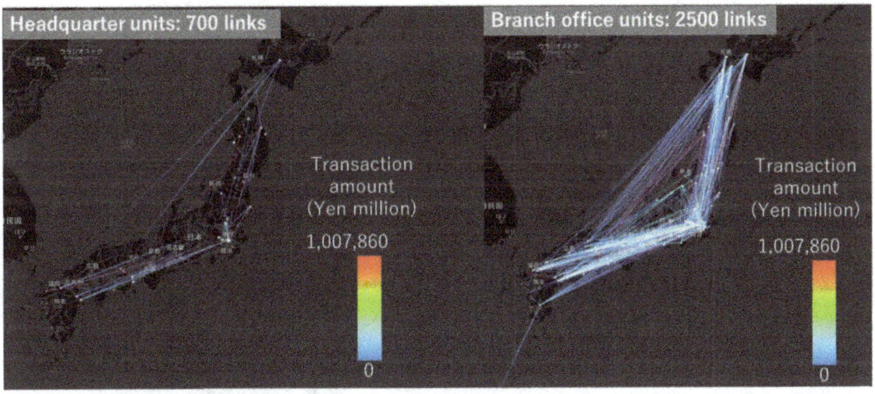

Fig. 19. Inter-firm business transaction in headquarter units (left) and in branch office units (right) for the mining sector.

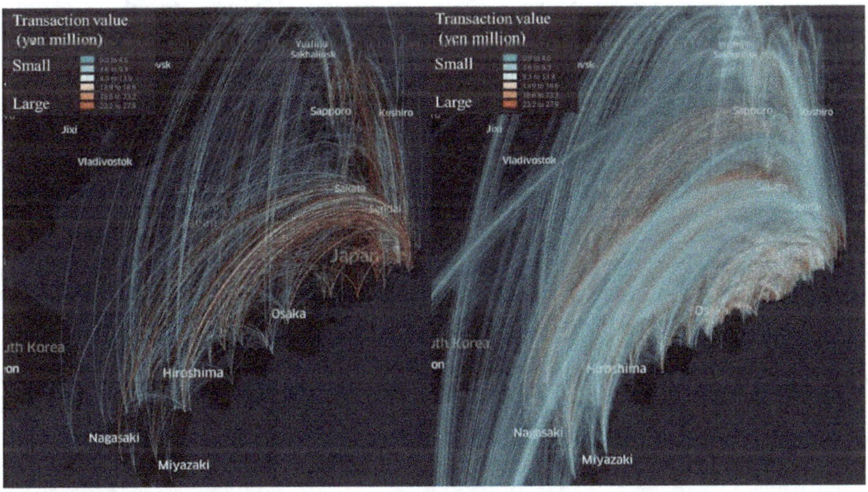

(a) (b)

Fig. 20. Example of estimation results of inter-establishments transactions. (a) Transactions in the agriculture, forestry, and fishery sector, 3,700 transactions, and (b) transactions in the retail and restaurant sector, 650,000 transactions.

undertakes transactions. When each establishment is considered, the number of transactions in the agriculture, forestry, and fishery sectors is 3,700, and the number of transactions in the retail and restaurant sectors is 650,000. In addition, compared to inter-establishments' transactions and inter-headquarters, transactions,

Table 10. Comparison of inter-company transactions based on headquarter units and branch office units.

Types	No. of headquarters	No. of establishments	No. of networks
Establishment-based data	147,000,000	580,000	550,000,000
Headquarter-based data	147,000,000	0	300,000,000

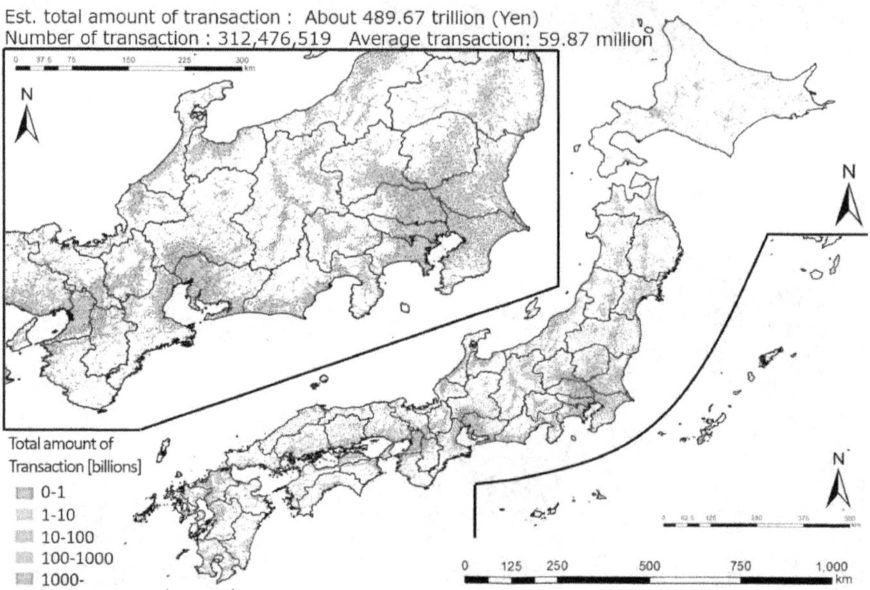

Fig. 21. Estimated total value of order and supply transactions per 1-km mesh.

the number of transactions is 550 million in headquarters inter-firm transactions and three billion transactions in establishment inter-firm transactions (Table 10). By estimating inter-establishment transactions, it is possible to capture transactions at the headquarters and establishment levels. Figure 21 shows the estimated total value of order and supply transactions per 1-km mesh in 2013 based on the inter-establishment transactions developed using Method 3, which had the highest correlation with the inter-regional IO table for each industry sector.

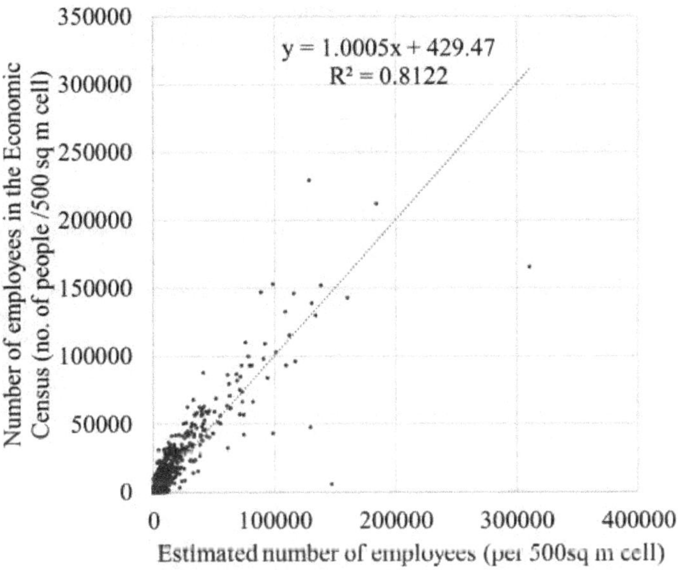

Fig. 22. Comparison between estimated number of employees and the economic census.

4.4. *Validation*

4.4.1. *Validation check of number of employees*

Regarding the reliability of the number of employees in each establishment, the values estimated using the model constructed for each sector and the true values were compared, and there was a comparatively high correlation for all sectors. A comparison of the results of this estimation with the number of people engaged per 500-m grid, obtained from the Economic Census (Japanese government), yielded a reliability coefficient of $R > 0.9$ (Fig. 22).

4.4.2. *Validation check of transaction value*

Reliability was verified by comparing the inter-regional transactions of all industries in the inter-regional IO table (64 networks among eight regions) with the inter-regional transactions estimated in this study. Table 12 shows the correlation coefficients between the inter-regional transactions in the inter-regional IO table and the transactions between head offices and business offices estimated using the four proposed methods in Table 9. All the methods showed

Table 11. Correlation coefficients between inter-regional IO table and inter-business transactions.

Method	Correlation
Headquarters transaction	0.9702
Method 1 of branch office transaction	0.9613
Method 2 of branch office transaction	0.9450
Method 3 of branch office transaction	0.9747
Method 4 of branch office transaction	0.9568

Table 12. Correlation coefficients between inter-firm transactional data and inter-regional IO tables in each method.

Sector	Headquarters	1	2	3	4
Agriculture, forestry, and fishery	0.155	0.129	0.108	0.154	0.153
Mining	0.541	0.425	0.526	0.46	0.517
Construction	0.965	−0.028	0.02	−0.022	0.018
Manufacturing	0.902	0.827	0.83	0.829	0.833
Wholesale, retail, and restaurants	0.961	0.372	0.814	0.456	0.804
Finance and insurance	0.885	0.846	0.895	0.948	0.956
Realty	0.968	0.963	0.969	0.96	0.973
Transport and telecommunications	0.788	0.785	0.773	0.831	0.839
Utilities	0.211	0.198	0.196	0.21	0.208
Service	0.943	0.94	0.946	0.942	0.947

extremely strong correlations between the two values. Next, we analyzed the correlation coefficients using the inter-regional IO table for transactions among branch offices in the same industry. The results for transactions between head offices only and the results for inter-establishment transactions were estimated using the four proposed distribution methods in Table 9. Table 11 shows the correlation between the results and the inter-regional IO table. Table 12 lists the correlation coefficients for transactions between the same industries. Transactions in agriculture, forestry, and fisheries showed almost no correlations, regardless of the method used. This may be due to the fact that the inter-industry transaction data did not sufficiently capture transaction records related to branch offices in these industries. However, strong correlations were observed in the manufacturing, finance, insurance, and real estate industries, regardless of the proposed method. This may be due to the fact that the value per transaction tends to be larger than in other industries, and that business offices in these industries tend to be concentrated in urban areas, making it easier to track transactions using business-to-business transaction data. Looking at the difference in correlation coefficients by method, the correlation was high in the construction industry for the method that considered only transactions between head offices and nearly uncorrelated for the other methods. This result suggests that most transactions among construction firms are conducted at the head office level rather than at the branch office level. On the other hand, some industries, such as the transportation and communications industry, were able to estimate more accurately than other methods by considering the distance between offices. We intend to continue to verify the estimation method for each industry to determine the most appropriate method for each industry and improve the accuracy of the distribution of transactions between branch office locations.

5. Advanced Simulations Through Data Linking

In Section 3, we introduced data linkages between social science simulations and IES systems using the building, human flow, and economic data introduced in the Section 4. Traditionally, simulations of earthquake tsunamis, buildings, and economic damages have been conducted independently. By linking social science simulations with data from earthquake and tsunami simulations, it is now possible to realize integrated simulations. In the next section, we introduce urban data linkage using economic recovery and evacuation as a case study.

5.1. *Economic simulation*

By linking social science simulations with the IES, we have realized advanced economic simulations that utilize inter-business transactions. By integrating these economic data with earthquake simulation results, integrated earthquake damage prediction, reflecting actual city and industrial structure models, has become possible.

Figure 23 shows the data linking the evacuation flow and economic simulation after earthquake tsunamis using advanced urban data.

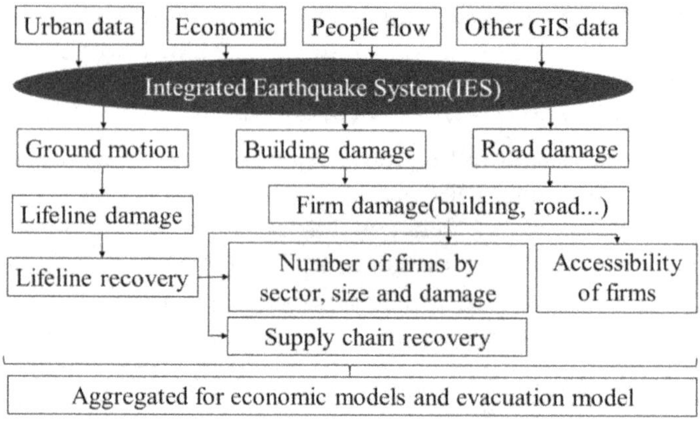

Fig. 23. Flowchart of data linking.

Data linking is achieved by inputting building data into the IES for earthquake tsunami simulation, receiving the calculation results of building damage, and spatially joining them with inter-business transaction data. The target earthquake was the Nankai trough earthquake. Since it was necessary to consider the impact of supply-chain disruptions in the economic simulation, we used the developed economic data to estimate which supply chain affected a company and for how long its production activities are affected. The various parameters required for the economic simulation, such as supply chains, capital, infrastructure, and lifelines, were developed by linking the data, as shown in Table 13. The area covered in this experiment was from Japan. By extending the data linkage to the entire country, it is possible to simulate the damage impact on companies and supply chains in Japan as a whole.

5.1.1. *Earthquake ground motion data*

The results from the earthquake simulation group were used for the seismic data. Data for the assumed seismic intensity distribution and inundation area of the Nankai Trough giant earthquake were also used. The data was sourced from the Cabinet Office's Nankai Trough giant earthquake model study group. Five cases of assumed seismic intensity data were used, which included the four strongest vibration waveform cases (the basic, east, west, and land cases), as well as the empirical method.

Table 13. List of variables for economic simulation.

Parameter type	Parameter	Description
Supply chain damage	$\Delta l_h^{ROK}, \Delta l_h^{ROJ}$	Labor damage rate by sector
	Company damage	Number of firms affected, and size of damage by industry, size, and region
	Supply chain damage rate and probability of production restoration	Supply chain damage rates and post-disaster time-series production restoration probabilities by industry, size, and region
Damage to capital stock	$\Delta k_h^{ROK+}, \Delta k_h^{ROJ+}$	Capital loss ratio
Damage to various infrastructures	$\Delta B_b^{ROk}, \Delta B_b^{ROJ}$	Infrastructure damage rate
Lifeline damage	Infrastructure damage rate and probability of restoration of supply	Infrastructure outage probability and post-disaster supply restoration probability

5.1.2. *Infrastructure damage*

(1) *Bayshore damage rates*
The amount of damage for the bay was determined using the bay damage function proposed by Ref. 37. Equation (3) was used to calculate the bay damage rate for each region from the corresponding measured seismic intensity for those regions. The average for each region was then used as the bay damage rate for that region:

$$P_{\text{Rank2}} = \Phi\left[\ln\left(\frac{I_{JMA}}{6.06}\right)/0.0546\right], \quad (3)$$

where P_{Rank2} is the X; and Φ is the Y, I_{JMA} is the Z, and 0.0546 is a constant.

(2) *Sewer damage*
Shoji *et al.*[38] used the earthquake damage rate curve to calculate the earthquake damage rate for buried sewer pipelines. Using the

damage rate curve function in Eq. 4, the standard damage rate $R(x)$ was calculated from the assumed measured seismic intensity for each region, and the average value for each region was used as the sewer damage rate for that region.

$$R(x) = 0.134 \cdot E\Phi\left[\frac{x - 5.000}{0.0546}\right] \quad (x : IJ). \tag{4}$$

(3) *Electricity, gas, and water supply damage*

The damage rates and post-disaster time-series restoration (or recovery) probabilities for the three services (electricity, gas, and water supply) were calculated. The method uses the seismic function evaluation model proposed by Nojima and Kato.[39] In this study, the probability of a supply service outage immediately after an earthquake was assumed to be the damage rate. The outage probability $P(I)$ was then calculated from the parameters b_0 and b_1 and the assumed measured seismic intensity I, as shown in Eq. (5):

$$P(I) = \frac{\exp[b_0 + b_1 \cdot EI]}{1 + \exp[b_0 + b_1 \cdot EI]}. \tag{5}$$

The lifeline and building damage were estimated from the results above (Fig. 24).

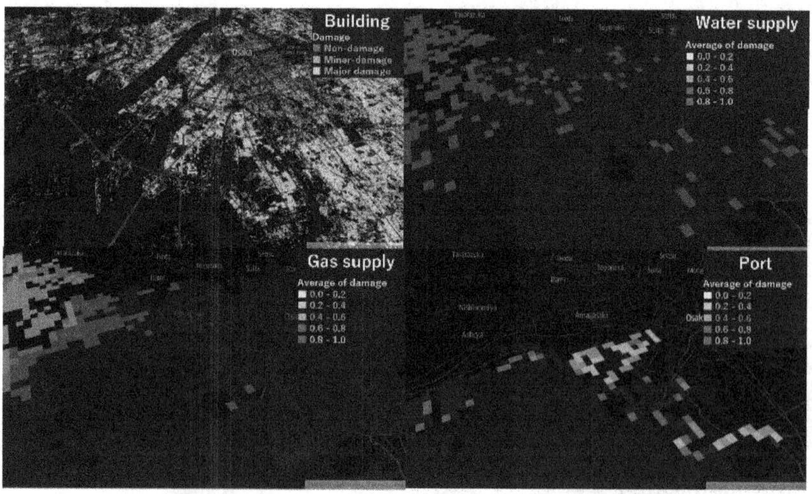

Fig. 24. Results of building and supply service damage (Earthquake scenario: landward case, with an amplitude of 1.5).

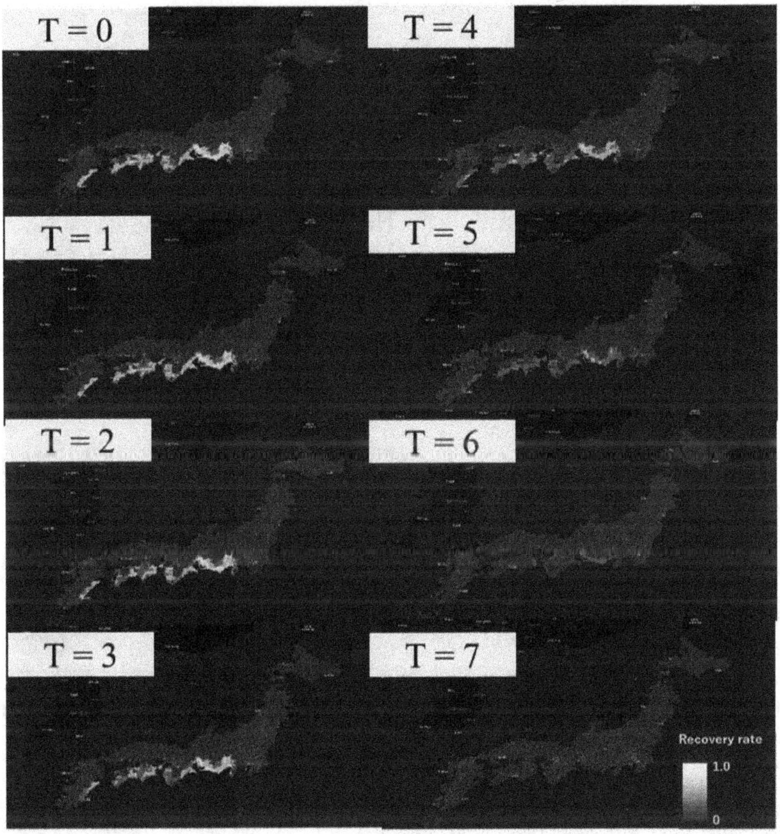

Fig. 25. Supply chain recovery time series.

Using the data aforementioned the restoration of supply chains in the event of a Nankai Trough earthquake was estimated. It was obtained from the damage rate and supply service damage rate of each company. The supply-chain recovery transition was calculated and is shown in Fig. 25. For details of the methodology, please refer to the Chapter 8 by Yokomatsu and Hori.

5.2. *Evacuation simulation*

An evacuation simulation that considers the population distribution over time during a tsunami was performed. This was done by inputting advanced people flow data, including demographics based on actual mobile phone GPS observed data, into an evacuation simulation model. In particular, the time when the tsunami occurred, attributes (age and gender) of evacuees in the target area, and their spatial distribution, may have a significant impact on lives lost. Conventionally, a Monte Carlo simulation is generally performed under initial conditions where people are randomly assigned using demographic data. However, by using people flow data based on mobile phone GPS movement histories, it is possible to perform simulations that consider the distribution of people more accurately.

Therefore, it is possible to simulate an evacuation according to the actual conditions of the region by considering the time of day and attributes of the evacuees.

Additionally, by visualizing the results of tsunami evacuation simulations using GIS, it is possible to easily understand the geographical trends in time series. To visualize the process of evacuation and create a video animation, the mobile data visualization software of Mobmap was used.[40] An example of a visualization of the evacuation simulation results is shown in Fig. 26. A video was created to visually understand the chronological evacuation behavior on a map, which aided in how behavior changes depending on

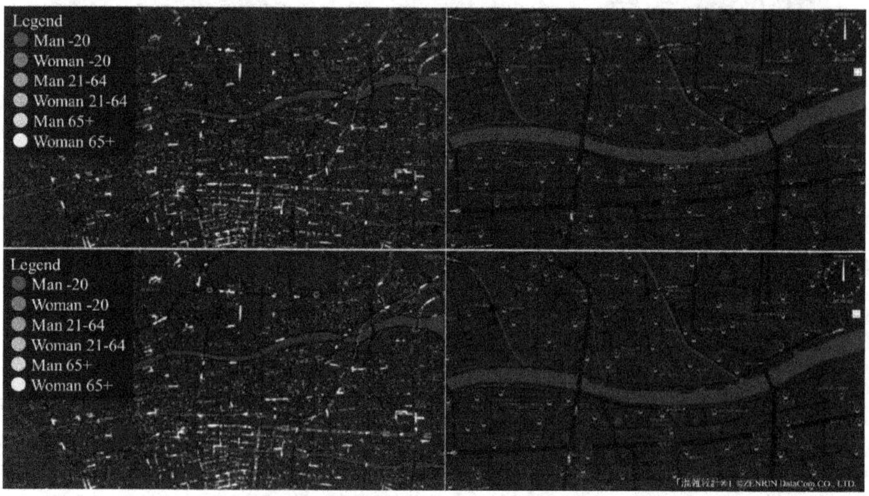

Fig. 26. Example of video visualization of evacuation simulation results.

the differences in input data and scenarios. This visualization will be useful for understanding data and to analyze results faster, in addition to considering regional vulnerability due to differences in simulation results.

6. Conclusion

In this study, urban data were developed to assess building risk, evacuation simulation, and economic damage caused by a large-scale earthquake. This was done by using a combination of various statistical information and large-scale spatial data such as digital residential maps for Japan. Although the data presented in this study are estimates, and thus contain errors compared to the true values, it is clear that results which are close to the true values can be obtained by aggregating the data. The data presented here were used for linking earthquake (or tsunami) simulations and social science simulations, and their usefulness was demonstrated empirically through evacuation and economic damage simulations.

It is worth noting that presently only basic data development has been completed, and the data developed in this study have important limitations. First, the amount of information in GIS and observed data is limited. For example, the development of 3D building models is urgently required because other factors, such as the placement of columns and weights, may also affect building damage. In addition, because people flow data are based on observation data from GPS observed data, it is also affected by, *inter alia*, the sample rate, observation error, weather, season, and time of day. The reliability of the people flow data also varied, depending on the sample rate. These are recognized as important limitations of the methodology. Further research is needed to verify whether these effects influence lives lost and evacuation simulation results or whether they are merely noise. Despite the limitations of the data, the potential applications of this method are vast.

It is expected that publication and sharing of these results will support and advance social science simulations in the future. People flow data have already been released as pseudo human flow data in JoRAS of the Center for Spatial Information Science, The University of Tokyo, and a community of researchers for this field has already been formed.[30] Building data are also being considered for open data for use in future research.

There are still various issues to be solved. However, the establishment of a database that can assess building damage, the effects on people and the economy of future earthquakes and tsunamis on a national scale, will contribute to the advancement of social science simulations in computer science. Additionally, it will contribute to disaster prevention policies and proactive disaster management.

References

1. T. Seto, Y. Sekimoto, K. Asahi, and T. Endo, Constructing a digital city on a web-3D platform: Simultaneous and consistent generation of metadata and tile data from a multi-source raw dataset, in: *Proceedings of the 3rd ACM SIGSPATIAL International Workshop on Advances in Resilient and Intelligent Cities*, pp. 1–9 (2020).
2. T. Horiuchi, K. Kajiwara, T. Yamashita, T. Aoki, T. Yashiro, Y. Sekimoto, and H. Koizumi, Study concept on the development of an urban cyber physical system for enhancing the capability to respond to large-scale earthquakes, *J. Disaster Res.* **16**(2), 287–297, (2021).
3. Mohammadi and Taylor, N. Mohammadi, and J. E. Taylor, Urban energy flux: Spatiotemporal fluctuations of building energy consumption and human mobility-driven prediction, *Appl. Energy* **195**, 810–818, (2017).
4. H. F. Nweke, Y. W. Teh, G. Mujtaba, and M. A. Al-Garadi, Data fusion and multiple classifier systems for human activity detection and health monitoring: Review and open research directions, *Info. Fusion* **46**, 147–170, (2019).
5. M. Hori, G. Sobhaninejad, T. Ichimura, and M. Lalith, Enhancement of integrated earthquake simulation with high-performance computing, *J. Earthq. Tsunami* **5**(03), 271–282, (2011).
6. G. Sobhaninejad, M. Hori, and T. Kabeyasawa, Enhancing integrated earthquake simulation with high performance computing, *Adv. Eng. Softw.* **42**(5), 286–292, (2011).
7. MILT (2020), https://www.mlit.go.jp/plateau.
8. J. D. Farmer, and D. Foley, The economy needs agent-based modelling. *Nature* **460**(7256), 685–686, (2009).

9. Z. Zhang, N. Li, P. Cui, H. Xu, Y. Liu, X. Chen, and J. Feng, How to integrate labor disruption into an economic impact evaluation model for postdisaster recovery periods, *Risk Anal.* **39**(11), 2443–2456, (2019).
10. M. Delalay, A. D. Ziegler, M. S. Shrestha, and V. Gopal, Methodology for future flood assessment in terms of economic damage: Development and application for a case study in Nepal, *J. Flood Risk Manag.* **13**(3), e12623, (2020).
11. E. Viegas, H. Goto, M. Takayasu, H. Takayasu, and H. J. Jensen, Assembling real networks from synthetic and unstructured subsets: The corporate reporting case, *Sci. Rep.* **9**(1), 1–10, (2019).
12. Y. Ogawa, Y. Akiyama, S. Yoshihide, and R. Shibasaki, Estimation of Transactional Network Data Between Branch Offices using Transactional Big Data Throughout Japan, in: *2019 IEEE International Conference on Big Data (Big Data)*, pp. 3916–3924, IEEE. (2019).
13. Y. Sekimoto, R. Shibasaki, H. Kanasugi, T. Usui, and Y. Shimazaki, Pflow: Reconstructing people flow recycling large-scale social survey data, *IEEE Pervasive Comput.* **10**(4), 27–35, (2011).
14. X. Song, Q. Zhang, Y. Sekimoto, R. Shibasaki, N. J. Yuan, and X. Xie, Prediction and simulation of human mobility following natural disasters, *ACM Trans. Intell. Syst. Technol. (TIST)*, **8**(2), 1–23, (2016).
15. M. Hori, *Introduction to Computational Earthquake Engineering*, 3rd edition, World Scientific Publishing Europe Ltd., Chapter 11, (2018).
16. M. Hori, T. Ichimura, and L. Maddegedara, *Integrated Earthquake Simulation.* CRC Press, (2022).
17. R. Shibasaki, Y. Akiyama, Y. Ogawa, Y. Hayashi, Y. Suzuki, S. Sato, and K. Tsukahara, Earthquake damage risk evaluation by micro geo data, in: *Disaster Resilient Cities-Concepts and Practical Examples* pp. 150–158, (2016). Butterworth-Heinemann.
18. Yamaguchi and Yamazaki, Fragility curves for buildings in Japan based on damage surveys after the 1995 Kobe earthquake, in: *Proceedings of the 12th World Conference on Earthquake Engineering, Auckland*, New Zealand, p. 2451, (2000).
19. Y. Ogawa, Y. Akiyama, and R. Shibasaki, Evaluation of catastrophic earthquake damage throughout Japan using estimated micro data, in: *CUPUM2013 Conference Papers* Vol. 103, pp. 1–30, (2013).
20. L. Breiman, Random forests, *Mach. Learn.* **45**(1), 5–32, (2001).
21. Y. Ogawa, T. Oki, S. Chen, and Y. Sekimoto, Joining street-view images and building footprint GIS data, in: *Proceedings of the 1st ACM SIGSPATIAL International Workshop on Searching and Mining Large Collections of Geospatial Data*, pp. 18–24, (2021, November).

22. A. Krizhevsky, I. Sutskever, and G. E. Hinton, Imagenet classification with deep convolutional neural networks, *Adv. in Neural Inform. Process. Syst.* **25**, 1097–1105, (2012).
23. I. Goodfellow, Y. Bengio, and A. Courville, *Deep Learning.* MIT Press, (2016).
24. Y. LeCun, Y. Bengio, and G. Hinton, Deep Learning, *Nature.* **521**(7553), 436–444, (2015).
25. S. H. Joo, Y. Ogawa, and Y. Sekimoto, Road-reconstruction after multi-locational flooding in multi-agent deep RL with the consideration of human mobility-Case study: Western Japan flooding in 2018, *Int. J. Disaster Risk Reduct.* **70**, 102780, (2022).
26. Q. Yu, H. Zhang, W. Li, X. Song, D. Yang, and R. Shibasaki, Mobile phone GPS data in urban customized bus: Dynamic line design and emission reduction potentials analysis, *J. Clean. Prod.* **272**, 122471, (2020).
27. Y. Qin, Y. Akiyama, Y. Ogawa, R. Shibasaki, and T. Sato, Study on the relationship between house rent and people congestion by time in Tokyo based on mobile phone GPS data, in: *2019 IEEE International Conference on Big Data (Big Data)*, pp. 5313–5320, IEEE. (2019).
28. T. Yabe, K. Tsubouchi, N. Fujiwara, T. Wada, Y. Sekimoto, and S. V. Ukkusuri, Non-compulsory measures sufficiently reduced human mobility in Tokyo during the COVID-19 epidemic, *Sci. Rep.* **10**(1), 1–9, (2020).
29. H. Zhang, J. Yan, Q. Yu, M. Obersteiner, W. Li, J. Chen, and R. Shibasaki, 1.6 million transactions replicate distributed PV market slowdown by COVID-19 lockdown, *Appl. Energy* **283**, 116341, (2021).
30. T. Kashiyama, Y. Pang, Y. Sekimoto, and T. Yabe, Pseudo-PFLOW: Development of nationwide synthetic open dataset for people movement based on limited travel survey and open statistical data, (2022), arXiv preprint arXiv:2205.00657.
31. Y. Ogawa, T. Sato, Y. Akiyama, R. Shibasaki, and Y. Sekimoto, Developing a model for estimating the home return of evacuees based on the 2011 Tohoku earthquake tsunami — Utilizing mobile phone GPS big data, in: *Advances and New Trends in Environmental Informatics*, pp. 227–240. Springer, Cham. (2018).
32. T. Horanont and R. Shibasaki, Nowcast of urban population distribution using mobile, phone call detail records and person trip data, in: *Proceedings of the International Conference Computers in Urban Planning and Urban Management, CD-ROM.* (2011).
33. M. Hadano, A. Wiayangkurn, Y. Akiyama, T. Horanont, and R. Sibasaki, A study on extracting characteristic of visitors at commercial area from GPS data, in: *IGU 2013 Kyoto Regional Conference*, 01435 (2013).

34. D. Acemoglu, V. M. Carvalho, A. Ozdaglar, and A. Tahbaz-Salehi, The network origins of aggregate fluctuations, *Econometrica* **80**(5), 1977–2016, (2012).
35. Y. Haimes and P. Jiang, Leontief-based model of risk in complex interconnected infrastructures, *J. Infrastruct. Sys.* **7**, 1–12, (2001).
36. J. Santos and Y. Haimes, Modeling the demand reduction input-output (I-O) inoperability due to terrorism of interconnected infrastructures, *Risk Anal.* **24**, 1437–1451, (2004).
37. Y. Akakura and K. Ono, Recovery curve of containerized cargo demand after a major earthquake and tsunami in shipping-dependent industries, *J. Jpn. Soc. Civ. Eng. D3, Infrastruct. Plann. Manage.* **70**(5), I_689–I_699, (2014).
38. M. Shoji, R. Terashima, and S. Nagata, Construction of earthquake damage rate curve for buried sewer pipelines, *J. Jpn. Soc. Civ. Eng.*, A1, *Struct. Eng. Earthquake Eng.* **70**(4), I_921–I_940, (2014).
39. N. Nojima and H. Kato, Vulnerability assessment of water distribution pipeline networks of national water utilities based on water supply statistics, A1, *Struct. Eng. & Earthquake Eng.* **70**(4), I_21–I_32, (2014).
40. S. Ueyama, Y. Akiyama, and R. Shibasaki, Integrated visual exploration tool for fusion of mass movement and static data, in: *Proceedings of the International Conference Computers in Urban Planning and Urban Management, CD-ROM.* (2015).

https://doi.org/10.1142/9781800614635_bmatter

Index

H

I

J

K

L

Printed in the USA
CPSIA information can be obtained
at www.ICGtesting.com
JSHW061930180724
66538JS00006B/5

9 781800 614628